PROTEIN AND PEPTIDE MASS SPECTROMETRY IN DRUG DISCOVERY

CONTENTS

PREFACE

For over a decade mass spectrometry (MS) has been one of the most highly utilized analytical technique for analysis of proteins and peptides. This is largely due to continuous refinement of ionization methods, including electrospray ionization (ESI) and matrix-assisted laser desorption ionization (MALDI), the improvement of MS instrumentation, and the growth in the data processing. Various niche applications in neuroproteomics and antigenic peptides could have important implications in drug discovery, and these developments are described in two chapters. Furthermore there has been considerable research activity focused on the development of new methodologies for the analysis of proteins and peptides; these methods have exploited ongoing instrumentation improvements and include both bottom-up and top-down protein sequencing. New approaches include those in imaging, ion mobility, and the use of microwave radiation to speed proteolysis, and these new ideas are covered in three chapters in this volume. Accompanying the analytical developments are new techniques for the determinations of protein structure, of their interactions with peptides, proteins, and ligands including drugs, and their folding and unfolding. These techniques are described in detail in this volume.

One of the important and immediate applications in protein and peptide MS is for pharmaceutical analysis throughout each stage of drug development process, ranging from drug discovery to manufacturing. MS-based technologies play critical roles in providing qualitative and quantitative information to characterization of target proteins and protein products for therapeutic use, as illustrated in this volume.

We are delighted to bring together the work of contributors from academe and industry in highlighting current analytical approaches, industry practices, and modern strategies for the characterization of proteins and peptides in drug discovery. Our goal is to present a compilation of the latest methodologies and applications to practitioners of protein and peptide MS with the focus on drug discovery efforts.

We would like to acknowledge the special efforts and patience of all the authors, who have made significant contributions to this book.

Michael L. Gross
Guodong Chen
Birendra N. Pramanik

CONTRIBUTORS

Christine L. Andrews, Merck Research Laboratories, Cambridge, MA

Catherine A. Srebalus Barnes, Eli Lilly and Company, Indianapolis, IN

Fotini N. Bazoti, The Goulandris Natural History Museum, Kifissia, Greece

Adriana Bora, University of Illinois, Urbana, IL

Ajay K. Bose (deceased), Stevens Institute of Technology, Hoboken, NJ

Ron Bose, Washington University, St. Louis, MO

Michael Boyne, Washington University, St. Louis, MO

Richard M. Caprioli, Vanderbilt University, Nashville, TN

Guodong Chen, Bristol-Myers Squibb, Princeton, NJ

Hao Chen, Ohio University, Athens, OH

Ismael Cotte-Rodriguez, Procter & Gamble, Loveland, OH

Peter O'Connor, University of Warwick, Coventry, UK

Surendra Dasari, Vanderbilt University Medical Center, Nashville, TN

Jeffrey R. Enders, Vanderbilt University, Nashville, TN

Jacob Galan, Purdue University, West Lafayette, IN

Michael L. Gross, Washington University, St. Louis, MO

Anton Iliuk, Purdue University, West Lafayette, IN

Sandra A. Kerfoot, Seattle Childrens Research Institute, Seattle, WA

Michal Kliman, Vanderbilt University, Nashville, TN

Amareth Lim, Eli Lilly and Company, Indianapolis, IN

Cheng Lin, Boston University, Boston, MA

John A. McLean, Vanderbilt University, Nashville, TN

Zhixin Miao, Ohio University, Athens, OH

Urooj A. Mirza, Merck Research Laboratories, Kenilworth, NJ

Elliott Nickbarg, Merck Research Laboratories, Cambridge, MA

Birendra N. Pramanik, Merck Research Laboratories, Kenilworth, NJ

Michelle L. Reyzer, Vanderbilt University, Nashville, TN

Henry Rohrs, Washington University, St. Louis, MO

Patrick J. Rudewicz, Elan Pharmaceuticals, South San Francisco, CA

Sevugarajan Sundarapandian, Vanderbilt University, Nashville, TN

Jonathan V. Sweedler, University of Illinois, Urbana, IL

David L. Tabb, Vanderbilt University Medical Center, Nashville, TN

W. Andy Tao, Purdue University, West Lafayette, IN

Anthony Tsarbopoulos, University of Patras, Greece

Fang Xie, Pacific Northwest National Laboratory, Richland, WA

Xianshu Yang, Merck Research Laboratories, Cambridge, MA

Yun Zhang, Ohio University, Athens, OH

Michael R. Ziebell, Merck Research Laboratories, Cambridge, MA

PART I

METHODOLOGY

CHAPTER 1

Ionization Methods in Protein Mass Spectrometry

ISMAEL COTTE-RODRIGUEZ, YUN ZHANG, ZHIXIN MIAO, and HAO CHEN

Mass spectrometry (MS) has become one of the most powerful and popular modern physical-chemical methods to study the complexities of elemental and molecular processes in nature. The advent of new methods of ion generation, novel mass analyzers, and new tools for data processing has made it possible to analyze almost all chemical entities by MS, ranging from small organic compounds, large biological molecules, to whole living cells/tissues. As proteins fulfill a plethora of biochemical functions within every living organism, equally spectacular efforts and advances have been seen for protein ionization methods. In particular, the invention of matrix-assisted laser desorption ionization (MALDI) [1] and electrospray ionization (ESI) technologies [2,3] allow one to measure protein molecular weights, to determine sequences, and to probe conformations and post-translational modifications of proteins. In addition the mass range of species amenable for MS analysis has been increased immensely, enabling the transfer into the gas phase of ionized noncovalent species with masses well over one million (e.g., a 100 MDa single DNA ion [4]). These advances move MS into the range of intact protein oligomers and functional machineries.

This chapter is an introduction to various ionization methods for proteins. As this is a broad topic with an immense literature coverage including many excellent books [5,6] and reviews [7–17], we will emphasize some types of spray or laser-based protein ionization techniques, including atmospheric pressure MALDI, surface-enhanced laser desorption/ionization (SELDI), nanostructure-initiator MS (NIMS), sonic spray ionization (SSI), electrosonic spray ionization (ESSI), desorption electrospray ionization (DESI), fused-droplet electrospray ionization (FD-ESI), electrospray-assisted laser desorption ionization (ELDI), and matrix-assisted laser desorption electrospray ionization (MALDESI). We begin with the introduction of some historic facts for the

Protein and Peptide Mass Spectrometry in Drug Discovery, Edited by Michael L. Gross, Guodong Chen, and Birendra N. Pramanik.

development of protein ionization methods, followed with the description of each method including the ionization principles, strengths, and analytical applications.

1.1 HISTORY OF THE DEVELOPMENT OF PROTEIN MASS SPECTROMETRY

MS originates from ninteen-century physics. The first known mass spectrometer was built by J. J. Thomson in the early 1900s to study and measure the mass (m)-to-charge (z) (m/z) values of the "corpusules" that make up "positive rays" [18], a type of radiation initially observed by German physicist Eugen Goldstein. Following the seminal work of Thomson, MS underwent countless improvements in instrumentation, ionization methods, and applications. The classical ionization method, electron ionization (EI), was devised by Dempster and improved later by Bleakney [19] and Nier [20], and became a widely used standard for ionization of volatile organic compounds. This ionization technique requires extensive derivatization and evaporation of a nonvolatile analyte into the ion source, and it involves numerous fragmentation and rearrangement reactions.

Applications of MS to peptides (derivatized via acylation) begun in the late 1950s by Biemann [21] and McLafferty [22] The first methods that allowed analysis of nonderivatized peptides were field desorption (FD) and chemical ionization (CI) developed in the 1960s [23,24]. Ionization by CI is achieved by interaction of its volatile molecules with reagent ions. CI allows ionization without significant degree of ion fragmentation but still requires gas-phase samples. Field desorption was reported by Beckey in 1969 [25], in which electron tunneling triggered by a very high electric field results in ionization of gaseous analyte molecules.

It was plasma desorption (PD) [26] and fast atom bombardment (FAB) [27] that opened the way to protein analysis. PD ionization, invented by R. D. Macfarlane in 1976 [28], a breakthrough in the analysis of solid samples, involves ionization of materials in the solid state by bombardment with ions or neutral atoms formed as a result of the nuclear fission of the Californium isotope ^{252}Cf. In 1982 Sundqvist and coworkers obtained the first spectrum of a protein, insulin (Figure 1.1), using bombardment with a beam of 90 MeV $^{127}I^{20+}$ ions from a tandem accelerator [26]. Later, FAB involving focusing the sample in liquid matrix with a beam of neutral atoms or molecules, was implemented for the ionization of proteins up to 24 kDa [29]. In 1983 Blakely and Vestal [30] introduced thermospray ionization (TSI) to produce ions from an aqueous solution sprayed directly into a mass spectrometer. Thermospray is a form of atmospheric pressure ionization in MS, transferring ions from the liquid phase to the gas phase for analysis. It was particularly useful in coupling liquid chromatography with mass spectrometry [31].

The breakthrough for large molecule laser desorption ionization came in 1987 when Tanaka combined 30-nm cobalt particles in glycerol with a 337-nm nitrogen laser for ionization and showed that singly charged protein molecular ions up to about 35 kDa can be introduced to a mass spectrometer [32]. During that time, MALDI [15,33], first reported in 1985 by Hillenkamp, Karas, and their colleagues,

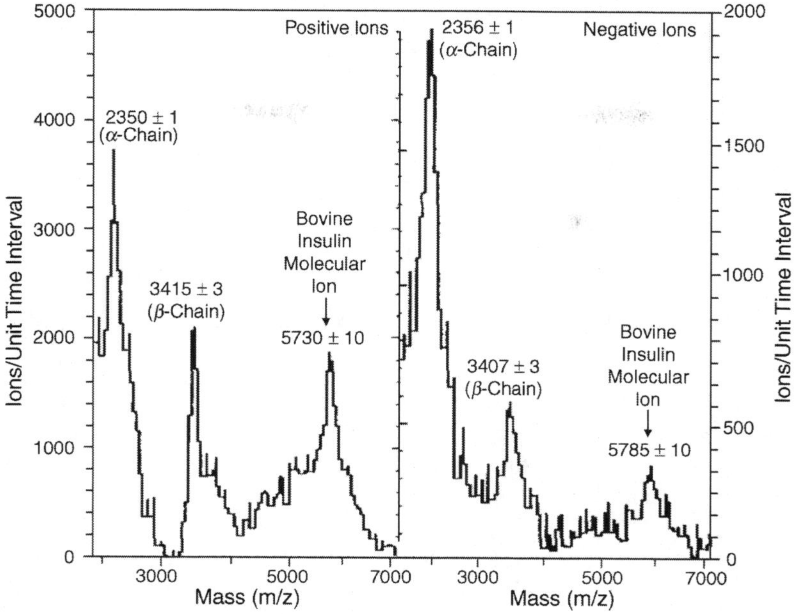

FIGURE 1.1 ^{127}I-PDMS spectra of bovine insulin recorded over a 1.5-h period with a 90-MeV ^{127}I (+20) beam current of 2000 s^{-1}. From [26]. Copyright permission was obtained from ACS.

emerged as the culmination of a long series of experiments using desorption ionization (DI). MALDI is a soft ionization technique for the analysis of biomolecules and large organic molecules and has gained wide success in protein analysis, particularly when coupled with time-of-flight (TOF) instruments [34,35].

Another breakthrough occurred in 1984 when Fenn and coworkers used electrospray to ionize biomolecules [2]; the first ESI analyses of biopolymers including proteins were published in 1989 [3]. MALDI and ESI have revolutionized protein mass spectrometry since their invention in 1980s, and they have triggered the explosion in application of mass spectrometry for protein studies [36].

1.2 LASER-BASED IONIZATION METHODS FOR PROTEINS

1.2.1 Matrix-Assisted Laser Desorption/Ionization (MALDI)

Investigations of the wavelength influence in ultraviolet-laser desorption [33] led to invention of ultraviolet-laser matrix-assisted laser desorption ionization (UV-MALDI) between 1984 and 1986 and summarized in a 1987 paper [37]. In 1988 Karas and Hillenkamp reported ultraviolet-laser desorption (UVLD) of bioorganic compounds in the mass range above 10 kDa [1]. As a soft desorption ionization method, MALDI handles thermolabile, nonvolatile organic compounds, especially those with

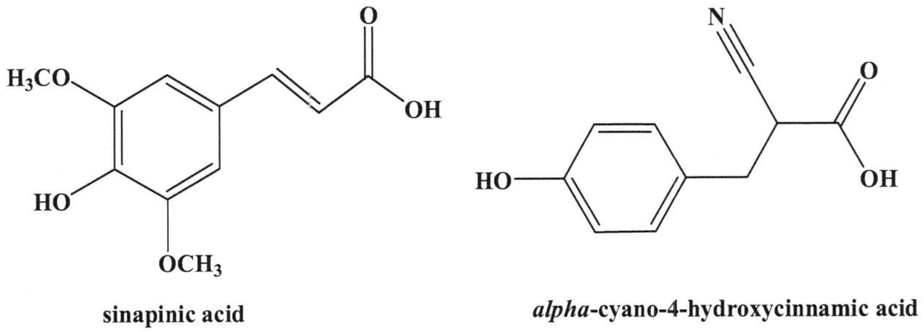

SCHEME 1.1 Structures of two common MALDI matrices.

high molecular weight and can be successfully used for the analysis of proteins, peptides, glycoproteins, oligosaccharides, and oligonucleotides. Its operation is relatively straightforward, although matrix preparation requires experience and perhaps some artistry.

MALDI is based on the bombardment of sample molecules with laser light, process that allows sample ionization [38]. It requires a specific matrix consisting of small organic compounds (e.g., nicotinic acid) that exhibit a strong resonance absorption at the laser wavelength used. The sample is premixed and diluted with the highly absorbing matrix and allowed to dry on a sample target. A range of compounds is suitable as matrices: sinapinic acid is a common one for protein analysis while *alpha*-cyano-4-hydroxycinnamic acid is often used for peptide analysis (the structures of matrices are shown in Scheme 1.1). This kind of acid serves well as a matrix for MALDI owing to the acid's ability to absorb laser radiation and also to donate protons (H^+) to the analyte of interest. Upon laser irradiation, energy is absorbed by the matrix in a localized region of the surface. As a result an explosive break up of the cocrystallized analyte/matrix sample occurs. The rapid expansion of the vaporized matrix in MALDI leads to the translational excitation of analyte molecules and the release of the analyte molecules from the surface of the condensed phase sample into vacuum. The analyte may be precharged (e.g., exist as a salt), and the intact analyte ion may simply be transferred as an ion from the solid to the vapor state upon laser irradiation of the matrix. Alternatively, a neutral analyte may be ionized through ion–molecule reactions (e.g., proton transfer reaction) occurring in the energized selvedge or interfacial region between the solid and gas phases.

MALDI has remarkable efficiency in producing intact molecular ions (often $[M + H]^+$, $[M + Na]^+$) of large biological compounds. MALDI ionization sensitivity is also extraordinary, and total amounts of sample loaded onto the target surface often are in the picomole to femtomole range. The method has tolerance to buffers and other additives and gives predominantly singly charged ions for large biomolecules [35]. TOF mass analyzers are ideal for use with this ionization technique because they are compatible with high-mass ions and pulsed-ion production [34,35]. TOF analyzers separate ions according to their *m/z* ratios by measuring the time it takes for ions, accelerated to the same kinetic energy, to travel through a field-free region known as the flight or drift tube. The heavier ions move slower than the lighter ones [6].

An important application of MALDI is chemical imaging, using a technique called matrix-assisted laser desorption/ionization imaging mass spectrometry (MALDI-IMS) [16]. Imaging combines parallel, high-throughput molecular analysis with location-specific information for the characterization of protein distributions directly from thin sections of intact biological tissue [39,40] and offers complementary information to two-dimensional (2D) gel electrophoresis and to shotgun proteomics for investigating proteomic differences. It is covered in Chapter 5 by Reyzer and Caprioli in this volume. Figure 1.2 illustrates the typical experimental process for MALDI imaging. Frozen tissue specimens are sectioned on a cryostat into about 5- to 20-mm thick sections. The sections are thaw-mounted onto conductive MALDI target plates. Matrix is applied to the sections, depending on the experiment to be performed: droplets (nL to pL) can be deposited in arrays or on discrete morphological areas, or a uniform coating of matrix can be applied to the entire tissue section.

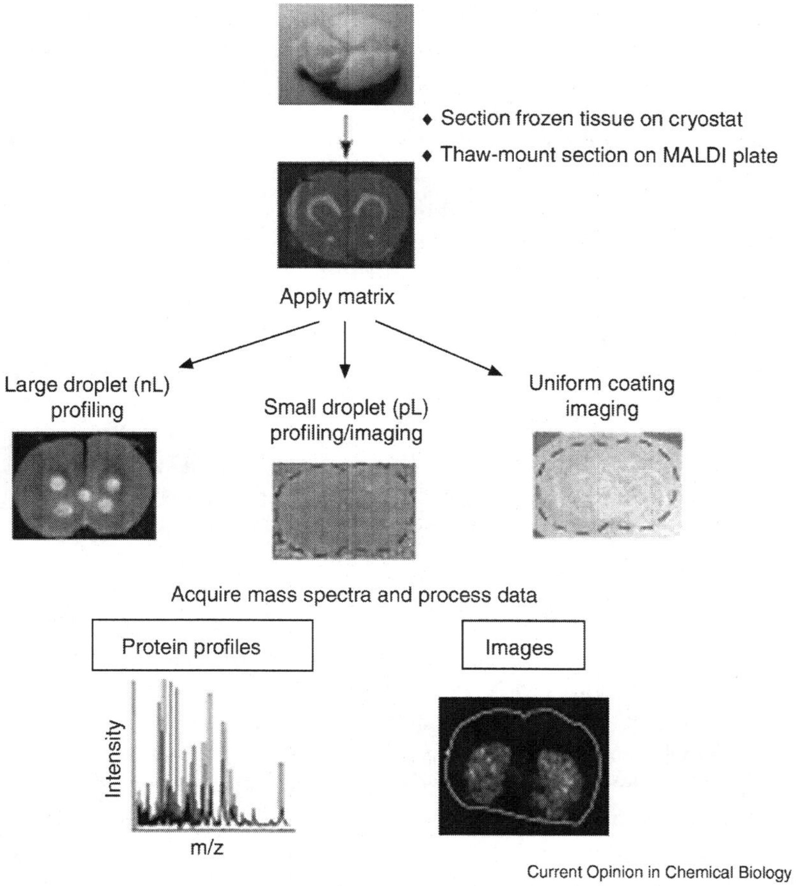

FIGURE 1.2 Typical experimental design for IMS. From [16]. Copyright permission was obtained from Elsevier. (*See the color version of this figure in Color Plates section.*)

Mass spectra are obtained from each spot or from across the entire tissue section in a defined raster pattern. The acquired spectra can then be examined and processed to form 2D molecule-specific ion images [16]. The power of MALDI-IMS technology is its capability to link reliably protein data with specific cellular regions within the tissue.

MALDI-IMS has been employed as an imaging technology in a wide variety of applications from the analysis of small molecules such as drugs and endogenous metabolites to high molecular weight proteins (e.g., MALDI-IMS of a mouse model of Parkinson's disease revealed a significant decrease in PEP-19 expression levels in the striatum after administration of the drug MPTP [41]).

1.2.2 Atmospheric Pressure Matrix-Assisted Laser Desorption/Ionization (AP-MALDI)

Atmospheric pressure matrix-assisted laser desorption/ionization (AP-MALDI) was first described by Laiko et al. [42]. In contrast to conventional vacuum MALDI, AP-MALDI can be operated at atmospheric pressure instead of high vacuum where ions are typically produced at 10 mTorr or less. During the ionization process, the solid-phase target material containing analyte sample and matrix is irradiated with a pulsed laser beam. The matrix absorbs the photon energy and undergoes fast heating and evaporation, which results in the formation of gaseous analyte ions [43]. Because the ionization of AP-MALDI occurs at atmospheric pressure, thermalization of the resulting ions takes place owing to collisions with the ambient gases used in AP-MALDI, accounting for the soft ionization nature of AP-MALDI [42].

The AP-MALDI source makes use of a high voltage potential that is applied between the target tip and the heated inlet transport capillary. The laser is focused onto the surface of the target plate. Ions are desorbed from the angled replaceable target tip and carried by the dry carrier nitrogen gas into a mass spectrometer [44]. The sensitivity of detection for AP-MALDI can be affected by the geometry of the target tip, and its position relative to the inlet orifice, the nitrogen gas flow rate, gas nozzle position, etc. [42]. Furthermore, when a Nd: YAG laser with high laser power rather than a nitrogen laser is used, the signal intensity can be improved [45].

Given that AP-MALDI and conventional vacuum MALDI share common ionization mechanisms, they have many similar features including simplicity of sample preparation and tolerance to interference from salts [42], which can be detrimental for biomolecule analysis. AP-MALDI is an extension of conventional MALDI, but it has some unique characteristics. First, samples are handled at atmospheric pressure. Second, AP-MALDI is a softer ionization technique than vacuum MALDI, which is favored for protein analysis. For example, the heavier peptides/glycopeptides from protein digestion are less likely to fragment by AP rather than vacuum MALDI; thus, more peptides can be detected (Figure 1.3) [42]. Third, AP-MALDI can employ liquid matrices to improve the ionization reproducibility, which otherwise results in source contamination in vacuum MALDI. Fourth, the AP-MALDI ion source is easy to exchange with other atmospheric ionization sources, allowing it to be easily coupled to different mass analyzers (e.g., quadrupole ion trap (QIT) [46], TOF [45],

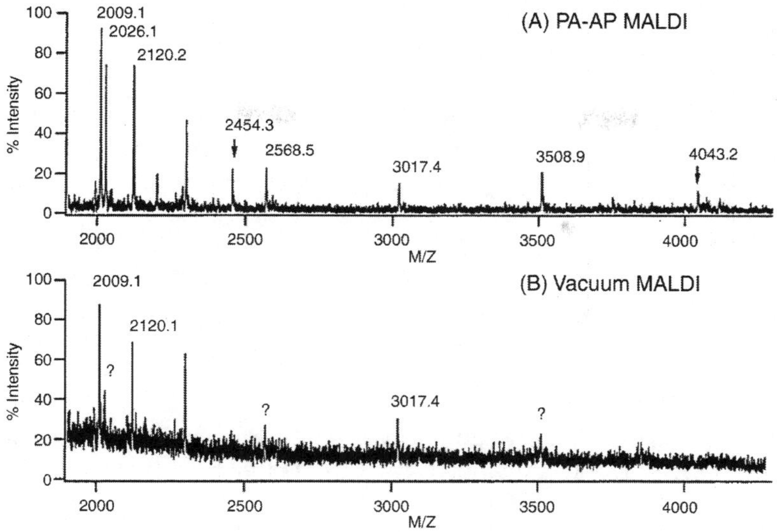

FIGURE 1.3 Partial mass spectra for tryptic digest of bovine fetuin. (*A*) AP-MALDI spectrum from 1.5-pmol deposition, showing eight identified peaks. (*B*) Vacuum MALDI spectrum for 1-pmol deposition, showing three identified peaks. From [42]. Copyright permission was obtained from ACS.

and Fourier transform ion cyclotron resonance (FT-ICR) [47]. Fifth, the analyte-matrix cluster ions in AP-MALDI caused by collisional cooling can be observed [43]. Nevertheless, the major disadvantage of AP-MALDI is the ion loss in atmospheric pressure interfaces, giving it lower sensitivity than conventional MALDI [46].

AP-MALDI MS has seen a variety of applications similar to those of conventional vacuum MALDI, including analysis in proteomics and determinations of oligosaccharides, DNA/RNA/PNA, lipids, bacteria, phosphopeptides, small molecules, and synthetic polymers [48]. The convenient and rapid exchange of the AP-MALDI source with other ionization sources and high throughput are attractive features. The major expected application for AP-MALDI is for the analysis of vacuum-incompatible samples like profiling of biological tissue samples, which requires the use of a wide range of liquid matrices at atmospheric pressure [43].

1.2.3 Surface-Enhanced Laser Desorption/Ionization (SELDI)

Surface-enhanced laser desorption/ionization (SELDI) as a prominent form of laser desorption/ionization (LDI) mass spectrometry was first described in 1993 by Hutchens and Yip [49]. It can be classified in three groups: surface-enhanced neat desorption (SEND), surface-enhanced affinity capture (SEAC), and surface-enhanced photolabile attachment and release (SEPAR) [50]. In SEND, analytes even for large molecules can be desorbed and ionized without adding matrix. This occurs because a compound with a chromophore to absorb laser energy is attached to the probe surface via physical adsorption or covalent modification [51]. In SEAC, the probe surface plays an active role in the extraction, fractionation, cleanup, and/or amplification of the sample of interest. Common are chemical surfaces such as H50 (hydrophobic surface, similar to C_6–C_{12} reverse-phase chromatography materials), CM10 (weak-positive ion exchanger), Q10 (strong anion exchanger), IMAC30 (metal-binding surface), and biochemical surfaces containing antibodies, receptors, enzymes or DNA [50,52]. In SEPAR, an energy-absorbing molecule promotes analyte desorption and ionization, making this approach a hybrid of SECA and SEND [50]. Furthermore SELDI is commercially embodied in Ciphergen's ProteinChip® Array System (Ciphergen Biosystemes, Palo Alto, CA, USA), which simplifies the sample preparation with on-chip binding and detection [50]. SELDI is typically coupled with TOF mass spectrometers and is applied to detect proteins in tissue, urine, blood, and other clinical samples.

SELDI-TOF-MS is the extended form of MALDI-TOF-MS. The differences are the sample preparation and the software tools for interpreting the acquired data. In executing the SELDI process (Figure 1.4) [53], the first step is to select a chromatographic and preactivated ProteinChip array. Next, the protein sample solution is applied and incubated on the spots of the ProteinChip array. Third, by allowing the proteins to interact with the chromatographic array surface, on-spot contaminants and salts of the sample can be washed away to ensure efficient sample cleanup. This binding step to the SELDI surface can be viewed as a separation step, purifying the proteins bound to the surface. Fourth, matrices are added for the formation of a homogeneous layer of cocrystallized target proteins. After that, a laser beam is used to irradiate the spot, causing desorption and ionization of the proteins. The laser beam raster can be applied to cover selectively the entire spot surface, affording an output of the entire spot. Finally, multiple spectra are averaged to yield a final spectrum that displays the protein ions. Protein quantification is achieved by the correlation between the signal intensities and analyte concentrations of proteins in the sample.

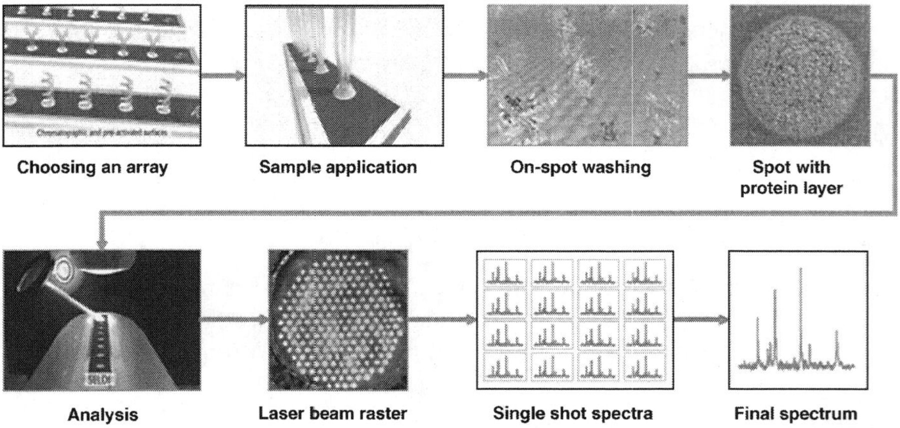

FIGURE 1.4 Experimental steps of SELDI-TOF-MS-based ProteinChip® System. From [53]. Copyright permission was obtained from Nature Publishing Group. (*See the color version of this figure in Color Plates section.*)

For the analysis of complicated biological systems containing hundreds of biological molecules together with salts (e.g., serum, blood, plasma, lymph, urine, whole cells, exudates) by MS, sample preparation and purification are necessary. Compared with some classic sample purification methods like liquid chromatography, electrophoresis, centrifugation, and immunoprecipitation, which are subject to losses of both analyte and minor components owing to nonspecific binding, SELDI can be directly and readily used to analyze the major and minor proteins in heterogeneous samples. This ionization method for analysis of macromolecules efficiently facilitates the investigation of biological molecules on-probe and simplifies sample purification and extraction steps in contrast to conventional LDI and MALDI [50]. Furthermore SELDI is rapid, highly reproducible, and offers good sensitivity for trace protein ($<$ fmol/mL using chemical arrays) analysis. It has had some impact in proteomics and drug discovery and can be used for discovery, analysis, and identification of post-translational modifications of disease-associated proteins [54].

1.2.4 Nanostructure-Initiator Mass Spectrometry (NIMS)

Nanostructure initiator mass spectrometry (NIMS) was introduced as a substitute to overcome typical limitations (sensitivity and spatial resolution) found with the use of matrices in laser methods such as MALDI. NIMS is a matrix-free, surface-based MS desorption/ionization technique that uses nanostructured surfaces or clathrates to trap liquid "initiator" materials (e.g., bis(tridecafluoro-1,1,2,2-tetrahydrooctyl)tetramethyl-disiloxane). These materials are released upon heating by laser irradiation, carrying with them absorbed analyte molecules (Figure 1.5*A*) [55–57]. The technique has been used in the characterization of proteolytic digests, single cells, tissues, biofluids (direct analysis of blood and urine), lipids, drugs, and carbohydrates. Imaging applications include peptide arrays, tissue (tissue/surface interface), and single cell. Some attributes of NIMS are minimal sample preparation, high-sensitivity/lateral resolution (ion-NIMS: 150 nm, as compared to MALDI and ESI), high salt tolerance, compatibility with standard laser based instruments, and reduced fragmentation (favored intact ion formation). The NIMS technique is also flexible, accommodating a variety of irradiation sources (laser or ion), surfaces, and initiator (depending on target analyte) compositions. In contrast to conventional MALDI, NIMS is capable of producing multiply charged proteins as ESI or cryo-infrared MALDI.

The nanostructured silicon surface in NIMS is composed of pores of approximately 10 nm in diameter (Figure 1.5*C*). Initiator molecules (Figure 1.5*B*), which are chosen depending on the target analyte, are trapped inside these pores. The initiator molecules are UV laser transparent (do not ionize) whereas the silicone nanostructure is an efficient UV absorber semiconductor. Analyte molecules are adsorbed on the initiator surface and desorbed upon initiator vaporization, caused by laser or ion irradiation. When ion irradiation is used, spatial resolution of approximately 150 nm can be achieved whereas laser-NIMS produces a spatial resolution of approximately 15 to 20 μm [55–57].

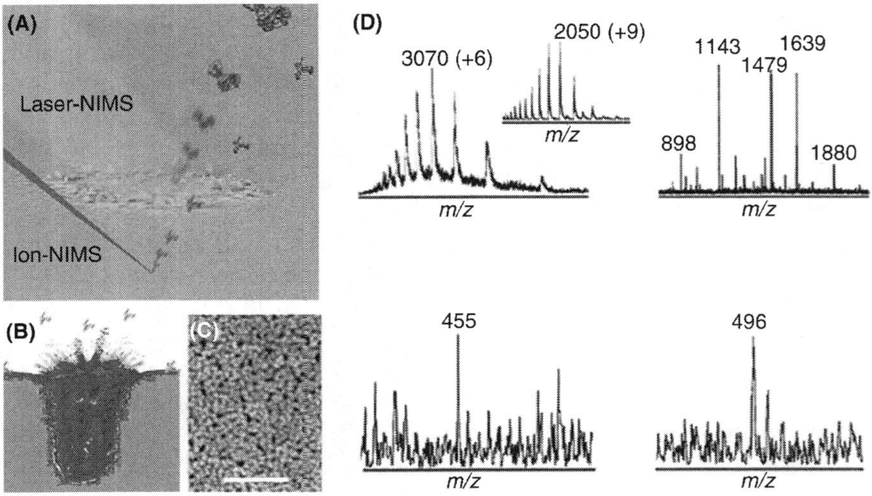

FIGURE 1.5 (*A*) Illustration superimposed on an SEM image of a NIMS surface after irradiation with a single laser shot (light grey), revealing localized surface distortion and destruction. By comparison, ion irradiation (dark grey red) allows a much higher lateral resolution. (*B*) Illustration of possible mechanism in which surface irradiation results in the vaporization or fragmentation of initiator (dark grey blue) trapped in a surface pore, triggering analyte desorption/ionization. (*C*) SEM image revealing that the NIMS surface is composed of 10-nm pores; scale bar, 100 nm. (*D*) Laser irradiation (wavelength 337 nm) of a NIMS surface. Upper left panel: detection of a multiply charged protein (50 nmol of β-lactoglobulin) in a similar manner to ESI (inset). Upper right panel: detection of a BSA tryptic digest (500 amol). Lower left panel: detection of the calcium antagonist verapamil (700 ymol). Lower right panel: detection of the endogenous metabolite 1-palmitoyllysophosphatidylcholine (50 amol). The initiator was bis(tridecafluoro-1,1,2,2-tetrahydrooctyl)tetramethyl-disiloxane; 0.5-µl drops were used. From [57]. Copyright permission was obtained from Nature Publishing Group.

Protein ionization via laser-NIMS generates ESI-like spectra showing multiply charged states (Figure 1.5*D*). In this specific case, laser-NIMS showed lower charges states than ESI for β-lactoglobulin, suggesting that the protein is less denatured by the NIMS ionization process. The superior sensitivity of laser-NIMS versus ESI or MALDI is also shown in Figure 1.5*D* for the detection of BSA peptide fingerprints at 500 amol (55% sequence coverage) [57]. Endogenous phospholipids can be detected from single metastatic breast cancer cells with less complexity than nano-ESI or MALDI [57]. Ion-NIMS, on the other hand, can be successfully used for high-resolving-power, label-free peptide array analysis, showing mass images and mass spectra collected for 1 fmol of peptide, representing a 1000-fold enhancement in sensitivity over TOF-SIMS strategies [57]. Typical laser energies for desorption/ionization with NIMS are approximately seven times lower than that of MALDI for analysis of a mixture of the tetrapeptide MRFA (50 fmol) and des-Arg9-bradikinin (25 fmol) when applying a laser energy of 110 mJ/cm^2 for MALDI and 15 mJ/cm^2 for NIMS. Better S/N ratios and less background ions are found for the collected NIMS mass spectra [58].

1.3 SPRAY-BASED IONIZATION METHODS FOR PROTEINS

1.3.1 Electrospray Ionization (ESI)

The principle of electrospray ionization was first described by Dole in 1968 [59] and coupled to MS in 1984 by Yamashita and Fenn [2]. ESI usually generates intact, multiply charged ions, generally in the form $[M + nH]^{n+}$ in both the positive (e.g., protonated) and negative (e.g., deprotonated) ion modes. In ESI-MS, "naked" ions form *via* progressive solvent evaporation from charged droplets of a liquid sample, sprayed in the presence of a strong electrical field. The formation of gaseous analyte ions by electrospray involves three steps: formation of charged droplets, shrinkage of the droplets owing to solvent evaporation, and transfer of ions to the gas phase. Although the macroscopic aspects of electrospray are generally well understood, the mechanisms for the final generation of desolvated (or nearly desolvated) ions from a charged droplet are not yet fully resolved. Two models describe this process. The charged residue model (CRM), conceived by Dole et al. [59], invokes successive cycles of solvent evaporation and coulombic fission at the Rayleigh limit until a droplet containing a single residual analyte ion remains. Complete evaporation of the solvent comprising this droplet eventually yields a "naked" analyte ion, the charged residue. The ion-evaporation model (IEM) proposed by Iribarne and Thomson [60] is based on transition-state theory and invokes, prior to complete desolvation of the droplet, sufficiently strong repulsion between the charged analyte ion and the other charges in the droplet that becomes to overcome solvation forces and the ion is ejected (field-desorbed) from the droplet surface into the gas phase [61].

With the advent of ESI, it became possible to study protein conformations. Different from traditional methods to investigate protein conformations such as circular dichroism (CD), NMR and X ray, ESI-MS offers several advantages for this purpose. First, ESI-MS is sensitive, requiring fmol and amol amounts of protein samples [12,62,63]. Second, ESI analysis makes use of a protein solution, which is important because most of biology and much of separations take place in solution. In traditional ESI experiments, organic compounds are often used as co-solvents; however, the use of highly organic solvents is no longer mandatory. This has lead to the birth of an emerging field in biomolecular MS, termed native ESI-MS [61,64–66]; the focus of this field is the analysis of intact proteins and protein complexes under near physiological conditions achieved by using neutral volatile buffer salts like ammonium acetate for protein sample preparation. The third is that gas-phase, multiply charged ions are generated from the protein sample [3]. This point plays a central role in protein studies, given that the charge-state distributions (CSDs) observable in protein ESI mass spectra are affected by the conformations that the protein held in solution at the moment of its transfer to the gas phase [12,67]. Typically, when a protein is in the folded structure, a narrow CSD in low-charge states is observed whereas the CSD is broadened and shifted to high-charge states after unfolding, probably because the unfolded protein has a greater capacity to accommodate charges on its surface because coulombic repulsions are reduced [62,68,69]. Therefore, information about the conformational states of the protein can often be extracted based on the structural interpretation of CSDs in ESI-MS, upon controlling other experimental conditions [12]. Another MS-based approach to protein conformation study is to monitor protein hydrogen/deuterium exchange reactions, which are sensitive to the conformational structure; thus the exchange level determined by MS can be related to protein conformation [70–76], and this subject is covered in Chapter 7 by Kerfoot and Gross in this volume.

In 1994 Wilm and Mann introduced an important variant of conventional ESI, termed nanoelectrospray (nESI) [77]. While this technique uses the same fundamental sequence of charged-droplet generation followed by solvent evaporation, coulombic fission events, and finally ion formation, it is distinguished from regular ESI in several ways. First, nESI is typically performed using glass or quartz capillaries that are pulled to a fine tip (~1-μm inner diameter) and given a metallic (usually gold) coating to hold the electric potential; these are used instead of the metallic capillary used for conventional ESI. Approximately 1 to 3 μL of sample is injected into the glass capillary and electrosprayed at flow rates in the range of around 1 nL/min to several tens of nL/min [78,79]. The spray is driven primarily by the approximately 0.5 to 1.5-kV potential applied to the capillary, although it is often necessary to provide an auxiliary backing gas pressure to the sample to initiate and/or maintain a steady stream of the solution through the tip [61]. Second, in comparison to conventional ESI, a smaller initial droplet size in nESI leads to less nonspecific aggregation (both protein–protein and protein–salt), and its gentler interface conditions, while still allowing adequate desolvation, lead to less dissociation and disruption of oligomeric and higher order structures (Figure 1.6 shows the contrast between nESI and ESI for the ionization of a GroEL complex). The benefits of nESI analysis include high ionization efficiency, well-resolved peaks corresponding to the protein assembly, narrow charge-state distributions, reduced nonspecific adduct formation, and high salt tolerance.

1.3.2 Sonic Spray Ionization (SSI)

Besides ESI, another spray technique that can be successfully used for the analysis of proteins and peptides is sonic spray ionization (SSI) [80–83]. This soft atmospheric pressure ionization (API) method was first introduced by Hirabayashi et al. [84] in the early 1990s as a method for interfacing capillary electrophoresis and liquid chromatography instrumentation to mass spectrometers. The source works at room temperature (no heating applied to capillary) [85,86]. Ions and charged droplets are produced under atmospheric pressure, and their abundances depend on the nebulization gas flow rate. Optimal ion abundances are obtained at Mach numbers of approximately 1, which corresponds to sonic velocity [84,85].

In SSI, a solution is infused through a fused-silica capillary, which is fixed by an external stainless steel capillary, allowing its accurate positioning in the source body (Figure 1.7*A*). The fused-silica capillary is then inserted into an orifice from which it protrudes approximately 0.6 mm [84]. Nitrogen gas is then passed through the orifice, coaxial to the fused-silica capillary, nebulizing the eluent at gas flow rates that match sonic velocities. The generated spray, composed of charged droplets and ions at

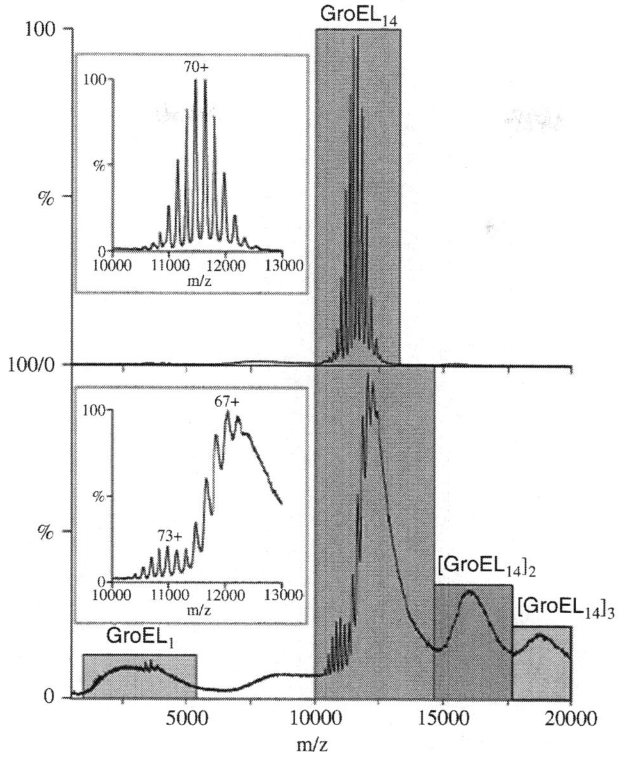

FIGURE 1.6 Conventional and nanoelectrospray MS of a protein complex. MS of the GroEL complex ionized by means of ESI (lower) and nESI (upper). Solution conditions were 200-mM ammonium acetate, pH 6.9, and a protein concentration of 2-μM tetradecamer. The nESI spectrum displays a series of peaks around 11,500 *m/z*, which correspond to the 800 kDa tetradecamer. Conventional ESI of the same solution results in poorly resolved "humps" centered on 12,500, 16,000, and 18,500 *m/z*. These are assigned to the tetradecamer, a dimer of tetradecamers, and a trimer of tetradecamers, respectively. There is also a signal at low *m/z* that corresponds to the GroEL monomer. From [61]. Copyright permission was obtained from ACS.

atmospheric pressure, is then introduced through a sampling orifice into the mass spectrometer for mass analysis.

The mechanism of ion formation by SSI is not yet well understood. Early studies on the charged droplet formation mechanism suggest that the origin of the charged species cannot be ascribed to the traditional models of friction electrification (between the capillary surface and the solution), electrical double layer or statistical charging model [85,87]. Instead, a charged droplet formation mechanism occurs based on the non-uniformity of positive and negative-ion concentrations near the solution surface (at a gas boundary), determined by the surface potential [83]. The charged droplet formation in SSI may be based on the statistical charging model (sudden evaporation of liquid into smaller equally sized droplets, which are charged owing to microscopic

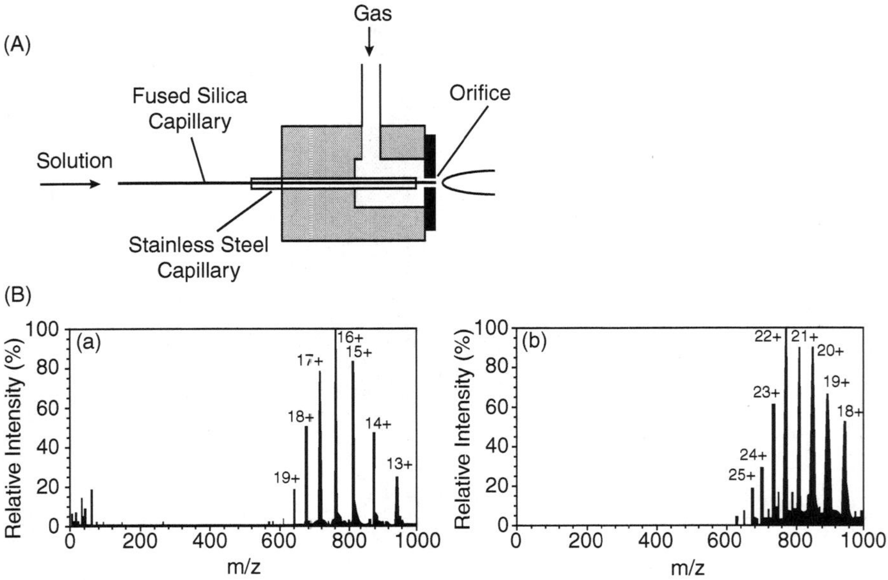

FIGURE 1.7 (*A*) Typical schematic of an SSI. From [84]. Copyright permission was obtained from ACS. (*B*) Mass spectra obtained from methanol/water/acetic acid (47.5/47.5/5.0%, v/v/v) solutions of (*a*) cytochrome *c* from sheep heart (MW 12, 300) and (*b*) myoglobin from horse skeletal muscle (MW-17,000 Da). A high voltage of 1 kV was applied to the source housing and the gas-flow rate was 3.0 L/min. From [83]. Copyright permission was obtained from Wiley.

fluctuations in the ion concentration in a bulk liquid). Gaseous ions are formed as a result of the charge residue model (continuous evaporation and fission cycles leading to droplets that contain on average one analyte or less); gas-phase ions are then formed after the remaining solvent has evaporated [82,84,85,88–92]. Given that no electrical field is applied to the solution in SSI, low charge-state ions are produced [83]. When one applies a high voltage to the source housing (solution surface), one can see increased charge density on the droplets and improved ion formation efficiencies [83].

An attribute of SSI is its simplicity because no high voltage or heating are used in ion formation. Ions are typically formed with low internal energies, making this technique promising for the study of thermal labile molecules, cluster ions, and fragile complexes (i.e., loosely bound metal-assembled cages) [83,93–95]. This attribute may be a disadvantage because excessive clustering makes data interpretation difficult. Given that low charge-state ions are typically generated by SSI, high voltages must be applied to the source housing to increase charge density on the droplets [83]. As an example of cluster formation and SSI gentle ionization character, the abundance for the protonated L-serine octamer is approximately 10–15 times higher when formed via SSI than by ESI, with virtually no oligomeric species, primarily attributed to the lower average internal energies of the ions produced as compared to ESI [91].

There are reported protein and peptide applications using SSI, and they include the analysis of RNase A, lysozyme, bovine serum albumin (BSA), myoglobin, cytochrome *c*, and carbonic anhydrase II [80–82]. Applications in other areas (e.g., drugs [96,97], oligosaccharides [98,99], phenolic compounds [100], oligonucleotides [101,102], and neurotransmitters [85]) have been reported, but these are beyond the scope of this chapter. The first spectra of proteins and the formation of multiply charged ions (Figure 1.7*B*) [83] by SSI were collected with a quadrupole mass spectrometer. The spectra show charge state distributions ranging from 13^+ to 19^+ for cytochrome *c* (MW $\approx$ 12,300) and from 18^+ to 25^+ for myoglobin (MW $\approx$ 17,000).

Capillary isoelectric focusing (CIEF) can be coupled with MS by using SSI for the analysis of proteins, as reported by Hirabayashi et al. [80,81]. An SSI interface setup with a buffer reservoir placed in between the sample introduction capillary of the ion source and the electrophoresis-separation capillary is required. This allows for online and one-step CIEF/MS analysis. Given that SSI uses a high-velocity gas to generate the spray, one can use a wide range of buffers solutions, solution flow rates, and high-polymer ampholytes (used in CIEF) without clogging the spray nozzle of the interface. Filling the buffer reservoir with acetic acid and introducing it through a pinhole into the sample introduction capillary of the SSI source reduces ion suppression caused by the ampholytes used in CIEF. Using this approach, one can detect down to 160 fmol of myoglobin and cytochrome *c* and separate the acidic and basic bands of myoglobin. In a recent application, SSI was used to obtain spectra of proteins with low charge states (as compared to ESI), hence decreasing overlap of peaks obtained from protein mixtures and facilitating mass spectral interpretation [82]. When contrasting SSI and ESI analysis of RNase A and lysozyme, one sees a dramatic reduction in charge states with SSI as the ionization method for these two proteins.

1.3.3 Electrosonic Spray Ionization (ESSI)

Electrosonic spray ionization (ESSI) is a hybrid between ESI and SSI; it uses a traditional micro ESI source and a supersonic gas jet similar to SSI [103]. The method can be used to study protein–ligand complexes owing to its gentle ionization character that allows formation of cold ions (low internal energies) [103–106]. Not only polymers [107] can be analyzed, also gas-phase basicities of proteins and peptides can be measured by this method [108–112]. The most distinctive characteristics of the ESSI method are the narrow charge-state distributions and narrow peak widths (efficient desolvation) as compared to those of ESI and nanospray [103]. The method can preserve solution protein and protein complex structures at physiological pH values, ionizing the systems with a charge-state distribution characteristic of its conformation in solution [103,104]. The formation of broader charge-state distributions is typically associated with unfolding of proteins during ionization whereas narrow and lower charge-states (as observed in ESSI) are associated with native-like or folded (which defines their biological role) ion structures in the gas phase [113]. Tolerance to high salt concentrations, tunable source potential, lack of arcing, and weak dependence on temperature are

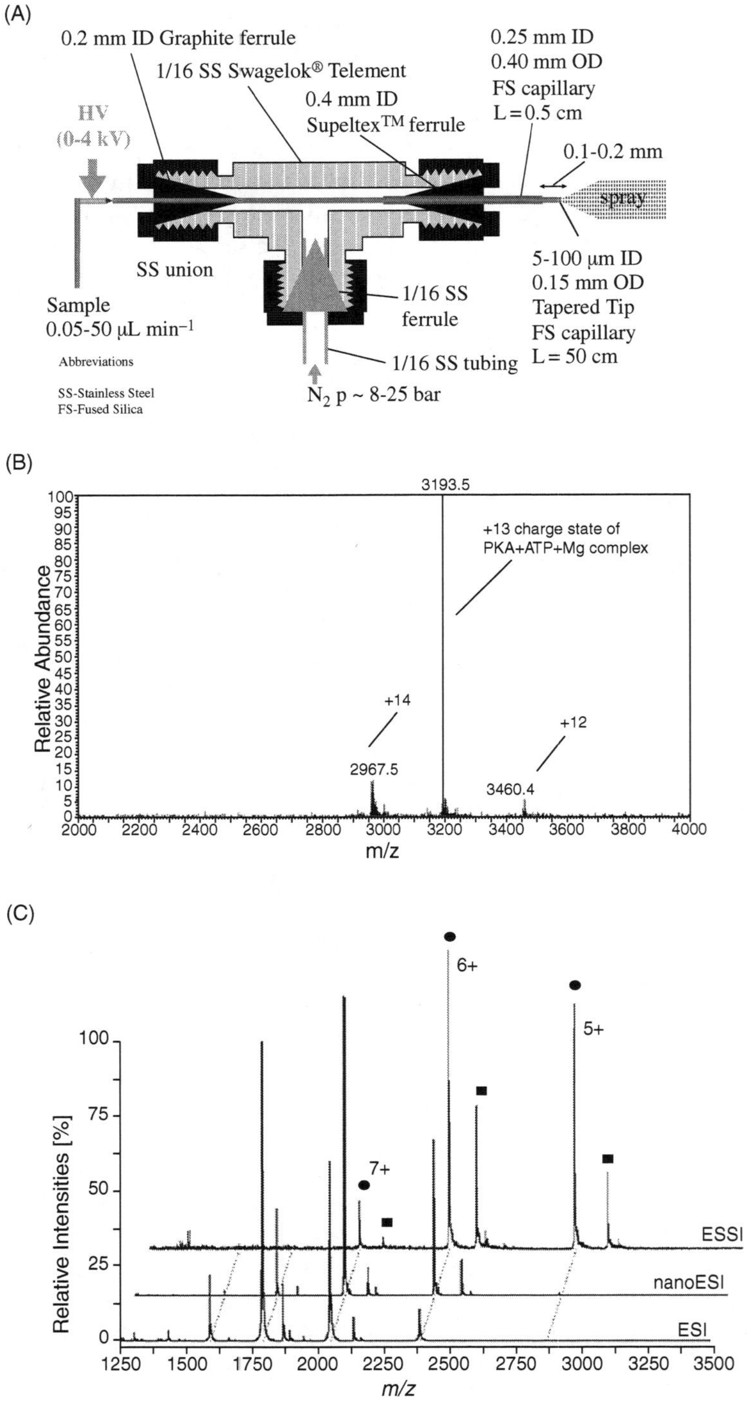
(A)
0.2 mm ID Graphite ferrule
1/16 SS Swagelok® Telement
0.4 mm ID
Supeltex™ ferrule
0.25 mm ID
0.40 mm OD
FS capillary
L = 0.5 cm
HV
(0-4 kV)
0.1-0.2 mm
spray
SS union
1/16 SS
ferrule
5-100 μm ID
0.15 mm OD
Tapered Tip
FS capillary
L = 50 cm
Sample
0.05-50 μL min−1
1/16 SS tubing
Abbreviations
SS-Stainless Steel
FS-Fused Silica
N2 p ~ 8-25 bar
(B)
3193.5
+13 charge state of
PKA+ATP+Mg complex
+14
2967.5
+12
3460.4
Relative Abundance
m/z
(C)
6+
5+
7+
ESSI
nanoESI
ESI
Relative Intensities [%]
m/z

other attributes of the technique [103]. ESSI can be successfully coupled to several mass analyzers, including hybrid quadrupole time of flight [105], triple quadrupoles, and linear ion traps [103,112], thus demonstrating that it can be used in any instrument that has an API interface.

The general design of an ESSI source consists of a gas nebulizer made of inner and outer deactivated fused silica capillaries (Figure 1.8*A*). Nitrogen (N_2) is used as the nebulizing gas at a typical flow rate of 3 L/min (sonic velocity). The solvent is sprayed under the influence of an applied high voltage, typically in the range 0 to 4 kV (0 kV would be "pure SSI"). The voltage can be applied to the liquid sample through a copper alligator clip that attaches to the stainless steel tip of the infusion syringe. The gas jet composed of electrosprayed aqueous microdroplets and free gas-phase ions is directed to the inlet of an atmospheric interface of the mass spectrometer [91].

The mechanism of ESSI ion formation is likely to be the charge residue model [88,114,115]. The main difference between ESSI and ESI or nano-ESI is that ESSI is a more efficient desolvation process, attributed to the production of initial ultrafine droplets (generated by the supersonic nebulizing gas). These droplets are easily desolvated in a short time [88,103]. The faster desolvation and low temperatures of the spray, caused by adiabatic expansion of the nebulizing gas, leads to the formation of ions with low internal energies (lower than those produced by ESI or nano-ESI), giving ESSI the required "softness" for the analysis of noncovalent interactions.

Electrosonic Spray Ionization for Protein Analysis ESSI can be a useful tool for the study of noncovalent interactions owing to its soft ionization character (a comparison of ESSI and nano-spray generated spectra recorded for trypsin was carried out, showing narrower peaks and lower charge states for ESSI) [103]. The ESSI spectrum is dominated by a single charged state, whereas the other charge states do not contribute to more than 5% relative abundance. Efficient transfer by ESSI of intact complexes to the mass spectrometer can be achieved (Figure 1.8*B* for kinase A after conversion to its ATP/Mg adduct by addition of excess ATP Mg salt) [103]. Other applications include the use of deprotonation reactions in an evaluation of gas-phase basisities of globular and denatured proteins [110] and the analysis of enzyme-substrate and enzyme-substrate inhibitor complexes [104]. ESSI can be utilized to measure dissociation constants (K_D) for protein–ligand systems, showing good agreement with solution results [105]. A comparison of K_D values obtained by

FIGURE 1.8 (*A*) Schematic of an ESSI source. From [103]. Copyright permission was obtained from ACS. (*B*) ESSI spectrum of bovine protein kinase A catalytic subunit (200 nM in 10-mM aqueous ammonium acetate, pH 7.8) in the presence of 100-μM ATP Mg salt. From [103]. Copyright permission was obtained from ACS. (*C*) Representative mass spectra of the noncovalent HEWL-NAG_3 complex (HEWL 10 μM, NAG_3 60 μM in 20-mM ammonium bicarbonate buffer) using different ionization techniques. The charge states of free protein signals (filled circle) and HEWL-NAG_3 complex signals (filled square) are given for ESSI. From [56]. Copyright permission was obtained from Elsevier.

ESSI with those from ESI or nano-ESI shows that ESSI can give K_D values that are most similar to those determined by solution methods. One reason may be that ESSI ion complexes are less prone to dissociation as compared to those formed by ESI or nano-ESI. An example is the HEWL-NAG3 complex (Figure 1.8*C*), demonstrating the softness of the method [105].

1.4 AMBIENT IONIZATION METHODS

In most applications, MS required moderate to extensive sample preparation followed by introduction of the sample into the high vacuum conditions prior to analysis, limiting in-situ analysis and increasing the possibility of contamination during sample handling. These drawbacks are overcome with the introduction of desorption electrospray ionization (DESI) and direct analysis in real time (DART), which can be viewed as ambient ionization methods. Samples can be examined in the open environment (natural or in the laboratory), and typically no sample preparation is required, allowing for in-situ analysis while preserving all attributes associated with MS analysis. These approaches should open a new era in mass spectrometry.

After the first reported applications using DESI and DART [116,117], a whole new family of ambient methods and variants emerged. DESI variants such as reactive-DESI (reactions accompanying desorption), nonproximate detection DESI (transport of sample ions at long distances), geometry-independent DESI, transmission-mode DESI, and liquid sample DESI were soon introduced either to increase selectivity and sensitivity for trace analysis or to facilitate direct sample analysis [118–124]. Another ionization method termed desorption atmospheric pressure chemical ionization (DAPCI) was also developed to study ionization mechanism for explosive compounds [118]. Other established ambient ionization methods include electrospray-assisted laser desorption/ionization (ELDI) [125], matrix-assisted laser desorption electrospray ionization (MALDESI) [126], extractive electrospray ionization (EESI) [127], atmospheric solid analysis probe (ASAP) [128], jet-desorption ionization (JeDI) [129], desorption sonic-spray ionization (DeSSI) [130], field-induced droplet ionization (FIDI) [131], desorption atmospheric pressure photoionization (DAPPI) [132], plasma-assisted desorption ionization (PADI) [133], dielectric barrier discharge ionization (DBDI) [134], liquid microjunction surface sampling (LMJ-SSP) [135], atmospheric pressure thermal desorption ionization (APTDI) [136], surface-sampling probe (SSP) [137], fused-droplet electrospray ionization (FD-ESI) [138], helium atmospheric pressure glow discharge ionization (HAPGDI) [139], neutral desorption extractive electrospray ionization (ND-EESI) [140], laser ablation electrospray ionization (LAESI) [141], low-temperature plasma (LTP) [82], and laser spray ionization (LSI) [142]. Although these methods can be used for ambient analysis, protein or peptide analysis has been achieved in a few cases owing to the ionization process involved (i.e., the amount of internal energy deposited into a protein). In the following subsection, we will focus on instrumentation, ionization mechanisms, and the successful applications on protein analysis of various ambient methods.

1.4.1 Desorption Electrospray Ionization (DESI)

DESI allows to record spectra of condensed-phase samples (pure, mixtures, or tissue) under ambient conditions, making the samples accessible during analysis for manipulation by ordinary physical or chemical means [118,143–146]. Analysis of small and large molecules, very short analysis time (high-throughput), high selectivity (reactive-DESI and MS/MS), and sensitivity are other attributes of this method. The DESI method is based on directing a pneumatically-assisted electrospray onto a surface (e.g., paper, metal, plastic, glass, and biological tissue), from which small organics and large biomolecules are picked up, ionized, and delivered as desolvated ions into the mass spectrometer. Ions are generated by the interaction of charged microdroplets or gas-phase ions derived from the electrospray with neutral molecules of analyte present on the surface [116,118]. DESI is a soft ionization method and shows ESI-like spectra of proteins, primarily attributed to some common features of the ionization process that produces low-energy intact molecular ions through fast collisional cooling under atmospheric conditions [145]. The method can be used for many types of compounds (polar/nonpolar, and low/high molecular weight) in forensics and homeland security (e.g., explosives, chemical warfare agents, bacteria) [118,120,121,146–148], biomedical (e.g., tissue imaging, proteomics, lipidomics, pathology) [26,149–153], pharmaceutical/industrial (e.g., drug analysis, pharmacokinetics, polymers, process monitoring, metabolomics, environmental analysis) [107,154–158], and other fields. Many of these applications can be implemented with various mass spectrometers, including triple quadrupoles [159], linear ion traps [160], Orbitrap [161], quadrupole time of flight (QTOF) [162], ion-mobility/TOF, and ion-mobility/QTOF hybrids [162], Qtraps [122], Fourier transform ion cyclotron resonance (FTICR) instruments [163], and miniature ion trap mass spectrometers [164].

DESI Ionization Source In a typical DESI setup (Figure 1.9*A*) the source consists of a solvent nebulizer made of deactivated fused-silica capillary, similar to the one used in ESSI [103]. Nitrogen (N_2) is used as the nebulizing gas at a linear velocity of approximately 350 m/s. The solvent (typically mixtures of methanol, water, and small amount of acetic acid) is sprayed under the influence of an applied high voltage (typically in the range 3–6 kV). The gas jet composed of electrosprayed aqueous microdroplets and free gas-phase ions is directed onto the analyte-containing surface at various incident angles (usually from as low as 25° up to 80° depending on the analyte) to the normal. The resulting droplets, ions, and neutrals are collected at a shallow angle from the surface. The ions are then transferred as a result of electrostatic and pneumatic forces to a mass spectrometer equipped with an atmospheric pressure interface. The source is typically mounted on an *xyz*-moving stage, allowing it to be positioned at any chosen point with respect to the sample. The moving stage also has a tangent arm drive miniature stage that allows precise angular adjustment from 0 to 90° (Figure 1.9*A*).

DESI Ionization Mechanisms Droplet pickup has been suggested as the primary ionization mechanism in DESI, although there is evidence for chemical

(A)

(B)

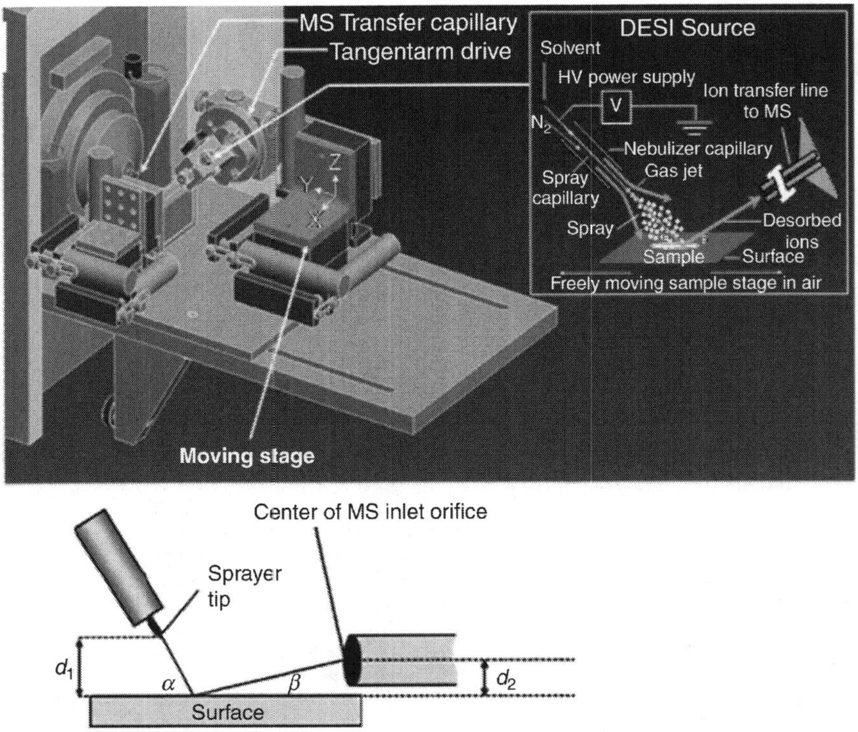

Term	Definition
α	Incident angle
β	Collection angle
d_1	Tip-to-surface distance
d_2	MS inlet-to-surface distance

FIGURE 1.9 (*A*) DESI source and moving stage used to position the source; an early prototype of the OmniSpray source of Prosolia, Inc. The source is fitted with an ion-transfer capillary. From [118]. Copyright permission was obtained from ACS. (*B*) Definitions of terms used in conjunction with DESI. From [144]. Copyright permission was obtained from Wiley.

sputtering (reactive ion–surface collisions) and gas-phase ionization processes (e.g., charge transfer, ion–molecule reactions, volatilization/desorption of neutrals followed by ionization) [116,118,144,165,166]. According to the droplet pickup mechanism, the surface is pre-wetted by initial droplets (velocities in excess of 100 m/s and diameters of less than 10 μm), forming a solvent layer that helps surface analytes become dissolved. These dissolved analytes are picked up by later arriving droplets that are impacting the surface, creating secondary droplets containing the dissolved analytes. Gas-phase ions are then formed from these secondary droplets by ESI-like mechanisms [144,165,166]. The resulting gas-phase ions have internal energy values similar to those in ESI and ESSI [167]. The formation of cold ions gives DESI its soft ionization character that affords ESI-like spectra, especially for proteins and polypeptides.

DESI Analytical Performance Signal intensity in DESI spectra depends on incident angle (β), collection angle (α), tip-to-surface distance (d_1), MS inlet-to-surface distance (d_2), and other geometric parameters, as defined in Figure 1.9*B*. Nebulization gas velocity, spray solvent flow rate, and spray potential also affect performance. The type of surface analyzed (its texture and electrical conductivity) is also a factor that affects the ionization process. The limits of detection (LODs) are in the low picogram to femtogram range for small molecules and some biopolymers [116,168]. The dynamic range is five orders of magnitude, and relative standard deviations (RSD) of 5% for quantitation (lower if using an internal standard) can be achieved [144]. For imaging applications, spatial resolution approaching 40 μm can be obtained [169]. Accuracies in the range of ±7% relative errors are possible [116,170].

DESI for Protein Analysis Protein and peptides show ESI-like spectra when analyzed by DESI, which is in part due to the ionization mechanism that takes place in DESI (droplet pickup or analyte microextraction into solution). Since the first reported applications of DESI for protein and peptide analysis [116,144], various research groups implemented applications ranging from solid-sample analysis (from surfaces) to direct analysis of liquid samples or liquid films [122,123,171–174]. An additional feature of liquid DESI is that it is easy to desorb large proteins directly from solution (Figure 1.10) [122]. For example, high mass proteins (e.g., BSA with MW of 66 kDa) appear to be relatively easily desorbed and ionized from solution than from dried samples on surface, probably due to less aggregation in solution than in the solid form [122]. Low detection limits and minimal sample preparation can apply to the

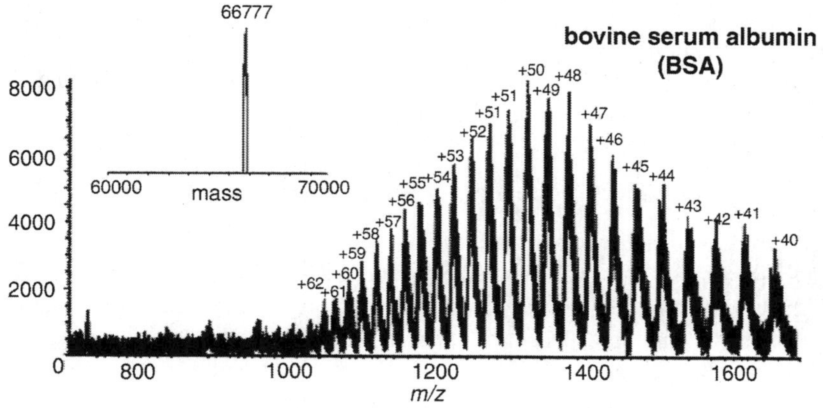

FIGURE 1.10 MS spectra showing the direct DESI-MS analysis of solutions containing bovine serum albumin (BSA). The insets show the corresponding deconvoluted spectra. From [122]. Copyright permission was obtained from Elsevier.

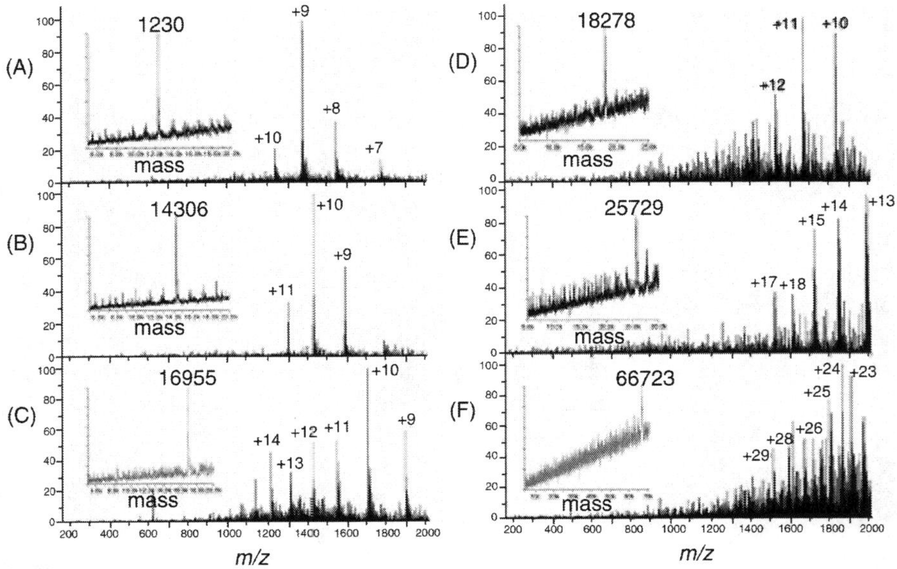

FIGURE 1.11 DESI-mass spectra and corresponding deconvoluted mass of intact proteins: (*A*) cytochrome *c* (> 10 ng/mm^2), (*B*) lysozyme (> 10 ng/mm^2), (*C*) apomyoglobin (> 50 ng/mm^2), (*D*) α-lactoglobulin B (> 100 ng/mm^2), (*E*) chymotrypsinogen A (> 100 ng/mm^2), and (*F*) BSA (> 4000 ng/mm^2). Proteins were deposited on a Plexiglas surface. From [159]. Copyright permission was obtained from ACS.

analysis of proteins from solid surfaces [159]. Basile et al. [159] evaluated the DESI response for the detection of proteins ranging in molecular mass from 12 to 66 kDa (Figure 1.11) and found detection limits that decrease with decreasing protein molecular mass. High mass resolving power can be obtained in protein and peptide identification by coupling DESI with Fourier transform ion cyclotron resonance mass spectrometry [163]. Other applications of peptide analysis can be envisioned for the direct identification of tryptic digests; examples are cytochrome *c* and myoglobin deposited on HPTL plates. After separation on the HPTL plates, the resulting bands are exposed to the DESI sprayer for peptide identification.

1.4.2 Fused-Droplet Electrospray Ionization (FD-ESI)

Fused-droplet electrospray ionization (FD-ESI) [138,175], a two-step electrospray ionization method [176,177], evolved from multiple-channel electrospray ionization (MC-ESI) [113,178–180]. In the multiple channel experiment, the analyte sample is introduced into one spray channel while other surrounding channels are used to generate the charged droplets that are fused with the analyte sample spray to form newly created droplets containing the analyte. Separating the ionization and nebulization process, Shiea and coworkers developed the newer ionization source, FD-ESI, in 2002 [175]. In the first step, the sample solution is ultrasonically nebulized to form fine aerosols that are transported to the skimmer of the mass spectrometer. These neutral

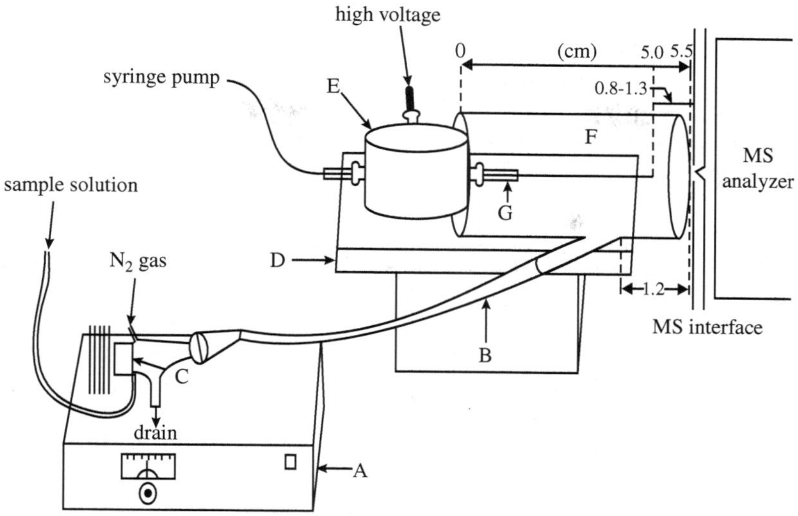

FIGURE 1.12 Schematic diagram of fused-droplet electrospray ionization mass spectrometry (FD-ESI-MS): (*A*) ultrasonic nebulizer, (*B*) Teflon tube, (*C*) piezoelectric transducer, (*D*) acrylic plate, (*E*) three way tee, (*F*) glass reaction chamber, (*G*) electrospray capillary in a Teflon tube. From [175]. Copyright permission was obtained from ACS.

aerosols are then fused in a reaction chamber with charged methanol droplets generated by electrospray. In the second step of a two-step process, ESI occurs for the newly created droplets, leading to the production of analyte ions [138,175].

FD-ESI Ionization Source A typical setup for FD-ESI (Figure 1.12) [175] consists of four parts, a traditional ESI source, a sample nebulizer assembly, a reaction chamber, and a mass spectrometer. The aqueous protein sample solution is pumped at an adjustable flow rate onto the surface of the piezoelectric transducer of an ultrasonic nebulizer to generate fine aerosols, which are subsequently transported with carrier nitrogen gas through the sidearm, a Teflon tube, into the reaction chamber. The end of the glass reaction chamber is positioned directly in front of the sampling skimmer of a quadruple mass spectrometer. The solvent such as methanol containing 1% acetic acid is electrosprayed continuously from a fused-silica capillary that is located at the center of the glass reaction chamber.

A modified FD-ESI apparatus [138] has advantages in providing salt tolerance for biological analysis. By replacing the ultrasonic nebulizer that generates the analyte aerosols with a pneumatic nebulizer from a commercial atmospheric pressure chemical ionization (APCI) probe [138], one can reduce sample consumption by 10 times compared to unmodified FD-ESI. To prevent the buildup of air pollutants in the methanol and fine acidic aerosols in the open air, an exhaust extractor is used in the fusion area.

FD-ESI for Protein Analysis FD-ESI can successfully ionize peptides and proteins dissolved in pure water [176,178]. Extremely high salt tolerance appears

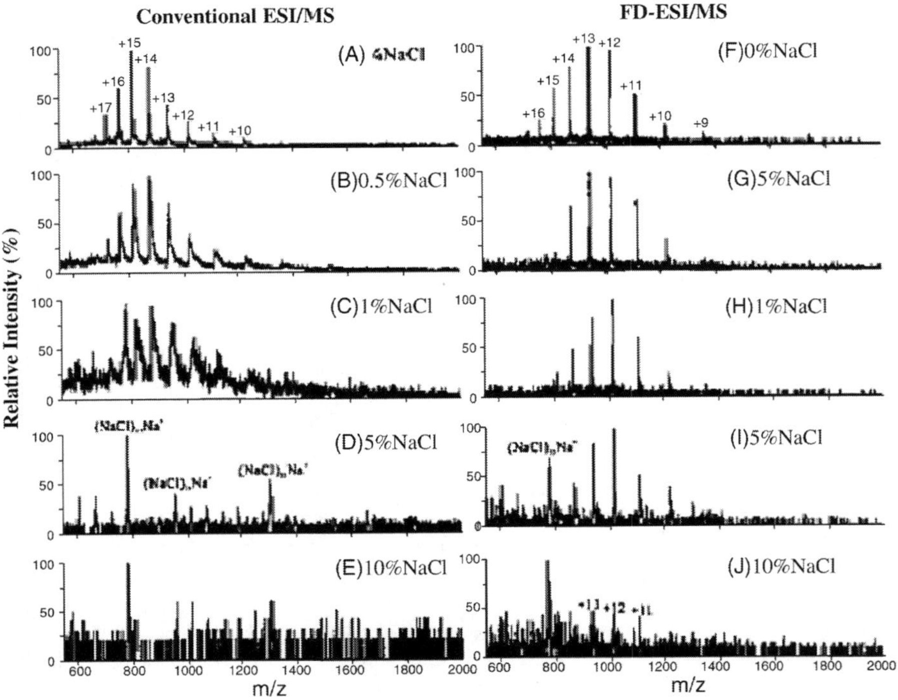

FIGURE 1.13 Positive ESI mass spectra of cytochrome *c* (10^{-6}M) that was dissolved in the aqueous solutions, which contained various amounts of NaCl (from 0 to 10% by weight). The mass spectra were obtained by conventional ESI-MS (*A–E*) and FD-ESI-MS (*F–J*). From [175]. Copyright permission was obtained from ACS.

to be one important advantage of FD-ESI compared to the traditional ESI in biological molecule analysis. As mentioned above, in ESI the moderate to large amounts of inorganic salts in the sample solution decrease the electrospray stability and sensitivity, owing to the formation of the various salt-protein adduct ions and the effect of ion suppression [138].

An example (Figure 1.13) is a comparison of conventional ESI and FD-ESI for the analysis of cytochrome *c* solution that contains NaCl at various concentrations. Using conventional ESI-MS, one finds that the mass spectra degrade with increasing NaCl concentration. When FD-ESI-MS is used to analyze these solutions, protonated proteins with nearly unchanged peak widths can still be observed, even when the NaCl concentration reaches 10% [175]. A similar desalting effect using organic spray solvent was seen in the liquid sample DESI experiment [181] whereby the liquid sample could be directly injected with no need of nebulization. The low solubility of inorganic salts in methanol spray solvent appears to exclude salt from the newly created fused droplets. Therefore, in FD-ESI the composition of the electrospray solvent is more important than that of sample solution in determining the ionization efficiency, and by adjusting the composition of the electrospray solvent, good quality mass spectra can be obtained [138], The disadvantage for FD-ESI, however, is that sample consumption exceeds that of the conventional ESI experiment.

1.4.3 Electrospray-Assisted Laser Desorption Ionization (ELDI)

Electrospray-assisted laser desorption ionization (ELDI) [125,161,182–188], another ambient ionization method, can be used to analyze protein samples both in the solid phase and in solution. Originating from the principle that the protein ionization can be achieved by mixing protein solutions with the ESI plume in FD-ESI, laser ablation was applied to desorb the protein samples. The created neutral proteins undergo post-ionization when merged with the ESI plume. Thus desorption and ionization are separate processes for protein analysis in ELDI.

Solid ELDI in Potein Analysis In the original setup for solid ELDI (Figure 1.14*A*) [125], the solid protein sample, from deposition on the mobile support plate, is desorbed by a laser beam. The resulting neutral protein droplets are ionized by the charged droplets generated from the ESI, giving multiply charged proteins that are detected by a mass spectrometer. By optimizing different distances and angles as well as the electrospray solvent composition, multiply charged cytochrome *c* ions can be successfully detected (Figure 1.14*B*).

Liquid ELDI in Potein Analysis Liquid ELDI allows the desorption and ionization of proteins from their native biological environment under ambient conditions [183]. In the liquid ELDI experiments, a small amount of protein solution, deposited onto the sample plate and mixed with the inert particles that serve as the matrix, is submitted to laser ablation. The laser energy is adsorbed by the inert particles and transferred to the surrounding solvent and analyte molecules for desorption. The desorbed neutral proteins are post-ionized by an ESI plume, producing multiply charged proteins. Given the high salt tolerance of ELDI and the effect of the ESI solvent, better-quality protein mass spectra (Figure 1.15) can be obtained in the analysis of proteins from human blood, tears, and bacteria extract than with traditional ESI and MALDI [183].

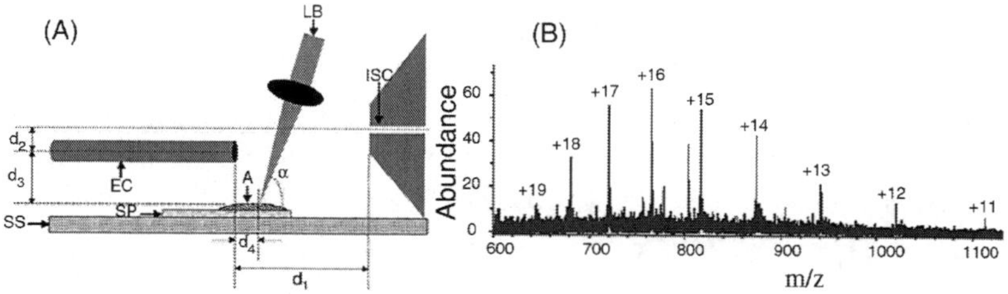

FIGURE 1.14 (*A*) Graphic representation of the geometry of the ELDI setup. analyte sample (A), sample support plate (SP), mobile sample stage (SS), laser beam (LB), electrospray capillary (EC), ion sampling capillary (ISC); (*B*) Solid ELDI mass spectrum of cytochrome *c*. From [125]. Copyright permission was obtained from Wiley.

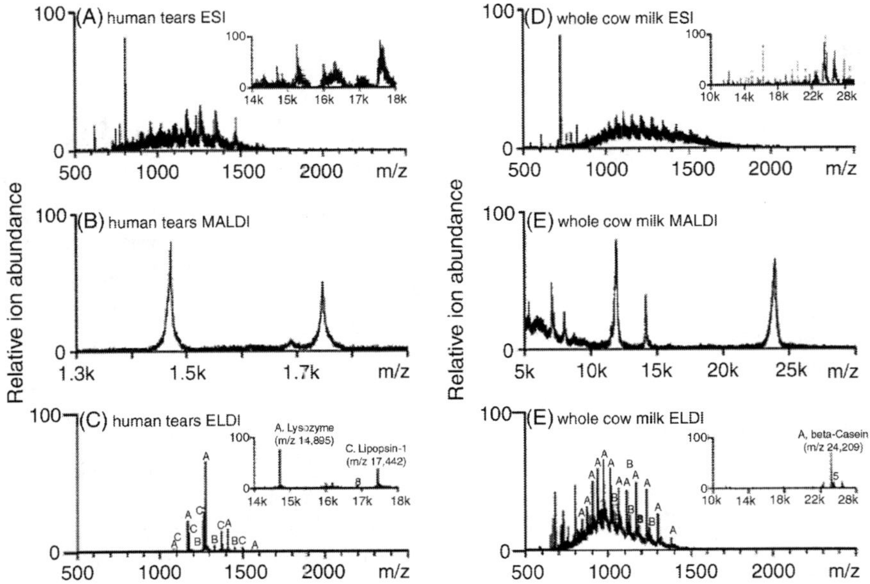

FIGURE 1.15 Positive ELDI mass spectra of human tears and whole cow milk; for comparison, conventional ESI and MALDI mass spectra of these biological fluids are also presented. From [183]. Copyright permission was obtained from ACS.

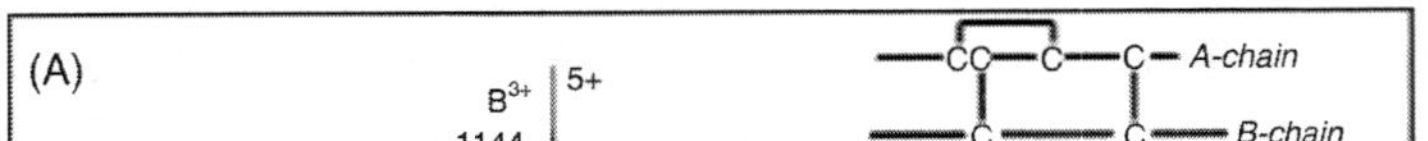

FIGURE 1.16 Reactive-ELDI experiments for online disulfide reduction of insulin with DTT. (*A*) Reactive-ELDI for disulfide reduction of insulin with DTT in the ESI solution yield new peaks at *m*/*z* 858 (insulin B-chain, 4^+) and 1144 (insulin B-chain, 3^+); (*B*) subsequent CADMS/MS of *m*/*z* 1144 confirmed the B-chain identity; (*C*) a solution of DTT was deposited onto the sample plate and desorbed by laser irradiation while insulin was electrosprayed. From [188]. Copyright permission was obtained from ACS.

Reactive-ELDI Similar to reactive DESI, some gas-phase reactions can be integrated in the ionization process of ELDI [187,188]. For reactive ELDI, the reactant ions are generated from either the ESI plume or the desorbed solution sample. Online disulfide bond cleavage of insulin can be achieved successfully via reactive ELDI in which the protein samples are either electrosprayed to react with laser-desorbed DTT or desorbed by laser irradiation followed by the reaction with sprayed DTT (Figure 1.16) [188]. Reactive ELDI can also be used to monitor other reactions including small-molecule reactions [187].

There are some interesting applications using ELDI, including coupling MS analysis with TLC [182] and chemically imaging different solid surfaces [189]. By

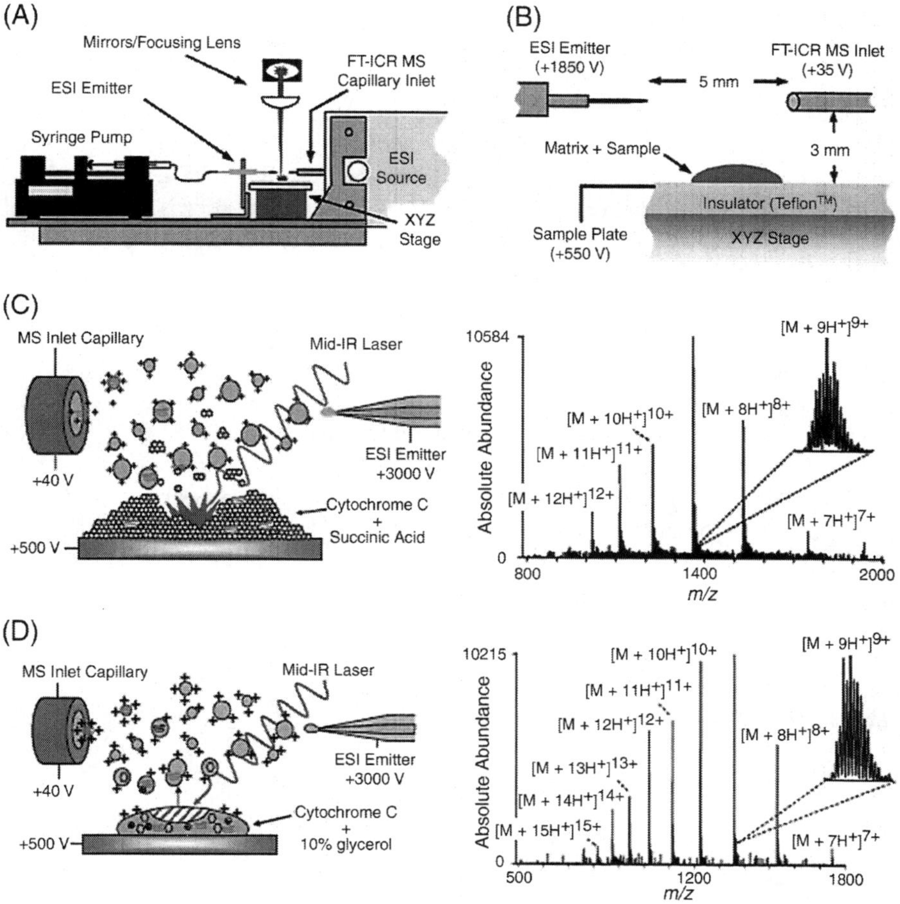

FIGURE 1.17 (*A*) Front and (*B*) side detailed views of the MALDESI Source. From [191]. Copyright permission was obtained from Elsevier. Schematic of (*C*) solid-state IR-MALDESI with ESI post-ionization and representative mass spectrum of bovine cytochrome *c* mixed with succinic acid. (*D*) liquid-state IR-MALDESI with ESI post-ionization and representative mass spectrum of bovine cytochrome *c* mixed with 10% glycerol. From [192]. Copyright permission was obtained from Elsevier.

optimizing the laser desorption energy, the composition of the electrospray solvent, and the matrix, intact protein ions can be observed with high sensitivity [184,185].

1.4.4 Matrix-Assisted Laser Desorption Electrospray Ionization (MALDESI)

MALDESI [126,190–193], a hybrid atmospheric pressure ionization method, combines the desirable attributes of ESI, MALDI, and ELDI into an integrated pulsed ionization source that generates multiply charged ions [191]. The MALDESI ion source is similar to the ELDI setup (Figure 1.17*A* and *B*), and the key to distinguishing between the two is that matrix is not necessary for ELDI whereas it is required for MALDESI [191]. For the latter, the analyte is deposited on the surface that is to be exposed to laser ablation, and the desorbed analyte then undergoes post-ionization by an ESI plume to generate multiply charged ions for MS detection.

MALDESI for Protein Analysis With minimal preparing of biological samples and by avoiding subjecting sensitive samples (i.e., tissue) to high vacuum, one can directly ionize samples in the solid or liquid states by using MALDESI (Figure 1.17*C* and *D*). For example, one can deposit a cytochrome *c* solution on a stainless steel target, dry it under open air, and analyze the sample by MALDESI to obtain multiply charged cytochrome *c* molecules.

1.5 CONCLUSIONS

One can see that a variety of protein ionization techniques based on MALDI or ESI are evolving. Because this field is rapidly developing, it is not possible to cover all protein ionization methods, and the authors apologize for any omissions. As the performance of the current MS ionization technologies, although highly effective, cannot meet all real-world demands in biochemistry and molecular biology, we can expect protein ionization methods to undergo further development; the only limitation seems to be our imagination [36].

ACKNOWLEDGMENTS

The preparation of this chapter was supported by NSF (CHE-0911160).

REFERENCES

1. Karas, M., Hillenkamp, F. (1988). Laser desorption ionization of proteins with molecular masses exceeding 10000 daltons. *Anal Chem 60*, 2299–2301.
2. Yamashita, M., Fenn, J. B. (1984). Electrospray ion source. Another variation on the free-jet theme. J Phys Chem *88*, 4451–4459.

3. Fenn, J. B., Mann, M., Meng, C. K., Wong, S. F., Whitehouse, C. M. (1989). Electrospray ionization for mass spectrometry of large biomolecules. *Science 246*, 64–71.
4. Chen, R., Cheng, X., Mitchell, D. W., Hofstadler, S. A., Wu, Q., Rockwood, A. L., Sherman, M. G., Smith, R. D. (1995). Trapping, detection, and mass determination of coliphage T4 DNA ions of 10^8 Da by electrospray ionization Fourier transform ion cyclotron resonance mass spectrometry. *Anal Chem 67*, 1159–1163.
5. Cole, R. B. (1997). *Electrospray Ionization Mass Spectrometry: Fundamentals, Instrumentation, and Applications.* Wiley -Interscience, New York.
6. Gross, J. H. (2004). *Mass Spectrometry: A Textbook.* Springer, New York.
7. Fenn, J. B. (2000). Mass spectrometric implications of high-pressure ion sources. *Int J Mass Spectrom 200*, 459–478.
8. Winston, R. L., Fitzgerald, M. C. (1997). Mass spectrometry as a readout of protein structure and function. *Mass Spectrom Rev 16*, 165–179.
9. Loo, J. A. (2000). Electrospray ionization mass spectrometry: A technology for studying noncovalent macromolecular complexes. *Int J Mass Spectrom 200*, 175–186.
10. Loo, J. A. (1997). Studying noncovalent protein complexes by electrospray ionization mass spectrometry. *Mass Spectrom Rev 16*, 1–23.
11. Kaltashov, I. A., Zhang, M., Eyles, S. J., Abzalimov, R. R. (2006). Investigation of structure, dynamics and function of metalloproteins with electrospray ionization mass spectrometry. *Anal Bioanal Chem 386*, 472–481.
12. Grandori, R. (2003). Electrospray-ionization mass spectrometry for protein conformational studies. *Curr Org Chem 7*, 1589–1603.
13. Calvo, E., Camafeita, E., Fernando Díaz, J., Antonio López, J. (2008). Mass spectrometry for studying the interaction between small molecules and proteins. *Curr. Proteomics 5*, 20–34.
14. Dalluge, J. J. (2002). *Matrix-assisted laser desorption ionization-mass spectrometry (MALDI-MS) Anal Bioanal Chem 372*, 18–19.
15. Harvey, D. J. (2003). Matrix-assisted laser desorption/ionization mass spectrometry of carbohydrates and glycoconjugates. *Int J Mass Spectrom 226*, 1–35.
16. Reyzer, M. L., Caprioli, R. M. (2007). MALDI-MS-based imaging of small molecules and proteins in tissues. *Curr Opin Chem Biol 11*, 29–35.
17. Najam-ul-Haq, M., Szabó, R. Z., Vallant, R., Huck, C. W., Bonn, G. K. (2007). Role of carbon nano-materials in the analysis of biological materials by laser desorption/ionization-mass spectrometry. *J Biochem Biophys Meth 70*, 319–328.
18. Thomson, J. J. (1913). *Rays of Positive Electricity and Their Applications to Chemical Analysis.* Longmans, London.
19. Bleakney, W. (1929). A new method of positive-ray analysis and its application to the measurement of iIonization potentials in mercury vapor. *Phys Rev 34*, 157–160.
20. Nier, A. O. (1947). A mass spectrometer for isotope and gas analysis. *Rev. Sci. Inst 18*, 398–411.
21. Biemann, K., Gapp, F., Seibl, J. (1959). Application of mass spectrometry to structure problems. I. Amino acid sequence in peptides. *J Am Chem Soc 81*, 2274–2275.
22. Senn, M., McLafferty, F. W. (1966). Automatic amino acid sequence determination in peptides. *Biochem Biophys Res Commun 23*, 381–385.

23. Munson, M. S. B., Field, F. H. (1966). Chemical ionization mass spectrometry. I. General introduction. *J Am Chem Soc 88*, 2621–2630.

24. Harrison, A. G. (1992). *Chemical Ionization Mass Spectrometry*. CRC Press, Boca Raton.

25. Beckey, H. D. (1969). Field ionization mass spectrometry. *Res Developm 20*, 26–29.

26. Hakansson, P., Kamensky, I., Sundqvist, B., Fohlman, J., Peterson, P., McNeal, C. J., Macfarlane, R. D. (1982). ^{127}I-plasma desorption mass spectrometry of insulin. *J Am Chem Soc 104*, 2948–2949.

27. Morris, H. R., Panico, M., Taylor, G. W. (1983). FAB-mapping of recombinant-DNA protein products. *Biochem Biophys Res Commun 117*, 299–305.

28. Macfarlane, R. D., Torgerson, D. F. (1976). Californium-252 plasma desorption mass spectroscopy. *Science 191*, 920–925.

29. Barber, M., Green, B. N. (1987). The analysis of small proteins in the molecular weight range 10-24 kDa by magnetic sector mass spectrometry. *Rapid Commun Mass Spectrom 1*, 80–83.

30. Blakely, C. R., Vestal, M. L. (1983). Thermospray interface for liquid chromatography/ mass spectrometry. *Anal Chem 55*, 750–754.

31. Koropchak, J. A., Veber, M., Browner, R. F. (1992). Thermospray sample introduction to atomic spectrometry. *Crit Rev Anal Chem 23*, 113–141.

32. Tanaka, K., Waki, H., Ido, Y., Akita, S., Yoshida, Y., Yoshida, T. (1988). Protein and polymer analyses up to *m/z* 100 000 by laser ionization time-of flight mass spectrometry. *Rapid Commun Mass Spectrom 2*, 151–153.

33. Karas, M., Bachmann, D., Hillenkamp, F. (1985). Influence of the wavelength in high irradiance ultraviolet laser desorption mass spectrometry of organic molecules. *Anal Chem 57*, 2935–2939.

34. Ingendoh, A., Karas, M., Hillenkamp, F., Giessmann, U. (1994). Factors affecting the mass resolution in matrix-assisted laser desorption-ionization mass spectrometry. *Int J Mass Spectrom Ion Processes 131*, 345–354.

35. Vestal, M. L., Juhasz, P., Martin, S. A. (1995). Delayed extraction matrix-assisted laser desorption time-of-flight mass spectrometry. *Rapid Commun Mass Spectrom 9*, 1044–1050.

36. Roepstoref, P. (2002). *Mass Spectrometry of Peptides and Proteins: A Personal Historical View*. Wiley New York.

37. Karas, M., Bachmann, D., Bahr, U., Hillenkamp, F. (1987). Matrix-assisted ultraviolet laser desorption of non-volatile compounds. *Int J Mass Spectrom Ion Processes 78*, 53–68.

38. Karas, M., Krüger, R. (2003). Ion formation in MALDI: The cluster ionization mechanism. *Chem Rev 103*, 427–439.

39. Cornett, D. S., Reyzer, M. L., Chaurand, P., Caprioli, R. M. (2007). MALDI imaging mass spectrometry: Molecular snapshots of biochemical systems. *Nat Meth 4*, 828–833.

40. Hardesty, W. M., Caprioli, R. M. (2008). In situ molecular imaging of proteins in tissues using mass spectrometry. *Anal Bioanal Chem 391*, 899–903.

41. Skold, K., Svensson, M., Nilsson, A., Zhang, X., K., N., Caprioli, R. M., Svenningsson, P., Andren, P. E. (2006). Decreased striatal levels of PEP-19 following MPTP lesion in the mouse. *J Proteome Res 5*, 262–269.

42. Laiko, V. V., Baldwin, M. A., Burlingame, A. L. (2000). Atmospheric pressure matrix-assisted laser desorption/ionization mass spectrometry. *Anal Chem 72*, 652–657.

43. Creaser, S. C., Ratcliffe, L. (2006). Atmospheric pressure matrix-assisted laser desorption/ionization mass spectrometry: A review. *Curr Anal Chem 2*, 9–15.

44. Moyer, S. C., Cotter, R. J. (2002). Atmospheric pressure MALDI. *Anal Chem 74*, 469A–476A.

45. McLean, J. A., Russell, W. K., Russell, D. H. (2003). A high repetition rate (1 kHz) microcrystal laser for high throughput atmospheric pressure MALDI-quardrupole-time-of-flight mass spectrometry. *Anal Chem 75*, 648–654.

46. Laiko, V. V., Moyer, S. C., Cotter, R. J. (2000). Atmospheric pressure MALDI/ion trap mass spectrometry. *Anal Chem 72*, 5239–5243.

47. Kellersberger, K. A., Tan, P. V., Laiko, V. V., Doroshenko, V. M., Fabris, D. (2004). Atmospheric pressure MALDI–Fourier transform mass spectrometry. *Anal Chem 76*, 3930–3934.

48. König, S., Kollas, O., Dreisewerd, K. (2007). Generation of highly charged peptide and protein Ions by atmospheric pressure matrix-assisted infrared laser desorption/ionization ion trap mass spectrometry. *Anal Chem 79*, 5484–5488.

49. Hutchens, T. W., Yip, T. T. (1993). New desorption strategies for the mass spectrometric analysis of macromolecules. *Rapid Commun Mass Spectrom 7*, 576–580.

50. Merchant, M., Weinberger, S. R. (2000). Recent advancements in surface-enhanced laser desorption/ionization time-of-flight mass spectrometry. *Electrophoresis 21*, 1164–1177.

51. Hutchens, T. W., Yip, T. T. (1999). Surface-enhanced neat desorption for disorption and detection of analytes. US Patent 5894063.

52. Tang, N., Tornatore, P., Weinberger, S. R. (2004). Current developments in SELDI affinity technology. *Mass Spectrom Rev 23*, 34–44.

53. Vorderwülbecke, S., Cleverley, S., Weinberger, S. R., Wiesner, A. (2005). Protein quantification by the SELDI-TOF-MS-based ProteinChip® System. *Nat Meth 2*, 393–395.

54. Wright, G. L., Cazares, L. H., Leung, S. M., Nasim, S., Adam, B. L., Yip, T. T., Schellhammer, P. F., Gong, L., Vlahou, A. (1999). Proteinchip® surface enhanced laser desorption/ionization (SELDI) mass spectrometry: A novel protein biochip technology for detection of prostate cancer biomarkers in complex protein mixtures. *Prostate Cancer Prostatic Dis 2*, 264–276.

55. Yanes, O., Woo, H. K., Northen, T. R., Oppenheimer, S. R., Shriver, L., Apon, J., Estrada, M. N., Potchoiba, M. J., Steenwyk, R., Manchester, M., Siuzdak, G. (2009). Nanostructure initiator mass spectrometry: Tissue imaging and direct biofluid analysis. *Anal Chem 81*, 2969–2975.

56. Woo, H. K., Northen, T. R., Yanes, O., Siuzdak, G. (2008). Nanostructure-initiator mass spectrometry: A protocol for preparing and applying NIMS surfaces for high-sensitivity mass analysis. *Nat Protoc 3*, 1341–1349.

57. Northen, T. R., Yanes, O., Northen, M. T., Marrinucci, D., Uritboonthai, W., Apon, J., Golledge, S. L., Nordstrom, A., Siuzdak, G. (2007). Clathrate nanostructures for mass spectrometry. *Nature 449*, 1033–1036.

58. Woo, H. K., Northen, T. R., Yanes, O., Siuzdak, G. (2008). Nanostructure-initiator mass spectrometry: A protocol for preparing and applying NIMS surfaces for high-sensitivity mass analysis. *Nat Protoc 3*, 1341–1349.

59. Dole, M., Mack, L. L., Hines, R. L. (1968). Molecular beams of macroions. *J Chem Phys 49*, 2240–2249.

60. Iribarne, J. V., Thomson, B. A. (1976). On the evaporation of small ions from charged droplets. *J Chem Phys 64*, 2287–2294.

61. Benesch, J. L. P., Ruotolo, B. T., Simmons, D. A., Robinson, C. V. (2007). Protein complexes in the gas phase: Technology for structural genomics and proteomics. *Chem Rev 107*, 3544–3567.

62. Kaltashov, I. A., Abzalimov, R. R. (2008). Do ionic charges in ESI MS provide useful information on macromolecular structure? *J Am Soc Mass Spectrom 19*, 1239–1246.

63. Przybylski, M., Glocker, M. O. (1996). Electrospray mass spectrometry of biomacromolecular complexes with non-covalent interactions—New analytical perspectives for supramolecular chemistry and molecular recognition processes. *Angewandte Chemie Int Ed 35*, 806–826.

64. van den Heuvel, R. H. H., Heck, A. J. R. (2004). Native protein mass spectrometry: From intact oligomers to functional machineries. *Curr Opin Chem Biol 8*, 519–526.

65. Kaddis, C. S., Loo, J. A. (2007). Native protein MS and ion mobility: Large flying proteins with ESI. *Anal Chem 79*, 1778–1784.

66. Benesch, J. L. P., Robinson, C. V. (2006). Mass spectrometry of macromolecular assemblies: Preservation and dissociation. *Curr Opin Struct Biol 16*, 245–251.

67. Chowdhurry, S. K., Katta, V., Chait, B. T. (1990). Probing conformational changes in proteins by mass spectrometry. *J Am Chem Soc 112*, 9012–9013.

68. Nemirovskiy, O., Giblin, D. E., Gross, M. L. (1999). Electrospray ionization mass spectrometry and hydrogen/deuterium exchange for probing the interaction of calmodulin with calcium. *J Am Soc Mass Spectrom 10*, 711–718.

69. Simmons, D. A., Konermann, L. (2002). Characterization of transient protein folding intermediates during myoglobin reconstitution by time-resolved electrospray mass spectrometry with on-line isotopic pulse labeling. *Biochemistry 41*, 1906–1914.

70. Winger, B. E., Light-Wahl, K. J., Rockwood, A. L., Smith, R. D. (1992). Probing qualitative conformation differences of multiply protonated gas-phase proteins via H/D isotopic exchange with D_2O. *J Am Chem Soc 114*, 5898–5900.

71. Campbell, S., Rodgers, M. T., Marzluff, E. M., Beauchamp, J. L. (1995). Deuterium exchange reactions as a probe of biomolecule structure: Fundamental studies of gas phase H/D exchange reactions of protonated glycine oligomers with D_2O, CD_3OD, CD_3CO_2D, and ND_3. *J Am Chem Soc 117*, 12840–12854.

72. Engen, J. R. (2009). Analysis of protein conformation and dynamics by hydrogen/deuterium exchange MS. *Anal Chem 81*, 7870–7875.

73. Zhang, Z., Smith, D. L. (1996). Thermal-induced unfolding domains in aldolase identified by amide hydrogen exchange and mass spectrometry. *Protein Sci 5*, 1282–1289.

74. Zhu, M., Rempel, D. L., Du, Z., Gross, M. L. (2003). Quantification of protein–ligand interactions by mass spectrometry, titration, and H/D exchange: PLIMSTEX. *J Am Chem Soc 125*, 5252–5253.

75. Zhu, M., Rempel, D. L., Zhao, J., Giblin, D. E., Gross, M. L. (2003). Probing Ca^{2+}-induced conformational changes in porcine calmodulin by H/D exchange and ESI-MS: Effect of cations and ionic strength. *Biochemistry 42*, 15388–15397.

76. McLafferty, F. W., Guan, Z., Haupts, U., Wood, T. D., Kelleher, N. L. (1998). Gaseous conformational structures of cytochrome *c*. *J Am Chem Soc 120*, 4732–4740.

77. Wilm, M. S., Mann, M. (1994). Electrospray and Taylor-cone theory, Dole's beam of macromolecules at last? *Int J Mass Spectrom Ion Processes 136*, 167–180.

78. El-Faramawy, Y. A., Siu, K. W. M., Thomson, B. A. (2005). Efficiency of nano-electrospray ionization. *J Am Soc Mass Spectrom 16*, 1702–1707.

79. Wilm, M., Mann, M. (1996). Analytical properties of the nanoelectrospray ion source. *Anal Chem 68*, 1–8.

80. Hirabayashi, Y., Hirabayashi, A. (2002). Capillary isoelectric focusing separation combined with mass spectrometry using sonic spray ionization for protein analysis. *J Mass Spectrom Soc Jap 50*, 21–23.

81. Hirabayashi, Y., Hirabayashi, A., Shimura, K. (2003). Protein and synthesized peptide pI marker analysis by capillary isoelectric focusing separation/mass spectrometry using sonic spray ionization. *J Mass Spectrom Soc Jap 51*, 67–71.

82. Harper, J. D., Charipar, N. A., Mulligan, C. C., Zhang, X., Cooks, R. G., Ouyang, Z. (2008). Low-temperature plasma probe for ambient desorption ionization. *Anal Chem 80*, 9097–9104.

83. Hirabayashi, A., Hirabayashi, Y., Sakairi, M., Koizumi, H. (1996). Multiply-charged ion formation by sonic spray. *Rapid Commun Mass Spectrom 10*, 1703–1705.

84. Hirabayashi, A., Sakairi, M., Koizumi, H. (1994). Sonic spray ionization method for atmospheric pressure ionization mass spectrometry. *Anal Chem 66*, 4557–4559.

85. Hirabayashi, A., Sakairi, M., Koizumi, H. (1995). Sonic spray mass-spectrometry. *Anal Chem 67*, 2878–2882.

86. Covey, T. R., Thomson, B. A., Schneider, B. B. (2009). Atmospheric pressure ion sources. *Mass Spectrom Rev 28*, 870–897.

87. Hirabayashi, A., Sakairi, M., Takada, Y., Koizumi, H. (1997). Recent progress in atmospheric pressure ionization mass spectrometry. *Trends Anal Chem 16*, 45–52.

88. Touboul, D., Jecklin, M. C., Zenobi, R. (2008). Ion internal energy distributions validate the charge residue model for small molecule ion formation by spray methods. *Rapid Commun Mass Spectrom 22*, 1062–1068.

89. Kebarle, P., Tang, L. (1993). From ions in solution to ions in the gas phase—The mechanism of electrospray mass spectrometry. *Anal Chem 65*, 972A–986A.

90. Hirabayashi, A., de la Mora, J. F. (1998). Charged droplet formation in sonic spray. *Int J Mass Spectrom 175*, 277–282.

91. Takats, Z., Nanita, S. C., Cooks, R. G., Schlosser, G., Vekey, K. (2003). Amino acid clusters formed by sonic spray ionization. *Anal Chem 75*, 1514–1523.

92. Dodd, E. E. (1953). The statistics of liquid spray and dust electrification by the hopper and laby method. *J Appl Phys 24*, 73–80.

93. Hirabayashi, Y., Hirabayashi, A., Takada, Y., Sakairi, M., Koizumi, H. (1998). A sonic spray interface for the mass analysis of highly charged ions from protein solutions at high flow rates. *Anal Chem 70*, 1882–1884.

94. Hirabayashi, Y., Hirabayashi, A., Koizumi, H. (1999). A sonic spray interface for capillary electrophoresis mass spectrometry. *Rapid Communi Mass Spectrom 13*, 712–715.

95. Gardner, J. S., Harrison, R. G., Lamb, J. D., Dearden, D. V. (2006). Sonic spray ionization mass spectrometry: A powerful tool used to characterize fragile metal-assembled cages. *New J Chem 30*, 1276–1281.
96. Bjorkman, H. T., Edlund, P. O., Jacobsson, S. P. (2002). Sonic spray ionization interface for liquid chromatography–mass spectrometry. *Anal Chim Acta 468*, 263–274.
97. Dams, R., Benijts, T., Gunther, W., Lambert, W., De Leenheer, A. (2002). Sonic spray ionization technology: Performance study and application to a LC/MS analysis on a monolithic silica column for heroin impurity profiling. *Anal Chem 74*, 3206–3212.
98. Deguchi, K., Takegawa, Y., Hirabayashi, A., Nakagawa, H., Nishimura, S. I. (2005). Characteristic pH and electrolyte concentration dependences of 2-aminopyridine-derivatized oligosaccharides (N-glycans) in sonic-spray ionization mass spectrometry. *Rapid Commun Mass Spectrom 19*, 2325–2330.
99. Takegawa, Y., Deguchi, K., Ito, S., Yoshioka, S., Nakagawa, H., Nishimura, S. (2005). Simultaneous analysis of 2-aminopyridine-derivatized neutral and sialylated oligosaccharides from human serum in the negative-ion mode by sonic spray ionization ion trap mass spectrometry. *Anal Chem 77*, 2097–2106.
100. Kovacs, Z., Dinya, Z., Antus, S. (2004). LC-SSI-MS techniques as efficient tools for characterization of nonvolatile phenolic compounds of a special Hungarian wine. *J Chromatogr Sci 42*, 125–129.
101. Huang, M., Hirabayashi, A. (2002). Multi-charged oligonucleotide ion formation in sonic spray ionization. *Anal Sci 18*, 385–390.
102. Huang, M., Hirabayashi, A., Okumura, A., Hirabayashi, Y. (2001). Matrix effect on the analysis of oligonucleotides by using a mass spectrometer with a sonic spray ionization source. *Anal Sci 17*, 1179–1182.
103. Takats, Z., Wiseman, J. M., Gologan, B., Cooks, R. G. (2004). Electrosonic spray ionization. A gentle technique for generating folded proteins and protein complexes in the gas phase and for studying ion–molecule reactions at atmospheric pressure. *Anal Chem 76*, 4050–4058.
104. Wiseman, J. M., Takats, Z., Gologan, B., Davisson, V. J., Cooks, R. G. (2005). Direct characterization of enzyme-substrate complexes by using electrosonic spray ionization mass spectrometry. *Angewandte Chemie–Int Ed 44*, 913–916.
105. Jecklin, M. C., Touboul, D., Bovet, C., Wortmann, A., Zenobi, R. (2008). Which electrospray-based ionization method best reflects protein–ligand interactions found in solution? A comparison of ESI, nanoESI, and ESSI for the determination of dissociation constants with mass spectrometry. *J Am Soc Mass Spectrom 19*, 332–343.
106. Czuczy, N., Katona, M., Takats, Z. (2009). Selective detection of specific protein-ligand complexes by electrosonic spray-precursor ion scan tandem mass spectrometry. *J Am Soc Mass Spectrom 20*, 227–237.
107. Nefliu, M., Venter, A., Cooks, R. G. (2006). Desorption electrospray ionization and electrosonic spray ionization for solid- and solution-phase analysis of industrial polymers. *Chem Commun* 888–890.
108. Wang, R., Allmendinger, P., Zhu, L., Gröhn, A. J., Wegner, K., Frankevich, V., Zenobi, R. (2011). The role of nebulizer gas flow in electrosonic spray ionization (ESSI). *J Am Soc Mass Spectrom 22*, 1234–1241.

109. Touboul, D., Jecklin, M. C., Zenobi, R. (2007). Rapid and precise measurements of gas-phase basicity of peptides and proteins at atmospheric pressure by electrosonic spray ionization–mass spectrometry. *J Phys Chem B 111* 11629–11631.

110. Touboul, D., Jecklin, M. C., Zenobi, R. (2008). Investigation of deprotonation reactions on globular and denatured proteins at atmospheric pressure by ESSI-MS. *J Am Soc Mass Spectrom 19*, 455–466.

111. Touboul, D., Jecklin, M. C., Zenobi, R. (2008). Exploring deprotonation reactions on peptides and proteins at atmospheric pressure by electro-sonic spray ionization–mass spectrometry (ESSI-MS). *Chimia 62*, 282–286.

112. Czuczy, N., Katona, M., Takats, Z. (2009). Selective detection of protein–ligand complexes by electrosonic spray-precursor ion scan tandem mass spectrometry. *J Am Soc Mass Spectrom 20*, 227–237.

113. Shiea, J., Wang, C. H. (1997). Applications of multiple channel electrospray ionization sources for biological sample analysis. *J. Mass Spectrom 32*, 247–250.

114. Nefliu, M., Jonell N. Smith, J. N., Andre Venter, A., Cooks, R. G. (2008). Internal energy distributions in desorption electrospray ionization (DESI). *J Am Soc Mass Spectrom 19*, 420–427.

115. Takats, Z., Cooks, R. G. Electrosonic spray ionization method and device for the atmospheric ionization of molecules. (2005). U.S. Pat. Appl. Publ., US 20050029442 A1 20050210.

116. Takats, Z., Wiseman, J. M., Gologan, B., Cooks, R. G. (2004). Mass spectrometry sampling under ambient conditions with desorption electrospray ionization. *Science 306*, 471–473.

117. Cody, R. B., Laramee, J. A., Durst, H. D. (2005). Versatile new ion source for the analysis of materials in open air under ambient conditions. *Anal Chem 77*, 2297–2302.

118. Cotte-Rodriguez, I., Takats, Z., Talaty, N., Chen, H. W., Cooks, R. G. (2005). Desorption electrospray ionization of explosives on surfaces: Sensitivity and selectivity enhancement by reactive desorption electrospray ionization. *Anal Chem 77*, 6755–6764.

119. Cotte-Rodriguez, I., Mulligan, C. C., Cooks, G. (2007). Non-proximate detection of small and large molecules by desorption electrospray ionization and desorption atmospheric pressure chemical ionization mass spectrometry: Instrumentation and applications in forensics, chemistry, and biology. *Anal Chem 79*, 7069–7077.

120. Cotte-Rodriguez, I., Hernandez-Soto, H., Chen, H., Cooks, R. G. (2008). In situ trace detection of peroxide explosives by desorption electrospray ionization and desorption atmospheric pressure chemical ionization. *Anal Chem 80*, 1512–1519.

121. Cotte-Rodriguez, I., Cooks, R. G. (2006). Non-proximate detection of explosives and chemical warfare agent simulants by desorption electrospray ionization mass spectrometry. *Chem Commun* 2968–2970.

122. Miao, Z. X., Chen, H. (2009). Direct analysis of liquid samples by desorption electrospray ionization-mass spectrometry (DESI-MS). *J Am Soc Mass Spectrom 20*, 10–19.

123. Chipuk, J. E., Brodbelt, J. S. (2008). Transmission mode desorption electrospray ionization. *J Am Soc Mass Spectrom 19*, 1612–1620.

124. Venter, A., Cooks, R. G. (2007). Desorption electrospray ionization in a small pressure-tight enclosure. *Anal Chem 79*, 6368–6403.

125. Shiea, J., Huang, M. Z., Hsu, H. J., Lee, C. Y., Yuan, C. H., Beech, I., Sunner, J. (2005). Electrospray-assisted laser desorption/ionization mass spectrometry for direct ambient analysis of solids. *Rapid Commun Mass Spectrom 19*, 3701–3704.

126. Sampson, J. S., Hawkridge, A. M., Muddiman, D. C. (2007). Direct characterization of intact polypeptides by matrix-assisted laser desorption electrospray ionization quadrupole Fourier transform ion cyclotron resonance mass spectrometry. *Rapid Commun Mass Spectrom 21*, 1150–1154.
127. Chen, H., Venter, A., Cooks, R. G. (2006). Extractive electrospray ionization for direct analysis of undiluted urine, milk and other complex mixtures without sample preparation. *Chem Commun* 2042–2044.
128. McEwen, C. N., McKay, R. G., Larsen, B. S. (2005). Analysis of solids, liquids, and biological tissues using solids probe introduction at atmospheric pressure on commercial LC/MS instruments. *Anal Chem 77*, 7826–7831.
129. Takáts, Z., Czuczy, N., Katona, M., Skoumal, R. (2006). Jet desorption ionization *Proc 54th ASMS Conf Mass Spectrometry,* Seattle, May 28–June 1.
130. Haddad, R., Sparrapan, R., Eberlin, M. N. (2006). Desorption sonic spray ionization for (high) voltage-free ambient mass spectrometry. *Rapid Commun Mass Spectrom 20*, 2901–2905.
131. Grimm, R. L., Beauchamp, J. L. (2005). Dynamics of field-induced droplet ionization: Time-resolved studies of distortion, jetting, and progeny formation from charged and neutral methanol droplets exposed to strong electric fields. *J Phys Chem B 109* 8244–8250.
132. Haapala, M., Pol, J., Saarela, V., Arvola, V., Kotiaho, T., Ketola, R. A., Franssila, S., Kauppila, T. J., Kostiainen, R. (2007). Desorption atmospheric pressure photoionization. *Anal Chem 79*, 7867–7872.
133. Ratcliffe, L. V., Rutten, F. J. M., Barrett, D. A., Whitmore, T., Seymour, D., Greenwood, C., Aranda-Gonzalvo, Y., Robinson, S., McCoustra, M. (2007). Surface analysis under ambient conditions using plasma-assisted desorption/ionization mass spectrometry. *Anal Chem 79*, 6094–6101.
134. Na, N., Zhang, C., Zhao, M., Zhang, S., Yang, C., Fang, X., Zhang, X. (2007). Direct detection of explosives on solid surfaces by mass spectrometry with an ambient ion source based on dielectric barrier discharge. *J Mass Spectrom 42*, 1079–1085.
135. Van Berkel, G. J., Kertesz, V., Koeplinger, K. A., Vavrek, M., Kong, A. T. (2008). Liquid microjunction surface sampling probe electrospray mass spectrometry for detection of drugs and metabolites in thin tissue sections. *J Mass Spectrom 43*, 500–508.
136. Chen, H., Ouyang, Z., Cooks, R. G. (2006). Thermal production and reactions of organic ions at atmospheric pressure. *Angewandte Chemie Int Ed 45*, 3656–3660.
137. Ford, M. J., Van Berkel, G. J. (2004). An improved thin-layer chromatography/mass spectrometry coupling using a surface sampling probe electrospray ion trap system. *Rapid Commun Mass Spectrom 18*, 1303–1309.
138. Shieh, I. F., Lee, C. Y., Shiea, J. (2005). Eliminating the interferences from TRIS buffer and SDS in protein analysis by fused-droplet electrospray ionization mass spectrometry. *J Proteome Res 4*, 606–612.
139. Andrade, F. J., Shelley, J. T., Wetzel, W. C., Webb, M. R., Gamez, G., Ray, S. J., Hieftje, G. M. (2008). Atmospheric pressure chemical ionization source. 1. Ionization of compounds in the gas phase. *Anal Chem 80*, 2646–2653.
140. Chen, H., Yang, S., Wortmann, A., Zenobi, R. (2007). Neutral desorption sampling of living objects for rapid analysis by extractive electrospray ionization mass spectrometry. *Angewandte Chemie Int Ed 46*, 7591–7594.

141. Nemes, P., Vertes, A. (2007). Laser ablation electrospray ionization for atmospheric pressure, in vivo, and imaging mass spectrometry. *Anal Chem 79*, 8098–8106.

142. Trimpin, S., Inutan, E. D., Herath, T. N., McEwen, C. N. (2010). Laserspray ionization, a new atmospheric pressure MALDI method for producing highly charged gas-phase ions of peptides and proteins directly from solid solutions. *Mol Cell Proteomics 9*, 362–367.

143. Venter, A., Nefliu, M., Cooks, R. G. (2008). Ambient desorption ionization mass spectrometry. *Trends Anal Chem 27*, 284–290.

144. Takats, Z., Wiseman, J. M., Cooks, R. G. (2005). Ambient mass spectrometry using desorption electrospray ionization (DESI): Instrumentation, mechanisms and applications in forensics, chemistry, and biology. *J Mass Spectrom 40*, 1261–1275.

145. Cooks, R. G., Ouyang, Z., Takats, Z., Wiseman, J. M. (2006). Ambient mass spectrometry. *Science 311*, 1566–1570.

146. Cotte-Rodriguez, I., Chen, H., Cooks, R. G. (2006). Rapid trace detection of triacetone triperoxide (TATP) by complexation reactions during desorption electrospray ionization. *Chem Commun* 953–955.

147. Ifa, D. R., Manicke, N. E., Dill, A. L., Cooks, R. G. (2008). Latent fingerprint chemical imaging by mass spectrometry. *Science 321*, 805–805.

148. Meetani, M. A., Shin, Y. S., Zhang, S., Mayer, R., Basile, F. (2007). Desorption electrospray ionization mass spectrometry of intact bacteria. *J Mass Spectrom 42*, 1186–1193.

149. Wu, C., Ifa, D. R., Manicke, N. E., Cooks, R. G. (2010). Molecular imaging of adrenal gland by desorption electrospray ionization mass spectrometry. *Analyst 135*, 28–32.

150. Dill, A. L., Ifa, D. R., Manicke, N. E., Costa, A. B., Ramos-Vara, J. A., Knapp, D. W., Cooks, R. G. (2009). Lipid profiles of canine invasive transitional cell carcinoma of the urinary bladder and adjacent normal tissue by desorption electrospray ionization imaging mass spectrometry. *Anal Chem 81*, 8758–8764.

151. Wiseman, J. M., Ifa, D. R., Zhu, Y., Kissinger, C. B., Manicke, N. E., Kissinger, P. T., Cooks, R. G. (2009). Desorption electrospray ionization mass spectrometry: Imaging drugs and metabolites in tissues. *Proc Nat Acad Sci (USA), 105*, 18120–18125.

152. Kertesz, V., Van Berkel, G. J., Vavrek, M., Koeplinger, K. A., Schneider, B. B., Covey, T. R. (2008). Comparison of drug distribution images from whole-body thin tissue sections obtained using desorption electrospray ionization tandem mass spectrometry and autoradiography. *Anal Chem 80*, 5168–5177.

153. Manicke, N. E., Wiseman, J. M., Ifa, D. R., Cooks, R. G. (2008). Desorption electrospray ionization (DESI) mass spectrometry and tandem mass spectrometry (MS/MS) of phospholipids and sphingolipids: Ionization, adduct formation, and fragmentation. *J Am Soc Mass Spectrom 19*, 531–543.

154. Kennedy, J. H., Wiseman, J. M. (2010). Evaluation and performance of desorption electrospray ionization using a triple quadrupole mass spectrometer for quantitation of pharmaceuticals in plasma. *Rapid Commun Mass Spectrom 24*, 309–314.

155. Harry, E. L., Reynolds, J. C., Bristow, A. W. T., Wilson, I. D., Creaser, C. S. (2009). Direct analysis of pharmaceutical formulations from non-bonded reversed-phase thin-layer chromatography plates by desorption electrospray ionisation ion mobility mass spectrometry. *Rapid Commun Mass Spectrom 23*, 2597–2604.

156. Luosujärvi, L., Laakkonen, U. M., Kostiainen, R., Kotiaho, T., Kauppila Tiina, J. (2009). Analysis of street market confiscated drugs by desorption atmospheric pressure photoionization and desorption electrospray ionization coupled with mass spectrometry. *Rapid Commun Mass Spectrom 23*, 1401–1404.

157. Soparawalla, S., Salazar, G. A., Perry, R. H., Nicholas, M., Cooks, R. G. (2009). Pharmaceutical cleaning validation using non-proximate large-area desorption electrospray ionization mass spectrometry. *Rapid Commun Mass Spectrom 23*, 131–137.

158. Williams, J. P., Scrivens, J. H. (2005). Rapid accurate mass desorption electrospray ionisation tandem mass spectrometry of pharmaceutical samples. *Rapid Commun Mass Spectrom 19*, 3643–3650.

159. Shin, Y. S., Drolet, B., Mayer, R., Dolence, K., Basile, F. (2007). Desorption electrospray ionization-mass spectrometry of proteins. *Anal Chem 79*, 3514–3518.

160. Myung, S., Wiseman, J. M., Valentine, S. J., Takats, Z., Cooks, R. G., Clemmer, D. E. (2006). Coupling desorption electrospray ionization with ion mobility/mass spectrometry for analysis of protein structure: Evidence for desorption of folded and denatured states. *J Phys Chem B110* 5045–5051.

161. Huang, M. Z., Hsu, H. J., Lee, J. Y., Jeng, J., Shiea, J. (2006). Direct protein detection from biological media through electrospray-assisted laser desorption ionization/mass spectrometry. *J Proteome Res 5*, 1107–1116.

162. Weston, D. J., Bateman, R., Wilson, I. D., Wood, T. R., Creaser, C. S. (2005). Direct analysis of pharmaceutical drug formulations using ion mobility spectrometry/quadrupole-time-of-flight mass spectrometry combined with desorption electrospray ionization. *Anal Chem 77*, 7572–7580.

163. Bereman, M. S., Nyadong, L., Fernandez, F. M., Muddiman, D. C. (2006). Direct high-resolution peptide and protein analysis by desorption electrospray ionization Fourier transform ion cyclotron resonance mass spectrometry. *Rapid Commun Mass Spectrom 20*, 3409–3411.

164. Ouyang, Z., Cooks, R. G. (2009) Miniature mass spectrometers, *An Rev Anal Chem 2*, 187–214.

165. Costa, A. B., Cooks, R. G. (2007). Simulation of atmospheric transport and droplet-thin film collisions in desorption electrospray ionization. *Chem Commun* 3915–3917.

166. Costa, A. B., Cooks, G. R. (2008). Simulated splashes: Elucidating the mechanism of desorption electrospray ionization mass spectrometry. *Chemical Phys Lett 464*, 1–8.

167. Nefliu, M., Smith, J. N., Venter, A., Cooks, R. G. (2008). Internal energy distributions in desorption electrospray ionization (DESI). *J Am Soc Mass Spectrom 19*, 420–427.

168. Takats, Z., Cotte-Rodriguez, I., Talaty, N., Chen, H., Cooks, R. G. (2005). Direct, trace level detection of explosives on ambient surfaces by desorption electrospray ionization mass spectrometry. *Chem Commun* 1950–1952.

169. Kertesz, V., Van Berkel, G. J. (2008). Improved imaging resolution in desorption electrospray ionization mass spectrometry. *Rapid Commun Mass Spectrom 22*, 2639–2644.

170. Ifa, D. R., Manicke, N. E., Rusine, A. L., Cooks, R. G. (2008). Quantitative analysis of small molecules by desorption electrospray ionization mass spectrometry from polytetrafluoroethylene surfaces. *Rapid Commun Mass Spectrom 22*, 503–510.

171. Mulligan, C. C., MacMillan, D. K., Noll, R. J., Cooks, R.G. (2007). Fast analysis of high-energy compounds and agricultural chemicals in water with desorption electrospray ionization mass spectrometry. *Rapid Commun Mass Spectrom 21*, 3729–3736.

172. Miao, Z., Chen, H. (2008). Analysis of continuous-flow liquid samples by desorption electrospray ionization-mass spectrometry (DESI-MS). *Proc 56th An Am Soc Mass Spectrom Conf on Mass Spectrometry*, Denver, CO, June 1–5.

173. Ma, X., Zhao, M., Lin, Z., Zhang, S., Yang, C., Zhang, X. (2008). Versatile platform employing desorption electrospray ionization mass spectrometry for high-throughput analysis. *Anal Chem 80*, 6131–6136.

174. Chipuk, J. E., Brodbelt, J. S. (2009). The influence of material and mesh characteristics on transmission mode desorption electrospray ionization. *J Am Soc Mass Spectrom 20*, 584–592.

175. Chang, D. Y., Lee, C. C., Shiea, J. (2002). Detecting large biomolecules from high-salt solutions by fused-droplet electrospray ionization mass spectrometry. *Anal Chem 74*, 2465–2469.

176. Lee, C. C., Chang, D. Y., Jeng, J., Shiea, J. (2002). Generating multiply charged protein ions via two-step electrospray ionization mass spectrometry. *J Mass Spectrom 37*, 115–117.

177. Chen, C. N., Jones, P. R., Shiea, J. (2005). Detection of preformed macrocyclic nickel complex ions using two-step electrospray ionization mass spectrometry. *J Mass Spectrom 40*, 413–415.

178. Shiea, J., Chang, D. Y., Lin, C. H., Jiang, S. J. (2001). Generating multiply charged protein ions by ultrasonic nebulization/multiple channel-electrospray ionization mass spectrometry. *Anal Chem 73*, 4983–4987.

179. Lee, C. Y., Shiea, J. (1998). Gas chromatography connected to multiple channel electrospray ionization mass spectrometry for the detection of volatile organic compounds. *Anal Chem 70*, 2757–2761.

180. Hong, C. M., Tsia, F. C., Shiea, J. (2000). A multiple channel electrospray source used to detect highly reactive ketenes from a flow pyrolyzer. *Anal Chem 72*, 1175–1178.

181. Zhang, Y., Chen, H. (2010). In-situ selective and sensitive detection of monosaccharides by reactive desorption electrospary ionization (DESI) using modified phenylboronic acids. *Int J Mass Spectrom 289*, 98–107.

182. Lin, S. Y., Huang, M. Z., Chang, H. C., Shiea, J. (2007). Using electrospray-assisted laser desorption/ionization mass spectrometry to characterize organic compounds separated on thin-layer chromatography plates. *Anal Chem 79*, 8789–8795.

183. Shiea, J., Yuan, C., Huang, M., Cheng, S., Ma, Y., Tseng, W., Chang, H., Hung, W. (2008). Detection of native protein ions in aqueous solution under ambient conditions by electrospray laser desorption/ionization mass spectrometry. *Anal Chem 80*, 4845–4852.

184. Huang, M., Jhang, S., Cheng, C., Cheng, S., Shiea, J. (2010). Effects of matrix, electrospray solution, and laser light on the desorption and ionization mechanisms in electrospray-assisted laser desorption ionization mass spectrometry. *Analyst 135*, 759–766.

185. Peng, I. X., Loo, R. O., Margalith, E., Little, M. W., Loo, J. A. (2010). Electrospray-assisted laser desorption ionization mass spectrometry (ELDI-MS) with an infrared laser for characterizing peptides and proteins. *Analyst 135*, 767–772.

186. Peng, I. X., Shiea, J., Loo, R. R. O., Loo, J. A. (2007). Electrospray-assisted laser desorption/ionization and tandem mass spectrometry of peptides and proteins. *Rapid Commun Mass Spectrom 21*, 2541–2546.

187. Cheng, C. Y., Yuan, C. H., Cheng, S. C., Huang, M. Z., Chang, H. C., Cheng, T. L., Yeh, C. S., Shiea, J. (2008). Electrospray-assisted laser desorption/ionization mass spectrometry for continuously monitoring the states of ongoing chemical reactions in organic or aqueous solution under ambient conditions. *Anal Chem 80*, 7699–7705.

188. Peng, I. X., Loo, R. R. O., Shiea, J., Loo, J. A. (2008). Reactive-electrospray-assisted laser desorption/ionization for characterization of peptides and proteins. *Anal Chem 80*, 6995–7003.

189. Huang, M., Hsu, H., Wu, C., Lin, S., Ma, Y., Cheng, T., Shiea, J. (2007). Characterization of the chemical components on the surface of different solids with electrospray-assisted laser desorption ionization mass spectrometry. *Rapid Commun Mass Spectrom 21*, 1767–1775.

190. Dixon, R. B., Muddiman, D. C. (2010). Study of the ionization mechanism in hybrid laser based desorption techniques. *Analyst 135*, 880–882.

191. Sampson, J. S., Hawkridge, A. M., Muddiman, D. C. (2006). Generation and detection of multiply-charged peptides and proteins by matrix-assisted laser desorption electrospray ionization (MALDESI) Fourier transform ion cyclotron resonance mass spectrometry. *J Am Soc Mass Spectrom 17*, 1712–1716.

192. Sampson, J. S., Murray, K. K., Muddimana, D. C. (2009). Intact and top-down characterization of biomolecules and direct analysis using infrared matrix-assisted laser desorption electrospray ionization coupled to FT-ICR mass spectrometry. *J Am Soc Mass Spectrom 20*, 667–673.

193. Sampson, J. S., Hawkridge, A. M., Muddiman, D. C. (2008). Development and characterization of an ionization technique for analysis of biological macromolecules: Liquid matrix-assisted laser desorption electrospray ionization. *Anal Chem 80*, 6773–6778.

CHAPTER 2

Ion Activation and Mass Analysis in Protein Mass Spectrometry

CHENG LIN and PETER O'CONNOR

In this chapter we consider the various methods of activation that can be used to fragment peptide and protein ions and thereby be used to determine their amino-acid sequences. After a brief introduction to the terms important in mass spectrometry (MS) analysis, we describe the methods used to activate peptide and protein ions for sequencing by MS. This section is followed by a discussion of mass analysis, particularly as it applies to the MS/MS experiment.

2.1 INTRODUCTION

A mass analyzer is the heart of a mass spectrometer, where ions are separated according to their mass-to-charge ratios (*m*/*z*). Although *m*/*z* has a dimension of mass over charge, it is often expressed as a dimensionless number in the MS literature, where the mass is measured in the unified atomic mass unit, *u*, or dalton (Da), 1 $u \approx 1.66 \times 10^{-27}$ kg, and the charge is measured as number of elementary charges, *e*, 1 $e \approx 1.602 \times 10^{-19}$ coulombs. The performance of a mass analyzer is characterized by a number of parameters, including its mass accuracy, mass resolving power, mass range, scan speed, and tandem MS analysis capability.

2.1.1 Mass Accuracy

Mass accuracy describes the ability of the mass analyzer to measure the correct mass (*m*/*z*) of an ion, and precision is a measure of the ability to reproduce the mass measurement. Mass accuracy may be expressed as an absolute number, typically in mDa's, representing the difference between the theoretical and the measured masses.

Protein and Peptide Mass Spectrometry in Drug Discovery, Edited by Michael L. Gross, Guodong Chen, and Birendra N. Pramanik.

It is also frequently given as the relative ratio of this mass difference to the theoretical mass value, in parts per million (ppm). Mass accuracy is closely related to the mass resolving power of the mass analyzer. For example, a low resolving power instrument, such as a quadrupole ion trap (QIT) mass spectrometer, can only provide a typical mass accuracy of approximately 100 ppm, whereas a high-end Fourier-transform ion cyclotron resonance (FTICR) mass spectrometer can routinely achieve a mass accuracy in the sub-ppm range. Other factors that affect the mass accuracy include the stability of the instrument, mass calibration, and the peak centroid determination.

2.1.2 Mass Resolving Power

Mass resolving power is the ability of a mass analyzer to separate ions with closely spaced *m/z* values. For an isolated peak the mass resolving power (RP) can be calculated using the formula

$$RP = \frac{m/z}{\Delta(m/z)_{FWHM}}, \qquad (2.1)$$

where $\Delta(m/z)_{FWHM}$ is the full width of the peak at its half maximum, but it can also be substituted by the peak width at other fractions of the peak maximum. Mass resolving power may also be calculated using adjacent overlapping peaks. In this definition the $\Delta(m/z)_{FWHM}$ of equation 2.1 is replaced by Δm, or the mass resolution, which is the smallest mass difference between two equal magnitude peaks so that the valley between them is a specific fraction of the peak height. For Gaussian shaped peaks, a 50% valley exists when Δm is approximately 141% of the $\Delta(m/z)_{FWHM}$ value. An immediate consequence of poor mass resolving power is the inability to determine the peak position accurately in the presence of nearby peaks. Figure 2.1 illustrates the effect of mass resolving power on the obtainable mass accuracy. For two Gaussian shaped peaks (Figure 2.1) of equal height for ions of *m/z* 1000 and 1001, a mass resolving power of 1000 results in close to a 100-mDa difference between the observed and actual peak positions (Figure 2.1*B*), whereas a slight increase of RP to 1410 improves the mass accuracy to around 4 mDa (Figure 2.1*A*). This overlapping problem is more severe when the nearby peak is of a higher intensity than the peak of interest. When the "interfering" peak at *m/z* 1000 is five times as intense as the one of interest at *m/z* 1001 and the RP is still 1410, the peak position of the latter is shifted by 26 mDa (Figure 2.1*C*); when that ratio increases to 10, the valley disappears, and the *m/z* 1001 peak cannot be identified in the spectrum (Figure 2.1*D*).

2.1.3 Mass Range

The mass range of a mass analyzer is the range of *m/z* values an ion can have to be detected. Quadrupole mass analyzers, magnetic sectors, and quadrupole ion traps can typically scan up to around *m/z* 4000, whereas FTICR mass analyzers can easily detect ions of *m/z* value over 10,000. A linear time-of-flight (TOF) analyzer has no upper mass limit in principle, but the practical upper mass limit of a reflectron TOF instrument is approximately 10,000; the limit is due to the tendency of large

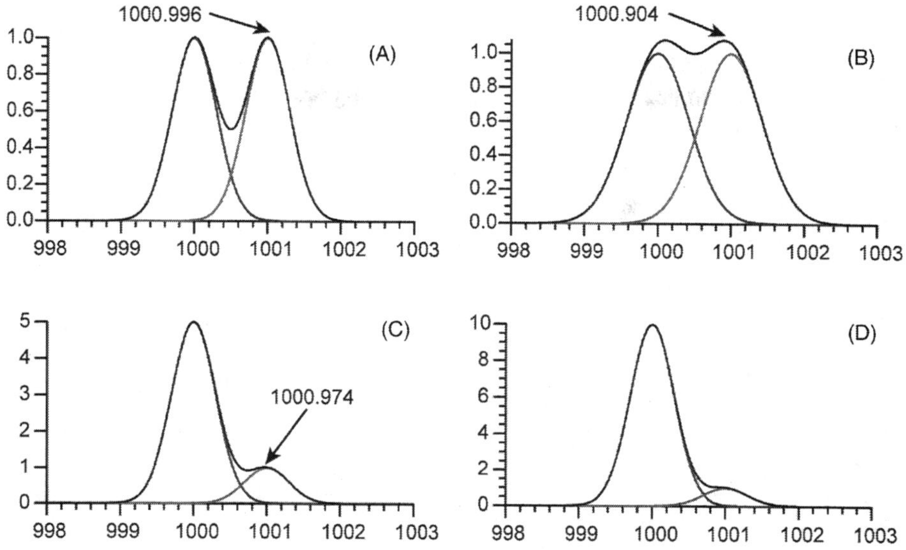

FIGURE 2.1 The impact of mass resolving power (R.P.) on the mass accuracy for two peaks separated by 1 Da. The arrow indicates the measured peak position of the *m/z* 1001 ion at the presence of another ion at *m/z* 1000. (*A*) R.P. = 1410, equal peak heights; (*B*) R.P. = 1000, equal peak heights; (*C*) R.P. = 1410, the peak height ratio is 5:1; (*D*) R.P. = 1410, the peak height ratio is 10:1.

biomolecules to undergo postsource decay (PSD) and not be detectable in a reflectron-based instrument because the precursor ions are lost to fragmentation, and the product ions are not refocused. *Low* mass limits also exist for some mass analyzers. For example, in both FTICR and Orbitrap instruments, the lowest *m/z* value detectable is limited by the sampling frequency applied, as determined by the Nyquist theorem, with typical values of 50 to 200 Da when the B field is large (e.g., >4.8 T). The mass range of a mass spectrometer may be further reduced by factors not determined by the mass analyzer. For example, multipole ion guides commonly used for ion transfer have their own mass cutoffs.

2.1.4 Scan Speed

Scan speed describes how fast a mass analyzer can acquire mass spectra. For scanning mass analyzers such as quadrupoles, QIT's, and sector instruments, this is literally the speed at which a mass analyzer scans through a certain *m/z* range. For other types of instruments (e.g., FTICR and orbitrap mass analyzers), ions of all *m/z* values are detected simultaneously, in which case the scan speed may be defined as the rate at which each individual mass spectrum is acquired. A TOF analyzer is the fastest analyzer, capable of acquiring thousands of spectra in one second. An FTICR mass spectrometer, on the other hand, is often operating at a much slower rate, taking a second or longer to acquire one high-resolving-power mass spectrum. Scan speed is particularly important when the mass analysis is performed in conjunction with online separation techniques such as high-performance liquid chromatography (HPLC) or ion mobility, where analytes of interest are eluting only for a short period of time. For a given mass analyzer, there is often a trade-off between scan speed and mass resolving power, and accuracy.

2.1.5 Tandem MS Analysis

Tandem MS analysis refers to the process where a selected ion of interest (called the precursor ion) is isolated and dissociated to generate fragment ions whose *m/z* values are then measured. The masses of the fragment ions can be used to elucidate the structure of the precursor ion or, as is relevant to this book, sequence peptides and proteins. Tandem MS experiments may be performed tandem in space, which requires the use of two separate, physically distinct mass analyzers, such as those done in a triple quadrupole instrument or in a TOF-TOF mass spectrometer. It may also be performed tandem in time, in which case isolation of the precursor ion and mass analysis of the fragment ions are achieved using the same mass analyzer, but the events of isolation, activation, and analysis are separated in time. This is usually done in trap instruments, such as a QIT or an FTICR mass spectrometer.

Tandem MS analysis may be performed once (MS/MS), or multiple times consecutively, with each of the MS/MS experiments done on a fragment ion generated in the previous MS/MS step (this is known as the MS^n experiment). MS^n experiments produce feature rich fingerprints of the precursor ion by providing detailed structural information on each of the isolated fragment ions from the product-ion spectrum acquired in an MS^{n-1} experiment, making them a valuable tool in metabolite identification in drug discovery. In addition, when a product-ion spectrum is dominated by just a few fragments resulting from facile cleavages, MS^3 experiments are often needed to generate more complete structural information of the precursor ion. MS^n experiments are also used to characterize carbohydrate structural isomers, based on sequential losses of different derivatized monosaccharide units that carry fragmentation "scars" throughout the MS^n tree [1]. MS^n ($n > 2$) experiments can best be performed in trapping instruments.

In tandem MS experiments, it is usually necessary to activate the precursor ion first to induce fragmentation. Ion activation can be achieved in many ways: via collisions with gases or surfaces, absorption of IR or UV photons, or activation by ion–electron interactions [2].

2.2 ION ACTIVATION AND TANDEM MS ANALYSIS

2.2.1 Introduction: Fragmentation in Protein MS

Before we get into the details of various ion activation methods, it is helpful to look at what fragment ions may be produced in tandem MS experiments of peptide ions, and how this information can be used for their structural characterization. Throughout the

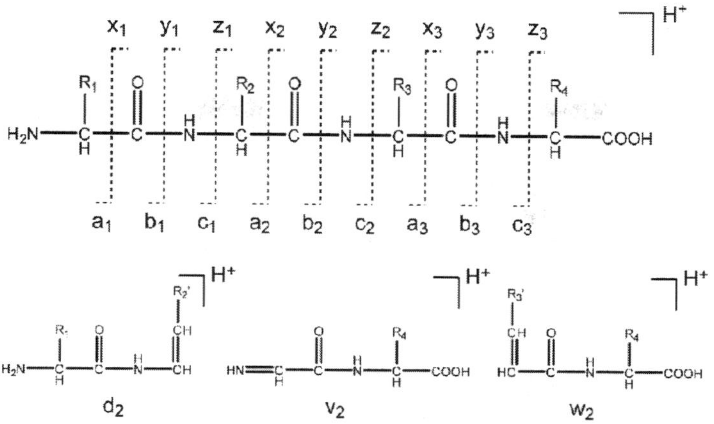

FIGURE 2.2 (*Top panel*) Nomenclature for sequence ions in peptide tandem mass spectra, as first proposed by Roepstorff and Fohlman (Roepstorff P., Fohlman J., *Biomedical Mass Spectrometry*, 1984, *11,* 601); (*bottom panel*) structure of satellite ions.

discussions in this section, the term peptide will be used in place of peptide/protein, as the two share similar fragmentation behavior. Figure 2.2 illustrates the common types of fragment ions observed in the tandem mass spectrum of a tetra-peptide ion [3]. These can be classified into two broad categories: the backbone fragment ions (a-, b-, c-, and x-, y-, z-ions), which result from the cleavage of a backbone bond, shown on the top panel; and the secondary side-chain fragment ions, or satellite ions (d-, v-, and w-ions), which are often generated by radical-driven, charge-induced, or even charge-remote fragmentations of the backbone fragment ions, shown on the bottom panel of the scheme.

Backbone fragment ions are also referred to as sequence ions because they are useful in peptide sequencing. Gas-phase peptide sequencing is based on the knowledge that adjacent backbone fragment ions of the same series are spaced by the masses of the amino-acid residues. For example, the mass difference between the b_n and b_{n+1} ion is the mass of the *n*th amino-acid residue, and that mass can be used to deduce the identity of the amino acid, with the exception of isomeric amino acids. Secondary side-chain fragment ions, on the other hand, contain important information on the identity of the side chain; these fragments are particularly useful for differentiation of isomeric amino-acid residues, (e.g., leucine and isoleucine). Other types of ions may also be produced in tandem MS experiments, including the immonium ions, internal fragment ions, and ions resulting from side-chain or small-molecule losses. Although these ions are not always structurally informative, and their presence can make the interpretation of the mass spectra more difficult, they can be useful (e.g., immonium ion formation and side-chain losses from the molecular ion are often used to identify the existence of certain amino-acid residues in the peptide).

Finally, because proteins undergo extensive post-translational modifications (PTMs), protein characterization should include the identification and location of PTMs in addition to the sequence determination. A PTM can be identified by the observation of the characteristic mass shift in molecular ions; however, PTM site location usually requires tandem MS experiments. Thus it is advantageous to retain PTMs during ion activation and backbone bond breakage so that the PTM mass "tags" the fragment ions to which it is attached, allowing successive localization in further stages of MS^n.

2.2.2 Collisional Activation Methods

Collisionally activated dissociation (CAD), or collision-induced dissociation (CID), is by far the most commonly applied ion activation method in tandem MS analysis. In a CAD experiment the precursor ion is allowed to collide with neutral gas molecules, resulting in energy transfer and ultimately internal excitation of the precursor ion [4–6]. Collisional activation can be achieved with a single high-energy (typically >1000 eV) collision, or with many low-energy (<1 to 100 eV) collisions.

Low-energy CAD is usually implemented in trapping instruments; examples are linear trapping quadrupoles (Q-CAD), quadrupole ion traps, and FTICR mass spectrometers (as in sustained off-resonance irradiation, or SORI-CAD) [7–10]. Although the peptide ion has been accelerated to a kinetic energy upward to 100 eV in the laboratory frame, it is the collisional energy in the center-of-mass frame (E_{COM}) that determines the maximum amount of energy that can be transferred to excite an ion's internal ro-vibrational modes. Because commonly used collision partners (e.g., He, N_2, or Ar) are much lighter than a typical peptide ion, E_{COM} is often orders of magnitude smaller, as calculated by equation 2.2, where m is the mass of the collision gas, M is the mass of the ion to be activated, and E_{Lab} is the laboratory energy:

$$E_{COM} = \frac{m}{m+M} E_{Lab}. \tag{2.2}$$

Thus, a heavier collision gas such as Ar or N_2 is frequently used in low energy CAD, as these gases allow a more efficient transfer of energy than does the lighter helium.

Low-energy CAD is generally considered an "ergodic" or "slow-heating" fragmentation method, where the term "slow" is used relative to the rate of intramolecular vibrational energy redistribution (IVR). In low-energy CAD experiments ion activation is achieved via multiple collisions, each depositing a small amount of energy into the precursor ion. Because the bond dissociation is preceded by the energy randomization, fragmentation rarely occurs at a site where the energy was first deposited in the collision. Instead, when the overall energy of the ion is raised above a certain dissociation threshold (activation barrier), fragmentation may occur, typically resulting in the rupture of the weakest bond within the molecule. For peptide ions, this is usually the amide bond, leading to the formation of b- and y-ions.

Given that direct amide bond cleavage requires the precursor ion to be excited to a substantially higher level than typically achievable in a low-energy CAD experiment, a "mobile proton" model can explain the b/y fragmentation pathway (Scheme 2.1) [11,12]. In essence, the fragmentation is initiated by the attachment or movement of a mobile proton to the oxygen or the nitrogen of the amide bond to be cleaved (the scheme shows the proton attachment to the amide nitrogen); the proton attachment not only weakens the amide bond but also increases the electrophilicity of

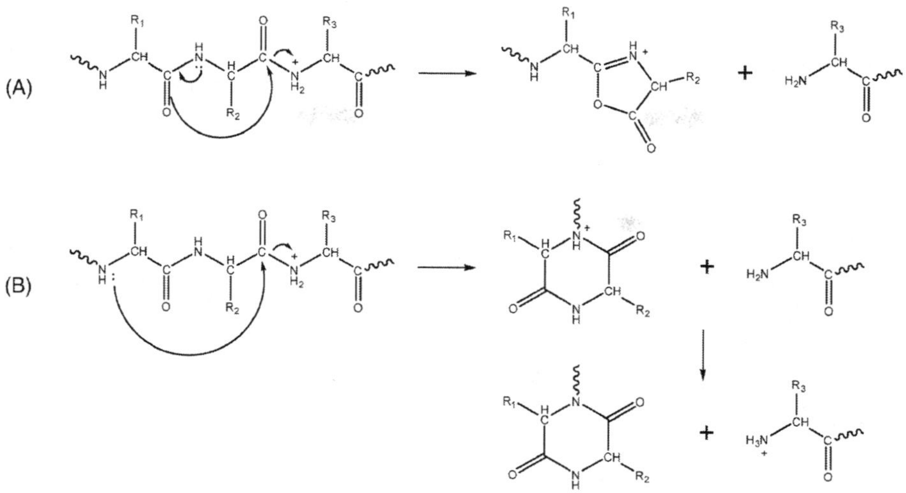

SCHEME 2.1 Mobile proton models for b/y cleavages. (*A*) the oxazolone pathway; (*B*) the diketopiperazine pathway.

the adjacent carbonyl, making it more susceptible to nucleophilic attack by either the amide oxygen (the oxazolone pathway, Scheme 2.1*A*) or the nitrogen (the diketopiperazine pathway, Scheme 2.1*B*) of its *N*-terminal neighbor. Thus peptide fragmentation proceeds via a low-energy rearrangement reaction, initiated by proton transfer, rather than direct accumulation of vibrational energy at an amide bond to induce its rupture. Because an arginine residue had the tendency to sequester a proton (it has high proton affinity), peptide ions with the number of arginine residues equal to or higher than the number of charges require higher collision energy to fragment. Although, in general, CAD cleaves the amide bonds relatively indiscriminately, selective cleavages are observed near particular amino-acid residues. For example, cleavage *N*-terminal to the proline residue is often enhanced, whereas cleavage *C*-terminal to the proline residue is suppressed. This "proline effect" may be due to its relatively high gas-phase proton affinity, which facilitates its amide nitrogen protonation, and to the hindered formation of an oxazolone b-ion containing the proline residue as its *C*-terminus owing to a strained bicyclic structure. Enhanced cleavage is also observed *C*-terminal to a protonated histidine residue, likely owing to its ability to transfer a proton to the backbone, allowing the side-chain nitrogen to attack the carbonyl and to form a resonance-stabilized cyclic b-ion. Finally, when no mobile proton is available, preferential cleavage at acidic residues occurs, with the acidic H of the aspartic or glutamic acid side chain serving to initiate cleavages at its *C*-terminal side.

High-energy CAD experiments are usually performed in magnetic sector [13,14] or TOF-TOF mass spectrometers [15] (beam instruments) where ions can be easily accelerated to have lab-frame translational energy of several thousand eV. Helium gas is the preferred collision partner in the high-energy CAD experiment, as it minimizes the scattering losses of both the precursor and the product ions; scattering is more severe in beam instruments because higher collision energy is employed, and focusing methods are sparse. These focusing methods are available when collisions are carried out in a multiple ion guide/trap. In high-energy CAD spectra, in addition to the b- and y-type ions and/or small-molecule losses, abundant immonium ions, internal ions and secondary fragments such as d- and w-type ions are also readily produced, the latter of which provide useful information for side-chain differentiations [13,16].

Despite its wide implementation, CAD also has several drawbacks. One of its major limitations is its poor applicability in PTM analysis. Many PTMs are more labile than the backbone amide bond and are the first to fall off (whether via direct scission or a rearrangement is not clear) when the ions are collisionally activated; this makes PTM location a challenging task. In addition, when a labile group is present in a peptide or protein, the CAD spectrum is often dominated by a fragment ion produced by loss of the PTM. This loss preempts peptide bond cleavages, causing fewer backbone fragments to form. Likewise, when a particularly labile dissociation channel exists, such as the b/y cleavage at the Asp-Pro sequence or loss of a phosphate, other fragmentation channels may also be suppressed, resulting in poor sequence coverage. Furthermore, sequence scrambling can occur in CAD experiments, in which an oxazolone b-ion can cyclize and reopen at different position; such processes give a product that, upon further activation, can produce misleading sequence ions [17,18]. The use of a collisional gas in CAD can compromise the high vacuum of the spectrometer, and the gas may need to be pumped away before mass analysis, particularly in an FTICR instrument. The pump-down time results in longer spectral acquisition time and reduced throughputs. Additionally CAD has limits in quadrupole ion traps and linear ion traps because the resonant excitation raises the low-mass cutoff of the instrument (discussed below in Section 2.3).

Alternatively, collisional activation can be achieved by ion/surface collisions without the use of collision gases, as implemented in surface-induced dissociation (SID) [19,20]. In general, SID produces product-ion spectra that are similar to those generated by CAD. Higher ratios of a- to b-ions and enhanced immonium-ion formation, however, also occur, and this is indicative of increased access to higher energy and secondary fragmentation channels. Unlike low-energy CAD, ion activation in SID is achieved in a single collision, rather than being slowly heated via multiple collisions until the dissociation threshold is reached. The effective neutral partner mass of the surface in SID (m in equation 2.2) is also much higher than that of collisional gases, leading to a higher center-of-mass collision energy available for ion excitation. The efficiency of translational-to-internal energy conversion for SID depends on the kinetic energy, the size of the precursor ions, and the nature of the surface. The most commonly used surface is metallic with a nonconducting fluorinated, self-assembled monolayer (SAM) to minimize ion neutralization. SID can be implemented in a variety of mass spectrometers, including the tandem quadrupole, TOF-reflectron, Q-TOF, and FTICR instruments [21–23].

2.2.3 Photodissociation

Peptide ions may be optically excited as well. Photodissociation (PD) is best applied to trapped ions to allow for sufficient ion/photon interaction time, hence indicating as instruments linear ion traps [24] or FTICR mass spectrometers [25], although photodissociation in TOF/TOF instruments [26] can also be useful. Photodissociation offers several advantages over CAD, including ease of implementation, better control of energy inputs, and selectivity based on the absorption spectra of precursor ions. For photodissociation in an FTICR instrument, the gas-free operation can dramatically reduce the MS/MS spectral acquisition time without deteriorating the vacuum, which is needed for high-resolving-power mass analysis. Compared to SORI-CAD experiments, which excite ions over a narrow *m/z* window over time, all precursor ions of interest and in different charge states are dissociated simultaneously in a single photodissociation event. For photodissociation in a linear or quadrupole ion trap, there is no low-mass cutoff because no translational excitation of the precursor ions is involved, and removing the need for translational activation also minimizes scattering. Finally, because all product ions are produced either on axis (as in linear ion traps) or in the center of the trap (as in FTICRs), MS^n experiments can be easily performed.

Infrared Multiphoton Dissociation The photons may originate from background blackbody irradiation, as in the blackbody infrared radiative dissociation (BIRD) experiment [27], or from a laser, with wavelengths ranging from the mid-infrared (IR) region to the vacuum ultraviolet (VUV) end of the spectrum. The most commonly used IR laser is the continuous wave (cw) CO_2 laser operating at 10.6 μm, which vibrationally excites peptide ions, for example. Given that a 10.6-μm photon has an energy of around 0.117 eV, or 11.3 kJ/mol, whereas a typical chemical bond has a bond dissociation energy (BDE) of around 400 kJ/mol, absorption of hundreds or even thousands of IR photons is necessary before fragmentation occurs. Thus, as for low-energy CAD, infrared multiphoton dissociation (IRMPD) also "heats" slowly the ions with IVR preceding bond dissociation. As a result IRMPD of peptide ions yields a fragment pattern similar to that of CAD, with the exception that, because fragment ions from IRMPD can continue to absorb photons and further fragment, secondary fragmentation is enhanced, for good or bad.

The IRMPD efficiency can be improved by covalently attaching to the peptides an IR-chromophore, such as a phosphonite or sulfonate group. *N*-terminal sulfonation affects the CAD and IRMPD spectra of the peptide YGGFLR; the spectrum was acquired in a linear ion trap (Figure 2.3) [28]. The sulfonate group increases the photoabsorptivity of the peptide at 10.6 μm and leads to extensive fragmentation at a shorter irradiation time. Furthermore the negative charge of the sulfonate neutralizes the *N*-terminal fragment charge, greatly simplifying the product-ion mass spectrum. Compared to the complex CAD spectrum showing both *N*- and *C*-terminal and other fragment ions, the IRMPD spectrum of the modified peptide is dominated by a y-ion series that makes de novo sequencing a much easier task. Finally, the low-mass cutoff in the CAD spectrum prevents observation of the y_1-ion, whereas a complete series of y-ions are formed upon IRMPD.

Ultraviolet Photodissociation UV lasers can also be used to fragment peptide ions in a variety of mass spectrometers, including the linear ion trap, tandem TOF, tandem sector, and FTICR instruments [26,29–32]. Unlike IR lasers, a UV laser

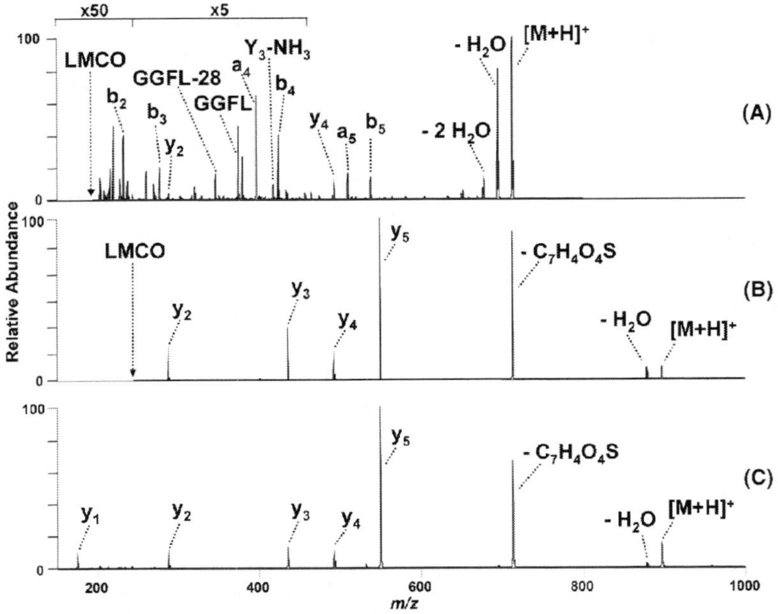

FIGURE 2.3 ESI-MS/MS mass spectra of YGGFLR. (*A*) CAD of unmodified peptide; (*B*) CAD of *N*-terminal sulfonated peptide; (*C*) IRMPD of *N*-terminal sulfonated peptide. Magnification scales apply to all spectra along the mass range indicated. Adapted with permission from American Chemical Society (Wilson J. J., Brodbelt J. S., *Anal. Chem.* 2006, *78*, 6855).

excites the peptide ions electronically. Electronic excitations are "vertical" excitations that occur on a femtosecond time scale, shorter than a vibrational period. Moreover, a UV photon contains much higher energy than an IR photon (e.g., a 193 nm photon has an energy of ~620 kJ/mol, enough to break most covalent bonds). Therefore, it is possible for dissociation induced by a single UV photon absorption to occur rapidly prior to the IVR. As such, UV photodissociation may produce spectra that are dramatically different from and complementary to either low-energy CAD or IRMPD spectra. Fast absorption of high energy also permits the UVPD method to be coupled to beam-type instruments.

Common UV lasers employed to fragment peptide ions include excimer lasers and various harmonics of the Nd:YAG laser. The 266 nm light (the third harmonic of the Nd:YAG laser output, or Y4) is absorbed strongly by the side chains of tryptophan, tyrosine, and phenylalanine and fragments peptide ions containing these chromophores [32]. A near-UV chromophore may also be covalently or noncovalently attached to the peptide ions, which enables photodissociation at longer wavelengths, such as the 355-nm light from the Nd:YAG laser (the second harmonic, Y3) [33]. In either case, as the absorption occurs locally at specific chromophores, the energy must be redistributed before extensive backbone fragmentations may occur, which results in a general fragmentation pattern similar to that observed in CAD, with enhanced fragmentation near the chromophores. Unusual fragmentations generating radical products sometimes occur, and are attributed to direct fragmentation near the chromophores.

A second UVPD approach is to choose a wavelength absorbed by a universal chromophore such as the backbone amide bond. The amide bond has several UV absorption bands centered at near 190 nm and near 160 nm, both of which are readily accessible with excimer lasers (ArF: 193 nm, F_2: 157 nm) [34]. Of the two wavelengths, the 193 nm is more convenient, primarily because it can travel through air with only small absorption [35]. Light of 157 nm must be transmitted either through vacuum or an inert-gas-protected environment because it is strongly absorbed by oxygen in the air. Figure 2.4 shows a typical UVPD spectrum of a tryptic peptide containing arginine at its *C*-terminus; the spectrum can be acquired by using either the 193- or the 157-nm light in a tandem TOF instrument [34]. Unlike IRMPD or UVPD at longer wavelengths, UVPD of peptide ions at these two wavelengths leads to extensive a/x cleavages and other secondary fragmentation. UV absorption at these

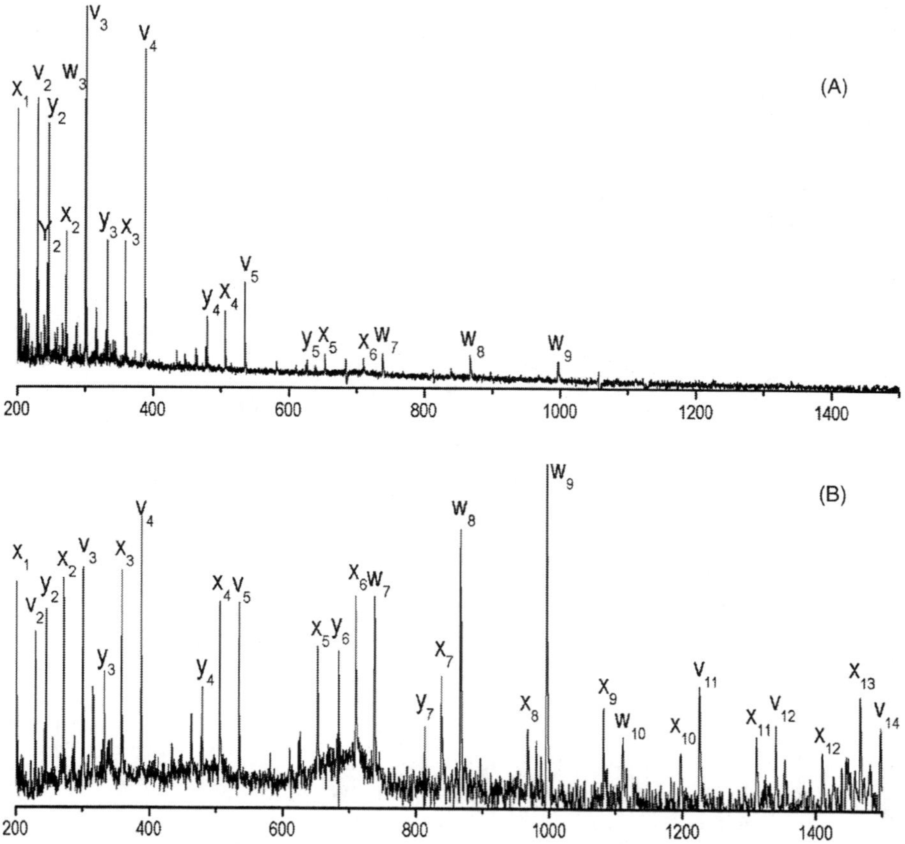

FIGURE 2.4 Tandem-TOF photodissociation of Glu-Fibrinopeptide B (EGVNDNEEGFFSAR) using (*A*) 193 and (*B*) 157 nm light. Reproduced with permission from Elsevier (Thompson M. S., Cui W., Reilly J. P., *J. Am. Soc. Mass Spectrom.* 2007, *18*, 1439).

wavelengths apparently results in the homolytic cleavage of the C_α–C(=O) bond, producing two radical species a + 1 and x + 1, both of which may either lose a hydrogen to form the a- or x-ions, depending on the location of the charge carrier(s), or undergo secondary radical-induced rearrangements to form d-, w-, and v-type ions. Some b- and y-ions are also seen in the UVPD spectra, particularly when the arginine residue is replaced by a lysine. This may be due to the higher mobility of a proton associated with the charged lysine residue, and that mobile proton facilitates the low-energy fragmentation processes. Finally, because UVPD may occur prior to energy randomization, a distinct advantage of UVPD over CAD and IRMPD is its ability to retain labile PTMs, such as phosphorylations and glycosylations, while backbone bonds break. Although this possibly non-ergodic or nonstatistical behavior may pertain to many electron-induced dissociation methods discussed in the next section, UVPD offers a unique benefit in that it is applicable to singly charged ions generated by MALDI, as it does not involve charge reduction.

Femtosecond Laser-Induced Dissociation Very recently a new tandem MS technique termed the femtosecond laser-induced ionization/dissociation, or fs-LID, was developed in which an ultrafast femtosecond laser with high peak power ($>10^{13}$ W/cm^2) and high repetition rate (kHz) fragments peptide ions in an ion trap. Although the fs laser is often a Ti:Al_2O_3 laser operated in the near IR region (~800 nm), the fs-LID spectra are very different from typical IRMPD spectra, as exemplified in Figure 2.5 [36]. In addition to the b- and y-ions associated with ergodic fragmentation pathways, abundant a-, c-, x-, and z-ions are produced along with secondary fragment ions. Phosphate groups are retained in many fragment ions produced by peptide-bond cleavages, and these ion series can be used to locate the

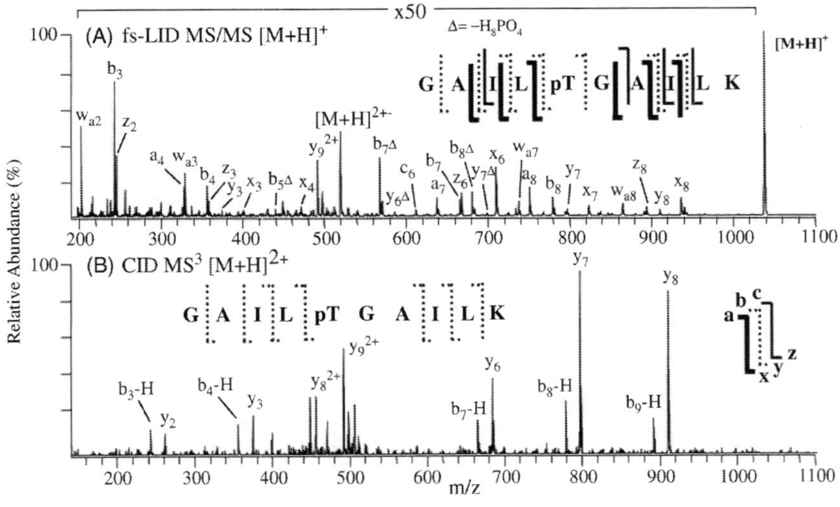

FIGURE 2.5 fs-LID MS/MS (200 msec irradiation) of the [M + H]$^+$ precursor ion of GAILpTGAILK. Adapted with permission from American Chemical Society (Kalcic C. L., et al., *J. Am. Chem. Soc.* 2009, *131*, 940).

phosphorylation site [37]. The mechanism of fs-LID involves tunneling ionization of the precursor ion in the presence of the strong electromagnetic field produced by the high-power femtosecond irradiation, which generates a radical species that may undergo fragmentation induced by vibrational or electronic excitation. This hypothesis is supported by the presence of a doubly charged radical species seen in the fs-LID spectrum of a singly charged precursor ion. Like the shorter wavelength UVPD, fs-LID also produces more nonstatistical fragmentations and is amenable to singly charged precursor ions. This approach complements conventional CAD, IRMPD, and electron-induced dissociation methods (next section).

2.2.4 Electron-Induced Dissociation

A third method to activate peptide ions is through ion–electron or ion–ion interactions. Electron-induced dissociation has been around for many decades, but it was not until the late 1990s with the implementation of electron-capture dissociation (ECD) that it found broad applications in the structural analysis of biomolecules [38–41]. The first ECD spectrum was actually acquired during a UVPD study in an FTICR instrument, where a misaligned 193-nm laser beam hit the ICR trap surface, generating photoelectrons that induced ECD of peptide ions. Since then, conventional electron sources (e.g., a directly heated filament, an indirectly heated dispenser cathode, or a cold field emitting device) are being used instead of the laser.

ECD spectra of multiply charged protein ions are usually dominated by the c- and z-ion series resulting from the N–C_α bond cleavage [38]. Preferential cleavage of disulfide bonds in ECD also occurs [42]. An important characteristic of ECD is its ability to generate extensive backbone cleavages while leaving the more labile PTMs and even noncovalent interactions intact [39–41,43–45]. The cause of this putative nonstatistical behavior of ECD is the center of an ongoing debate on the primary ECD mechanism. Some propose that ECD is a non-ergodic process initiated by the electron capture at a charge site, followed by hydrogen transfer to the backbone carbonyl inducing N$-$$C_\alpha$ bond cleavages. Others argue against the non-ergodic premise and propose that the electron capture first occurs at the backbone carbonyl, generating an anion-radical super base stabilized by a remote charge; the newly formed species then undergoes facile N$-$$C_\alpha$ bond cleavage prior to proton transfer, leading to the formation of c- and z-ions. A general mechanism (Scheme 2.2) proposes that the electron capture puts the peptide ion in a Rydberg state (with ~4–6 eV of excess energy owing to recombination of opposite charges) that may sample a number of electronic states of the charge-reduced ion as it "rattles" down the energy ladder. Both mechanisms could be at work, depending on the electronic state of the peptide ions from which the dissociation occurs [46].

ECD is highly complementary to the conventional CAD method, as illustrated by Figure 2.6, which is a heat map showing the frequency of fragmentation occurrence as a function of the neighboring amino acid residues [47]. One feature that stands out is that whereas cleavage *N*-terminal to proline is enhanced in CAD, it is rarely observed in ECD, because the N$-$$C_\alpha$ bond cleavage at the proline site still leaves the *N*- and *C*-terminal "fragments" connected by a covalent bond owing to proline's ring

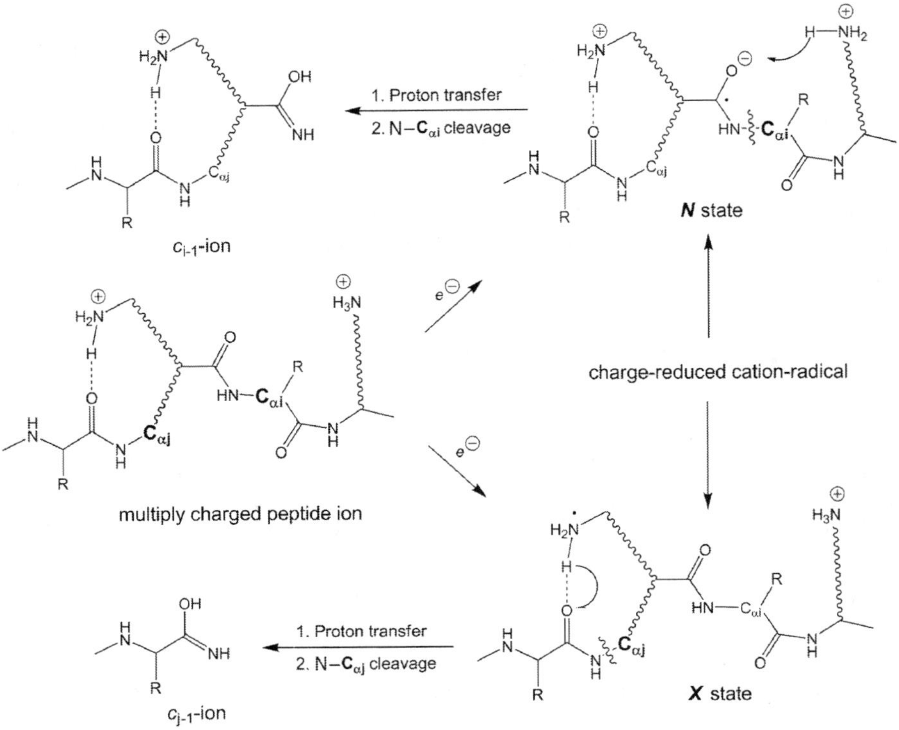

SCHEME 2.2 General mechanism for ECD in ground and excited electronic states of peptide ions. Reproduced with permission from Elsevier (Syrstad E. A., Turecěk F., *J. Am. Soc. Mass Spectrom.* 2005, *16*, 208).

structure. Figure 2.6 suggests that it is often advantageous to perform ECD and CAD as a duet to take advantage of the complementary nature of the two methods. Moreover, obtaining both the ECD and CAD spectra of the same sample increases the confidence of peak assignment by utilizing the "golden pair,"[†] and consequently lead to more reliable protein identifications in both database searching method and de novo sequencing [48].

It is worth noting that N−C$_\alpha$ bond cleavage may also occur on the N-terminal side of the affected carbonyl. Although this is generally disfavored because the N-radical that is formed is unstable, it may be important when unusual amino-acid residues are involved. For instance, C$_\alpha$-C$_\beta$ cleavage at the isoaspartic acid residue site will lead to the formation of c + 57 and z − 57 diagnostic ions for the differentiation of aspartic and isoaspartic acid residues; this differentiation is an important goal in protein deamidation studies [49,50].

[†] The "golden pair" refers to a b- (or y-) ion from the CAD spectrum that is also observed with a correlated c- (or z-) ion from the ECD spectrum. These pairs allow determination of the directionality of the cleavage, thus inherently labeling the *N*-terminal and/or *C*-terminal fragment ion series.

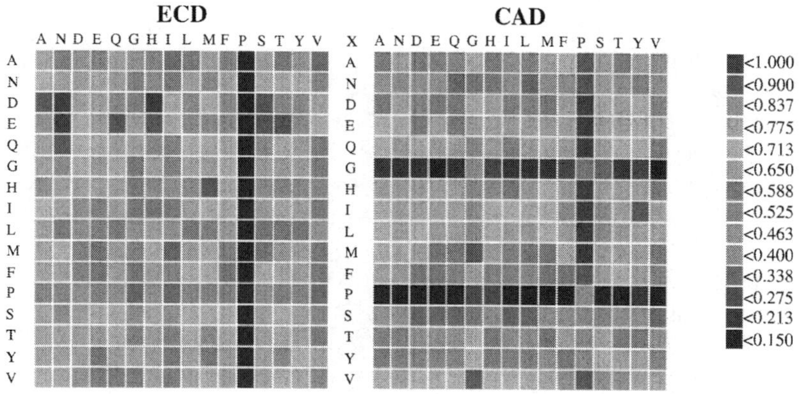

FIGURE 2.6 Amino acid preferences in ~15,000 tandem mass spectra CAD and ECD. Reproduced with permission from Elsevier (Zubarev R. A., et al. *J. Am. Soc. Mass Spectrom.* 2008, *19*, 753). (*See the color version of this figure in Color Plates section.*)

In ECD, the initial N$-$C$_\alpha$ bond cleavage produces an even-electron c ion and an odd electron z$^\bullet$ ion. The radical on the alpha carbon of the z$^\bullet$ ion may propagate via radical-driven rearrangements, including hydrogen transfer between the c and z species before they separate to produce a radical c$^\bullet$ ion and an even-electron z ion [51–53]. Hydrogen abstraction within the z$^\bullet$ fragment can lead to the formation of secondary w ions, internal fragments, or z-ions with additional partial or complete side-chain losses [54–56]. This free-radical cascade may explain the formation of backbone fragments in cyclic-peptide ECD, where the capture of a single electron results in multiple backbone cleavages [57].

The ECD efficiency and sequence coverage can often be improved by vibrational excitation of the precursor ions, particularly for larger proteins or peptides with extensive noncovalent interactions [58] such as those between a phosphate group of a phosphopeptide and other positive charges [59]. The vibrational excitation breaks noncovalent bonds that hold the nascently formed c and z fragments together and facilitates formation of individual fragment ions. Furthermore there may be an increase in the conformational heterogeneity of the precursor ion, leading to a more extensive fragmentation pattern or increased secondary fragmentation. Ion activation can be achieved via collisions with gases as done in plasma ECD [60] via IR irradiation, as implemented in the activated ion or AI-ECD [58], or simply by increasing the electron energy as employed in hot ECD [61,62].

The success of ECD has led to the implementations of other electron-induced dissociation methods [63] including electron-detachment dissociation (EDD) [64,65], electron-ionization dissociation (EID) [66], and electron-transfer dissociation (ETD) [67], which are collectively known as the ExD methods. In EDD, instead of capturing a low-energy electron, a negatively charged precursor ion, when bombarded by high-energy electrons, experiences loss of an electron from its valence shell and forms a charge-reduced radical anion. This radical anionic species may undergo backbone cleavages leading to the formation of a-, c-, and z-ions, much like those observed in ECD. EDD is particularly useful for fragmenting peptides that can easily generate multiply charged anions in ESI; examples are phosphorylated, sulfated peptides, or peptides with multiple acidic residues.

Given that both ECD and EDD are initiated by a charge-reduction process, they are only applicable to multiply charged ions. Singly charged precursor ions, as those produced by MALDI, would be neutralized by either electron capture in the positive-ion mode, or electron detachment in the negative-ion mode, preventing the detection of fragments. EID, however, can be applied to singly charged precursor ions. The exact mechanism of EID is not well understood, but it probably involves first the ionization of the precursor ion $[M + nH]^{n+}$ by interaction with a high-energy electron to produce a radical $[M + nH]^{(n+1)+\bullet}$, which may dissociate directly or capture a low-energy electron and then dissociate. EID spectra are often complex; they show ions produced both by ergodic processes and by radical pathways.

In the early stages of development, ExD was used exclusively with FTICR instruments; the presence of a magnetic field is beneficial for trapping thermal electrons to allow efficient ion–electron interactions. More recently the implementation of ECD has been successfully extended to other types of mass spectrometers (e.g., a linear ion trap) with a superimposed magnetic field generated by a permanent magnet [68].

ExD in an RF-only trap remained an elusive goal until 2004, when electron transfer dissociation was developed. ETD takes advantage of ion–ion interactions, where electron transfer from an anion radical, rather than the capture of an unbound electron, initiates the bond dissociation. The reagent anions are generated in a negative chemical ionization source (*n*CI), and introduced into the same ion trap where positive peptide ions are stored. Commonly used anion reagents include aromatic compounds with low electron affinities, such as azobenzene and fluoranthene, which also have favorable Franck–Condon factors for transition from the ground vibronic state of the anion to the low–lying vibrational states of the ground electronic state of the neutral molecule. An important competing reaction in ETD of peptides is a proton-transfer reaction (PTR) that involves the movement of a proton from the multiply protonated peptide precursor or fragment ion to the anion radical [69,70]. Although PTR is often an undesired competition, it does have utility in ETD experiments performed in low–resolving-power mass spectrometers, particularly for top–down analysis of large protein ions. For the latter, PTR can reduce the charge state of the highly charged fragment ions, enabling the accurate determination of their charge states, and thus the mass values, which are otherwise difficult to obtain because achieving isotopic resolution at higher charge states is difficult [71].

ETD shares many similarities with ECD, including the preferential and extensive $N{-}C_\alpha$ bond cleavages, preservation of labile modifications, and ability to differentiate certain isomeric amino-acid residues via secondary, radical-induced rearrangements [72–74]. ETD is considered to give even "colder" fragmentation than ECD, capable of retaining even sulfations, the most labile of PTMs. Its success stems from the smaller amount of energy deposited than by ECD. One difference is that some energy is needed to overcome the electron affinity of the anion reagent; another is the collisional cooling afforded by a higher pressure ion trap. Like for ECD, the initially formed c/z ion pair in ETD may still be held together by noncovalent interactions; these need to be ruptured before the products can be detected individually. Overcoming these noncovalent interactions is usually achieved by post-ETD collisional activation, abbreviated as ETcaD [75]. ETD efficiency also increases with the charge state of the precursor ion. ETD efficiency and sequence coverage may be improved by introducing fixed charge tags to the peptides, via, for example, amidation of the carboxylic groups or derivatization of cysteines [76].

To date, there has been only one report of the ETD analog of EDD, where xenon radical cations react with multiply deprotonated peptide anions to generate EDD-like spectra for which a- and x-ions are the major fragments [77]. Current research efforts using radical cations from polycyclic aromatic hydrocarbons (PAHs) as the reverse ETD reagent (electron acceptor) for negatively charged peptide and carbohydrates suggest promise for reverse ETD (rETD) in the structural analysis of biomolecules.

2.2.5 Other Radical-Induced Fragmentation Methods

ExD and UVPD are just two of several classes of fragmentation methods that involve radical-induced reactions. Other methods that generate reactive radical peptide ions include collisional activation and interaction with metastable atoms or other free radicals. In free-radical-initiated peptide sequencing (FRIPS), a free-radical initiator is conjugated to the *N*-terminus of a peptide; the initiator can be cleaved by CAD to leave the radical on the peptide [78]. Subsequent collisional activation induces fragmentation of the peptide, generating abundant a- and z-type ions. Two advantages of FRIPS over ExD are its ability to fragment singly charged peptide ions and the possibility of generating radicals with different reactivities for selective gas-phase fragmentations.

In metastable-atom fragmentation (MAF) [79] or metastable-atom dissociation (MAD) [80], the radical on the peptide is generated by collisions with metastable atoms, usually electronically excited He*, Ne*, Ar*, or Kr*. Metastable atom beams may be produced by electron impact, DC plasma discharge, or RF discharge. The MAF process likely involves Penning ionization of the precursor ion by collision with the metastable atom, generating a radical cation that undergoes ExD-like fragmentations. A typical MAF spectrum (Figure 2.7) of a peptide shows promise as an alternative to ExD for odd-electron ion fragmentation, capable of producing extensive backbone cleavages without the loss of labile PTMs. Further, MAF is not subject to the charge-state limitation of ECD/ETD, and is applicable to singly charged and negatively charged ions. The MAF source can also generate reactive radical species, such as $CH_3^{\bullet}$ or $OH^{\bullet}$, simply by doping the reagent rare gas with methane or water; these radical species may abstract a hydrogen from the peptide precursor and initiate fragmentation.

2.3 MASS ANALYZERS

Tandem MS analysis usually requires selection of a precursor and mass analysis of the products; both steps employ mass analysis. In this section we consider the principles

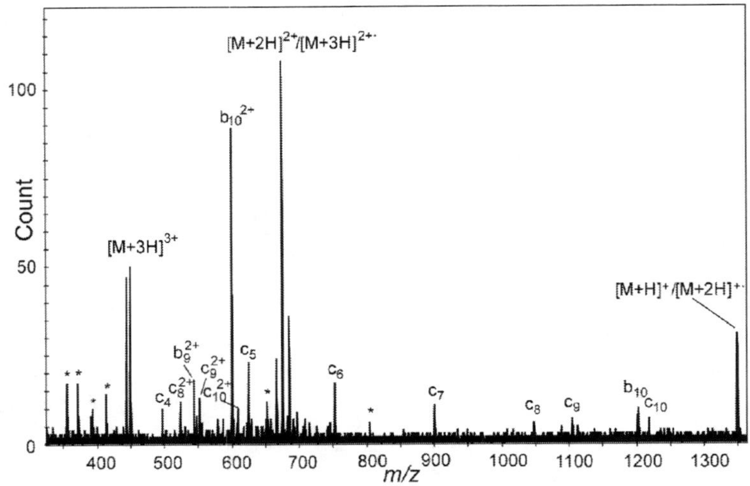

FIGURE 2.7 Fragmentation spectrum of substance P obtained via interaction with a low kinetic energy beam of metastable argon atoms (* denotes contaminant peaks). Adapted with permission from American Chemical Society (Berkout V. D., *Anal. Chem.* 2006, *78*, 3055).

of various mass analyzers that are used in MS/MS experiments in peptide and protein MS. Mass analyzers determine the *m/z* value of a charged particle based on its trajectory in electric or magnetic field or both. Different mass analyzers influence the ion motions in different ways; some through the application of electrostatic or electrodynamic fields, others via the use of a combination of electric and magnetic fields. The fundamental law that governs the ion motion is Newton's second law of mechanics:

$$\mathbf{F} = m\frac{d^2\mathbf{r}}{dt^2}, \tag{2.3}$$

and the force an ion experienced in electromagnetic fields can be expressed as

$$\mathbf{F} = q\mathbf{E} + q\mathbf{v} \otimes \mathbf{B}, \tag{2.4}$$

where the two terms on the right-hand side represent the electric and magnetic components, respectively. Many different types of mass analyzers have been developed and applied to the structural analysis of proteins; the most common are the time of flight, quadrupole, quadrupole ion trap, orbitrap, and Fourier-transform ion cyclotron resonance mass spectrometers.

2.3.1 Time-of-Flight Mass Analyzer

A time-of-flight instrument [81] separates ions of different *m/z* based on their flight times through a field-free drift region. TOF analyzers are well suited to pulsed ion sources such as MALDI because they operate in a pulsed ion-counting mode. A linear

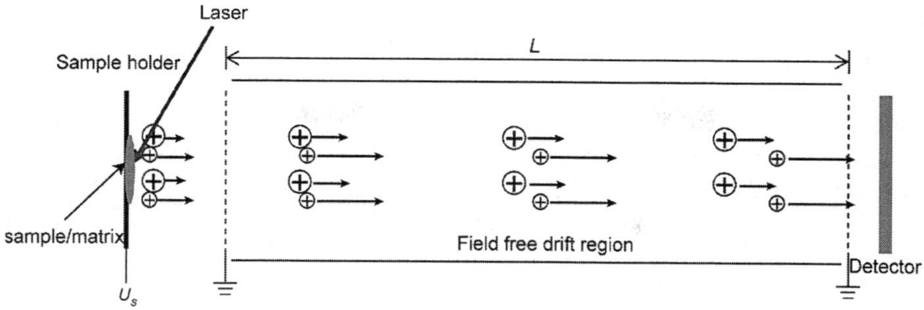

FIGURE 2.8 Principles of a linear time-of-flight mass analyzer.

TOF analyzer is conceptually the simplest type of mass analyzer, whose principles are illustrated in Figure 2.8. All ions with the same number of charges z are accelerated by an electrical potential U_s applied between the sample plate and an extraction electrode to the same kinetic energy ($E_k = zeU_s$), where e is the elementary charge, but with different velocities v, as determined by

$$v = \sqrt{\frac{2zeU_s}{m}}. \qquad (2.5)$$

The flight time of an ion through the drift region is given by

$$t = \frac{L}{\sqrt{2eU_s}}\left(\frac{m}{z}\right)^{1/2}. \qquad (2.6)$$

After exiting the flight tube, the ions strike a detector successively in the order of their m/z values, from low to high. The m/z value of a given ion arriving at time t can be calculated by using the calibration equation

$$\left(\frac{m}{z}\right)^{1/2} = A^*t + B, \qquad (2.7)$$

where A is determined by the flight tube length and the acceleration voltage, and B is a correction term for time zero offset, which may be caused by trigger delay or propagation delay in the detector circuit.

A simple detector for TOF instruments is a secondary emission multiplier (SEM) consisting of a series of dynodes held at decreasing negative potentials. Ions striking the surface of the first dynode cause an emission of secondary electrons, which are then accelerated toward the next dynode held at a less negative potential, generating more secondary electrons upon impact. This process may continue as the electrons travel toward the ground potential, leading to a cascade of electrons. The final electron flow out of the last dynode will be orders of magnitude higher than the initial one emitted from the first dynode; the amplified current can be converted to a voltage that is easily detected by a conventional electronic amplifier.

Alternatively, a microchannel plate (MCP) may be employed as the detector. An MCP is essentially a glass plate with many channels, whose surfaces are coated to achieve a high ion/electron conversion and electron-multiplication yield. The inner surface of each channel resembles a continuous array of dynodes; the potentials vary from high to low negative values from near the front to the back surface of the MCP, as sustained by applying an around 1 kV voltage difference between the two sides of the plate. Ions hitting the front surface of the MCP will induce emission of electrons that cascade down the channel, much like what happens in the SEM. Compared to SEMs, MCPs have the advantage of a faster response time, but they can be easily saturated because their recovery time is long.

The analog electron current signal can be digitized by using either a standard analog-to-digital converter (ADC) or a time-to-digital converter (TDC). An ADC samples the analog detector voltage at discrete intervals and stores the digitized value in a memory from which the signal can be reconstructed or read out by the computer. ADCs for modern TOF instruments can operate at a sampling rate of 1–4 GHz, but only with an 8-bit board, which limits the dynamic range of stored signal amplitudes to a maximum of 256. Moreover, saturation may occur in both the digitizer and the upstream analog current-to-voltage amplifier, leading to flat-topped peaks and erroneous ion-abundance measurements. A TDC is like a 1-bit ADC, which records the arrival time of ions not as an analog signal, but as an array of 1's and 0's. The ion abundance information is recovered by summing over a large number of spectra. A TDC has the advantage of ultra-fast response and data-transfer rate. A TDC, however, suffers from its unit dynamic range; it is particularly undesirable when multiple ions strike the detector at the same time, which results in missing the signal from the slower arriving ions. Further, these detectors suffer from dead-time issues, which occurs when one ion strikes the detector so closely following another that the detector cannot respond to the second ion. Generally, TDCs are used in orthogonal TOF instruments, whereas the ion-extraction optics can operate at a very fast repetition rate (several kHz), and each extracted ion packet contains only a small number of ions.

The mass resolving power of a TOF instrument scales with the length of the flight tube. There is usually a practical limit on how long a flight tube can be, and a longer flight tube is also associated with decreased sensitivity caused by ion loss due to angular dispersion of the ion beam. TOF mass resolving power is also limited by the time width of the ion packet arriving at the detector, which is determined by variations in when and where they are formed, as well as their kinetic-energy spread.

The TOF broadening caused by the kinetic energy spread can be partially corrected by using a delayed extraction (DE) scheme [82–84], as shown in Figure 2.9. With continuous extraction, all ions are extracted and accelerated by the same electric potential U_s, and the ions with a higher initial velocity will arrive at the detector earlier than the ions (with the same *m/z*) with a lower initial velocity (Figure 2.9*A*). With delayed extraction, ion extraction and acceleration are done in two stages (Figure 2.9*B*). The extraction potential U_e is not applied until after a certain delay time following the ion formation, during which time the ion packet will expand in space. Ions with lower initial forward velocity will not move as far down the extraction field as those with higher initial forward velocity, and consequently will experience more acceleration when the extraction voltage is turned on, allowing them to catch up with the faster moving ions. Careful selection of the delay as well as

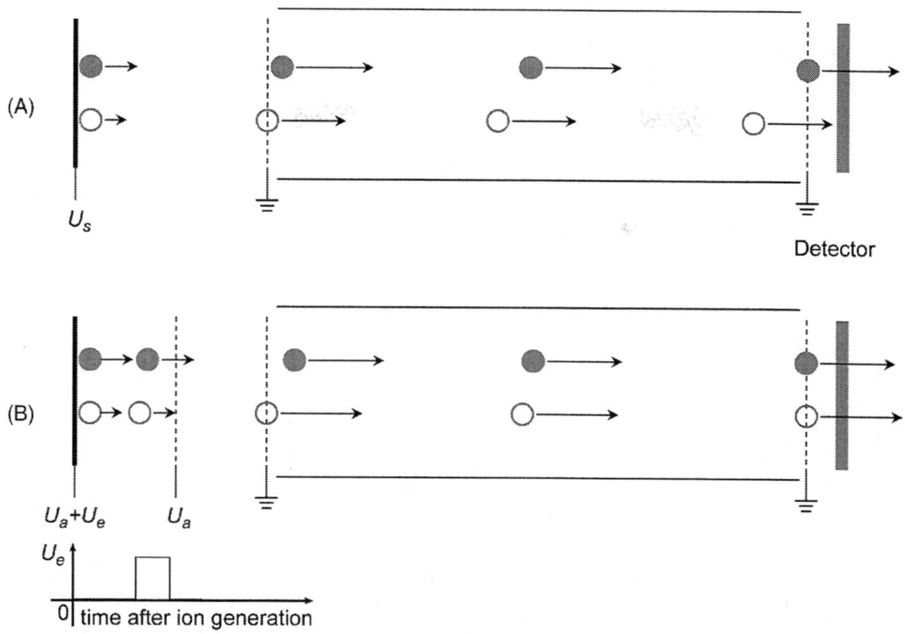

FIGURE 2.9 Schematic of a linear TOF instrument with (*A*) continuous ion extraction and (*B*) delayed pulsed ion extraction showing that TOF broadening caused by the kinetic energy spread of ions can be reduced by employing a delayed ion extraction scheme.

the extraction and acceleration voltages, U_a, permits all ions of a specific *m/z* value to be "time focused" to arrive at the detector at nearly the same time regardless of their initial velocities. It should be noted that the DE focusing condition is mass dependent, and the resulting improvement in mass resolving power drops significantly when applied to a broad *m/z* range.

TOF broadening resulting from the ion kinetic energy spread may also be reduced by employing an electrostatic ion mirror, known as the reflectron [85], as shown in Figure 2.10. The reflectron is usually a stack of ring electrodes, located at the end of the flight tube; when the electrodes are "turned on," they create a constant electric field, usually through a linear voltage gradient, that slows down the ions and turns them around toward the detector located at the other end of the flight tube. For ions of a given *m/z* value, those with higher initial kinetic energy will penetrate more deeply the reflectron field, spending more turn-around time inside the reflectron, thus partially compensating their shorter flight time outside of the reflectron. With proper setting of the reflectron voltage, U_r, ions with both high and low initial kinetic energies can be focused at the detector. The best reflectron-focusing condition is typically achieved when the ion spends an equal amount of time inside and outside of the reflectron. Modern TOF instruments equipped with both DE and a reflectron can routinely achieve mass resolving powers of >20,000 and mass accuracies in the 2 to 5-ppm range.

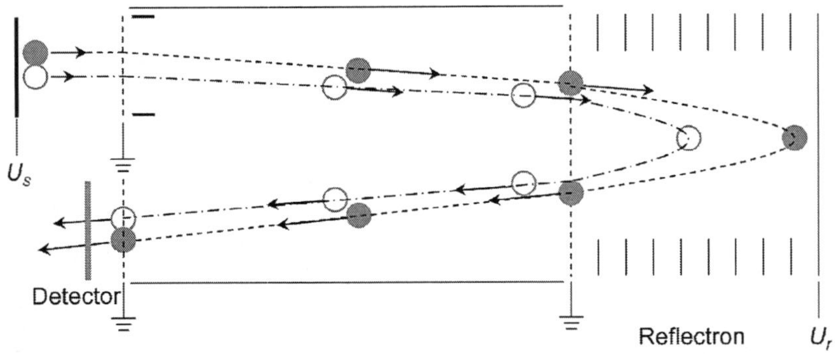

FIGURE 2.10 Schematic of a reflectron TOF instrument illustrating the focusing effect of the reflectron on ions of the same *m/z* but with different initial kinetic energy.

TOF mass resolving power may also be improved even more by adopting an orthogonal TOF (oTOF) arrangement [86,87], a schematic of which is shown in Figure 2.11. In an oTOF instrument, instead of being accelerated along its axis of motion, the ion packet is extracted and accelerated sideways by a pulsed voltage applied to the deflector. Because the resulting TOF axis is perpendicular to the original axis of ion motion, the initial ion kinetic energy spread will not signficantly compromise the achievable mass resolving power. oTOF instruments often include a high-pressure quadrupole for ion cooling and focusing, as well as additional ion optics to squeeze the ion packet both radially and axially for improved mass resolving

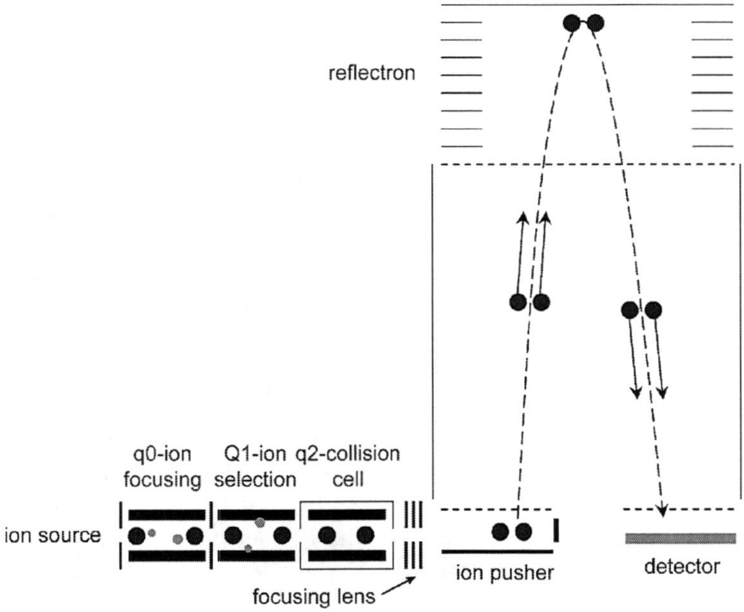

FIGURE 2.11 Schematic of an orthogonal TOF instrument.

power. A distinct advantage of oTOF instruments is that they can work with continuous ion sources, such as ESI because the pulsed TOF analyzer is decoupled from the ion source.

Tandem MS in TOF Instruments Tandem MS experiments cannot be performed on a simple linear TOF mass spectrometer because the ion velocity is already established as the ion enters the field-free flight tube; thus a fragment ion formed in the drift region, via postsource decay (PSD), will have the same velocity as its precursor ion, hence the same TOF. On the other hand, PSD fragments can be mass analyzed on a reflectron TOF instrument because they have kinetic energies that are proportional to their *m/z* [88]. This change in kinetic energy comes about because the kinetic energy of the precursor must be conserved; thus it is partitioned between the fragments as a function of their masses. Despite having the same initial velocity as the precursor ion and the same flight time outside of the reflectron, a PSD fragment with its lower kinetic energy will not penetrate the reflectron as deeply as the precursor ion. Lighter fragments will spend less time inside the reflectron, and arrive at the detector earlier. It is important to note that the focusing condition for each PSD fragment is different, and a complete PSD spectrum generally requires piecing together multiple spectra obtained at several different reflectron voltages, each covering only a fraction of the mass range. Precursor ion selection is usually achieved by placing a pair of electrodes outside of the source region to deflect unwanted ions, although a Bradbury–Nielsen gate, consisting of a set of alternatively biased wires with voltages applied at high frequency, which allows ion passing only in certain voltage phase, is sometimes used. These PSD spectra do not have high mass resolving power and are not extensively used today in peptide sequencing.

Tandem MS experiments can also be performed more effectively on a TOF/TOF instrument, which consists of two TOF analyzers in tandem, with a typical configuration as shown in Figure 2.12 [15]. The first TOF analyzer is usually a short linear drift tube, separating ions according to their *m/z* values. A timed ion selector in front of the collision cell is switched open at a proper delay to select a small *m/z* range including those precursor ions of interest for CAD. The collisional energy can be adjusted by changing the offset potential of the collision cell. The fragment ions formed can be re-accelerated into the second TOF region, typically a high–resolving

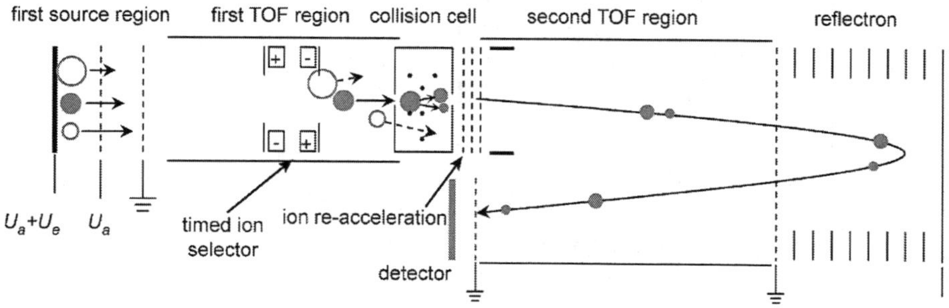

FIGURE 2.12 Schematic of a tandem TOF instrument.

power reflectron for mass analysis. A TOF/TOF mass spectrometer is one of the only two instruments (the other one being tandem sector instruments, which are not widely used for protein analysis) that are used for conducting high-energy CAD experiments (high-energy activation can also be carried out with an FTICR instrument, but the efficiency of product-ion detection is poor). UVPD can also be implemented in TOF/TOF instruments, as the pulsed laser can both time-select and optically excite the precursor ions [26].

2.3.2 Quadrupole Mass Analyzer and Quadrupole Ion Trap

Quadrupole Mass Analyzer A quadrupole mass analyzer, or quadrupole mass filter (QMF), separates ions of different *m/z* based on the stability of their trajectories inside an RF field [89]. An ideal quadrupole contains four highly parallel metal rods, of hyperbolic cross section, arranged in a square configuration. Each pair of the opposing rods is connected electrically. An RF voltage is applied between two pairs of rods, with a DC voltage superimposed on it, creating a quadrupolar field inside the rod arrangement; this field can be expressed as

$$\phi_{x,y} = \frac{(U + V\cos\Omega t)}{r_0^2}(x^2 - y^2) + C, \qquad (2.8)$$

where U is the DC voltage, V is the RF amplitude, Ω is the RF frequency, r_0 is the radius of the circle inscribed in the inner surface of the quadrupole, x and y are the cartesian coordinate positions of the ions, and C is a constant voltage offset. In practice, most quadrupoles use circular rods because they are easier to construct than hyperbolic electrodes. With proper choice of the rod diameter and inter-rod distance, a cylindrical quadrupole can produce an electric field that closely approximates a quadrupolar field (see Figure 2.13 for the geometry of a quadrupole constructed with circular rods).

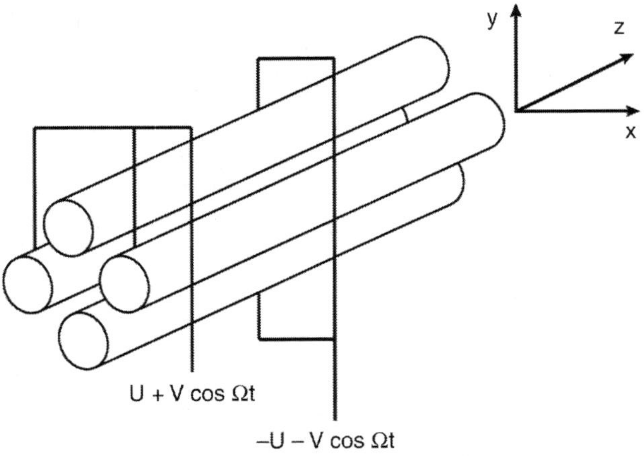

FIGURE 2.13 Schematic of a quadrupole mass analyzer with cylindrical rods.

The ion motion inside a quadrupolar field can be described by the following equations:

$$m\frac{d^2x}{dt^2} = -e\frac{d\phi}{dx} = -2ze\frac{(U + V\cos\Omega t)}{r_0^2}x \qquad (2.9a)$$

and

$$m\frac{d^2y}{dt^2} = -e\frac{d\phi}{dy} = 2ze\frac{(U + V\cos\Omega t)}{r_0^2}y, \qquad (2.9b)$$

where z is the number of charges the ion carries, rather than the cartesian coordinate.

Equations (2.9) can be rewritten in the more familiar form of the Mathieu equation by expressing the instrumental parameters as four dimensionless numbers:

$$a_x = -a_y = \frac{8zeU}{mr_0^2\Omega^2} \qquad (2.10a)$$

and

$$q_x = -q_y = \frac{-4zeV}{mr_0^2\Omega^2}. \qquad (2.10b)$$

The stability of the ion trajectory depends on the ion's a and q values. A stable ion trajectory is one where the ion motion is bound in both x- and y-dimensions (i.e., $|x| < r_0$ and $|y| < r_0$ at all times). Although many stability regions exist in the a/q space, nearly all commercial quadrupoles operate in the first stability region, as outlined in Figure 2.14. It is evident from equations (2.10) that the stability diagram in the a/q space can be directly translated into the U/V space with a scaling factor that is proportional to the ion's mass-to-charge ratio. Figure 2.15 depicts the stability diagrams of several ions of different m/z values. Normally, when the quadrupole is used as a mass analyzer, the DC potential and the RF amplitude are ramped together while the ratio of U/V is kept constant, along the operating line as shown in Figure 2.15 (dashed line). The operating line scans across the tips of the stability region of each m/z, allowing the sequential passage of ions of different m/z values, from low to high. Near the vertex of the stability diagram of a given m/z ion, only ions within a small window of that m/z can have stable trajectories and be transmitted. Given that it takes a finite time for an ion to "fly" through the quadrupole, the U and V values must be kept within its stability region for the whole length of the ion residence time inside the quadrupole, limiting the scan speed. The scan speed can be increased by decreasing the slope of the operating line, thus increasing the m/z window of transmission at any given instant. A broader transmission window, however, also means a reduced mass resolving power. Modern quadrupole mass analyzers can achieve a mass resolving power of up to nearly 10,000 in a high–resolving power mode, and a scan speed well over 10,000 u/sec at unit mass resolving power (i.e., sufficient to separate adjoining m/z ions). The upper mass limit of commercial quadrupoles is typically between m/z 4000 and 6000, which is limited by the maximum amplitude the RF power supply can provide, and by its frequency, which is normally a constant and cannot be readily varied by the operator.

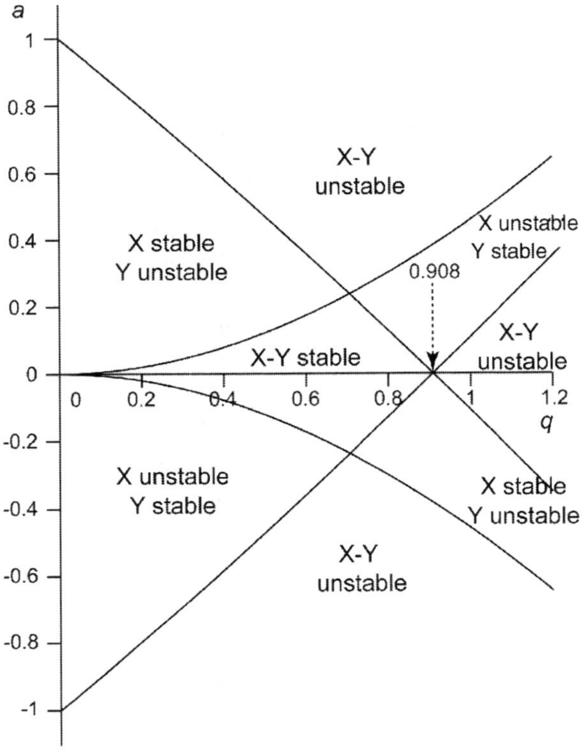

FIGURE 2.14 Stability diagram of a quadrupole in *a/q* space.

As suggested by the stability diagram, quadrupoles can be used as ion guides when operated in the RF-only mode. With no DC potential applied ($a = 0$), all ions above a certain *m/z* values can be transmitted. This low mass cutoff may be decreased by applying lower RF amplitude, *V*, allowing, in principle, ion transmission across a

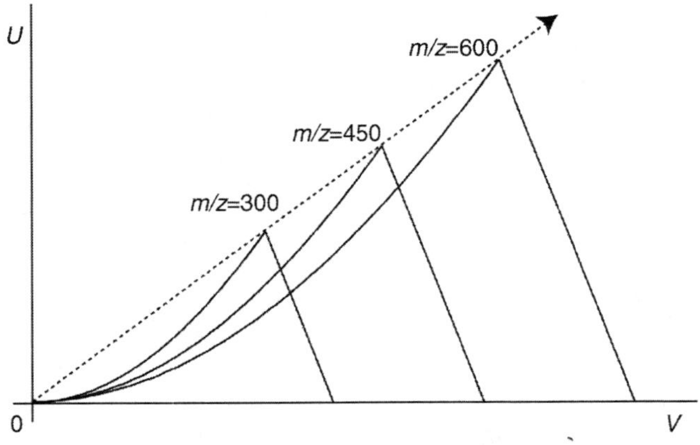

FIGURE 2.15 Principles of a quadrupole mass analyzer.

wider mass range. In practice, however, the ion transmission efficiency is also limited by the focusing ability of the quadrupole. Ion dispersion in an angular way toward the quadrupole rods may be caused by a number of reasons, including the initial ion velocity, which often contains a radial component, space-charge effects, which originate from the mutual repulsion of ions of like charge within the ion packet, and scattering collisions with residual or background gas molecules. The focusing ability of the quadrupole is characterized by the depth of the effective radial trapping potential well, which is proportional to V^2. Ions with a higher *m/z* value are less affected by the electric field, requiring a deeper trapping well to be efficiently focused and transferred. When transferring ions across a broad mass range, there is often a compromise in choosing the RF amplitude to avoid the loss of low-mass ions, owing to their unstable trajectories, and to reduce the loss of high mass ions, owing to inefficient focusing.

Tandem MS in Triple-Quadrupole Mass Spectrometers Tandem MS experiments cannot be performed on a stand-alone quadrupole mass analyzer because it lacks ion-trapping capability. Spectrometers employing multiple quadrupoles in series, on the other hand, are widely used for tandem MS analysis [90–92]. The most common type is the triple quadrupole instrument, as shown in Figure 2.16. The first and third quadrupoles (Q1 and Q3) are used as mass filters, and the second quadrupole (q2) is operated in the RF-only mode and used as a collision cell.

Given that Q1 and Q3 can be either tuned to a fixed DC/RF value for mass selection, or scanned to perform mass analysis, a triple quadrupole mass spectrometer can be operated under four different modes. The most commonly used mode is the product-ion scan, where Q1 is tuned to select precursor ions of a specific *m/z*, which are further subjected to CAD in q2. Q3 is scanned to measure the fragment ion masses. The product-ion scan is frequently used in liquid chromatography LC-MS/MS, and is very useful for deducing the structure of precursor ions. Alternatively, Q1 can be scanned, while Q3 is held at a constant DC/RF value to allow detection of a particular fragment ion of interest. This precursor-ion scan mode is particularly useful for identifying all precursor ions that produce a common product ion of interest. The third scanning mode is the neutral loss scan, where Q3 is scanned at the same rate as Q1, but with an offset, Δm. The value of Δm is usually negative, and the resulting mass spectrum displays all precursor ions that undergo the same neutral loss upon collisional

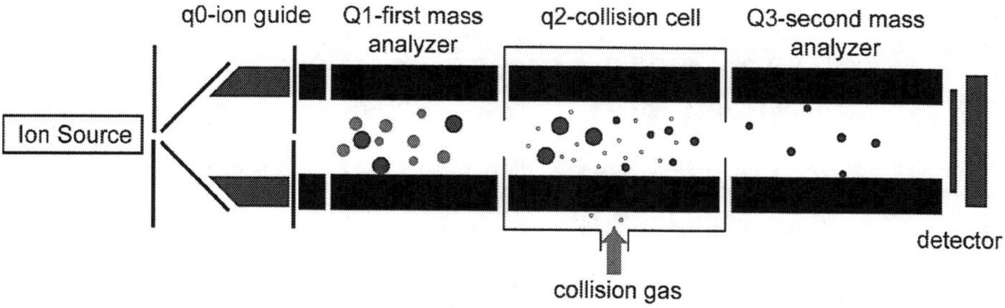

FIGURE 2.16 Schematic of a triple quadrupole mass spectrometer.

activation in q2. The neutral loss scan is often used to screen for a class of compounds that contain a similar labile group (e.g., a phosphate group that is eliminated as neutral phosphoric acid from phosphorylated peptides). The final mode is the selected reaction monitoring (SRM), sometimes also referred to as the multiple reaction monitoring (MRM) mode, where Q1 and Q3 are both operated at fixed DC/RF values, with Q1 allowing passage of a particular precursor ion and Q3 allowing transmission of an expected fragment ion from the selected precursor [93]. SRM is usually performed when the instrument is coupled with LC, producing a chromatogram of a specific precursor ion corresponding to a specific analyte of interest. Because both mass analyzers are operated at fixed *m/z* values, SRM is a highly specific and sensitive method for identifying compounds, provided that their fragmentation behaviors are known.

Linear Quadrupole Ion Trap A quadrupole can be turned into an ion-storage device by adding end trapping electrodes or by segmenting the quadrupole (see Figure 2.17 for a linear quadrupole ion trap (LIT) with sectioned rods) [94,95]. The segmented quadrupole has the advantage of minimizing potential ion losses caused by the fringe field near the end of the center quadrupole when the main RF voltage used to confine the radial ion motion is also applied to the end quadrupole segments. A DC potential is applied to the end segments to confine the axial ion motion, and collision gas is also used to facilitate the ion trapping. Two of the four center quadrupole rods (the X-pair in the diagram) have open slits in the middle for radial ion ejection during mass analysis. A conversion dynode with an electron multiplier is placed on each side of the ion trap to allow detection of most ejected ions. Compared with the 3D quadrupole ion trap (discussed below), a linear ion trap has the advantage of a higher space-charge limit as it allows ion clouds to expand axially over a larger volume.

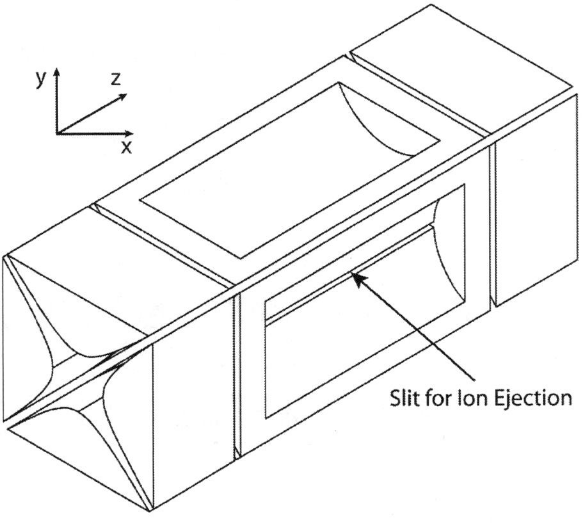

FIGURE 2.17 Schematic of a linear quadrupole ion trap.

Unlike a quadrupole mass analyzer which uses a combination of RF and DC voltages to achieve mass selection, an LIT performs mass analysis by scanning the RF amplitude alone, without any DC potential applied between the two sets of quadrupole rods. As evident from the stability diagram of a quadrupole trap, Figure 2.14, only ions with $q < 0.908$ have stable trajectories along the q-axis (where $a = 0$). Recall that q is proportional to $V/(m/z)$; as the RF amplitude is ramped up, low m/z ions will reach the edge of the stability region first, and they are ejected from the trap. Although conceptually simple and easy to implement, such mass-selective instability scan has several drawbacks, including slow ion ejection near the edge of the stability region and low detection efficiency because only a portion of the ions actually exit from the slits.

An improved way to perform mass analysis using an LIT is through resonance ejection. Each ion has its unique oscillation frequency inside a quadrupolar field, known as the secular frequency f_s, which differs from the fundamental RF frequency. Because heavier ions have slower response to the change of the electric field, they oscillate at lower frequencies than those of lighter ions. The secular frequency is related to the ion q value (defined in equation 2.10), and f_s increases as q increases. During the mass analysis, a small auxiliary AC voltage at a fixed frequency is applied between the two X-rods only. As the main RF amplitude is ramped up, all ions experience an increase in their q values and consequently their secular frequencies. When the secular frequency of a certain ion reaches the frequency of the auxiliary AC, it will be resonantly ejected from the trap and detected. The AC frequency chosen allows ions to reach resonance at a q value slightly below 0.908, which permits faster ejection than allowing the ions reach the edge of the stability region; the outcome is improved mass resolving power. Furthermore, the sensitivity is also improved via resonance ejection because ions exit the trap almost exclusively along the x-axis, with minimal scattering loss.

Given that an LIT is an ion-trapping device, tandem MS analysis is possible using a single LIT. Precursor ion selection is achieved by applying a tailored RF waveform to the X-rods, which contains RF power at the secular frequencies of all ions except for the ion of interest. Selected ions are then activated by applying a resonant dipolar excitation waveform along the X-rods, with the amplitude kept low to avoid ion ejection. Collision energy can be increased by performing resonance excitation at a higher q. A higher q, however, means a smaller m/z range for detectable fragment ions. For example, for a precursor ion of around m/z 908 excited at around q 0.30, all fragment ions with m/z less than 300 will have a q value above 0.908, and thus be unstable and undetectable. A pulsed Q dissociation (PQD) scheme can be implemented to alleviate this problem. In PQD, precursor ions are first resonantly excited at a higher q value and held there for a short period of time for collisional excitation, but not long enough for significant dissociation to occur. After this short excitation period, the main RF amplitude is dropped to bring the q values down before or as the ions dissociate, and the fragment ions are trapped at low q values.

Alternatively, photodissociation, either UVPD or IRMPD, or ETD can be used to circumvent the compromise between ion activation and fragment-ion trapping. The geometry of an LIT allows easy axial introduction of the laser beam, and the extended ion-storage time permits extensive ion–photon interactions for efficient precursor-ion excitation [24]. The elevated pressure in an LIT is sometimes undesirable for IRMPD as the rapid collisional cooling can compete with ion activation. For ETD, simultaneous axial trapping of analyte cations and ETD reagent anions is achieved by applying an RF trapping voltage to the end lenses of the linear trap.

3D Quadrupole Ion Trap A 3D quadrupole ion trap (QIT), or Paul trap, works under the same principle that governs the operation of an LIT [96–98]. A QIT consists of a cylindrical ring electrode and two end caps. Conceptually a QIT can be considered as a quadrupole that bends around itself to form a closed loop. As the inner radius of the resulting "donut" shrinks, the inner rod is reduced to a point, the outer rod becomes the ring electrode, and the top and bottom rods become the end caps (see Figure 2.18 for a schematic). Externally generated ions enter the trap through an opening in one end cap, and are trapped by a combination of the alternating electric field and collisions with buffer gas that refocus the ions to the center of the trap. Ions are ejected through an opening on the other end cap during mass analysis and ion detection.

Ion motion inside the QIT is also described by the Mathieu equation, although the expressions of a and q are slightly different from those for the linear quadrupole ion trap because they have different geometries:

$$a_z = -2a_r = \frac{-16zeU}{m(r_0^2 + 2z_0^2)\Omega^2} \tag{2.11a}$$

and

$$q_z = -2q_r = \frac{8zeV}{m(r_0^2 + 2z_0^2)\Omega^2}. \tag{2.11b}$$

Like its 2D cousin, a 3D QIT works along the q-axis, with no DC potential difference applied between the ring electrode and end caps. Most of the time the main RF is only applied to the ring electrode, while the end caps are primarily used

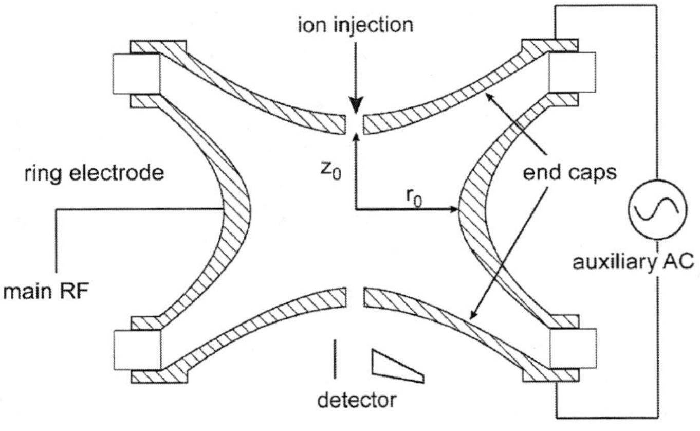

FIGURE 2.18 Schematic cross-sectional view of a 3D quadrupole ion trap.

for resonant-ion ejection and excitation. Mass analysis can be performed either by mass-selective instability scan or by resonance ejection. The upper mass limit is once again limited by the ability of the RF power supply to provide a sufficiently high voltage to drive the q_z over 0.908. Although a reduction in trap dimension can lead to an increase in its upper mass limit, it is also associated with reduced ion-storage capacity due to coulomb repulsion between like charges. Alternatively, the upper mass limit can be increased by performing resonance ejection at a frequency corresponding to a lower q.

CAD experiments can be similarly carried out on a 3D QIT, but loss of low mass fragment ions can occur as the V is increased for precursor-ion activation. Alternative fragmentation methods employing ion/ion interactions may benefit because a 3D QIT is a charge-sign-independent trapping device, which makes it particularly suitable for ETD tandem MS analysis. Consecutive tandem MS analysis (MS^n) can be easily performed on a QIT, by alternating the ion-selection and ion-fragmentation steps. The low cost, compact size, rapid analysis time, and MS^n capability of a QIT make it one of the most common instruments for LC/MS-MS analysis.

2.3.3 Fourier-Transform Ion Cyclotron Resonance Mass Spectrometer

The Fourier-transform ion cyclotron resonance (FTICR) MS was developed in the 1970s by Comisarow and Marshall [99,100]. An FTICR mass spectrometer determines the *m/z* value of ions based on their cyclotron frequencies in a homogeneous magnetic field. The Lorentz force an ion experienced in a magnetic field of strength B is normal to its velocity v and the magnetic field lines, causing the ion to undergo cyclotron motion, with the Lorentz force balancing the centrifugal force:

$$F = zevB = \frac{mv^2}{r}. \tag{2.12}$$

Thus the unperturbed angular cyclotron frequency ω_c of an ion of a given *m/z* value in a fixed magnetic field is given by

$$\omega_c = \frac{v}{r} = \frac{Be}{m/z} \tag{2.13}$$

where e is the elemental charge, and z is the charge in integers.

In practice, because a homogeneous magnetic field can only confine the ion motion in the radial direction (i.e., the direction perpendicular to the magnetic field line), an inhomogeneous electrostatic field is also applied to trap ions in the axial direction. The axial electrical trapping results in an axial oscillation with the frequency ω_z, given by

$$\omega_z = \sqrt{\frac{2zV_{trap}\alpha}{ma^2}}, \tag{2.14}$$

where V_{trap} is the trapping potential applied, a is the dimension of the ICR trap, and α is a constant that depends on the geometry of the trap. The electric field and the

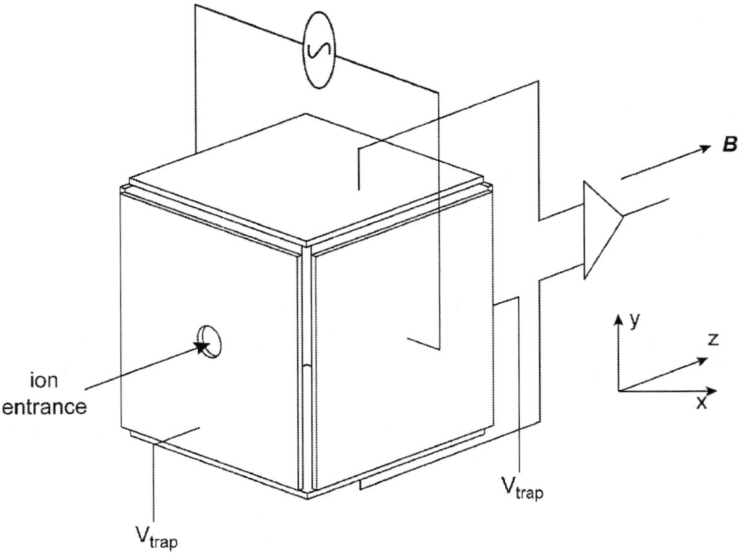

FIGURE 2.19 Schematic of a cubic FTICR trap.

resulting axial harmonic motion reduce the cyclotron frequency and introduce a second radial motion called the magnetron motion. The natural angular frequencies of the ion motions are now

$$\omega_{\pm} = \frac{\omega_c}{2} \pm \sqrt{\left(\frac{\omega_c}{2}\right)^2 - \left(\frac{\omega_z}{2}\right)^2}, \tag{2.15}$$

where ω_+ is the reduced cyclotron frequency and ω_- is the magnetron frequency. It is this reduced cyclotron frequency ω_+ that is measured in FTICR. Thus a proper calibration equation is needed to correct for the trapping term when calculating the m/z value of an ion based on its measured reduced cyclotron frequency [101].

The simplest ICR trap is of cubic shape, which consists of three pairs of metal plates orthogonal to each other (Figure 2.19). The pair of plates that is perpendicular to the magnetic field is used as trapping plates, to which a small DC voltage is applied to confine the ion motions along the z-axis. The other two pairs of plates are used as excitation and detection plates, respectively. Although this is an informative trap for explaining the principles, most modern traps are cylindrical in design.

Figure 2.20 illustrates the principle of operation for an FTICR mass spectrometer. The initially trapped ions are confined radially to very small cyclotron radii owing to their thermal velocities. For example, at room temperature a singly charged ion of $m = 100$ Da in a magnetic field of 12 T has a thermal ICR orbital radius of around 0.02 mm. This small-amplitude thermal cyclotron motion is not useful for ion detection because it is neither coherent nor can it induce significant image currents on the detection plates. The ion packet may be excited to a larger orbit by applying an azimuthal (i.e., perpendicular to the magnetic field) spatially uniform field that is

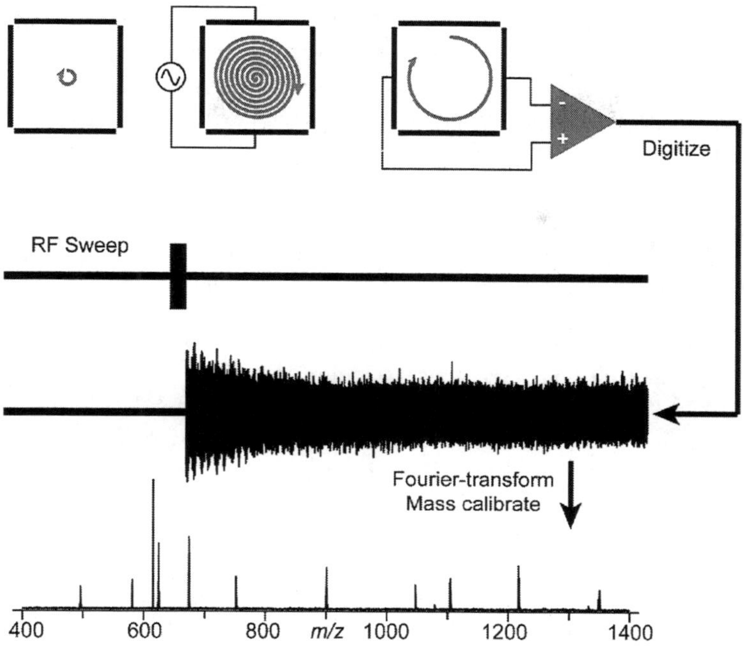

FIGURE 2.20 Principles of FTICR mass spectrometry.

oscillating sinusoidally with the same angular frequency, ω_c, as the ion's characteristic angular cyclotron frequency. After the excitation all ions of the same *m/z* value will move coherently as a tight packet in a larger cyclotron orbit, inducing alternating image charges on the opposing detection plates. The induced alternating image current has the same frequency as the ion cyclotron frequency. Multiple ion packets of different *m/z* values can be excited to the same cyclotron radius, albeit with different frequencies, by applying a swept *rf* excitation waveform with equal magnitude for all frequencies (also known as the "chirp" excitation). Image current induced by all ion packets can be detected simultaneously as a superposition of many sine waves, which are amplified, digitized, and stored as a time-domain transient. This transient is Fourier-transformed to give a frequency-domain spectrum, and finally mass-calibrated to produce the mass spectrum.

The Fourier-transform limited mass resolving power of an FTICR mass analyzer is roughly equal to $f^*t/2$, where f is the cyclotron frequency and t is the transient length. Thus, it is very important to maintain an ultra-high vacuum (typically $\sim 10^{-10}$ torr) in the ICR region; otherwise, collisions of ions with background gas may lead to rapid transient decay and poor mass resolving power. One may calculate that an FTICR mass spectrometer with a 7 T magnet can provide around 100,000 resolving power at *m/z* 500 with a one-second transient. The high resolving power may also be appreciated by considering the distance an excited ion traverses in its orbital motion in a short time. In the example above, if the ion were excited to an orbit of 5 cm in radius, it would travel a distance of around 63 km during a one-second observation time. Given that the frequencies can be measured with high accuracy, their corresponding *m/z* can also be calculated with high accuracy. With a well-constructed ICR trap and careful control of experimental conditions, a modern FTICR mass analyzer can routinely achieve mass accuracy of <2 ppm with external mass calibration, and into the ppb region with internal calibration.

Tandem MS in FTICR Mass Spectrometers Because an FTICR trap can store ions, tandem MS analysis can be easily performed with an FTICR mass spectrometer. A precursor ion may be isolated by applying tailored excitation waveforms, such as the SWIFT (stored waveform inverse Fourier transform) [102]. Precise ion isolation down to a 0.1 Da *m/z* window as well as multiple precursor-ion selection can be achieved with SWIFT. The ICR trap is particularly well suited for performing several tandem MS experiments, including IRMPD and ECD. IRMPD in an ICR trap benefits from its long ion-storage time, which allows extensive ion–photon interaction, and its ultra-high vacuum, which minimizes collisional cooling. Until very recently an FTICR instrument is the only type of mass spectrometer that is capable of performing ECD analysis, primarily because of its ability to guide and trap electrons with magnetic field.

CAD can also be used to fragment ions in an FTICR mass spectrometer, but generally not with resonant excitation because ions can be lost owing to the high-energy collisions, and collisional damping is required if one to bring the product ions back to the center of the trap for re-excitation for mass analysis. With resonant excitation, fragment ions tend to be formed off-axis and, given large magnetron motion amplitudes, produce poor spectra and extensive fragment-ion losses. Instead, selected ions are usually excited by a slightly off-resonance waveform that periodically excites and de-excites the ions, ensuring ample collisions while keeping the ions relatively close to the center of the ICR trap [9]. Although such sustained off-resonance irradiation (SORI) can produce product-ion spectra similar to those obtained by other low energy CAD methods, direct introduction of the collision gas into the ICR trap is undesirable, as a long pump-down delay is usually needed after the fragmentation event to achieve the low pressure that is suitable for high-resolving-power mass analysis. The extra delay leads to low duty cycles, and thus SORI-CAD is not suitable for high-throughput tandem MS analysis, particularly when the mass spectrometer is coupled with LC.

Nearly all modern commercial FTICR instruments are hybrid instruments that employ either a linear ion trap or a QMF-collision cell setup as the front end. Given that CAD performed in the front end does not compromise the vacuum in the ICR trap, these hybrid instruments are ideal for performing high-throughput, high-mass-accuracy LC-MS/MS analyses. In addition, the front end can also be used to isolate ions, without inducing significant magnetron motions. The LIT offers the possibility of automatic gain control (AGC) for maintaining constant ion populations in the ICR trap, which is crucial for achieving high mass accuracies. The QMF-collision cell setup, on the other hand, allows selected ion accumulation, which is beneficial for analysis of low-abundance ions, leading to dramatically increased dynamic range.

The excellent mass resolving power and accuracy achievable on an FTICR and its versatile tandem MS analysis capability make FTICR optimal for many applications (e.g., top-down proteomics [103] where whole intact protein ions are fragmented in the gas phase for identification and characterization, and complex mixture analysis in petroleomics research [104]). The analysis time on FTICR mass spectrometers, however, is long, relative to times of chromatographic separation, with a typical acquisition time around one second to achieve a reasonably high mass resolving power. In addition, FTICR instruments are usually expensive, owing to the cost for the magnet, further limiting their application in routine sample analysis.

2.3.4 Orbitrap

Ion trapping by a pure electrostatic field is also possible. The first electrostatic ion trap was developed by Kingdon in the early 1920s; it consists of a central wire and an outer cylindrical electrode to produce a radial electrostatic trapping field. An ion circles around the central wire (orbital motion) in a Kingdon trap, with the centrifugal force balanced by an attractive coloumbic force. The outer electrode was later modified to include an axial quadrupolar term for axial ion-motion confinement. Neither configuration was reported to produce mass spectra. The breakthrough came in the late 1990s with the development of the orbitrap by Makarov. Aside from redesigning the electrodes to generate a quadro-logarithmic electric field, Makarov also devised a way to introduce externally generated ions into the orbitrap, and more important, with the same initial phase with regard to their axial motion [105–107].

The schematic cross section view of an orbitrap (Figure 2.21) shows a trap consisting of an inner spindle-shaped electrode and an outer barrel-shaped electrode,

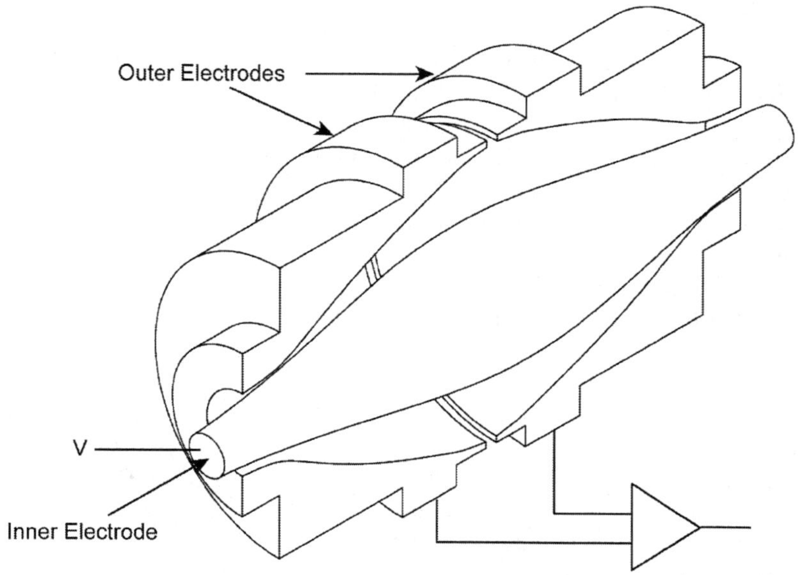

FIGURE 2.21 Schematic cross-section view of an orbitrap mass spectrometer.

which is sectioned in the middle. The electrostatic field inside the orbitrap can be described by a quadro-logarithmic distribution:

$$U(r,z) = \frac{k}{2}\left(z^2 - \frac{r^2}{2}\right) + \frac{k}{2}R_m^2 \ln\left(\frac{r}{R_m}\right) + C, \tag{2.16}$$

where z and r are cylindrical coordinates, with $z = 0$ being the plane that bisects the outer electrode, k is the field curvature, R_m is the characteristic radius of the trap, and C is a constant voltage offset. Stable ion trajectories inside the orbitrap combine orbital rotations around the inner electrode and harmonic oscillations along it. Several *m/z*-dependent characteristic frequencies exist, including the frequency of rotation ω_ϕ, the frequency of radial oscillation ω_r, and the frequency of axial oscillation ω_z. Recall that the restoring force along the z-axis can be calculated as

$$F_z = -q\frac{\partial U}{\partial z} = -qkz = m\frac{d^2z}{dt^2}, \tag{2.17}$$

which describes a simple harmonic oscillator, with the frequency of oscillation being

$$\omega_z = \sqrt{\frac{k}{m/q}} = \sqrt{\frac{ke}{m/z}}, \tag{2.18}$$

where e is the elemental charge.

Ion oscillation along the z-axis induces an image current between the two parts of the sectioned outer electrode. As for FTICR MS, this image current can be Fourier-transformed to generate the frequency domain spectrum, and further mass-calibrated to produce the mass spectrum.

Successful generation of mass spectra by using an orbitrap hinges on the ability to introduce the externally generated ions as a tight packet into the orbitrap, so that the axial motion of ions of the same *m/z* is coherent. The original orbitrap design adopted a high voltage pulsed ion deflector to achieve this purpose, although only a small fraction of the ions made their way into the orbitrap. The latest commercial orbitrap, the LTQ-Orbitrap, employs a C-trap, which is a curved quadrupole that can be pulsed to push all trapped ions into the orbitrap, all with the same initial phase axial motion (Figure 2.22) [108,109].

As for all Fourier-transform mass analyzers, the Fourier-transform limited mass resolving power of an orbitrap is approximately equal to the product of the measured frequency (in this case the frequency of ion oscillation along the z-axis) and the transient length. The commercial orbitrap can routinely achieve a mass resolving power of around 100,000 at *m/z* 400 for a 1.25 s transient. This mass resolving power is similar to that obtainable on an FTICR mass analyzer with a 4.3 T magnet at the same *m/z* and with the same transient length. Because the z-axis oscillation frequency of an ion inside an orbitrap has a weaker dependence on *m/z* (equation 2.18) than the cyclotron frequency of an ion inside an ICR trap (equation 2.13), the mass resolving power of an orbitrap does not decrease as fast as that of an FTICR as the *m/z* increases, and can exceed that of a commercial FTICR with a higher field (7 or 9.4 T) magnet for

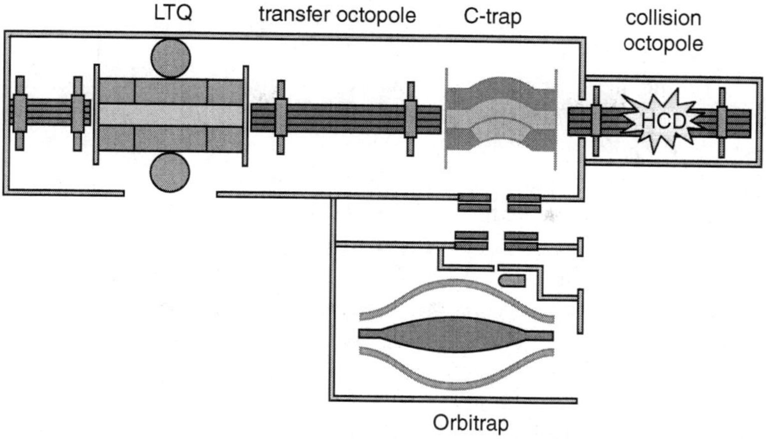

FIGURE 2.22 Schematic of the LTQ-Orbitrap XL instrument. Adapted with permission from Macmillan Publishers Ltd: [Nature Methods] (reference [108]), copyright (2007).

ions of higher *m/z* for the same acquisition time. The longest attainable transient produced by an orbitrap (currently at approximately 2 s), however, is significantly shorter than that by an FTICR (> 100 s); thus the ultimate mass resolving power achievable on an orbitrap is also significantly lower. Transient decay in an orbitrap results from ion loss and dephasing, which are caused by several factors including ion collisions with background gas, field imperfections, instability of power supplies, and space-charge effects.

The LTQ-Orbitrap can achieve a mass accuracy in the low ppm range with external calibration. A "lock-mass" standard, such as polycyclodimethylsiloxane (PCM-6) ions (*m/z* 445.1200) generated as the background ions during the ESI process from a siloxane contaminant in the atmosphere, can be added to the analyte ion packet via sequential filling of the C-trap. The presence of the lock-mass ion in the same ion population as the analyte of interest provides an internal standard for mass calibration, allowing mass measurement with better than 1-ppm mass accuracy. Scan speeds of up to 5 scans/s are possible, albeit with reduced mass resolving power and mass measurement accuracy.

Tandem MS in LTQ-Orbitrap An ion packet of a specific *m/z* in the orbitrap can be selectively excited or de-excited by applying a resonant dipolar AC signal to each half of the outer electrodes. Although such ability to manipulate confined ion populations allows precursor-ion selection and excitation, tandem MS analysis inside the orbitrap has not been demonstrated to date. In a commercial orbitrap, ion fragmentation is typically done in the front end linear ion trap (called an LTQ in the commercial instrument). The LTQ can be used to perform low-energy CAD and ETD analyses, as well as a full range of MS^n experiments. Fragment ions can be mass analyzed in the LTQ for high-throughput analysis, or in the orbitrap if high mass resolving power and accuracy are desired. For ETD experiments, reagent anions can be introduced into the LTQ either from the front or from the back. The original front-loading design employs a dual, pulsed nanoESI (nESI) source, with the second nESI emitter operating in the negative-ion mode to generate deprotonated molecules (anions) [110]. These even-electron anions can only react with analyte cations by proton transfer. Thus it is necessary to trap these anions in a separate region in the LTQ, where they first undergo a charge-selective CAD process to generate ETD-inducing anion radicals, before they are allowed to react with analyte ions. This front-end ETD approach has a major drawback in that a long acquisition time is required to accommodate the additional CAD step and the switching delay between two pulsed nESI sources. The newer commercial design employs an nCI source mounted on the back of the orbitrap to generate radical reagent anions that can be brought into the LTQ from its rear entrance. High-throughput ETD analyses on the chromatographic time scale are readily achieved with this rear-end ETD design [111].

Ion fragmentation can also be achieved outside of the LTQ. For example, CAD can be performed in the C-trap, by accelerating the isolated precursor ions from the LTQ through the transfer octopole toward the C-trap [108]. The higher energy C-trap dissociation can generate additional fragment ions that are unobtainable in the low-energy CAD performed in the LTQ. During the C-trap CAD, however, a higher RF amplitude is needed to trap efficiently the high-mass incoming precursor ions, leading to an increase of the low-mass cutoff of the fragment ions. To overcome this difficulty, the newer LTQ-Orbitrap XL instrument employs a dedicated collisional octopole attached to the rear end of the C-trap for higher energy, collision-induced dissociation (HCD). After dissociation in the collisional octopole, fragment ions are sent back to the C-trap, and injected into the orbitrap for mass analysis. The C-trap CAD and HCD tandem mass spectra often contain more structural information than low-energy CAD spectra obtained in the LTQ, particularly in the low *m/z* region. These low-*m/z* ions are useful in a number of applications. For example, the presence of phosphotyrosine immonium ions in higher energy CAD spectra provides information on PTMs [108]. As another example, quantitative proteomics studies employing iTRAQ labels cannot be performed on ion trap instruments because the reporter ions are usually of too low an *m/z* to be trapped efficiently, but iTRAQ can be performed on an LTQ-Orbitrap with HCD capability [112].

The availability of these tandem MS tools, along with orbitrap's superior mass analysis performance and its compatibility to LC-MS/MS analysis, make the LTQ-Orbitrap an extremely versatile and powerful mass spectrometer for a wide variety of applications.

2.3.5 Ion-Mobility Instruments

To conclude this discussion on mass analysis and MS/MS in various instrument configurations, we bring to the readers' attention a new capability for MS. Ion mobility has been known for some time, and when an ion-mobility device is combined with a mass spectrometer, it gives a new dimension to peptide and protein analysis. While it is still too early to determine the ion-mobility device's impact, we know that it offers the capability to do fast separations prior to mass analysis and to investigate ion conformations, which are of key importance in the biophysics of peptides and proteins.

Most mass spectrometers operate at 10^{-6} to 10^{-9} mbar background gas pressure so that the perturbations on ion motion caused by collisions with neutral gas molecules do not greatly affect their trajectories. Ion-mobility devices operate on a different principle, at around 1 mbar. Ions in an ion-mobility device are subjected to a constant electric field at this high pressure so that they accelerate and quickly achieve a terminal velocity, relying on the "drag" force from background gas collisions to separate ions based on the balance between their acceleration (dependent on their mass/charge ratio) and the drag (dependent on their cross-sectional area and the mass of the background gas). Thus ion-mobility devices are not strictly mass analyzers, the separation of the masses is also dependent on average cross-sectional area [113,114]. Ion-mobility devices, however, have become quite useful as quick gas-phase separation tools when combined with traditional mass analyzers such as time-of-flight instruments [114–116].

For example, recently a mass-spectrometer manufacturer, Waters, released a new instrument, called the Synapt, that employs an ion-mobility separator prior to a quadrupole/time-of-flight instrument. In this case the ion-mobility separator uses a stacked-ring geometry ion guide to keep the beam radially confined, but is still clearly capable of separating ions by their rotationally averaged cross-sectional area. When combined with a high-speed quadrupole/time-of-flight instrument, there are substantial improvements in peak capacity, baseline chemical noise levels, and the ability to perform conformation-dependent MS and tandem MS experiments, thus providing interesting new capabilities. Other instrument manufacturers are also currently working to develop similar, competing instruments.

REFERENCES

1. Ashline, D. J., Lapadula, A. J., Liu, Y.-H., Lin, M., Grace, M., Pramanik, B., Reinhold, V. N. (2007). Carbohydrate structural isomers analyzed by sequential mass spectrometry. *Anal Chem 79*, 3830–3842.
2. Sleno, L., Volmer, D. A. (2004). Ion activation methods for tandem mass spectrometry. *J Mass Spectrom 39*, 1091–1112.
3. Roepstorff, P., Fohlman, J. (1984). Proposal for a common nomenclature for sequence ions in mass-spectra of peptides. *Biomed Mass Spectrom 11*, 601–601.
4. Shukla, A. K., Futrell, J. H. (2000). Tandem mass spectrometry: Dissociation of ions by collisional activation. *J Mass Spectrom 35*, 1069–1090.
5. Papayannopoulos, I. A. (1995). The interpretation of collision-induced dissociation tandem mass spectra of peptides. *Mass Spectrom Rev 14*, 49–73.
6. Hayes, R. N., Gross, M. L. (1990). Collision-induced dissociation. *Meth Enzymol, 193*, 237–263.
7. Yost, R. A., Enke, C. G., McGilvery, D. C., Smith, D., Morrison, J. D. (1979). High efficiency collision-induced dissociation in an RF-only quadrupole. *Int J Mass Spectrom Ion Phys 30*, 127–136.

8. Louris, J. N., Brodbelt-Lustig, J. S., Graham Cooks, R., Glish, G. L., van Berkel, G. J., McLuckey, S. A. (1990). Ion isolation and sequential stages of mass spectrometry in a quadrupole ion trap mass spectrometer. *Int J Mass Spectrom Ion Processes 96*, 117–137.
9. Gauthier, J. W., Trautman, T. R., Jacobson, D. B. (1991). Sustained off-resonance irradiation for collision-activated dissociation involving Fourier-transform mass-spectrometry: Collision-activated dissociation technique that emulates infrared multi-photon dissociation. *Anal Chim Acta 246*, 211–225.
10. Laskin, J., Futrell, J. H. (2003). Collisional activation of peptide ions in Ft-Icr mass spectrometry. *Mass Spectrom Rev 22*, 158–181.
11. Wysocki, V. H., Tsaprailis, G., Smith, L. L., Breci, L. A. (2000). Special feature: Commentary—Mobile and localized protons: A framework for understanding peptide dissociation. *J Mass Spectrom 35*, 1399–1406.
12. Paizs, B., Suhai, S. (2005). Fragmentation pathways of protonated peptides. *Mass Spectrom Rev 24*, 508–548.
13. Fabris, D., Michele, K., Constance, M., Zhuchun, W., Catherine, F. (1993). High-energy collision-induced dissociation of multiply charged polypeptides produced by electro-spray. *J Am Soc Mass Spectrom 4*, 652–661.
14. Medzihradszky, K. F., Burlingame, A. L. (1994). The advantages and versatility of a high-energy collision-induced dissociation-based strategy for the sequence and structural determination of proteins. *Methods 6*, 284–303.
15. Medzihradszky, K. F., Campbell, J. M., Baldwin, M. A., Falick, A. M., Juhasz, P., Vestal, M. L., Burlingame, A. L. (2000). The characteristics of peptide collision-induced dissociation using a high-performance MALDI-TOF/TOF tandem mass spectrometer. *Anal Chem 72*, 552–558.
16. Falick, A. M., Hines, W. M., Medzihradszky, K. F., Baldwin, M. A., Gibson, B. W. (1993). Low-mass ions produced from peptides by high-energy collision-induced dissociation in tandem mass spectrometry. *J Am Soc Mass Spectrom 4*, 882–893.
17. Harrison, A. G., Young, A. B., Bleiholder, C., Suhai, S., Paizs, B. (2006). Scrambling of sequence information in collision-induced dissociation of peptides. *J Am Chem Soc 128*, 10364–10365.
18. Bleiholder, C., Osburn, S., Williams, T. D., Suhai, S., Van Stipdonk, M., Harrison, A. G., Paizs, B. (2008). Sequence-scrambling fragmentation pathways of protonated peptides. *J Am Chem Soc 130*, 17774–17789.
19. Dongré, A. R., Somogyi, Á., Wysocki, V. H. (1996). Surface-induced dissociation: An effective tool to probe structure, energetics and fragmentation mechanisms of protonated peptides. *J Mass Spectrom 31*, 339–350.
20. Wysocki, V. H., Joyce, K. E., Jones, C. M., Beardsley, R. L. (2008). Surface-induced dissociation of small molecules, peptides, and non-covalent protein complexes. *J Am Soc Mass Spectrom 19*, 190–208.
21. Wysocki, V. H., Ding, J.-M., Jones, J. L., Callahan, J. H., King, F. L. (1992). Surface-induced dissociation in tandem quadrupole mass spectrometers: A comparison of three designs. *J Am Soc Mass Spectrom 3*, 27–32.
22. Williams, E. R., Fang, L., Zare, R. N. (1993). Surface-induced dissociation for tandem time-of-flight mass spectrometry. *Int J Mass Spectrom Ion Processes 123*, 233–241.

23. Laskin, J., Denisov, E. V., Shukla, A. K., Barlow, S. E., Futrell, J. H. (2002). Surface-induced dissociation in a Fourier transform ion cyclotron resonance mass spectrometer: Instrument design and evaluation. *Anal Chem 74*, 3255–3261.

24. Brodbelt, J. S., Wilson, J. J. (2009). Infrared multiphoton dissociation in quadrupole ion traps. *Mass Spectrom Rev 28*, 390–424.

25. Little, D. P., Speir, J. P., Senko, M. W., O'Connor, P. B., McLafferty, F. W. (1994). Infrared multiphoton dissociation of large multiply charged ions for biomolecule sequencing. *Anal Chem 66*, 2809–2815.

26. Zhang, L., Reilly, J. P. (2009). Peptide photodissociation with 157 Nm light in a commercial tandem time-of-flight mass spectrometer. *Anal Chem 81*, 7829–7838.

27. Dunbar, R. C. (2004). Bird (blackbody infrared radiative dissociation): Evolution, principles, and applications. *Mass Spectrom Rev 23*, 127–158.

28. Wilson, J. J., Brodbelt, J. S. (2006). Infrared multiphoton dissociation for enhanced de novo sequence interpretation of *N*-terminal sulfonated peptides in a quadrupole ion trap. *Anal Chem 78*, 6855–6862.

29. Kim, T.-Y., Thompson, M. S., Reilly, J. P. (2005). Peptide photodissociation at 157 Nm in a linear ion trap mass spectrometer. *Rapid Commun Mass Spectrom 19*, 1657–1665.

30. Bowers, W. D., Delbert, S. S., Hunter, R. L., McIver, R. T. (1984). Fragmentation of oligopeptide ions using ultraviolet laser radiation and Fourier transform mass spectrometry. *J Am Chem Soc 106*, 7288–7289.

31. Martin, S. A., Hill, J. A., Kittrell, C., Biemann, K. Photon-induced dissociation with a four-sector tandem mass spectrometer. *J Am Soc Mass Spectrom 1*, 107–109.

32. Oh, J. Y., Moon, J. H., Kim, M. S. (2004). Sequence- and site-specific photodissociation at 266 Nm of protonated synthetic polypeptides containing a tryptophanyl residue. *Rapid Commun Mass Spectrom 18*, 2706–2712.

33. Wilson, J. J., Brodbelt, J. S. (2007). Ms/Ms simplification by 355 Nm ultraviolet photodissociation of chromophore-derivatized peptides in a quadrupole ion trap. *Anal Chem 79*, 7883–7892.

34. Reilly, J. P. (2009). Ultraviolet photofragmentation of biomolecular ions. *Mass Spectrom Rev 28*, 425–447.

35. Madsen, J. A., Boutz, D. R., Brodbelt, J. S. Ultrafast ultraviolet photodissociation at 193 Nm and its applicability to proteomic workflows. *J Proteome Res 9*, 4205–4214.

36. Kalcic, C. L., Gunaratne, T. C., Jones, A. D., Dantus, M., Reid, G. E. (2009). Femtosecond laser-induced ionization/dissociation of protonated peptides. *J Am Chem Soc 131*, 940–942.

37. Smith, S. A., Kalcic, C. L., Safran, K. A., Stemmer, P. M., Dantus, M., Reid, G. E. (2010). Enhanced characterization of singly protonated phosphopeptide ions by femtosecond laser-induced ionization/dissociation tandem mass spectrometry (fs-LID-MS/MS). *J Am Soc Mass Spectrom 21*, 2031–2040.

38. Zubarev, R. A., Kelleher, N. L., McLafferty, F. W. (1998). Electron capture dissociation of multiply charged protein cations: A nonergodic process. *J Am Chem Soc 120*, 3265–3266.

39. Cooper, H. J., Håkansson, K., Marshall, A. G. (2005). The role of electron capture dissociation in biomolecular analysis. *Mass Spectrom Rev 24*, 201–222.

40. Zubarev, R. A. (2004). Electron-capture dissociation tandem mass spectrometry. *Curr Opin Biotechnol 15*, 12–16.

41. Meng, F., Forbes, A. J., Miller, L. M., Kelleher, N. L. (2005). Detection and localization of protein modifications by high resolution tandem mass spectrometry. *Mass Spectrom Rev 24*, 126–134.

42. Zubarev, R. A., Kruger, N. A., Fridriksson, E. K., Lewis, M. A., Horn, D. M., Carpenter, B. K., McLafferty, F. W. (1999). Electron capture dissociation of gaseous multiply-charged proteins is favored at disulfide bonds and other sites of high hydrogen atom affinity. *J Am Chem Soc 121*, 2857–2862.

43. Kelleher, N. L., Zubarev, R. A., Bush, K., Furie, B., Furie, B. C., McLafferty, F. W., Walsh, C. T. (1999). Localization of labile posttranslational modifications by electron capture dissociation: The case of γ-carboxyglutamic acid. *Anal Chem 71*, 4250–4253.

44. Shi, S. D. H., Hemling, M. E., Carr, S. A., Horn, D. M., Lindh, I., McLafferty, F. W. (2000). Phosphopeptide/phosphoprotein mapping by electron capture dissociation mass spectrometry. *Anal Chem 73*, 19–22.

45. Mirgorodskaya, E., Roepstorff, P., Zubarev, R. A. (1999). Localization of *O*-glycosylation sites in peptides by electron capture dissociation in a Fourier transform mass spectrometer. *Anal Chem 71*, 4431–4436.

46. Syrstad, E. A., Turecek, F. (2005). Toward a general mechanism of electron capture dissociation. *J Am Soc Mass Spectrom 16*, 208–224.

47. Zubarev, R. A., Zubarev, A. R., Savitski, M. M. (2008). Electron capture/transfer versus collisionally activated/induced dissociations: Solo or duet? *J Am Soc Mass Spectrom 19*, 753–761.

48. Savitski, M. M., Nielsen, M. L., Kjeldsen, F., Zubarev, R. A. (2005). Proteomics-grade de novo sequencing approach. *J Proteome Res 4*, 2348–2354.

49. Cournoyer, J. J., Pittman, J. L., Ivleva, V. B., Fallows, E., Waskell, L., Costello, C. E., O'Connor, P. B. (2005). Deamidation: Differentiation of aspartyl from isoaspartyl products in peptides by electron capture dissociation. *Protein Sci 14*, 452–463.

50. Cournoyer, J. J., Lin, C., Bowman, M. J., O'Connor, P. B. (2007). Quantitating the relative abundance of isoaspartyl residues in deamidated proteins by electron capture dissociation. *J Am Soc Mass Spectrom 18*, 48–56.

51. O'Connor, P. B., Lin, C., Cournoyer, J. J., Pittman, J. L., Belyayev, M., Budnik, B. A. (2006). Long-lived electron capture dissociation product ions experience radical migration via hydrogen abstraction. *J Am Soc Mass Spectrom 17*, 576–585.

52. Savitski, M. M., Kjeldsen, F., Nielsen, M. L., Zubarev, R. A. (2007). Hydrogen rearrangement to and from radical Z fragments in electron capture dissociation of peptides. *J Am Soc Mass Spectrom 18*, 113–120.

53. Tsybin, Y. O., He, H., Emmett, M. R., Hendrickson, C. L., Marshall, A. G. (2007). Ion activation in electron capture dissociation to distinguish between *N*-terminal and *C*-terminal product ions. *Anal Chem 79*, 7596–7602.

54. Cooper, H. J., Hudgins, R. R., Hakansson, K., Marshall, A. G. (2003). Secondary fragmentation of linear peptides in electron capture dissociation. *Int J Mass Spectrom 228*, 723–728.

55. Savitski, M. M., Nielsen, M. L., Zubarev, R. A. (2007). Side-chain losses in electron capture dissociation to improve peptide identification. *Anal Chem 79*, 2296–2302.

56. Li, X., Lin, C., Han, L., Costello, C. E., O'Connor, P. B. (2010). Charge remote fragmentation in electron capture and electron transfer dissociation. *J Am Soc Mass Spectrom 21*, 646–656.

57. Leymarie, N., Costello, C. E., O'Connor, P. B. (2003). Electron capture dissociation initiates a free radical reaction cascade. *J Am Chem Soc 125*, 8949–8958.

58. Horn, D. M., Ge, Y., McLafferty, F. W. (2000). Activated ion electron capture dissociation for mass spectral sequencing of larger (42 kDa) proteins. *Anal Chem 72*, 4778–4784.

59. Creese, A. J., Cooper, H. J. (2008). The effect of phosphorylation on the electron capture dissociation of peptide ions. *J Am Soc Mass Spectrom 19*, 1263–1274.

60. Sze, S. K., Ge, Y., Oh, McLafferty, F. W. (2003). Plasma electron capture dissociation for the characterization of large proteins by top down mass spectrometry. *Anal Chem 75*, 1599–1603.

61. Kjeldsen, F., Haselmann, K. F., Budnik, B. A., Jensen, F., Zubarev, R. A. (2002). Dissociative capture of hot (3–13 Ev) Electrons by polypeptide polycations: An efficient process accompanied by secondary fragmentation. *Chem Phys Lett 356*, 201–206.

62. Tsybin, Y. O., Witt, M., Baykut, G., Håkansson, P. (2004). Electron capture dissociation Fourier transform ion cyclotron resonance mass spectrometry in the electron energy range 0–50 Ev. *Rapid Commun Mass Spectrom 18*, 1607–1613.

63. Zubarev, R. A. (2003). Reactions of polypeptide ions with electrons in the gas phase. *Mass Spectrom Rev 22*, 57–77.

64. Budnik, B. A., Haselmann, K. F., Zubarev, R. A. (2001). Electron detachment dissociation of peptide di-anions: An electron-hole recombination phenomenon. *Chem Phys Lett 342*, 299–302.

65. Wolff, J. J., Amster, I. J., Chi, L., Linhardt, R. J. (2007). Electron detachment dissociation of glycosaminoglycan tetrasaccharides. *J Am Soc Mass Spectrom 18*, 234–244.

66. Fung, Y. M. E., Adams, C. M., Zubarev, R. A. (2009). Electron ionization dissociation of singly and multiply charged peptides. *J Am Chem Soc 131*, 9977–9985.

67. Syka, J. E. P., Coon, J. J., Schroeder, M. J., Shabanowitz, J., Hunt, D. F. (2004). Peptide and protein sequence analysis by electron transfer dissociation mass spectrometry. *Proc Nat Acad Sci USA 101*, 9528–9533.

68. Satake, H., Hasegawa, H., Hirabayashi, A., Hashimoto, Y., Baba, T., Masuda, K. (2007). Fast multiple electron capture dissociation in a linear radio frequency quadrupole ion trap. *Anal Chem 79*, 8755–8761.

69. Gunawardena, H. P., He, M., Chrisman, P. A., Pitteri, S. J., Hogan, J. M., Hodges, B. D. M., McLuckey, S. A. (2005). Electron transfer versus proton transfer in gas-phase ion/ion reactions of polyprotonated peptides. *J Am Chem Soc 127*, 12627–12639.

70. Xia, Y., Gunawardena, H. P., Erickson, D. E., McLuckey, S. A. (2007). Effects of cation charge-site identity and position on electron-transfer dissociation of polypeptide cations. *J Am Chem Soc 129*, 12232–12243.

71. Coon, J. J., Ueberheide, B., Syka, J. E. P., Dryhurst, D. D., Ausio, J., Shabanowitz, J., Hunt, D. F. (2005). Protein identification using sequential ion/ion reactions and tandem mass spectrometry. *Proc Nat Acad Sci USA 102*, 9463–9468.

72. Mikesh, L. M., Ueberheide, B., Chi, A., Coon, J. J., Syka, J. E. P., Shabanowitz, J., Hunt, D. F. (2006). The utility of ETD mass spectrometry in proteomic analysis. *Biochim Biophys Acta (BBA)–Proteins Proteomics 1764*, 1811–1822.

73. O'Connor, P. B., Cournoyer, J. J., Pitteri, S. J., Chrisman, P. A., McLuckey, S. A. (2006). Differentiation of aspartic and isoaspartic acids using electron transfer dissociation. *J Am Soc Mass Spectrom 17*, 15–19.

74. Wiesner, J., Premsler, T., Sickmann, A. (2008). Application of electron transfer dissociation (ETD) for the analysis of posttranslational modifications. *Proteomics 8*, 4466–4483.
75. Swaney, D. L., McAlister, G. C., Wirtala, M., Schwartz, J. C., Syka, J. E. P., Coon, J. J. (2006). Supplemental activation method for high-efficiency electron-transfer dissociation of doubly protonated peptide precursors. *Anal Chem 79*, 477–485.
76. Vasicek, L., Brodbelt, J. S. (2009). Enhanced electron transfer dissociation through fixed charge derivatization of cysteines. *Anal Chem 81*, 7876–7884.
77. Coon, J. J., Shabanowitz, J., Hunt, D. F., Syka, J. E. P. (2005). Electron transfer dissociation of peptide anions. *J Am Soc Mass Spectrom 16*, 880–882.
78. Hodyss, R., Cox, H. A., Beauchamp, J. L. (2005). Bioconjugates for tunable peptide fragmentation: Free radical initiated peptide sequencing (FRIPS). *J Am Chem Soc 127*, 12436–12437.
79. Berkout, V. D. (2006). Fragmentation of protonated peptide ions via interaction with metastable atoms. *Anal Chem 78*, 3055–3061.
80. Cook, S. L., Collin, O. L., Jackson, G. P. (2009). Metastable atom-activated dissociation mass spectrometry: Leucine/isoleucine differentiation and ring cleavage of proline residues. *J Mass Spectrom 44*, 1211–1223.
81. Weickhardt, C., Moritz, F., Grotemeyer, J. (1996). Time-of-flight mass spectrometry: State-of-the-art in chemical analysis and molecular science. *Mass Spectrom Rev 15*, 139–162.
82. Wiley, W. C., McLaren, I. H. (1955). Time-of-flight mass spectrometer with improved resolution. *Rev Sci Instr 26*, 1150–1157.
83. Brown, R. S., Lennon, J. J. (1995). Mass resolution improvement by incorporation of pulsed ion extraction in a matrix-assisted laser desorption/ionization linear time-of-flight mass spectrometer. *Anal Chem 67*, 1998–2003.
84. Vestal, M. L., Juhasz, P., Martin, S. A. (1995). Delayed extraction matrix-assisted laser desorption time-of-flight mass spectrometry. *Rapid Commun Mass Spectrom 9*, 1044–1050.
85. Mamyrin, B. A., Karataev, V. I., Shmikk, D. V., Zagulin, V. A. (1973). The mass-reflectron, a new nonmagnetic time-of-flight mass spectrometer with high resolution. *Soviet Phys JETP 37*, 45–48.
86. Morris, H. R., Paxton, T., Dell, A., Langhorne, J., Berg, M., Bordoli, R. S., Hoyes, J., Bateman, R. H. (1996). High sensitivity collisionally-activated decomposition tandem mass spectrometry on a novel quadrupole/orthogonal-acceleration time-of-flight mass spectrometer. *Rapid Commun Mass Spectrom 10*, 889–896.
87. Chernushevich, I. V., Loboda, A. V., Thomson, B. A. (2001). An introduction to quadrupole–time-of-flight mass spectrometry. *J Mass Spectrom 36*, 849–865.
88. Spengler, B. (1997). Post-source decay analysis in matrix-assisted laser desorption/ionization mass spectrometry of biomolecules. *J Mass Spectrom 32*, 1019–1036.
89. Dawson, P. H. (1995) *Quadrupole Mass Spectrometry and Its Applications.* Springer, New York.
90. Hunt, D. F., Shabanowitz, J., Giordani, A. B. (1980). Collision activated decompositions in mixture analysis with a triple quadrupole mass spectrometer. *Anal Chem 52*, 386–390.
91. Perchalski, R. J., Yost, R. A., Wilder, B. J. (1982). Structural elucidation of drug metabolites by triple-quadrupole mass spectrometry. *Anal Chem 54*, 1466–1471.

92. Hunt, D. F., Yates, J. R., Shabanowitz, J., Winston, S., Hauer, C. R. (1986). Protein sequencing by tandem mass-spectrometry. *Proc Nat Acad Sci USA 83*, 6233–6237.

93. Lange, V., Picotti, P., Domon, B., Aebersold, R. (2008). Selected reaction monitoring for quantitative proteomics: A tutorial. *Mol Syst Biol 4*, 222.

94. Schwartz, J. C., Senko, M. W., Syka, J. E. P. (2002). A two-dimensional quadrupole ion trap mass spectrometer. *J Am Soc Mass Spectrom 13*, 659–669.

95. Douglas, D. J., Frank, A. J., Mao, D. (2005). Linear ion traps in mass spectrometry. *Mass Spectrom Rev 24*, 1–29.

96. March, R. E. (1992). Ion trap mass spectrometry. *Int J Mass Spectrom Ion Processes 118–119*, 71–135.

97. March, R. E. (1997). An introduction to quadrupole ion trap mass spectrometry. *J Mass Spectrom 32*, 351–369.

98. March, R. E. (2000). Quadrupole ion trap mass spectrometry: A view at the turn of the century. *Int J Mass Spectrom 200*, 285–312.

99. Amster, I. J. (1996). Fourier transform mass spectrometry. *J Mass Spectrom 31*, 1325–1337.

100. Marshall, A. G., Hendrickson, C. L., Jackson, G. S. (1998). Fourier transform ion cyclotron resonance mass spectrometry: A primer. *Mass Spectrom Rev 17*, 1–35.

101. Zhang, L.-K., Rempel, D., Pramanik, B. N., Gross, M. L. (2005). Accurate mass measurements by fourier transform mass spectrometry. *Mass Spectrom Rev 24*, 286–309.

102. Guan, S., Marshall, A. G. (1996). Stored waveform inverse fourier transform (SWIFT) ion excitation in trapped-ion mass spectometry: Theory and applications. *Int J Mass Spectrom Ion Processes 157–158*, 5–37.

103. Bogdanov, B., Smith, R. D. (2005). Proteomics by FTICR mass spectrometry: Top down and bottom up. *Mass Spectrom Rev 24*, 168–200.

104. Marshall, A. G., Rodgers, R. P. (2003). Petroleomics: The next grand challenge for chemical analysis. *Accounts Chem Res 37*, 53–59.

105. Hardman, M., Makarov, A. A. (2003). Interfacing the orbitrap mass analyzer to an electrospray ion source. *Anal Chem 75*, 1699–1705.

106. Hu, Q., Noll, R. J., Li, H., Makarov, A., Hardman, M., Graham Cooks, R. (2005). The orbitrap: A new mass spectrometer. *J Mass Spectrom 40*, 430–443.

107. Perry, R. H., Cooks, R. G., Noll, R. J. (2008). Orbitrap mass spectrometry: Instrumentation, ion motion and applications. *Mass Spectrom Rev 27*, 661–699.

108. Olsen, J. V., Macek, B., Lange, O., Makarov, A., Horning, S., Mann, M. (2007). Higher-energy C-trap dissociation for peptide modification analysis. *Nat Meth 4*, 709–712.

109. Olsen, J. V., Schwartz, J. C., Griep-Raming, J., Nielsen, M. L., Damoc, E., Denisov, E., Lange, O., Remes, P., Taylor, D., Splendore, M., Wouters, E. R., Senko, M., Makarov, A., Mann, M., Horning, S. (2009). A dual pressure linear ion trap orbitrap instrument with very high sequencing speed. *Mol Cell Proteomics 8*, 2759–2769.

110. McAlister, G. C., Phanstiel, D., Good, D. M., Berggren, W. T., Coon, J. J. (2007). Implementation of electron-transfer dissociation on a hybrid linear ion trap-orbitrap mass spectrometer. *Anal Chem 79*, 3525–3534.

111. McAlister, G. C., Berggren, W. T., Griep-Raming, J., Horning, S., Makarov, A., Phanstiel, D., Stafford, G., Swaney, D. L., Syka, J. E. P., Zabrouskov, V., Coon, J. J. (2008). A

proteomics grade electron transfer dissociation-enabled hybrid linear ion trap-orbitrap mass spectrometer. *J Proteome Res 7*, 3127–3136.

112. Bantscheff, M., Boesche, M., Eberhard, D., Matthieson, T., Sweetman, G., Kuster, B. (2008). Robust and sensitive iTRAQ quantification on an LTQ-orbitrap mass spectrometer. *Mol Cell Proteomics 7*, 1702–1713.
113. Clemmer, D. E., Jarrold, M. F. (1997). Ion mobility measurements and their applications to clusters and biomolecules. *J Mass Spectrom 32*, 577–592.
114. Kanu, A. B., Dwivedi, P., Tam, M., Matz, L., Hill, H. H. (2008). Ion mobility–mass spectrometry. *J Mass Spectrom 43*, 1–22.
115. Tang, K., Shvartsburg, A. A., Lee, H.-N., Prior, D. C., Buschbach, M. A., Li, F., Tolmachev, A. V., Anderson, G. A., Smith, R. D. (2005). High-sensitivity ion mobility spectrometry/mass spectrom using electrodynamic ion funnel interfaces. *Anal Chem 77*, 3330–3339.
116. Pringle, S. D., Giles, K., Wildgoose, J. L., Williams, J. P., Slade, S. E., Thalassinos, K., Bateman, R. H., Bowers, M. T., Scrivens, J. H. (2007). An investigation of the mobility separation of some peptide and protein ions using a new hybrid quadrupole/travelling wave IMS/oa-TOF instrument. *Int J Mass Spectrom 261*, 1–12.

CHAPTER 3

Target Proteins: Bottom-up and Top-down Proteomics

MICHAEL BOYNE and RON BOSE

3.1 MASS SPECTRAL APPROACHES TO TARGETED PROTEIN IDENTIFICATION

The ability to use mass as a feature to identify proteins and peptides has undergone a revolution over the past 20 years. The development of electrospray ionization (ESI) [1], matrix-assisted laser desorption/ionization (MALDI) [2,3], and related methods (see Chapter 1 by Coffee-Rodriguez, Zhang, Miao, and Chen in this volume) in the late 1980s made direct measurement of the mass of proteins and peptides routinely possible. The development of mass spectrometers with increasing mass accuracy, higher sensitivity, and faster duty cycles combined with the coupling of these instruments to protein and peptide separation techniques has produced a number of highly sophisticated approaches for the identification and characterization of proteins. Mass spectrometry (MS) has become the dominant analytical tool for identifying proteins, whether in their purified form or within a complex mixture, pushing aside Edman sequencing owing to the increased sensitivity (<femtomoles of material) and throughput (100–1000 proteins/day) of the MS techniques.

The two main mass spectral approaches to proteomics and targeted protein identification are termed "bottom-up" and "top-down" [4,5]. Bottom-up commonly refers to exhaustive digestion of proteins either pre-fractionated (commonly by one- or two-dimensional polyacrylamide gel electrophorisis [6,7]) or proteolyzed in a whole-cell extract *en masse* (shotgun digestion) [6,8] followed by mass measurement of the peptides in the resulting mixture (Figure 3.1). Top-down proteomics removes the proteolytic digestion step described above, focusing instead on sequence coverage (routinely 100%) and complete protein characterization to tease out more biologically relevant information by integrating all the protein complexity into an array of forms

Protein and Peptide Mass Spectrometry in Drug Discovery, Edited by Michael L. Gross, Guodong Chen, and Birendra N. Pramanik.

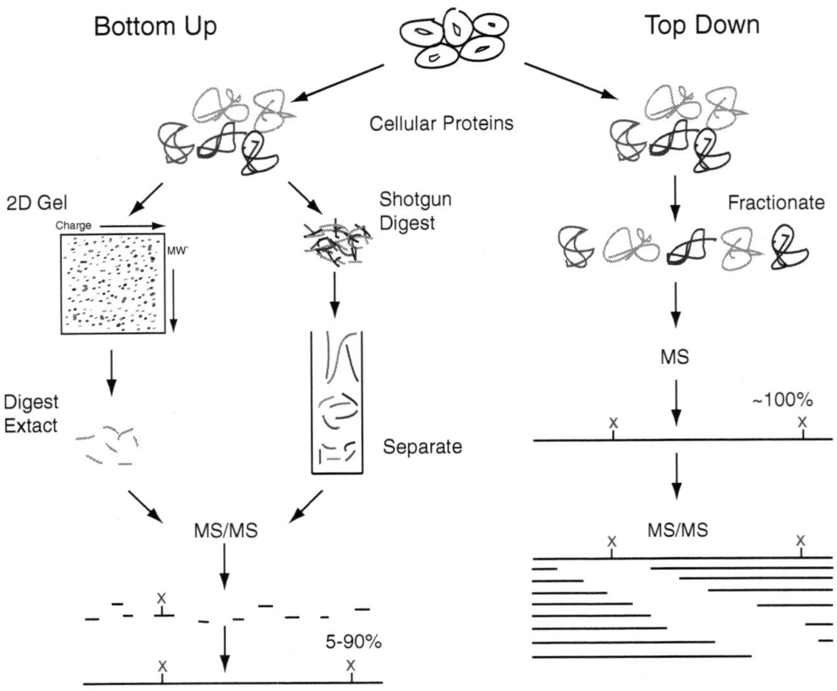

FIGURE 3.1 Comparison of bottom-up and top-down approaches to protein identification. Bottom-up approaches use proteolytic digestions to create a mixture of peptides, which are then introduced to the mass spectrometer. Protein identification is inferred from two or more confident peptide identifications. Top-down approaches differ in that proteins are fractionated and then introduced into the MS instrument while still intact. Subsequent fragmentation produces a series of fragment masses, which are then used in combination with the intact mass to identify the protein.

from a single gene (Figure 3.1) [9]. The appropriate technique to use depends on the capabilities of the mass spectrometer available, the limitations of each approach, and ultimately the biological question to be answered.

3.2 BOTTOM-UP PROTEOMICS

Bottom-up proteomics is predicated on the generation of 1- to 3 kDa peptides from the protein in question. These peptides can easily be measured in many types of mass spectrometers including ion traps, triple quadrupoles, time of flight (TOF), and Fourier-transform instruments. Most often the regioselective endopeptidase trypsin, which cleaves the *C*-terminal to lysine and arginine residues, is employed to generate peptides of the desired size. These peptides are then identified, and these identifications are used to infer the protein precursors. Peptide mass fingerprinting, GeLC-MS, and shotgun digestion are the three most widely used experimental approaches in bottom-up protein identification.

3.2.1 Peptide Mass Fingerprinting

Peptide mass fingerprinting (PMF) is a protein identification process where by a single protein or simple mixture of proteins is digested into peptides whose absolute masses are subsequently measured in a mass spectrometer, typically an ESI-TOF, ESI-trap, ESI-FTMS, or MALDI-TOF. Developed in the early 1990s, this method's analytical identification power comes from the generated list of peptide masses. PMF is based on the idea that each protein in an organism will generate a unique set of peptides, upon proteolytic digestion, whose masses will provide a molecular signature (i.e. a fingerprint) identifying that particularly protein. In practice, these peptide masses are searched against a protein or translated sequence database that has been digested in silico (in a computer), knowing the specificity of the digesting enzyme. Proteins are identified based on the number of peptide masses matching to the theoretical generated list within a mass tolerance.

In a typical experiment a single protein or protein mixture is loaded onto a SDS-PAGE or 2D gel and resolved. An individual band or spot is excised from the gel, and the isolated protein is reduced in the gel to break disulfide bonds, and alkylated to prevent crosslinking of peptides. This protein is then digested for 12 to 18 hours (50:1 protein to enzyme ratio) before the peptides are extracted from the gel with acetonitrile and evaporated to dryness. The dried peptides are subsequently re-suspended and desalted to prepare them for MS analysis.

This method's advantages are speed and low performance requirements for the MS instruments, but it is often unable to handle mixtures of proteins. In the tryptic peptide mass range of 600 to 2500 Da, 0.2- to 0.3-ppm relative mass error is need to determine the amino-acid composition with 99% confidence [10]. This performance is beyond the capabilities of all but the most expensive, highest resolving power instruments and limits PMF to all but the simplest of protein mixtures. Database search algorithms to support this technique are available on Mascot (http://www.matrixscience.com) and ProFound (http://prowl.rockefeller.edu/) as well as others.

Figure 3.2*A* shows PMF analysis of bovine serum albumin (BSA) collected on a MALDI-TOF instrument, which we commonly use as a positive control. The resulting mass peak list from the TOF spectrum was searched with Mascot, and the only significant identification was BSA (Mowse Score [11]). Sheep serum albumin was the next closest match but did not reach the confidence threshold. Approximately 80% sequence coverage was obtained (Figure 3.2*B*) from matching peptide masses.

The growing capabilities of tandem MS instruments and computer algorithms to automate peptide product-ion spectral (from MS/MS) interpretation have largely replaced this technique. As this example shows, however, selective and careful use of PMF makes it an adequate approach to the identification of proteins, particularly when speed and ease of sample preparation are of utmost importance or when tandem MS is not available.

3.2.2 Bottom-up Proteomics Using Tandem MS: GeLC-MS/MS and Shotgun Digests

The evolution of sensitive ion traps and fast hybrid instruments helped tandem MS experiments replace peptide mass fingerprinting as the default technique for protein

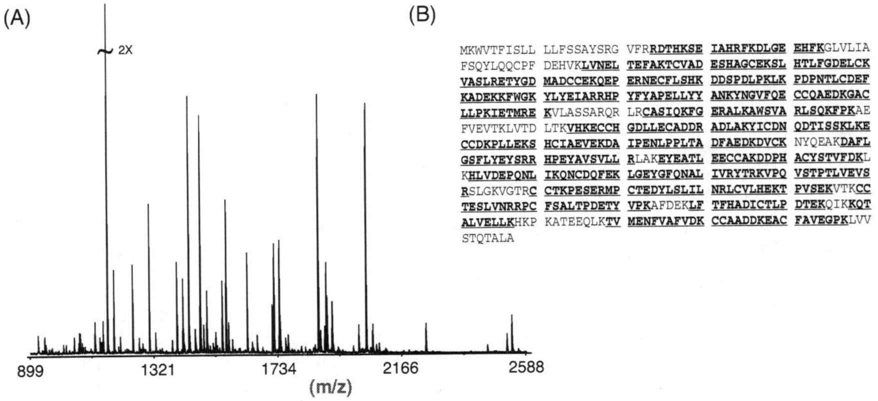

FIGURE 3.2 Peptide fingerprint mapping of bovine serum albumin. (*A*) MALDI-TOF spectrum of digested, desalted BSA. (*B*) Sequence coverage map after Mascot search against the Swissprot other mammalia database. The underlined text highlights the observed sequence coverage (~80%). (*See the color version of this figure in Color Plates section.*)

identification. In a tandem MS experiment, a precursor scan is acquired, and all of the peptide masses in the spectrum are recorded. Subsequently individual peptide ion populations are isolated and fragmented within the mass spectrometer and a second mass scan is taken to measure the *m/z*'s of the generated fragments. Threshold dissociation of peptides (collisional-induced/activated dissociation—CID or CAD, infrared multiphoton dissociation—IRMPD) yields a mixture of predominantly b- and y-ions, whereas electron-based dissociation (electron capture dissociation – ECD or electron transfer dissociation—ETD) yields a mixture of predominantly c- and z-ions (Figure 3.3). These two sets of data (the precursor mass and its corresponding

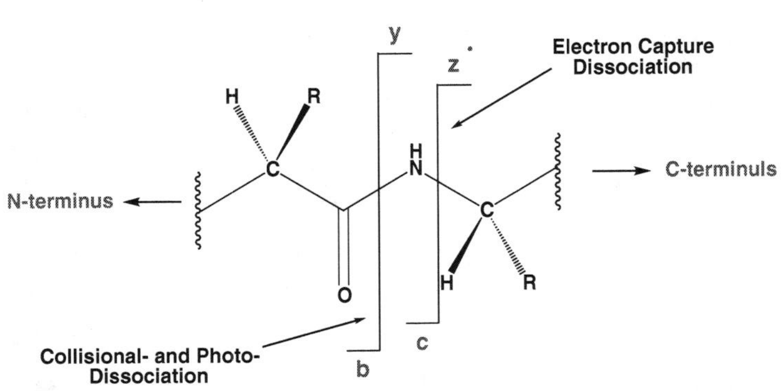

FIGURE 3.3 Fragmentation of peptides. The Roepstorff nomenclature [27] of peptide fragment ions is shown with b- and c-ions representing fragments to the *N*-terminal side of fragmented bond and y- and z$^{\bullet}$ ions representing those to *C*-terminal side. Cleavage of the C–C carbonyl bond would result in a- and x-ions, but these are rarely used in peptide identifications.

fragment ion masses) can then be used to identify more confidently a peptide and thereby a protein.

These complex product-ion (MS/MS) spectra can be interpreted in three ways: (1) by statistical matching of observed fragment-ions to predicted fragment-ion spectra or masses generated from a genomic database (e.g., Sequest, Mascot, OMSSA), (2) by de novo sequencing, which relies on identifying a ladder of sequential fragment ions and deducing the amino acid residue masses from it, or (3) by the sequence tag approach, where a limited amount of the amino acid sequence in the peptide is obtained by de novo sequencing and is coupled with the mass of the peptide and the location of the "tag" within the peptide to conduct a database search. In practice, the first approach of statistically matching observed fragmentation data to predicted data is the dominant methodology in bottom-up proteomics, given the extensive amount of genome sequence information that is now publically available. The statistical considerations involved in these database search approaches require detailed consideration and are discussed in greater detail in Chapter 9. The advantages of this technique include its speed and relatively high throughput as well as the standardization and robustness of the methods involved.

3.2.3 GeLC-MS/MS

In a GeLC-MS/MS experiment the protein or proteins of interests are separated on a SDS-PAGE gel, and individual protein bands or gel slices are digested with trypsin, typically after a reduction and alkylation step. The resulting peptides are most often analyzed via liquid chromatography coupled to a mass analyzer via ESI (LC-MS/MS). The infused peptides' masses are measured with at least unit resolution followed by a fragmentation step where ions are activated by threshold dissociation techniques and the subsequent fragment ions are also measured at unit resolution in an automated data-dependent manner. Usually, ion traps or hybrid time-of-flight instruments are used to collect tandem mass spectral data. These data are then searched against curated peptide databases created from the target organism's genome sequence to retrieve a list of observed peptides and thereby a list of identified proteins.

GeLC-MS/MS approaches are particularly suited for targeted protein analysis. The use of a gel matrix simultaneously traps the proteins of interest while removing detergents and other buffer containments from the sample, eliminating a common limitation in MS analysis. Moreover the gel acts as a stage of fractionation at the protein level, which can greatly reduce the sample complexity and thereby aid the MS analysis.

Many robust methods are available in core proteomic facilities that routinely perform this experiment, but a GeLC-MS/MS experiment suffers a few limitations. A GeLC-MS/MS approach is often less sensitive than other bottom-up strategies, often requiring enough material (>50 ng protein) to be visualized using Coomassie Blue staining. Although analysis of smaller quantities of material that can visualized by using silver or sypro ruby stain levels is possible, the chances of success are greatly

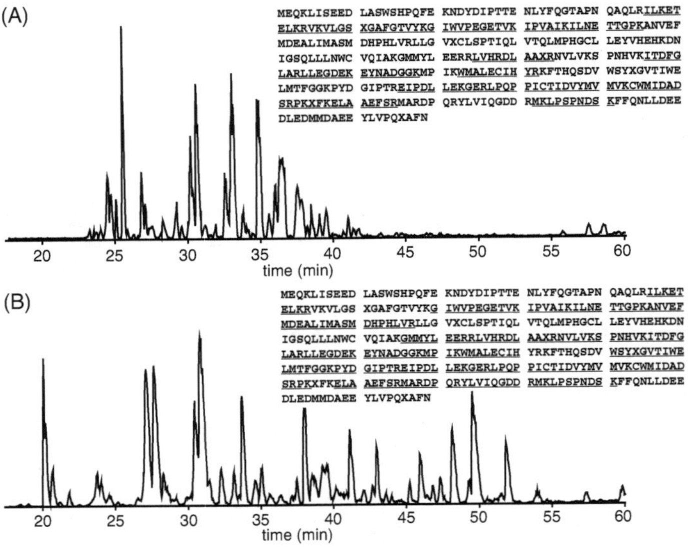

FIGURE 3.4 GeLC-MS/MS versus a shotgun digest. Recombinant Her4 kinase domain was analyzed by GeLC-MS/MS (panel *A*) and Shotgun digest (panel *B*) approaches. The greater sequence coverage and denser base peak chromatograph emphasize the enhanced sensitivity of a shotgun approach (~60% sequence coverage vs. 40% sequence coverage). (*See the color version of this figure in Color Plates section.*)

diminished owing to incomplete extraction and the sensitivity limits of the instruments. GeLC-MS/MS is also time-consuming. Each band must be excised, reduced, alkylated, washed, swelled with trypsin, digested, and extracted prior to MS analysis. The published versions of this protocol take a minimum of 4 h and can range up to 24 h [12], depending on the number of bands to excise and the length of digestion employed. The multiple steps and long time also increase the likelihood of keratin contamination, which confounds analysis and may mask low-abundance species. Even with all of these caveats, GeLC-MS/MS is often the fastest way for nonexperts to confidently identify a targeted protein.

As an example, we show in Figure 3.4*A*, a recombinant construct of the Her4 kinase domain (~40 kDa) that was subjected to a GeLC-MS/MS approach. Panel *A* shows a base peak chromatogram where the intensity of the signal for most abundant species in the mass spectrometer is plotted against time and is a measure of the chromatography performance and instrument sensitivity. In total, around 40% of the construct was sequenced using this method.

3.2.4 Shotgun Digest

For targeted protein analysis, the gel separation step may be unnecessary, particularly for recombinant proteins or immunoprecipitated proteins. In a shotgun digest approach, the target protein is digested along with any and all contaminating proteins and the resulting peptide mixture is loaded onto a reverse phase column coupled to a mass spectrometer. The eluting peptides are analyzed as described above. This setup provides the greater sensitivity and often improves sequence coverage because there is not a gel extraction step. In Figure 3.4*B*, the same recombinant construct of the Her4 kinase domain was shotgun digested and analyzed by LC-MS/MS. Panel *B* shows a base peak chromatogram. In total, around 60% of the construct was sequenced using this method, demonstrating the enhanced sensitivity of this approach (Figure 3.4).

If the target protein is isolated in a mixture, however, the increase in complexity of the sample may offset the gains in sensitivity and may require more extensive fractionation. MudPIT (Multidimensional Protein Identification Technology) is a multi-phase chromatography-based approach to protein identification where two orthogonal stationary phases (usually strong cation exchange and reverse phase) are used to enhance binding capacity, increasing both the resolution of the separation and the sensitivity of the analysis. A mixture of proteins is precipitated, and the pellet is washed and then digested, typically with trypsin. After digestion, the entire mixture is then loaded onto the first dimension column, which is then sequentially eluted with increasing amounts of salt. After each salt "bump," the eluted peptides are separated by reverse-phase chromatography coupled directly to a mass spectrometer. In general, this setup has the highest sensitivity and greatest dynamic range but also demands the greatest amount of instrument time and expertise. For targeted proteomics, a MudPIT-based approach may not be warranted except in certain cases like the isolation and identification of members of protein complexes.

Despite the success, growth and expansion of mass spectrometry based approaches, scientists and laboratories employing these methods should employ great technical care and a cautious scientific approach to both the methodological and broader conceptual issues involved here. Major protein contaminants such as keratins, which are commonly operator-introduced during sample handling, and chemical contaminants, including plasticizers, detergents, and trypsin autolysis products, can obscure the spectra of the desired proteins to be characterized. Limitations of the bottom-up approach include mixture complexity, a stochastic data-acquisition process (discussed below) and the problems of false negatives (the peptide is detected, but no identification can be made) and false positives (the peptide sequence identification or protein identification assigned is incorrect).

By proteolytically digesting mixtures of proteins, the complexity of a protein mixture is dramatically increased. The median size of a tryptic peptide is 10 amino-acid residues, so a single protein of 40 kD can give rise to over 40 possible tryptic peptides. When missed cleavage events and protein modifications are considered, the peptide mixture arising from a single protein is even more complex. Translating this to a proteomic scale, we see that an in silico of human proteome database from the International Protein Index (May 2006), containing 58,099 protein sequences, produces 15.5 million theoretical peptides. Even after two dimensions of separation commonly associated with bottom-up, the sheer number of peptides eluting during an LC-MS/MS experiment overwhelms all currently available mass spectrometers [13] and leads to a stochastic data-acquisition process [14,15] where multiple injections of the same sample only have around 30% overlap in the peptides identified.

Other significant limitations exist in a shotgun digestion analysis, particularly in the context of a targeted protein. The sequence coverage (i.e., the fraction of an identified protein actually detected) is usually less than 100%, limiting the ability to detect and characterize post-translational modifications, such as phosphorylation, or biological variation, such as coding SNPs. Moreover shotgun digestion methods make an already complex mixture proteins [16] into an even more complex mixture of peptides.

3.3 TOP-DOWN APPROACHES

Traditionally top-down experiments involve high resolving power (>50,000), accurate measurement of the intact protein mass followed by its isolation, and fragmentation within the mass spectrometer [17]. The fragmentation data are then used to retrieve the correct protein isoform from an *annotated* database of predicted protein forms [18]. In those cases where a mass discrepancy (Δm) is observed between the predicted and observed forms, the MS/MS data can localize the Δm to a specific region (or even a single amino acid) of the protein. This data analysis logic–the use of high resolving power, mass accurate MS, and MS/MS—becomes increasingly important for multicellular eukaryotes where there are a large number of protein modifying events (nonsynonymous coding polymorphisms, alternative splicing, post-translational modifications, etc.) that can cause the mass of a protein to differ from that predicted by the gene sequence. By incorporating known and predicted modifying events into a single organism database, one can greatly reduce the difficulty in identifying, characterizing, and distinguishing multiple protein isoforms from one another; these isoforms would otherwise be collapsed into a single protein identification in a bottom-up experiment.

Top-down approaches are not nearly as widespread as bottom-up, owing to the lack of available software, the difficulty in obtaining robust, automated MS/MS fragmentation from proteins, and the increased cost and decreased availability of high resolving power mass spectrometers. For protein targets below 60 kDa that are highly modified or that contain other biological variations, top-down's potential 100% sequence coverage often yields information that is missed by bottom-up.

In a typical top-down experiment, a mixture of proteins is fractionated by using one or more dimensions of separation, and then individual fractions are desalted and infused directly into the mass spectrometer. Although online approaches to top-down are available [9], these methods are limited to the highest performing mass spectrometers and often require nonstandard methods to ensure detection of targeted species. An offline approach to top-down is highlighted in Figure 3.5, which demonstrates the significant differences between top-down and bottom-up. Multiple scans are typically summed to reach the signal-to-noise level necessary to obtain the high-quality isotopic distributions needed to make accurate mass measurements. This also means a single MS/MS experiment takes much longer than in the corresponding bottom-up experiments. For targeted species, this time gap is often inconsequential, but for proteome-wide studies, this decrease in speed is often prohibitive. The ability to sum

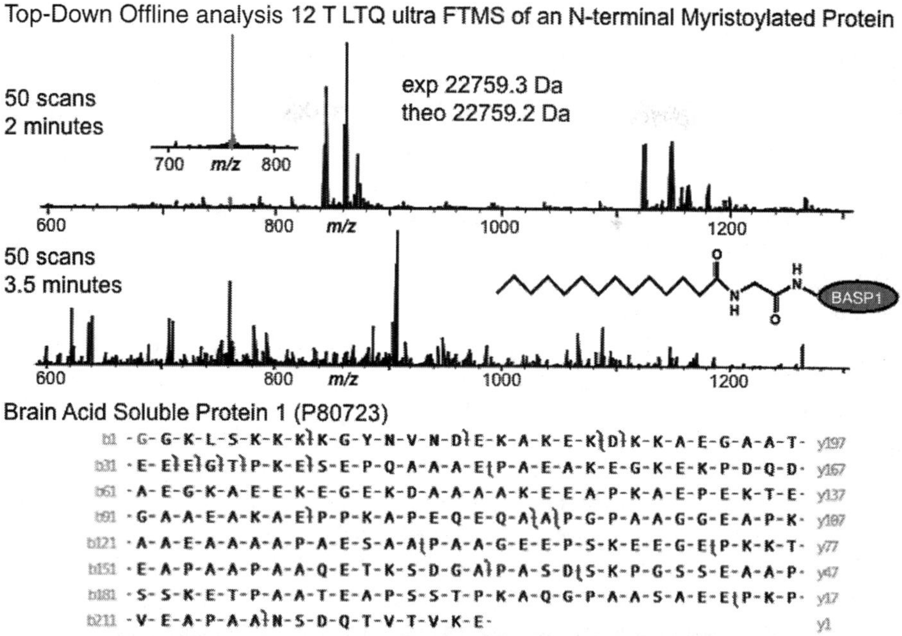

Length	Mass	Mass Diff	PPM Diff	B Ions	Y Ions	Total Ions	PDE Score	Expectation	P Score
226	22759.2000	0.0694	3.0510	14	8	22	68.2000	1.76E-13	7.98E-20

FIGURE 3.5 Top-down identification of brain acid soluble protein 1 (BASP1). A single charge state of BASP1 was isolated (inset) and fragmented on a 12 Tesla LTQ-FT. The intact mass and fragmentation data were searched against an annotated human protein database and BASP1 was identified with high confidence (8E-20 expectation value) and characterized as myristoylated.

scans, however, allows for the collection of robust MS/MS data, which improves identification confidence.

Figure 3.5 shows the identification and characterization of brain acid-soluble protein (P80723) from a HeLa cell lysate. The component corresponding to the peak shown in red was isolated and accumulated in the mass spectrometer, and fragmentation data were collected. The resulting intact mass of 22759.3 Da and its corresponding fragment-ion list were searched with ProSightPC. The identified protein was myristolylated with fragment ions, partially localizing it to the *N*-terminus. Bottom-up LC-MS/MS approaches might have identified the protein but would have missed the *n*-terminal modification because myristolylation is not a frequently searched modification and a three-residue peptide is often hard to identify, given the lack of fragmentation data and loss in the liquid chromatography.

While the common distinction between top-down and bottom-up is the digestion step, the more fundamental difference lies in their approaches to data acquisition and analysis. Bottom-up experiments use fast, sensitive, lower resolution mass spectrometers in an effort to increase proteome coverage and measure quantitative dynamics at the expense of peptide and protein characterization and confidence in the identifications. Top-down experiments use the highest performing mass spectrometers available, Fourier-transform instruments whose mass resolving power (>50,000) and mass accuracy (routinely <5 ppm) dramatically increase the confidence in identifications. When top-down data are combined with well-annotated databases, the output often simultaneously characterizes biological variation, greatly clarifying the outcome for researchers who are not mass spectrometrists. This depth of knowledge comes at the expense of throughput and sensitivity. Fourier-transform mass spectrometers are an order of magnitude slower and less sensitive than ion traps and time-of-flight instruments, since they measure a frequency rather than an electron multiplier response, but the gap in speed and sensitivity is narrowing.

3.4 NEXT-GENERATION APPROACHES

The evolution of higher performance hybrid instruments (Q-FTMS, LTQ-FT, LTQ-Orbitrap) has spawned a new generation of data acquisition and data analysis techniques [19,20] that blur the distinction between bottom-up and top-down. The higher mass resolving power and greater mass accuracy that these instruments confer, at both the MS and MS/MS level, allow researchers to identify more proteins, faster and with greater confidence [16].

In PMF experiments, where the a list of measured masses is compared against the in silico digests of a given database, the number of masses matching within a given tolerance and the number of masses searched are the key components in obtaining confident scores [21]. Higher mass resolving power and mass accuracy help in two ways. More accurate mass measurement eliminates a significant portion of the peptides with the same nominal mass but different amino-acid compositions [22]. Eliminating possible candidate peptides to search increases the confidence in an identification, while correspondingly decreasing the incident of false positives [23]. Moreover search speed is increased as fewer candidates must be considered by identification software. In the extreme, accurate mass measurements can be used to discriminate between nonpeptide and peptide signals in the mass spectrometer, since the monoisotopic mass of all peptides must be in a predictable range of values. By only submitting those species that can arise from peptides, one can exponentially increase the confidence in which a protein is identified while gaining a linear increase in search speed and specificity [24].

The benefits to data-dependent MS/MS experiments are also significant. The use of a high mass resolving power, accurate mass precursor scan, followed by a lower resolution data-dependent MS/MS, has been shown to provide more identifications at higher confidence levels than traditional low-resolution experiments with increased characterization rates for post-translational modifications in pull-down experiments [25,26]. This increase in data depth and quality can be attributed to a multitude of factors: the exclusive selection of multiply charged peptides, improved identification power when spectrum quality is poor, and the reduction in the number of peptides considered by the search algorithm [25]. Moreover mass accuracy can be independently used to validate peptide spectral masses; true positives' mass deviations tend to cluster together whereas false positives' mass deviations are evenly distributed across mass space [25].

Even with these increases in performance, targeted protein studies are also benefiting from more focused MS techniques that attempt to minimize the stochastic nature of data-dependent acquisition processes and direct the mass spectrometer to "look" for the target proteins and their peptides. For data-dependent LC/MS-MS processes this means the use of inclusion lists and some form of quantification (see Chapter 4) where the mass spectrometer is programmatically set to fragment peptides of a particular *m/z* (or mass in some acquisition software) eluting during a particular time window. Here the precursor mass, fragment-ion masses, and elution time are used to confirm the identity of the peptide. The combination of targeted analysis and quantitative information produces a data set that monitors protein changes during the experiment, which is often the goal of biological studies. In addition to inclusion lists, selected reaction monitoring or multiple reaction monitoring on triple quads and Q-traps are being developed as focused MS approaches. These techniques move mass spectrometer based identification into the realm of specific assays, where both the detection and quantification of a target protein can be assessed in one experiment, and offer a fast and sensitive alternative when the target protein is known and well described and multiple samples need to be tested.

This evolution of mass spectrometry techniques means both bottom-up and top-down approaches for target proteins are being continually optimized. In this chapter we have provided an overview of the most established practices in the field: peptide mass fingerprinting, GeLC-MS/MS, shotgun digestion for bottom-up, and direct infusion for top-down, while keeping an eye toward the future and the use of increasingly accurate data and focused mass spectrometry techniques. Bottom-up, top-down, mass accurate and focused, mass spectrometry based analysis will remain as valuable tools for identifying and characterizing target proteins.

REFERENCES

1. Fenn, J. B., et al. (1989). Electrospray ionization for mass spectrometry of large biomolecules. *Science 246*, 64–71.
2. Karas, M. Hillenkamp, F. (1988). Laser desorption ionization of proteins with molecular masses exceeding 10,000 Daltons. *Anal Chem 60*, 2299–2301.
3. Tanaka, K., et al. (1988). Protein and polymer analyses up to m/z 100 000 by laser ionization time-of-flight mass spectrometry. *Rapid Commun Mass Spectrom 2*, 151–153.
4. Kelleher, N. L. (2004). Top down proteomics. *Anal Chem 76*, 197A–203A.
5. Kelleher, N. L., et al. (1999). Top down versus bottom up protein characterization by tandem high-resolution mass spectrometry. *J Am Chem Soc 121*, 806–812.
6. McCormack, A. L., et al. (1997). Direct analysis and identification of proteins in mixtures by LC/MS/MS and database searching at the low-femtomole level. *Anal Chem 69*, 767–776.
7. O'Farrell, P. H. (1975). High resolution two-dimensional electrophoresis of proteins. *J Biol Chem 250*, 4007–4021.

8. Washburn, M. P., Wolters, D., Yates, J. R. (2001). Large-scale analysis of the yeast proteome by multidimension protein identification technology. *Nat Biotechnol 19*, 242–247.
9. Roth, M. J., et al. (2008). "Proteotyping": Population proteomics of human leukocytes using Top Down mass spectrometry. *Anal Chem 80*, 2857–2866.
10. Zubarev, R., Mann, M. (2007). On the proper use of mass accuracy in proteomics. *Mole Cell Proteomics 6*, 377–381.
11. Pappin, D. J. C., Hojrup, P., Bleasby, A. J. (1993). Rapid identification of proteins by peptide-mass fingerprinting. *Curr Biol 3*, 327–332.
12. Shevchenko, A., et al. (2007). In-gel digestion for mass spectrometric characterization of proteins and proteomes. *Nat Proto 1*, 2856–2860.
13. MacCoss, M. J. (2005). Computational analysis of shotgun proteomics data. *Curr Opin Chem Biol 9*, 88–94.
14. Elias, J. E., et al. (2005). Comparative evaluation of mass spectrometry platforms used in large-scale proteomics investigations. *Nat Meth 2*, 667–675.
15. Liu, H., Sadygov, R. G., Yates, J. R. (2004). A model for random sampling and estimation of relative protein abundance in shotgun proteomics. *Anal Chem 76*, 4193–4201.
16. Liu, T., et al. (2007). Accurate Mass Measurements in Proteomics. *Chemical Reviews 107*, 3621–3653.
17. Meng, F., et al. (2002). Processing complex mixtures of intact proteins for direct analysis by mass spectrometry. *Anal Chem 74*, 2923–2929.
18. Pesavento, J. J., et al. (2004). Shotgun annotation of histone modifications: A new approach for streamlined characterization of proteins by top down mass spectrometry. *J Am Chem Soc 126*, 3386–3387.
19. Gorshkov, M. V., Zubarev, R.A. (2005). On the accuracy of polypeptide masses measured in a linear ion trap. *Rapid Commun Mass Spectrom 19*, 3755–3758.
20. Frank, A. M., et al. (2007). De novo peptide sequencing and identification with precision mass spectrometry. *J Proteome Res 6*, 114–123.
21. Perkins, D. N., et al. (1999). Probablility based protein identification by searching sequence databases using mass spectrometry data. *Electrophoresis 20*, 3551–3567.
22. He, F., et al. (2004). Theoretical and experiemental prospects for protein identification based solely on accurate mass measurement. *J Proteome Res 3*, 61–67.
23. Dodds, E. D., et al. (2007). Systematic characterization of high mass accuracy influence on false discovery and probability scoring in peptide mass fingerprinting. *Anal Biochem 372*, 156–166.
24. Dodds, E. D., et al. (2006). Enhanced peptide mass fingerprinting through high mass acccuracy: Exclusion of non-peptide signals based on residual mass. *J Proteome Res 5*, 1195–1203.
25. Bakalarski, C. E., et al. (2007). The effects of mass accuracy, data acquisition speed, and search algorithm choice on peptide identification rates in phosphoproteomics. *Anal Bioanal Chem 389*, 1409–1419.
26. Wu, S. L., et al. (2005). Extended range proteomic analysis (ERPA): A new and sensitive LC-MS platform for high sequence coverage of complex proteins with extensive post-translational modifications-comprehensive analysis of beta-casein and epidermal growth factor receptor (EGFR). *J Proteome Res 4*, 1155–1170.
27. Roepstorff, P., Fohlman, J. (1984). Proposal for a common nomenclature for sequence ions in mass spectra of peptides. *Biomed Mass Spectrom 11*, 601.

CHAPTER 4

Quantitative Proteomics by Mass Spectrometry

JACOB GALAN, ANTON ILIUK, and W. ANDY TAO

4.1 INTRODUCTION

Modern drug research and development depend heavily on the ability to target and analyze compounds of interest on a large scale from a wide variety of complex sources. The ability to identify the proteins and other molecules that change as a consequence of the progression of an illness or a disorder is the vital first step in drug development. This stage is often referred to as biomarker discovery. Biomarkers are biologically relevant proteins, peptides, and other molecules that exhibit variation from homeostasis during the onset or progression of a certain disease [1,2]. Commonly, their deviations from the steady state include differential expression of the proteins and/or their post-translational modifications (PTMs), including phosphorylation and glycosylation (PTMs are discussed in Chapter 12 by Tsarbopolous and Bazoti in this volume). Identification of such biomarkers not only can provide molecules necessary for disease prognosis and diagnosis but also can offer potential therapeutic targets.

Many approaches have been utilized to study the changes in protein expression or its PTM levels. The usual techniques such as Western blot or ^{32}P labeling are regarded as efficient and robust methods that provide important protein information with good reproducibility. They are mainly used, however, on a small scale provided that we know or can speculate reasonably on the identity of the protein. Systems biology emerges as a new direction for biomarker discovery, allowing identification of multiple markers for a single disease and providing *unequaled* specificity for diagnosis. Mass spectrometry (MS)-based proteomics is a choice for large-scale and unbiased analyses. Presently, improvements in the accuracy, sensitivity, and throughput capabilities have advanced mass spectrometry to be a major tool for

Protein and Peptide Mass Spectrometry in Drug Discovery, Edited by Michael L. Gross, Guodong Chen, and Birendra N. Pramanik.

complex sample analyses [3]. Although the identification of proteins from a mixture has become routine, the mass spectrometer itself does not allow confident biomarker recognition because there are serious issue for quantitation. For this reason quantitative proteomics has been identified as an important subject and activity and efforts to develop it further are on the rise. A successful outcome will provide reliable means to obtain relative or absolute quantitation of proteins in various complex samples.

The implementation of quantitative proteomics using MS originally utilized stable isotope dilution, where an isotopically labeled isotopomer is used as an internal standard for quantification. The isotope dilution method has been a "workhorse" in modern pharmaceutical research [4–6]. To allow for accurate measurements of small molecules by using MS, references or standards are required to be added to the unknown sample. Such materials should have the same chemical properties as, but be distinguishable from, the analytes. A reference or a standard is commonly prepared by incorporating stable isotopes (e.g., ^{13}C, ^{15}N, and ^{2}H), allowing for mass discrimination of target analytes from the standards. The reference molecules are spiked or "diluted" with the sample before the MS analysis. Quantitative MS-based proteomics uses essentially the same concept to analyze two or more proteomic samples, but unlike much target compound analysis, there is a need to measure a large number of proteins in one experiment.

A number of attempts have been made to discover disease markers, motivating the development of a variety of methods for quantitative proteomics [7–9]. Although the overall strategy is promising and good progress has been made, major obstacles still remain, particularly when working with a complex sample. A common difficulty with working on the whole-cell level is the wide dynamic range of protein abundances contained in the cell, spanning over 6 to 8 orders of magnitude [10]. This severely reduces the efficiency of biomarker identification, primarily because the typical candidates are present in relatively low concentrations and are, therefore, masked by the highly abundant "housekeeping" proteins [2]. This dilemma is even more noticeable for serum and plasma samples, which can exhibit close to 12 orders of magnitude in protein abundances [11]. Some setbacks that are typically encountered during quantitative proteomic analyses include incomplete quantitative label incorporation, variation between samples during preparation, and insufficient bioinformatics tools. Several recently developed quantitation methods attempt to address these issues.

This chapter is intended to provide a general description of the approaches for quantitative proteomics and an in-depth catalog of the currently available methods. Each method has its own unique advantages and shortcomings, depending on its characteristics and application. We do not believe that there is a general method that is best for quantitation, but that rather each analytical approach has advantages with respect to the others depending on the sample specifications and desired results. Therefore selection of the most effective quantitation technique for a specific set of experiments requires care and may even be critical for success. Our goal is to provide the necessary information on major quantitation approaches that should aid in this selection process.

TABLE 4.1 Selected Methods for Quantitative Proteomics via Stable Isotope Labeling on Proteins

Reactive Group	Tagging Method†	Stable Isotope	Stage of Labeling	References
Biological incorporation				
	SILAC	^{15}N	Protein	12
	CDIT	^{2}H, ^{13}C, ^{15}N	Protein	13
	^{15}N labeling (ammonium salt)	^{15}N	Protein	14
	BONCAT	No isotope	Protein	15
	SILIP	^{15}N	Protein	16
Enzymatic incorporation				
	^{18}O labeling	^{18}O	Protein	17
	QCET	^{18}O	Protein	18
Small chemical labeling				
Cysteine	ICAT	^{2}H & ^{13}C	Protein	19
	HysTag	^{2}H	Protein	20
	N-t-butylidoacetimide	^{2}H	Protein	21
	2-vinyl-pyridine	^{2}H	Protein	22
	CAR	^{13}C	Protein	23
	Iodoacetanilide	^{2}H	Protein	21
	SoPIL	^{13}C	Protein	24
	MeCAT	None	Protein	25
	FCAT	^{13}C & ^{15}N	Protein	26
	methyl iodide	^{13}C & ^{2}H	Protein	27
	VICAT	^{13}C, ^{14}C, ^{15}N	Protein	28
N-Terminus/Lysine	ISIL	^{2}H	Protein	29
	ANIBAL	^{13}C	Protein	30
	Urea	^{13}C	Protein	31
	Guanidination (O-methyl-isourea)	^{13}C & ^{15}N	Protein	32
	ICPL	^{13}C	Protein	33
Carboxyl	ANIBAL	^{13}C	Protein	30
Tryptophan	NBSCl	^{13}C	Protein	34

†SILAC - Stable isotope labeling of amino acids in cell culture, CDIT- Cell culture derived isotope labeling, BONCAT - Bioorthogonal noncanonical amino acid tagging, SILIP - Stable isotope labeling in *planta*, QCET - Quantitative cysteinyl-peptide enrichment technology, ICAT – Isotope-coded affinity tagging, HysTag – Histidine tag, CAR - Catch and release, SoPIL - Soluble polymer-based isotope label, MeCAT - Metal-coded affinity tag, FCAT - Fluorescent isotope-coded affinity tag, VICAT - Visual isotope-coded affinity tag, ISIL – in-gel isotope labeling, ANIBAL - Aniline and Benzoic acid labeling, ICPL – Isotope-coded protein label, NBSCl – Nitrobenzenesulfonyl chloride.

The number of quantitative techniques is large, ranging from stable isotope labeling to label-free approaches and biological incorporation, from small-molecule labeling to innovative multifaceted chemical structural designs combining purification mechanisms and isotope incorporation (Tables 4.1 and 4.2). Depending

TABLE 4.2 Selected Reagents for Stable Isotope-based Quantitative Proteomics on Peptides

Reactive Group	Tagging Method†	Stable Isotope	Stage of Labeling	References
Small chemical labeling:				
Cysteine	ICAT	^{2}H & ^{13}C	Peptide	19
	SoPIL	^{13}C	Peptide	24
	FCAT	^{13}C, ^{15}N	Peptide	26
	ALICE	^{2}H	Peptide	35
	Solid-phase tagging	^{13}C	Peptide	36
	ICROC	^{2}H	Peptide	37
N-Terminus/Lysine	TMT	^{13}C	Peptide	38
	Succinic anhydride	^{2}H	Peptide	39
	N-acetoxysuccinamide	^{2}H	Peptide	40
	Acetic anhydride	^{2}H	Peptide	41
	Propionic anhydride	^{2}H	Peptide	42
	Nic-NHS	^{13}C	Peptide	43
	iTRAQ	^{13}C	Peptide	44
	MCAT	No isotope	Peptide	45
	Guanidination (O-methyl-isourea)	^{13}C & ^{15}N	Peptide	32
	Phenyl isocyanate	^{2}H	Peptide	46
	QUEST	No isotope	Peptide	47
	SPITC	^{13}C	Peptide	48
	2-Methoxy 4,5-1H-imidazole	^{2}H	Peptide	49
	Formaldyde	^{2}H	Peptide	50
	N-terminal stable-isotope labeling of tryptic peptides (pentafluorophenyl-4-anilino-4-oxbutanoate)	^{2}H or ^{13}C	Peptide	51
Tyrosine	CILAT	^{13}C	Peptide	52
Carboxyl	Methyl esterification	^{2}H	Peptide	53
	Ethyl esterification	^{2}H	Peptide	54
Glycosyl	QUIBL	^{13}C	Glycan	55
	Methyl iodide	^{13}C	Glycan	56

†ICAT - Isotope coded affinity tagging, SoPIL - Soluble polymer-based isotope label, FCAT - Fluorescent isotope-coded affinity tag, ALICE - Acid-labile isotope-coded extractant, ICROC – Isotope-coded reduction off column, TMT - Tandem mass tag, Nic-NHS - Nicotinoyloxy-succinamide, iTRAQ - Isobaric tag for relative and absolute quantitation, MCAT - Mass-coded abundance tagging, QUEST - Quantitation using enhanced sequence tags, SPITC - Differential isotope-coded N-terminal protein sulphonation, CILAT - Cleavable isobaric labeled affinity tag, QUIBL - Quantitation by isobaric labeling.

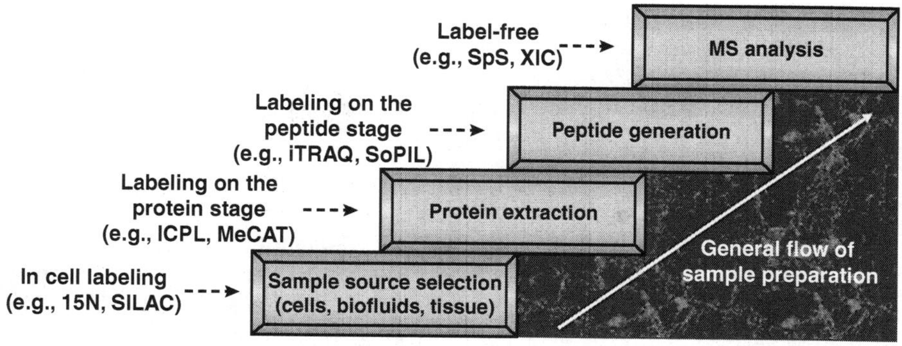

FIGURE 4.1 General workflow of quantitative proteomic approach. Samples can come from various cell types, biofluids, or plant tissue. The proteins can be metabolically labeled as they are synthesized using isotopically differentiated amino acids. Alternatively, samples can be labeled on the protein stage using a variety of chemical reagents that can target specific residues. Chemical labeling can also be incorporated on the peptide stage after protein digestion with proteases. This usually requires protein or peptide samples to be purified (IP, IMAC, density gradient fractionation (organelle)), or fractionated using HPLC, simplifying and reducing the sample to subproteomic scale, followed by MS-based analysis. After MS analysis, each MS spectrum can be integrated (relying on software), allowing for relative quantitation.

on the stage in which the stable isotope is introduced, we group them as follows: (1) in-cell incorporation of isotope label, (2) introduction of label at the protein level, (3) introduction of label at the peptide level, and (4) label-free quantitation (Figure 4.1).

4.2 IN-CELL LABELING

In-cell labeling, usually referred to as metabolic labeling, is a quantitation approach that relies on normal biological functions of a cell, specifically protein synthesis. Isotopically coded molecules can be introduced during translation to achieve their incorporation into nascent proteins. Perhaps the most well-known strategies of in-cell labeling include ^{15}N metabolic labeling and SILAC (stable isotope labeling by amino acids in cell culture), and we discuss them in the following.

4.2.1 ^{15}N Metabolic Labeling

One of the first quantitative labeling methods developed for in-cell isotope incorporation was ^{15}N labeling; ^{15}N-enriched media can be used to introduce the "heavy" isotopic form of nitrogen into the proteins. By this approach, two samples of interest (a control and a test) can be grown separately in either ^{14}N- or ^{15}N-enriched media. After nitrogen incorporation, the samples are pooled and treated together to minimize the amount of variation introduced during further manipulations. Finally, the mixture is analyzed by MS, which allows the relative quantitation to be carried out on the basis of isotopically differentiated protein labels. Although nitrogen is a ubiquitous element used by every organism, there are limitations as to which tissues and cell cultures can efficiently incorporate its "heavy" form. Often ^{15}N labeling can be used in intact plants owing to good incorporation [57,58].

Although the approach has also been attempted in microorganisms [59], *C. elegans* [60], fruit flies [61], and cell cultures [13,14], the amount of ^{15}N labeling in peptides may be lower than in the amount of ^{14}N-incorporation in their counterparts. Needless to say, this incomplete incorporation can translate into challenges during proper peak selection, peptide sequencing, and quantitation at the MS stage [62]. An example problem that arises from inefficient labeling is the presence of ^{14}N-containing isotopologs corresponding to peaks that appear before the monoisotopic peak of the ^{15}N-containing peptide. These peaks can be mistakenly judged as monoisotopic ^{15}N-containing, thus resulting in missed or false identification and quantitation.

To address these issues, improved labeling experiments and quantitation software have been designed. Some of the enhanced approaches include the introduction of an internal standard into both samples, thereby canceling out any errors between the samples [63], or utilizing ^{15}N enrichment in combination with precursor mass correction to increase the level of identification and quantitation [62]. Another novel approach designed to improve ^{15}N incorporation is termed SILIP (stable isotope labeling *in plants*) [16]. As the name suggests, the approach is designed for plant research and is based on optimization of soil-based medium, thus promoting 98–99% incorporation of "heavy" nitrogen.

Despite continuing efforts to improve metabolic labeling using ^{15}N-enriched media, a number of limitations prevent its broad usage. Primary reasons for its lack of broad usage are the high cost (need an excess amount of isotopically labeled nuclei), difficulty in interpretation of results, and incomplete isotope incorporation, as discussed above. To overcome some of these drawbacks, a new metabolic labeling approach, developed by Mann and coworkers, involves substitution of ^{15}N-rich media with isotopically differentiated amino acids [12,64]. The new method, named SILAC, has shown great promise toward efficient protein labeling in vivo.

4.2.2 Stable Isotope Labeling by Amino Acid (SILAC)

Quantitation using SILAC is based on the introduction of isotopically labeled amino acids into a growing peptide chain during translation. During the experiment two or more sets of cell cultures are grown using depleted media designed for SILAC that are enriched with amino acids in their "light" or "heavy" forms (usually through ^{2}H, ^{13}C, or ^{15}N isotopes) [65]. After a number of cell divisions, nearly 100% of the proteins will be newly synthesized, and thus would incorporate the labeled amino acids, leading to a very efficient process. Full isotopic inclusion, however, would not be possible unless the amino acids chosen for the reaction are essential, thus creating only a single source of amino acids. For this reason leucine [12], lysine [66], and methionine [67] are the best candidates. Recently arginine was added to the list of amino acids that exhibit efficient incorporation [68]. Although arginine is not an essential amino acid, it is for the most part obtained by many organisms through diet, thus making it a viable labeling source [69]. The combination of arginine and lysine has become perhaps the most often used labeling mixture because, after trypsin digestion, each peptide theoretically will have a single label, enhancing protein identification coverage and improving quantitation.

In-cell quantitative labeling offers many advantages compared to tagging of samples during later stages. During SILAC labeling, as in ^{15}N labeling, an isotopic molecule is localized based on the protein sequence. Such sequence dependence aids in protein identification because the differences in masses generated can also be used as evidence for peptide sequence confirmation. This is particularly true when isotopic incorporation is nearly 100%, which can be achieved with many cell lines [12]. Nevertheless, the major reason why many groups employ SILAC as their method of choice is that it introduces early in the experiment the quantitative label into the proteins. This benefit allows researchers to combine the samples before any preparative manipulations. Since its introduction, SILAC quantitation has been successfully utilized by many research groups and applied to various cell types for both focused studies [70–74] and large-scale biomarker discovery [8,75].

Although the advantages of SILAC are many, there are some drawbacks that researchers need to understand before using this quantitation method. First, like ^{15}N labeling, it is expensive. Second, given that the labeled amino acids need to be incorporated during cell culture growth, no complete quantitative labeling is possible in plants, clinical samples, biofluids, or tissues, thus limiting the broad application of the technique. An approach to address potentially this latter restriction may be to use culture-derived isotope tags (CDITs) such that an in vivo isotope incorporated sample can be employed as an internal standard in the quantitation of tissue proteins [13]. Previously this labeling limitation also held true for primary neurons because of their inability to divide. This problem in neurons can also be addressed by growing primary cortical and hippocampal neurons for a prolonged period of time in derived media and demonstrating efficient incorporation of SILAC amino acids [76]. A setback that can occur during the SILAC labeling is the metabolic conversion of isotope-coded arginine to proline. This conversion issue can be addressed by using *L*-proline [77] or a mathematical correction [78]. Owing to these drawbacks, a large number of alternative methods are available and introduce quantitative tagging at a later stage of sample preparation.

4.3 QUANTITATION VIA ISOTOPIC LABELING OF PROTEINS

Introduction of stable isotopes at the protein isolation stage is another useful option in quantitative proteomics. The approach still allows relatively early incorporation of the label while offering a wider variety of methods than metabolic labeling. Major approaches of isotopic labeling on proteins include 2D PAGE-based quantitation, proteolytic labeling by ^{18}O and chemical isotope tagging.

4.3.1 2D PAGE-Based Quantitation

Two-dimensional polyacrylamide gel electrophoresis (2D PAGE) has remained a method of choice for protein separation and quantitation for many years. It allows separation of complex mixtures based on the protein isoelectric point (*pI*) in the first dimension and molecular mass (M_r) in the second. 2D PAGE is considered perhaps the earliest method for quantitative proteomics. The protocol is as follows: two samples of interest are run side-by-side by using 2D PAGE, the proteins stained, and the band sizes quantitatively compared to each other according to the exact positions on the gels (a 2D PAGE-based quantitation flowchart is in Figure 4.2). Here the quantitation is done at an early stage before any identification is accomplished. Recently with the development of improved instrumentation and software, the quantitative differentiation is much more accurate. Quantitation using 2D PAGE is particularly effective when comparing proteins that have undergone post-translational modifications [79]. Usually phosphorylation, glycosylation, and some other modifications can be detected on a gel as discrete band trains.

Despite the unique nature of quantitation using 2D PAGE in comparison to other methods, it too has a number of drawbacks. First, separation and quantitation by 2D electrophoresis discriminates against the proteins with molecular masses and *pI*s outside the range, thus making complete proteome analyses difficult. Second, the reproducibility of the method has room for improvement; the lack of precision is due to gel heterogeneities introduced during polymerization and fluctuations during the run. Furthermore, because it is virtually impossible to identify proteins from a

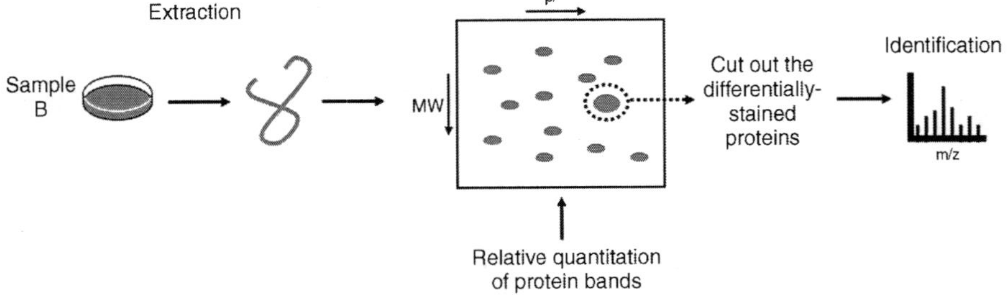

FIGURE 4.2 Workflow for using 2D-PAGE for quantitation. Protein samples are extracted from a cell lysate or other biologically important sources and run on two separate 2D-PAGE gels. In the first dimension, the proteins are separated according to their *pI*, followed by molecular weight (MW)-based separation in the second dimension. After staining, each protein can be visualized and quantified based on the intensity of each band for relative comparison. The proteins of interest are excised, digested with proteases, and identified using MS.

complex mixture by electrophoresis, an additional identification step is required. This step was traditionally Edman degradation, but recently MS has become the method of choice. Therefore many research groups now use MS-based quantitation to combine the quantitation and identification steps. The most critical downside of 2D PAGE, however, is its low sensitivity. Because the concentration range of proteins in a cell range is between five and seven orders of magnitude, it is necessary to distinguish proteins in low abundance (usually interesting signaling molecules) from the high background of high-abundance proteins (usually "housekeeping" molecules). Regrettably, most current gel staining techniques are not capable of achieving this. In addition, 2D PAGE is biased toward membrane and other less soluble proteins. Finally, the 2D PAGE approach is relatively labor-intensive and needs great care for large-scale and frequent studies.

To address some of the drawbacks of the traditional 2D PAGE, one can turn to a novel enhanced method called difference gel electrophoresis (DIGE) [80]. The unique design of the approach uses fluorescent dyes, which can label proteins in complex mixtures. Because fluorescent tags generally have good sensitivity, they add to proteins detectability over a wide dynamic range. Besides sensitivity, reproducibility (precision) has also been improved thanks to the ability to run both labeled samples on a single gel, thus eliminating any gel-to-gel variations [81]. Nonetheless, protein discrimination, its time-consuming nature, and the inability to identify the bands of interest without further manipulations still remain as problems.

4.3.2 Proteolytic Labeling Using ^{18}O Water

Proteolytic labeling in ^{18}O water is a unique method of quantitation because the tagging occurs not precisely at the protein stage but rather during the digestion step. Here the two protein samples to be analyzed undergo proteolysis in either ^{16}O or ^{18}O water, thus permitting one of the samples to incorporate an isotopic tag on the carboxy termini of its peptides [82,83]. This results in a mass difference of 2 to 4 Da between the two peptides that are identical in every other way. At this stage the peptides can be combined, and the remainder of the protocol can be carried out as usual owing to the excellent stability of the C–^{18}O bond. Finally, quantitation is carried out at the MS stage by determining the $^{16}O/^{18}O$ ratio.

The exact number of "heavy" atoms introduced during labeling depends on the completeness of the proteolysis and the enzyme selected. Trypsin is usually the first choice owing to its capacity to produce relatively uniform peptides of manageable length that are easier to analyze by MS [84]. Along with a number of other proteases, like Glu-C [82] and chymotrypsin, [85] trypsin incorporates two ^{18}O atoms during digestion. On the other hand, it is possible to introduce only a single ^{18}O isotope during proteolysis if metalloendopeptidase Lys-N is used [86].

Whether one or two oxygen atoms are added during proteolysis, the result of the reaction is still a small mass difference (2 or 4 Da), making overlap between test and control likely and quantitation difficult, particularly when low resolving power mass spectrometers are utilized. Another disadvantage of a proteolytic labeling method is the lack of automated software for efficient quantitation. Back-exchange of the ^{18}O

with solvent during proteolysis is also a substantial problem, resulting in poor quantitation because the continuous activity of trypsin can sometimes substitute the incorporated ^{18}O atom with an ^{16}O isotope when the two samples are pooled together [87]. Usually it helps to inactivate the enzyme by lowering the pH and/or the temperature of the digest solution [88]. These approaches only reduce the problem, not eliminate it. Lastly, poor labeling efficiency and high cost of the technique work against the method's widespread adoption.

Regardless of the pitfalls and disadvantages, proteolysis in ^{18}O water remains a useful technique for quantitative labeling primarily owing to its simplicity. No extra steps are necessary for incorporating the isotope tag, and the labeling can be done on any biological sample. To address the low efficiency and high cost issues, a new approach was introduced involving two sets of digestion—the first in "normal" water and the second in "heavy" water—to incorporate the label at the peptide stage [85,89]. This possibility stems from the ability of trypsin to recognize peptides as pseudo-substrates. This strategy optimizes the labeling efficiency and reduces the amount of ^{18}O water used, thus lowering the cost.

4.3.3 Quantitative Labeling by Chemical Tagging

Quantitation by chemical labeling has become one of the most prevalent approaches in proteomic research. An extensive selection of various chemical tags is available, and the ability to label different residues and PTMs within the samples provide an opportunity for any research group to select the most effective and appropriate design for individual experiments. Although the majority of chemical tags are designed to be incorporated at the peptide stage, there are a few methods that introduce quantitative capabilities by selectively labeling at the protein level (selected chemical tagging reagents are illustrated in Figure 4.3).

Isotope-coded affinity tag (ICAT) technology is perhaps the most popular and widely used chemical labeling method on proteins [19]. The design consists of three functional groups: a chemical tagging element that is capable of reacting and binding to specific residues within a protein (in this case, cysteine residues), an isotopically modified linker that incorporates a "light" or "heavy" label on the protein, and an affinity tag, usually biotin that allows for efficient enrichment of the tagged peptides, thus reducing sample complexity (Figure 4.3*a*). Although the original ICAT version was designed for protein tagging, recently, a number of peptide labeling versions of ICAT have become available and used (to be discussed later).

Since the introduction of the original ICAT, many deviations of isotope tagging reagents for protein labeling have been developed, each having its own advantages. An example of one such reagent is isotope-coded protein label (ICPL), introduced in 2005 [33]. The foundation of the ICPL method is an isotopically differentiated Nic-NHS tag (nicotinoyloxy succinamide) that is capable of attaching to all free amine groups, thus introducing a much higher labeling coverage than ICAT (Figure 4.3b). The method does not require an affinity tag.

Another chemical tagging method, fluorescent isotope-coded affinity tag (FCAT), developed in 2008, introduces a new dimension to quantitative proteomics—the

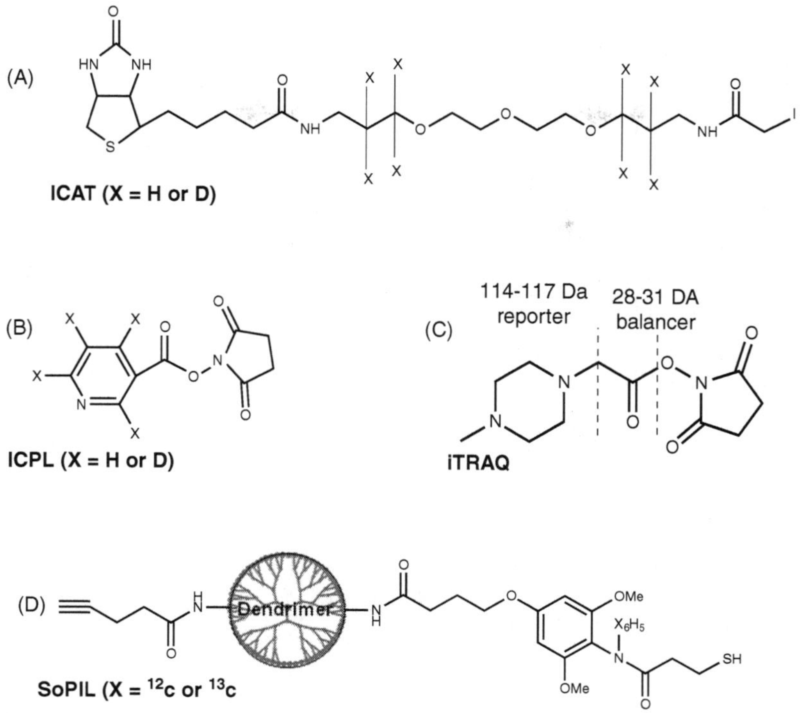

FIGURE 4.3 Schematic elemental illustrations of selected chemical labeling reagents and their chemical compositions.

ability to combine relative and absolute quantitation [26]. Like ICAT, the reagent is comprised of sulfhydryl-reactive chemical binding group and an isotopically labeled linker. Its unique features, however, are the base-labile group, which is designed to allow cleavage of a part of the tag to make it smaller, and a fluorescent tag, which can be used for antibody-based purification or absolute quantitation by fluorescence detection.

A different tagging method, developed also in 2008 and termed aniline benzoic acid labeling (ANIBAL), aims to use a combined "symmetric" chemistry approach [30]. The method takes advantage of two residue groups found commonly in proteins—amino and carboxylic functionalities. Both tagging reactions rely on carbodiimide chemistry to introduce ^{13}C isotopic tags. This dual labeling offers wider proteome coverage for more efficient quantitation. The relatively low cost of the reagents, however, is a real advantage of the method.

The final strategy for protein labeling discussed here is the metal-coded affinity tag (MeCAT) approach [25]. The method utilizes a macrolytic metal chelate complex loaded with a lanthanide as its central metal ion. In this case lanthanide (III) ions (e.g., Lu(III), Tm(III), or Tb(III)) are the source of differential labeling (metal coding) that allows for quantitation. The reagent also contains a thiol-specific maleimido group and a biotin molecule for enrichment. With the availability of various lanthanides, this method is capable of achieving simultaneous relative quantitation for more than two samples.

Despite the different strategies that are available for protein labeling, there are a number of drawbacks to tagging at this stage. Primarily it is sometimes difficult to combine the tagging at this early stage with the subsequent manipulations need for analysis; this is due to the poor compatibility of a tag with many subsequent steps in a protocol. Often a chemical tag introduces variations in refolding, hydrophobicity, or net charge within the labeled proteins, which could hinder ensuing chromatography or fractionation steps. For example, the labeling of myoglobin with ICPL reduces the *pI* and altered the migration toward the acidic side in 2D PAGE [33]. In addition, a complete labeling of all of the resulting peptides is virtually impossible. For these reasons peptide tagging has become a widely used approach in quantitative proteomics.

4.4 QUANTITATION VIA ISOTOPIC LABELING ON PEPTIDES

In this section we review the common labels that have routinely been used for labeling peptides from complex mixtures. These chemical tags could be used for protein labeling but, to date, have not been utilized or reported for that purpose.

4.4.1 ICAT

ICAT reagents were first applied at the protein stage and used to quantify changes in glucose-mediated protein expression in yeast. Since then, newer versions of ICAT have become available, primarily focusing on peptide-stage labeling. The biotin affinity tag was a key limitation for the first generation ICAT, and modified ICAT reagents have a photo- or acid-cleavable linker to help remove the biotin tag before MS analysis [90]. The adaptation of a solid-phase format allowed stable isotopic labeling along with isolation but with elimination of extra sample cleanup steps [55]. This tactic [35,91], termed reagent acid-labile isotope-coded extractant (ALICE), also targets cysteine-containing peptides. ALICE, though similar to ICAT, in principle, was designed with a maleimido reactive group instead of an iodoacetyl group, and a linker that can incorporate up to 10 deuterium atoms for quantitation by MS. The use of the inert nonbiological acid-labile linker (Sieber amide polymer matrix) is the first significant contribution that greatly decreases nonspecific binding and eliminates extra cleanup steps, thus reducing sample loss. A solid-phase reagent that is specific toward higher amino groups [92] can also be used. Other improvements to ICAT include the replacement of ^{2}H with ^{13}C for stable isotope labeling [93]. Using deuterium for labeling of peptides may lead to a lack of co-elution of standard and unknown (lack of resolved chromatographic isotopic peaks), and introduce errors in quantitation [94]. This deuterium effect becomes more pronounced in peptides with a higher number of deuterium atoms. In contrast, using heavy isotopes ^{13}C or ^{15}N as replacements does not lead to observable changes in isotopic resolution with contemporary LC chromatography.

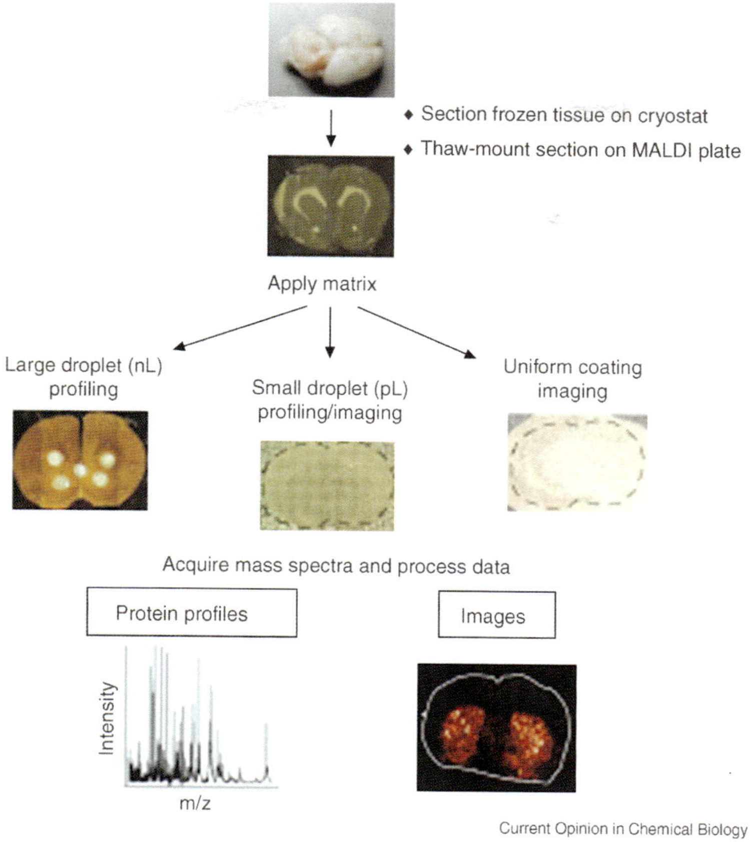

FIGURE 1.2 Typical experimental design for IMS. From [16]. Copyright permission was obtained from Elsevier.

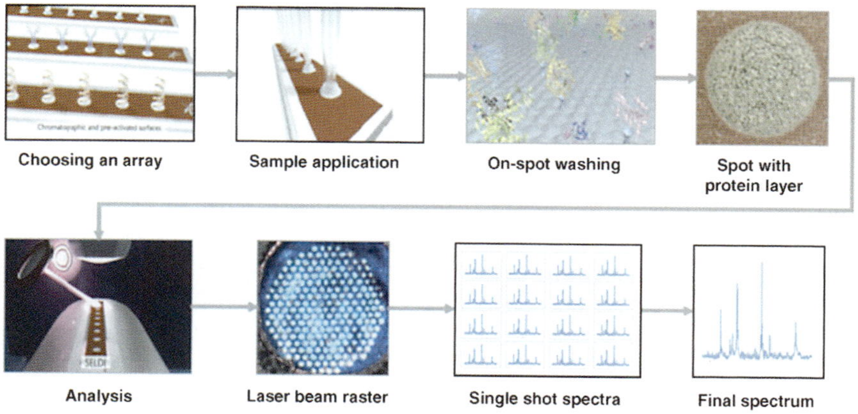

FIGURE 1.4 Experimental steps of SELDI-TOF-MS-based ProteinChip® System. From [53]. Copyright permission was obtained from Nature Publishing Group.

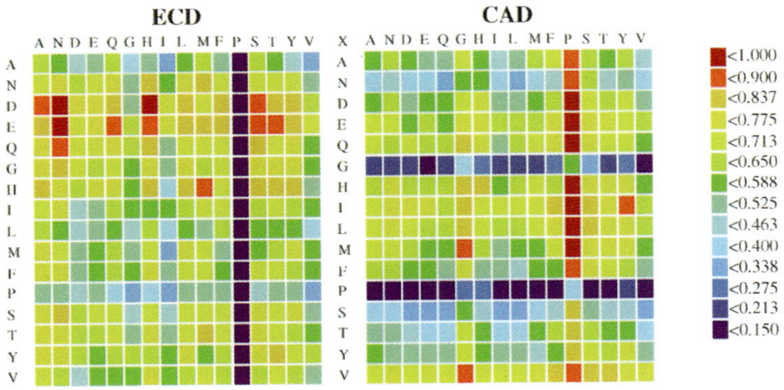

FIGURE 2.6 Amino acid preferences in ~15,000 tandem mass spectra CAD and ECD. (*See text for full caption.*)

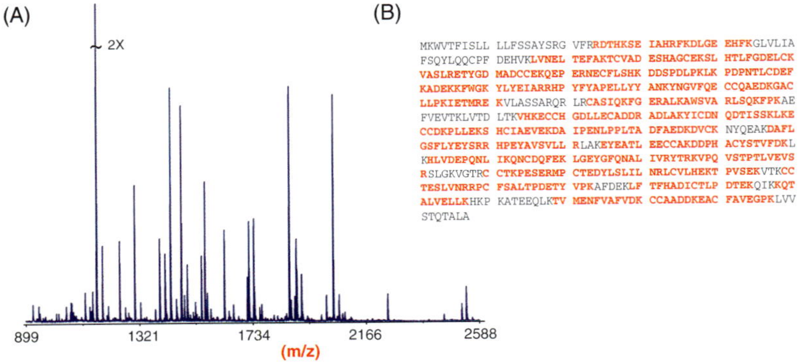

FIGURE 3.2 Peptide fingerprint mapping of bovine serum albumin. (*See text for full caption.*)

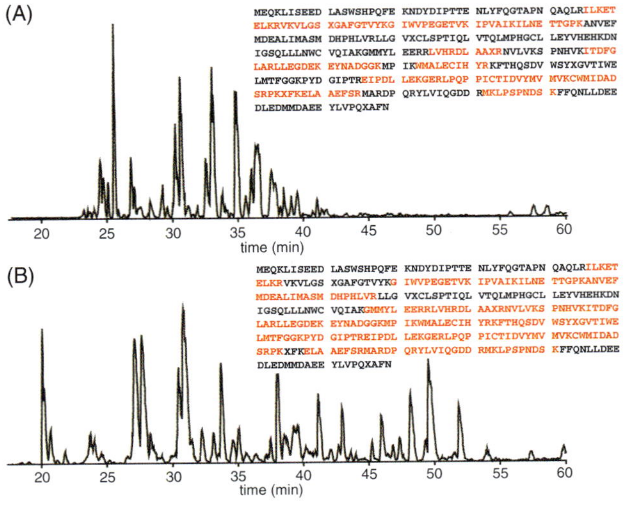

FIGURE 3.4. GeLC-MS/MS versus a shotgun digest. (*See text for full caption.*)

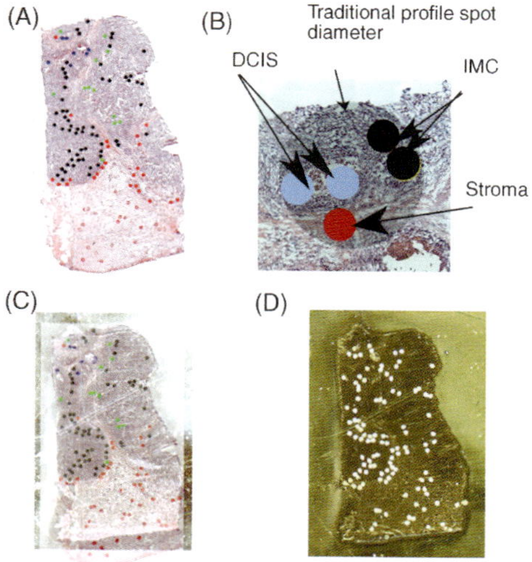

FIGURE 5.1 Histology-directed protein profiling for comparative proteomics. (*A*) H&E stained section of human breast cancer specimen annotated by a pathologist to locate regions of interest: *red,* peritumoral stroma; *black,* IMC; *blue,* DCIS; and *green,* non-tumor epithelium. (*B*) Illustration of the different surface areas profiled by the histology-directed strategy (colored spots) and traditional profiling (shaded area). (*C*) Overlay of the aligned H&E image with the section on the MALDI target plate for matrix spotting. (*D*) Optical image of the section on the MALDI target plate after robotic deposition of matrix onto the designated sites. Reproduced with permission from [15].

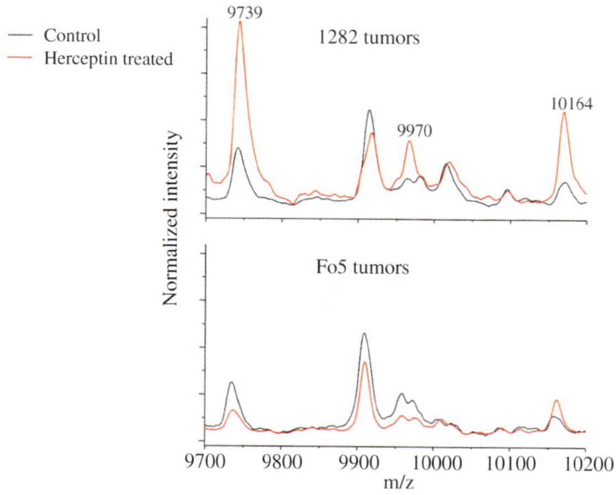

FIGURE 5.2 Drug-induced changes in the proteome predict for therapeutic resistance. Mice bearing Fo5 (Herceptin-resistant) and 1282 (Herceptin-sensitive) tumors were treated with a single dose of Herceptin (30 mg/kg i.p). Tumors were harvested after dosing and subjected to mass spectral proteomic analysis. An example of a statistically significant change observed after Herceptin-treatment in the 1282 tumors that is not observed in the Fo5 tumors is shown (an increase in *m/z* 9212). The sensitive tumor line traces consist of untreated tumors (average of 20 spectra from 6 tumors) and Herceptin-treated tumors (average of 13 spectra from 4 tumors). The resistant tumor line traces consist of untreated tumors (average of 11 spectra from 3 tumors) and Herceptin-treated tumors (average of 20 spectra from 4 tumors). Reproduced with permission from [16].

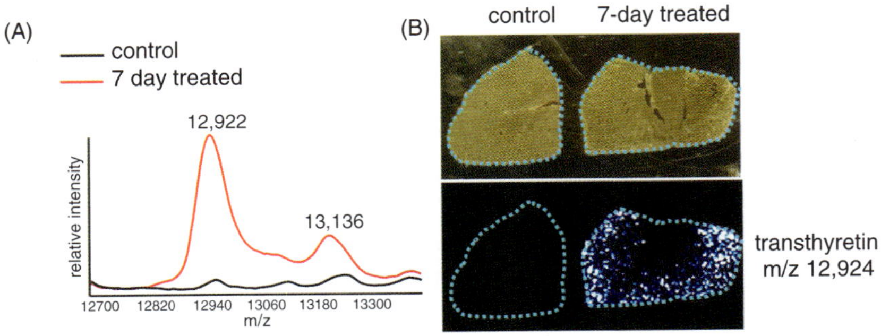

FIGURE 5.3 Drug-induced changes in the proteome correlate with drug-induced toxicity. Monkeys were dosed with a combination of the known nephrotoxicant gentamicin (10 mg/kg) and everninomicin (30 mg/kg) for 7 days. Kidneys were harvested and subjected to mass spectral proteomic analysis. (*A*) A signal at *m/z* 12,922 (subsequently identified as transthyretin) was found to be significantly increased in the dosed kidneys compared to controls. (*B*) High-resolution image analysis of kidneys from one control and one dosed monkey show the transthyretin ion is localized to the cortex of the dosed kidneys.

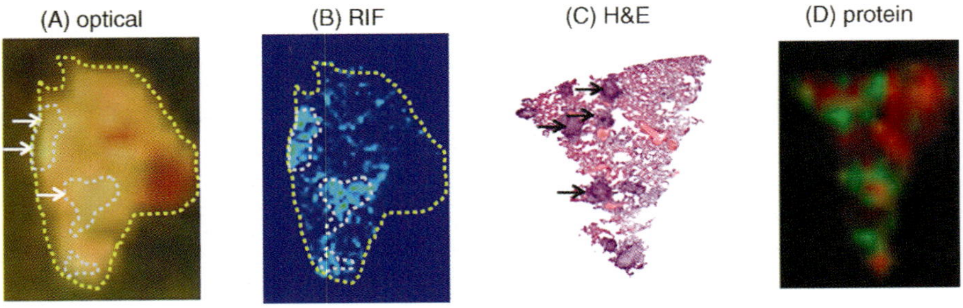

FIGURE 5.4 Examining drug distribution in the granuloma microenvironment in a rabbit model of tuberculosis infection. Rabbits were infected with *M. tuberculosis* and orally dosed with a combination of antituberculosis drugs, including rifampin at 30 mg/kg for 5 days. (*A*) Optical image of an infected rabbit lung section on a gold-coated MALDI target plate. This animal was sacrificed 1 h 5 min after the final dose. Granulomas are indicated with white arrows. (*B*) MALDI MS image of the distribution of rifampin (MS/MS 821→397 + 722) in the lung section shown in *A*. Rifampin appears to localize to granulomas compared to surrounding lung. (*C*) H&E stained serial section of the lung tissue shown in *A*, with granulomas indicated by black arrows. (*D*) MALDI MS protein image showing the localization of *m/z* 11,345 (green) to the granuloma areas and *m/z* 15,787 (red) to adjacent uninvolved tissue.

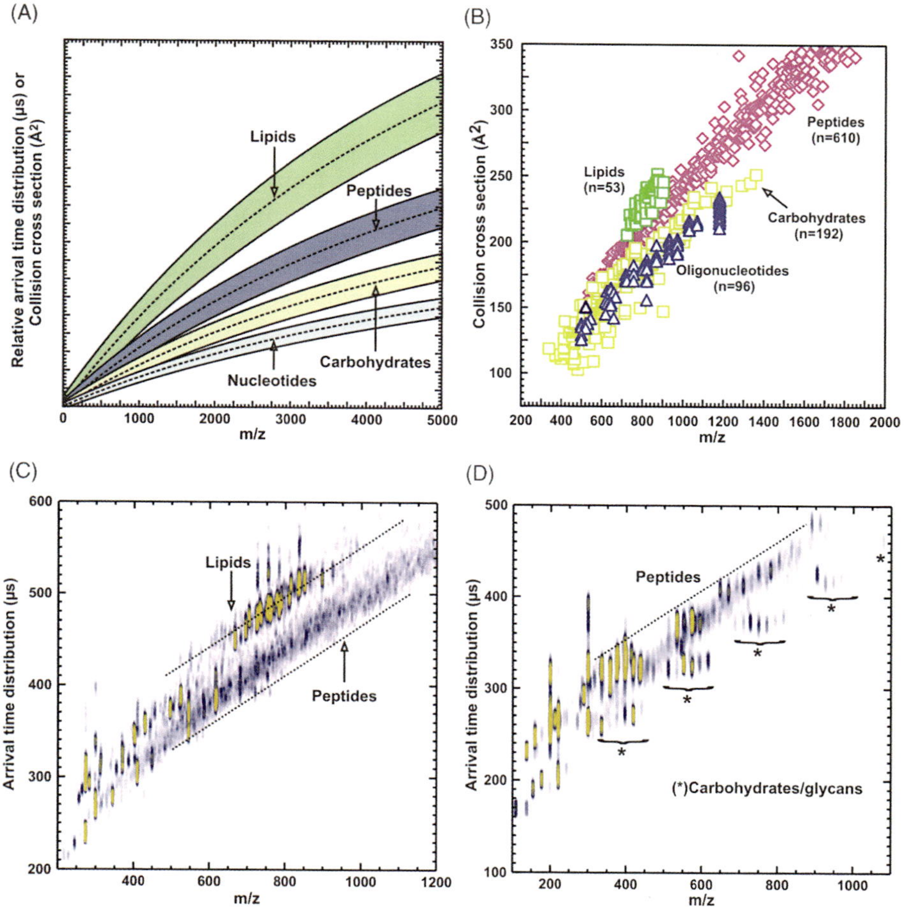

FIGURE 6.4 (*A*) A hypothetical plot highlighting where particular biomolecular classes are expected to appear in conformation space based on differing gas-phase packing efficiencies. (*B*) A plot showing the calculated collision cross-sectional data collected from these biomolecular classes, including oligonucleotides ($n = 96$), carbohydrates ($n = 192$), peptides ($n = 610$), and lipids ($n = 53$). All species correspond to singly charged ions generated using MALDI, where error $\pm 1\sigma$ is generally within the data point. Values for peptide species are from [73]. (*C*) A plot of conformation space illustrating the simultaneous separation of peptides and lipids. (*D*) A plot of conformation space illustrating the simultaneous separation of peptides and carbohydrates. Part (*a*) is adapted with kind permission from Springer Science + Business Media: *Anal. Bioanal. Chem.*, Biomolecular structural separations by ion mobility–mass spectrometry, 391, 2008, 906, L. S. Fenn and J. A. McLean, Fig. 2(a). Part (*b*) is adapted with kind permission from Springer Science + Business Media: *Anal. Bioanal. Chem.*, Characterizing ion mobility–mass spectrometry conformation space for the analysis of complex biological samples, 2009, in press, L. S. Fenn, M. Kliman, A. Mahsutt, S. R. Zhao, and J. A. McLean, Fig. 1(a).

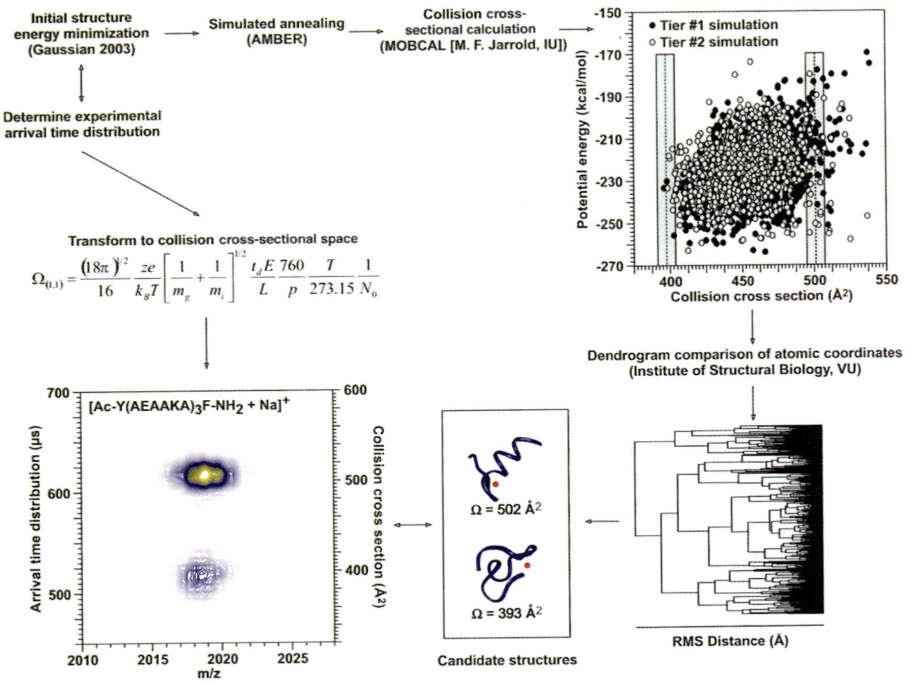

FIGURE 6.5 Modeling protocol used to interpret peptide and protein structure based on the absolute collision cross-sectional measurements acquired from IM-MS data.

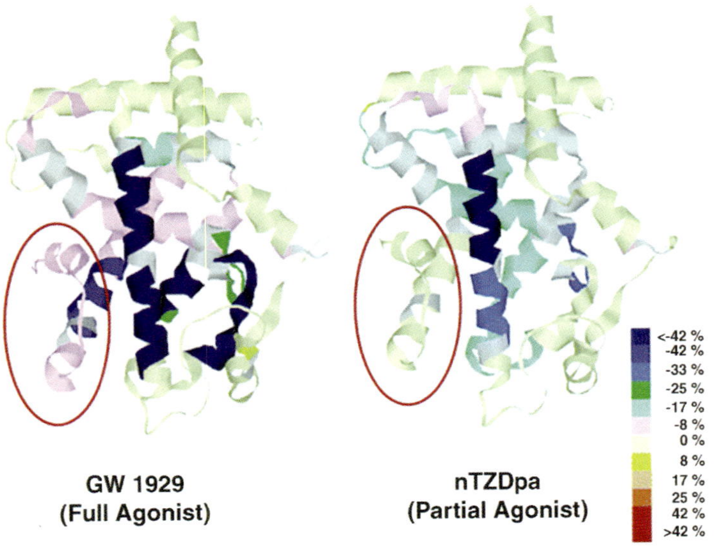

FIGURE 7.4 Change in HDX rate constants as a result of binding a full agonist (*left*) and a partial agonist (*right*) to PPARγ. From M. J. Chalmers et al., Probing protein ligand interactions by automated hydrogen/deuterium exchange mass spectrometry. *Anal Chem* 78(4), 1005–1014. Copyright © 2006 by American Chemical Society. Reprinted by permission of American Chemical Society.

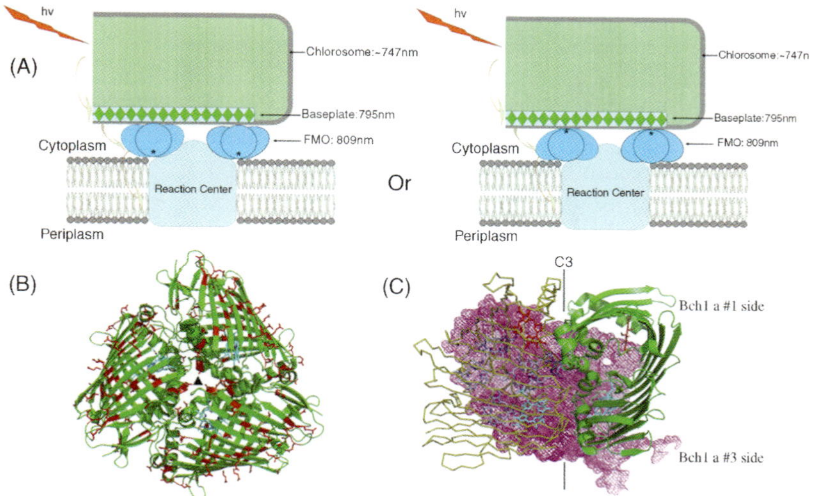

FIGURE 7.18 Photosystem from *C. tepidum* and structure of FMO. (*A*) Model architecture of photosystem from *C. tepidum*. The two possible orientations of FMO on the CM are presented. Bchl a #3 is shown as a star. (*B*) Top view of the FMO trimer with the Bchl a #3 side shown. All the pigments are omitted except Bchl a #3 which is colored cyan. (*C*) Side view of the FMO trimer shown as cartoon, ribbon, and mesh for clarity. Positions of Bchl a #3 (cyan) and Bchl a #1 (red) are labeled in the monomer. From J. Wen et al., Membrane orientation of the FMO antenna protein from Chlorobaculum tepidum as determined by mass spectrometry-based footprinting. *Proc Nat Acad Sci USA* 106(15), 6134–6139.

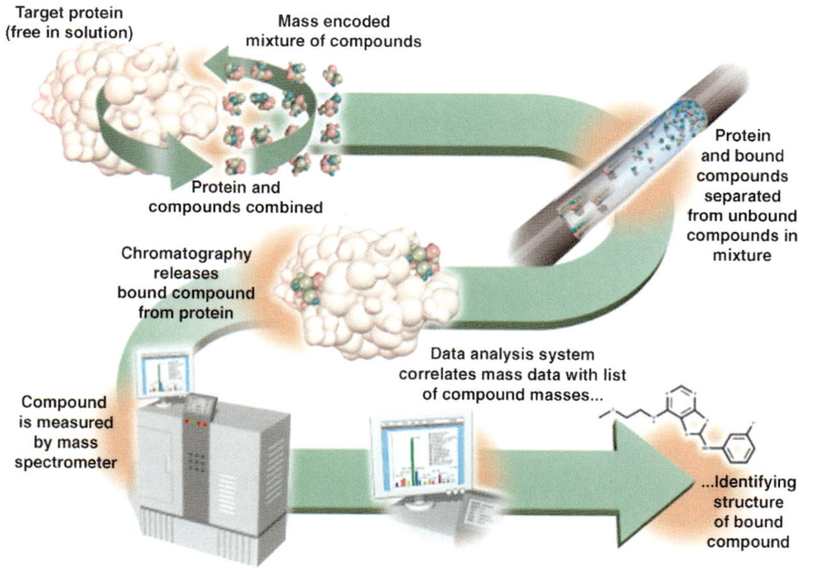

FIGURE 10.1 Schematic of ALIS, the automated ligand identification system that utilizes size-exclusion chromatography coupled online to LC-MS for the study of protein–ligand interactions. Reprinted with permission from [27].

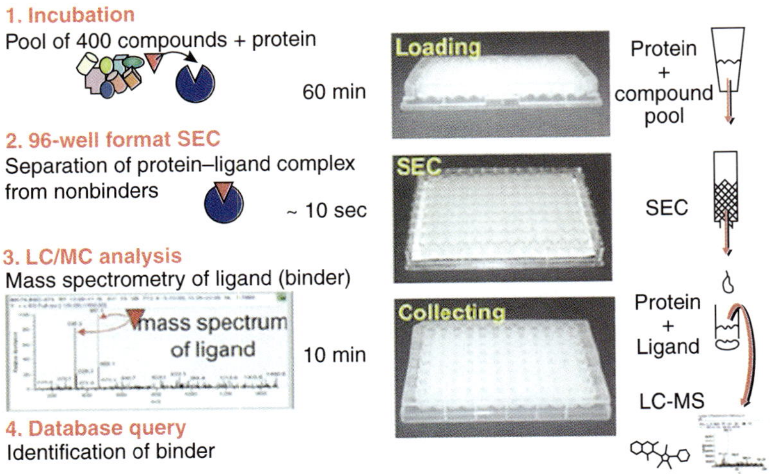

FIGURE 10.3 The basic principles of SpeedScreen technology. The left panel describes the four process steps of incubation, 96w-SEC, LC/MS-analysis, and database query. The right panel depicts the material used for these process steps. Reprinted with permission from [31].

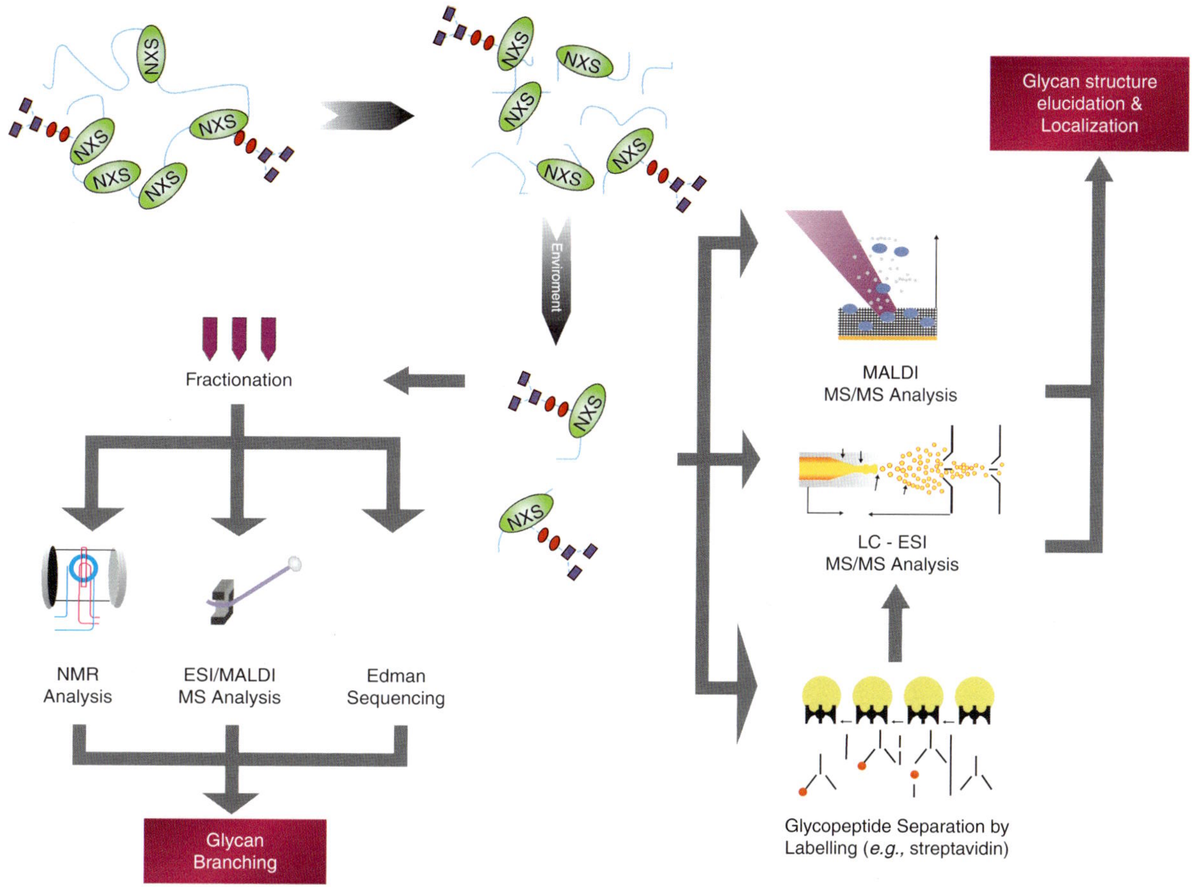

FIGURE 12.6 Scheme of the different analytical approaches employed for the separation and analysis of glycoproteins by LC-ESI and MALDI tandem MS.

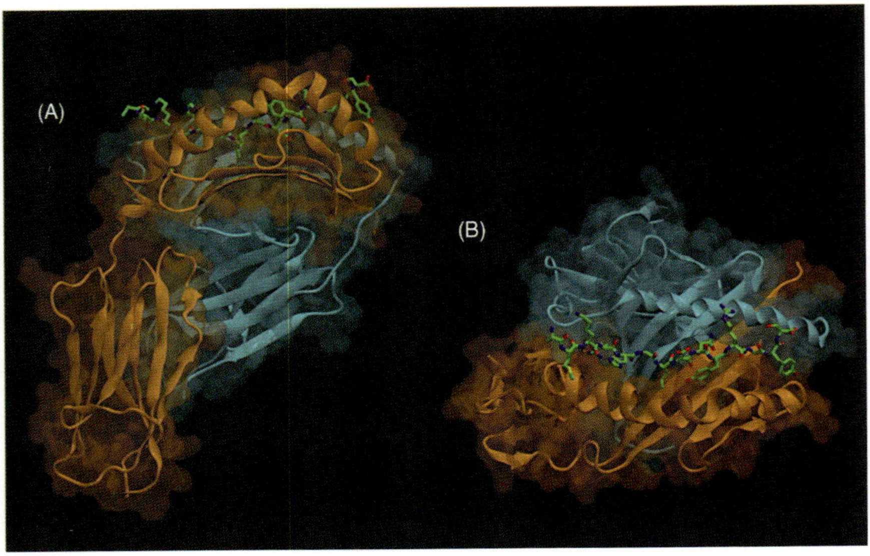

FIGURE 13.1 Murine class II MHC, IAg^7, with an antigenic peptide, HEL11-25, in the binding cleft viewed form the side (*A*) and from above (*B*). The alpha and beta chains of the protein are shown in turqouise and gold, respectively. Note the groove formed by the two alpha helices and the underlying beta sheet. This is structure 1F3J from the Protein Data Bank (www.pdb.org) and was rendered with VMD (www.ks.uiuc.edu/Research/vmd/) by Dr. Manolo Plasencia.

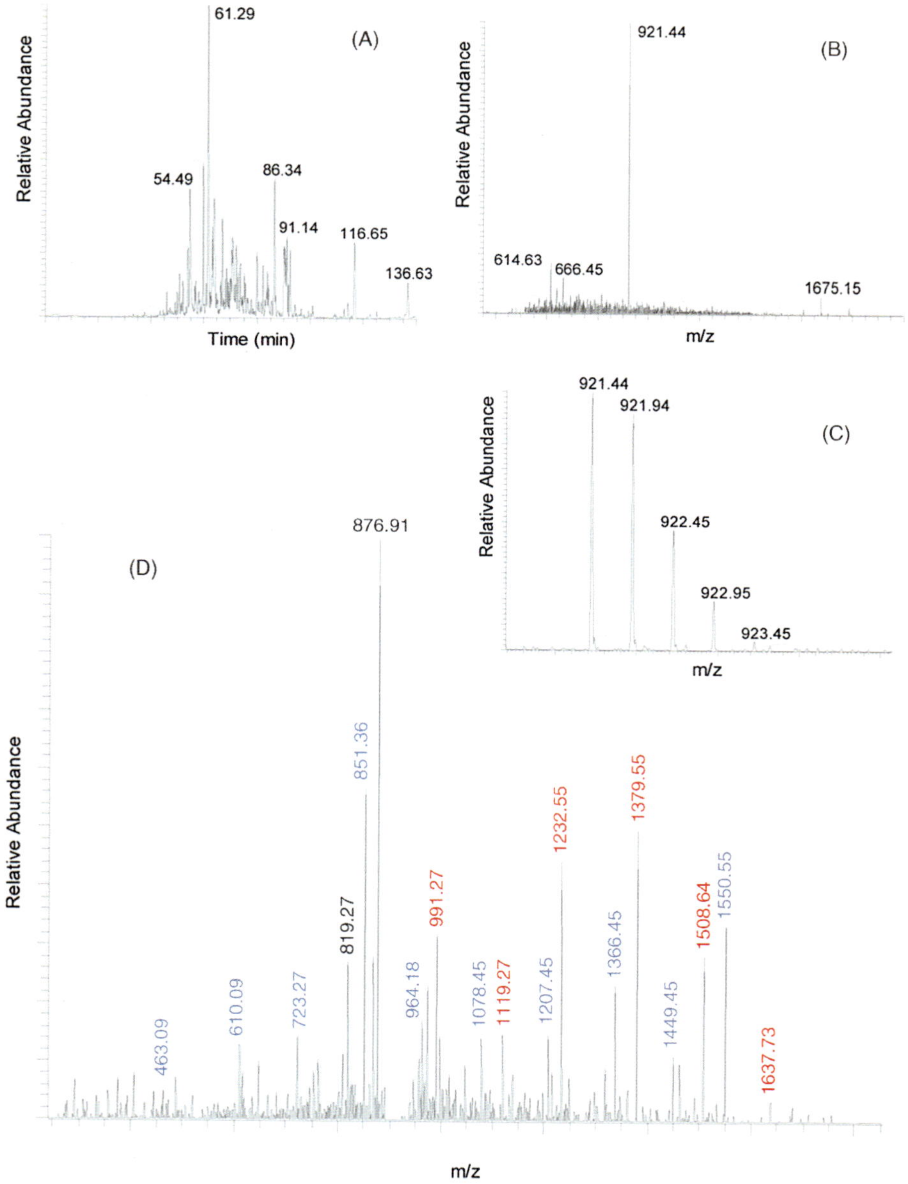

FIGURE 13.2 Proteomics applied to antigenic peptides. (*A*) The base peak chromatogram from a 150-min gradient separation of peptides eluted from the class II MHC IAg[7]. (*B*) A full mass spectrum from 75.11 min in the chromatogram. (*C*) The isotopically resolved peaks in an expanded view of the mass spectrum shown in *B*. (*D*) The MS2 spectrum from the peak at m/z 921.44. A Mascot database search determined that best match to the spectrum was YQTIEENIKIFEEDA from the murine protein ITM2B. The b-ions are shown in red and the y-ions in blue. The other two ions are doubly charged *b*-ions.

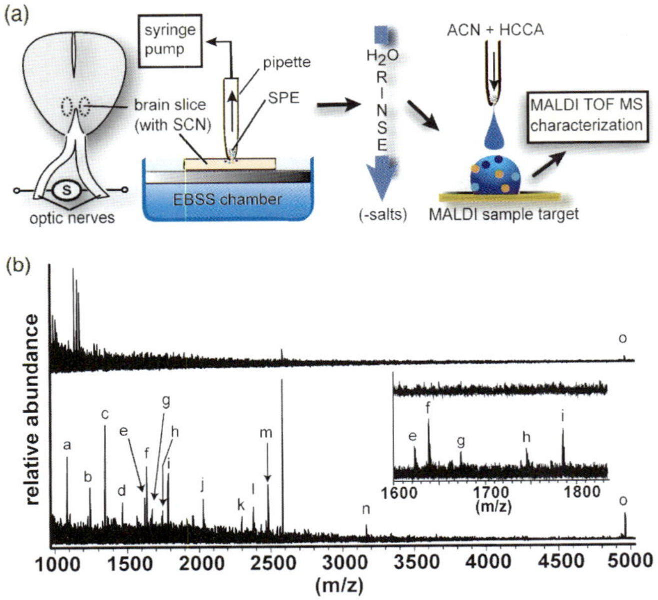

FIGURE 14.4 Schematic workflow of releasate collection and characterization. (*See text for full caption.*)

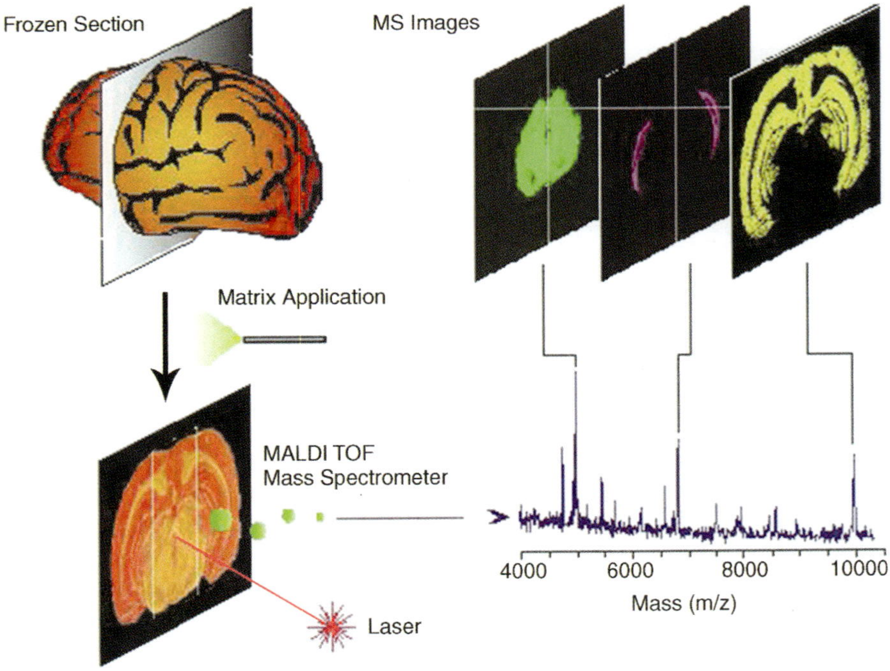

FIGURE 14.5 Schematic workflow of the spatial analysis by MALDI TOF mass spectrometry. (*See text for full caption.*)

4.4.2 iTRAQ

The majority of peptide labeling methods allow for a direct comparison of peptides that have the same sequence but are differentiated by isotopic masses at the MS stage. The development of isobaric tags for relative and absolute quantitation (iTRAQ) is a response; it takes a completely different approach than older methods [44]. Although iTRAQ is based on a chemical tag, like many others created before it, the addition of different iTRAQ reagents to the peptides results in an identical mass after tagging. It is during the MS/MS fragmentation stage that iTRAQ quantitation is achieved by liberating reporter ions that are differentiating because they contain either the "light" or "heavy" label, thus allowing for relative quantitation. This analytical advance is achieved by the unique construction and structure of the iTRAQ labeling reagent and is based on the concept that iTRAQ reagents consist of reporter, balancer, and target groups (Figure 4.3*c*). The target group is *N*-hydroxysuccinimide, which reacts specifically with the ε-amino group of lysines and the *N*-termini of peptides. The reporter group can contain up to eight differentially tagged sites, allowing for detection of mass differences of 1 to 8 Da. These 1-Da differential mass shifts increase the capability for high-throughput analyses, where up to eight samples can be quantified in a single MS experiment. The balancer group is designed to offset the differential reporter masses. The iTRAQ-labeled peptides with the same sequence have the same LC retention times, thus increasing the quantitation accuracy. During the MS/MS stage, peptides containing iTRAQ reagents fragment to generate reporter ions in the *m/z* 114 to 121 range. This range is ideal for identification of reporter ions because there is little background noise and essentially no peptide fragment ions.

Another isobaric chemical tagging reagent, called tandem mass tags (TMT), was designed and tested at approximately the same time as iTRAQ [38]. TMT can be used as a multiplexing strategy allowing up to six samples to be analyzed simultaneously. Cleavable isobaric labeled affinity tag (CILAT) is another chemical tag that is part of a strategy similar to that of iTRAQ, but the reagent contains an affinity tag to allow enrichment of labeled peptides [52]. iTRAQ and ICAT are, at this time, perhaps the most widely used chemical labeling tags in proteomics research and by investigators in other biomedical fields who seek novel protein identification and quantitation.

4.4.3 SoPIL

SoPIL is another approach to quantitative proteomics [24]. The new reagent uses a water-soluble nanopolymer as a support for both chemical derivatization and subsequent capture of cysteine-containing peptides from a complex mixture. This appears to be the first use of nanopolymers for proteomics, where a dendrimer can be employed as the reactant and a soluble support to capture phosphorylated peptides [95]. The nanopolymer has many unique features that can be useful in biological applications; those features include high structural and chemical homogeneity, compact spherical shape, high branching, controlled surface functionalities, and capacity to permeate cells [96]. The solubility of the dendrimer allows for a complete and efficient capture due to the homogenous reaction conditions.

The foundation of the SoPIL reagent is a polyamino-amine (PAMAM) generation-4 dendrimer functionalized with a bromo-aceto reactive group for cysteine-specific capture, 5-(2-formyl-3,5-dimethoxyphenoxy)pentanoic acid as the acid-cleavable linker, and aniline ^{12}C or ^{13}C as the isotope tag (Figure 4.3*d*). Once the cysteine peptides are tagged by SoPIL, the complexes (SoPIL/peptide) are captured by azide, solid-phase beads. This is achieved by functionalizing the dendrimer with a pentyl group that serves as a handle to allow the use of the click chemistry (copper-catalyzed azide/alkyne cycloaddition) and give a fast and efficient reaction. The labeled peptides can be liberated by elution with 90% trifluoric acetic acid (TFA). The majority of methods based on small chemical labeling require additional purification procedures (e.g., reagent removal or desalting steps) that can result in severe sample loss. In contrast, most solid-phase methods result in nonlinear kinetics caused by the heterogeneous (two-phase) reaction conditions. SoPIL combines efficient labeling and isolation of target cysteine peptides in a homogeneous environment with no need for extra desalting steps. The usefulness of SoPIL in complex sample quantitation, is demonstrated in the expression-level quantification of highly complex protein mixtures from the venoms of *Crotalus scutulatus scututlaus* type A and B, *Crotalus oreganus helleri*, and *Bothrops colombiensis* [97]. The use of SoPIL permits the identification, in a single experiment, of Mojave toxin, a known neurotoxin, and quantification of the increased expression levels of hemorrhagic proteases, which are found exclusively in the venoms of *C. s. scutlatus* type A and type B, respectively.

4.4.4 Absolute Quantitation

Absolute quantitation of proteins has become essential for biomarker discovery and is now emerging as a promising opportunity for proteomics. The current methods are based on the same principles as the stable isotope dilution method wherein internal standards are used as reference markers for quantification by mass spectrometry. In absolute quantitation, standard peptides are synthesized de novo, and stable isotopes are incorporated to allow differential detection in MS. A standard curve is generated by using known amounts of internal standards (represented peptide) (Figure 4.4).

Several strategies can be used to afford a standard procedure for absolute quantitation (Table 4.3). In one approach, called AQUA for proteins and posttranslational modifications [98], protein of interest and one of the resulting peptides to represent it in the MS analyses are selected. The peptide is synthesized by using a standard solid-phase based peptide synthesis; this peptide is a mimic of a peptide digested from the sample protein. A leucine is used for stable isotope incorporation, allowing a 7-Da shift (six ^{13}C and one ^{15}N). The AQUA standard is added to the cell lysate before the digestion procedure, and the peptide/protein of interest is quantified according to the amount of the synthesized or internal-standard peptide introduced. The method was validated by AQUA synthesis of myoglobin in a yeast background.

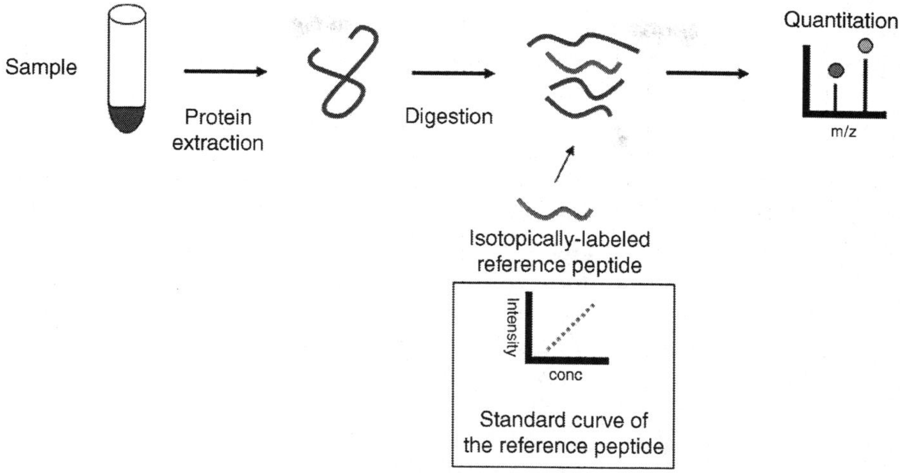

FIGURE 4.4 Example of a sample workflow for absolute quantitation. Proteins are extracted from a biologically significant source (e.g., serum or plasma). Samples are pre-run or known protein targets are selected. A reference peptide mimicking a peptide from the target protein is synthesized de novo and used to generate a statistical linear relationship of concentration in regard to ion intensity. The proteins are digested, and the reference peptide is "spiked" into the sample for quantitation by MS.

Although the AQUA approach is robust and potentially universal to quantify any protein, its ability to quantify multiple proteins in complex mixtures in high-throughput analyses is limited owing to the need to synthesize all of the peptides of interest de novo and analyze them independently. A multiplexed approach called QCAT (concatenations of Q peptides (standard peptides)) circumvents this limitation [99,100]. QCAT is a strategy that uses engineered artificial proteins that are concatenations of Q peptides for the proteins of interest. The protein of interest is recombinantly synthesized and metabolically labeled in selected media in *E. coli*, then grown and isolated. The isotopically labeled proteins are mixed with the real

TABLE 4.3 Selected Methods for Absolute Quantitation

	Method†	Stable Isotope	References
Absolute Quantitation			
	AQUA	^{2}H, ^{13}C, ^{15}N	98
	QCAT	^{2}H, ^{13}C, ^{15}N	99
	QconCAT	^{2}H, ^{13}C, ^{15}N	100
	SISCAPA	^{2}H, ^{13}C, ^{15}N	101
	iTRAQ	^{13}C	44

†AQUA - Absolute quantification, QCAT - Multiplexed absolute quantification, QCONCAT - Multiplexed absolute quantification using concatenate signature, SISCAPA - Stable isotope standards and capture by anti-peptide antibodies, iTRAQ - isobaric tag for relative and absolute quantification.

sample and quantified accordingly. As a proof of principle, twenty proteins in skeletal muscle cells from the organism *Gallus gallus* were quantified.

Owing to the wide concentration range of proteins in cells and biofluids, an enrichment step can be added to remove unwanted, high-abundance proteins. For this purpose a novel absolute quantitation technique called SISCAPA (stable isotope standard with capture by antipeptide antibodies) is now available [101]. This method relies on four steps: (1) digestion of proteins, (2) addition of an internal standard that is labeled with stable isotopes, (3) enrichment of low-abundant peptides using immobilized antibodies, and (4) quantitation by MS.

Absolute quantitation has become a robust and useful approach for proteomic quantitation by MS. The strategy relies heavily on the use of de novo generation of peptide standards for quantitative comparisons. Therefore, not only must target molecules be selected, but also the correct peptide representation needs to be used for analyses. Protein digestion by trypsin or other proteolytic enzymes must generate the correct *N* and *C* termini after digestion so that accurate quantitative comparison with standard peptides can be achieved. This can be a limitation in real sample analyses where the predicted digestion may depend on the nature of the protein and sample conditions. One should consider a pre-run of the complex mixture proteins to determine the peptides that should be synthesized de novo. This limits absolute quantitation as an unbiased comprehensive global-scale proteomics approach. In addition, unknown, or even known, post-translational modifications of target molecules may disallow accurate quantitation. To address this issue, one can analyze "proteotryptic" peptides according to their physiochemical properties (e.g., charge, hydrophobicity, length, and amino-acid composition) to identify a number of parameters that help the selection of representative peptides for MS-based absolute quantification [102].

4.5 LABEL-FREE QUANTITATION

Although quantitative proteomics research has produced novel and versatile chemical-tagging strategies for determination of protein expression changes in organisms, the obvious shortcomings of isotope labeling based methods are the introduction of extra steps during sample preparation and the small number of samples that can be analyzed simultaneously. A number of quantitative and semiquantitative strategies are now available; they are based on and include, but not limited to, statistical treatment and chromatographic retention time, or peak area, to generate direct determination of protein expression changes from multiple sample runs (Figure 4.5). These methods, referred to as label-free quantitation, are alternative methods to isotopic-labeling approaches for quantitative proteomics [103] (Table 4.4). The current detection schemes are based on measuring ion signal intensity [104], counting spectra, counting number of fragments for a particular peptide [105], and/or using abundance indicators for normalization [106–109].

Ion intensity measurement, or extracted ion counting (XIC), is the systematic counting of selected ions of a particular *m/z* plotted over elution time, resulting in a

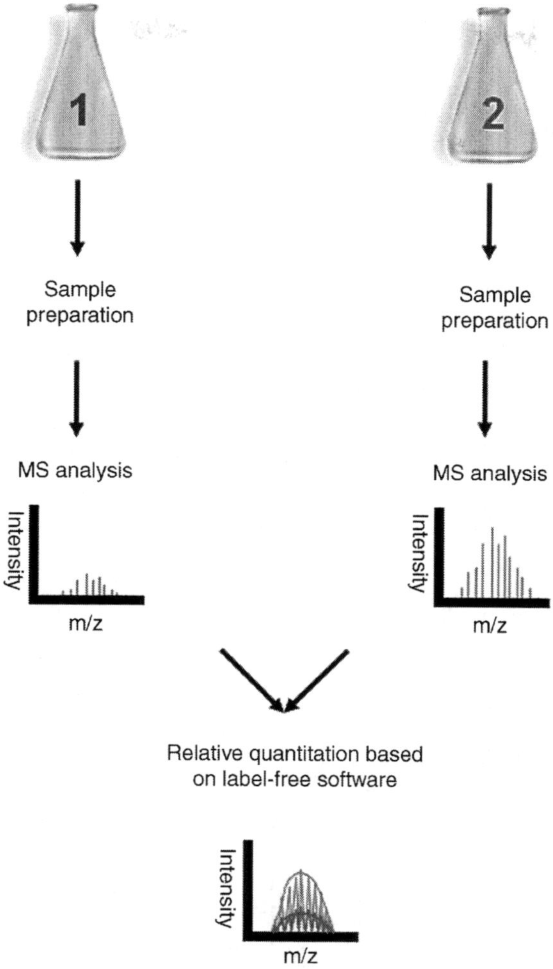

FIGURE 4.5 Example of a sample workflow for label-free quantitation. Two biological samples are selected. Separately, proteins are extracted and digested with proteases, followed by MS analyses. Relative quantitation of the two samples is achieved by integrating the ion current (or spectral counting using MS/MS) of each MS generated form a peptide and its represented protein.

peak area. The ion signal intensity can then be compared to that of a control ion with the same *m/z*. To achieve this in a complex mixture, the peptides must be separated by chromatography methods. This allows for ion extraction and peak selection from individual LC runs to be integrated over the time of the chromatographic separation (Figure 4.5) [110–112]. Ion peak intensities do correlate with protein abundances in complex mixtures [113]. Because variation can occur between multiple sample runs owing to differences in sample handling and loading, performance, and MS ionization efficiency, normalization is required [114]. Correction for variability can be reduced

TABLE 4.4 Selected Label-free Strategies Currently Used in Quantitative Proteomics

	Method†	References
Label-Free Quantitation		
	SpS	105
	XIC	111
	PAI	106
	emPAI	115
	PMSS	116

†SpS - Spectra sampling, XIC - Extracted ion current, PAI - Protein abundance index, emPAI - Exponentially modified protein abundance index, PMSS - Probabilistic peptide scores.

by normalizing each peak intensity by the sum or median of all peak intensities over the chromatographic run [117]. Normalizing the peptide signal intensity by the sum of the total ion signal intensities and correcting the signal intensity in various fractions can reduce variability for the population mean, thus reducing the standard deviation [114]. One disadvantage of signal intensity measurements is the balance of MS acquisition that allows for identification and quantitation of ions during a run. Therefore it is necessary for equal ion acquisition of survey (MS) and fragments (MS/MS) to be optimized for maximum high-throughput usage. Ion trap based instruments have a greater advantage compared to quadrupole instruments because they are able to acquire a spectrum every 0.2 s or even less, whereas quad-TOF could take up to 3 s/spectrum [108,118–121].

A new statistical method called spectrum sampling (SpS) (often referred to as spectral counting) [105] relies on counting the identifying spectra for each protein. Spectral counting analysis is of the number of product-ion (MS/MS) spectra obtained for each peptide. It is based on the assumption that the higher number of product-ion spectra produced by a peptide from a single run infers higher protein abundance. By comparing the spectral numbers from two samples, it generates a semiquantitative strategy for the analysis of protein-expression changes [121]. Spectral counting has fair reproducibility and shows good correlation with protein changes [105]. In addition an increase in spectral number could result in an increase in accuracy and a decrease in false positives. An advantage of spectral counting is the correlation with protein abundance is linear [114], and this is also observed for ion intensity peak area or peak height. One additional benefit for spectral counting when compared to ion intensity-based quantitation is the the MS/MS acquisition over a single chromatographic time allows for both identification and quantification. A comparison study demonstrates that spectral counting correlates well with relative protein abundances [114]. Spectral counting is also sufficiently sensitive to determine a small, 2-fold change in protein expression levels. By using the spectral counting approach, one can measure more drastic abundance changes throughout a wider range of protein expression levels than by using any other quantitative approach (2–100-fold). Ideally, spectral counting can provide a simple way for relative protein quantitation, but despite the advantages of the method, there are a number of drawbacks that remain to be addressed. For instance, high-abundance proteins that provide many spectra during the analyses can cause saturation effects and impair quantification attempts. Another disadvantage is that high-abundance proteins can mask low-abundance proteins, severely limiting the number of quantifiable proteins, particularly when comparing, for example, signaling (low-abundance) versus "housekeeping" proteins [105,114,122].

Steady efforts have been made to identify a more accurate label-free protein/peptide quantitation strategy. For example, similar to SpS, the peptide-matching score summation (PMSS) is a label-free technique that assumes ideal scoring for proteins as the summative of the identification scores of their constituent peptides freed upon digestion. A higher score represents a more abundant protein [116]. Another method, PAI or Protein Abundance Indices, is semiquantitative but a more reliable indicator for abundance than PMSS. The approach is also based on the assumption that an increasing number of identified peptides indicate an increasing protein abundance [106]. This method can be modified by developing emPAI (exponentially modified protein abundance index), which uses a logarithmic relationship between the number of signature peptides and the protein abundance represented by these peptides [115]. Probabilistic peptide identification scores can be used for differential proteomic analyses by combining PMSS and SpS with a statistical validation method called LPET (local pool error test) [109]. This method is able to quantify differences in purified proteins and serum samples with 2- to 5-fold changes with 90-95% confidence.

Currently label-free methods are still a controversial means for protein quantitation using MS, especially if follow-up experiments are not performed to validate the initial quantitative data. Chemical labeling approaches are generally more accurate than label-free methods, but label-free quantitation provides a potentially simpler approach to analyze a large number of samples in parallel [123]. With the continued development of statistics and bioinformatics approaches, label-free quantitation is expected to continue improving steadily.

4.6 CONCLUSIONS

Quantitative proteomics has evolved immensely, contingent on novel tagging strategies, new high resolving power mass spectrometers, and advanced software [124]. Progress toward the goal of uncovering the vast genomes and proteomes of lower and higher order organisms is on a rapid pace. MS-based proteomics started with high promise, but it has not yet fulfilled it. A main limitation is the lack of quantitative capability. Applying a successful quantitative proteomic strategy to make accurate measurements for novel discoveries is challenging owing to varying circumstances, including the species investigated (e.g., the swine genome has not been sequenced to date), sample state (e.g., cell, tissue, or serum), reagents used (e.g., ICAT or iTRAQ), and mass spectrometers utilized. Each circumstance has some advantages and drawbacks. In addition, due to the dynamic range of cell and serum samples and detection limits of current mass spectrometers, enrichment, depletion, or multidimensional chromatography must be implemented beforehand. To detect such low-abundance proteins, the aforementioned fractionation strategies must be applied, often at the cost of accurate quantitation. Multifractionation may distort the naturally existing ratios between two samples, resulting in false or biased information. Integrated efforts on the development of technologies, chemistries, methods, and software will continue to advance the young field of quantitative proteomics and allow proteomics to live up to its promise.

ACKNOWLEDGMENT

This work has been funded in part by an NSF CAREER award, a 3M general fund (WAT), and by National Institutes of Health Grants, S10RF025044 and R21RR025802.

REFERENCES

1. Tao, W. A., Aebersold, R. (2003). Advances in quantitative proteomics via stable isotope tagging and mass spectrometry. *Curr Opin Biotechnol 14*, 110–118.
2. Tao, W. A. (2007). Soluble polymer-based isotopic labeling (SoPIL): A new strategy to discover protein biomarkers? *Expert Rev Proteomics 4*(5), 603–607.
3. Aebersold, R., Mann, M. (2003). Mass spectrometry-based proteomics. *Nature 422*, 198–207.
4. Schoenheimer, R., Rittenberg, D. (1935). Deuterium as an indicator in the study of intermediary metabolism. *Science 82*, 156–157.
5. Schoenheimer, R., Rittenberg, D., Foster, G. L., Keston, A. S., Ratner, S. (1938). The application of the nitrogen isotope N15 for the study of protein metabolism. *Science 88*, 599–600.
6. Arsene, C. G., Ohlendorf, R., Burkitt, W., Pritchard, C., Henrion, A., O'Connor, G., Bunk, D. M., Guttler, B. (2008). Protein quantification by isotope dilution mass spectrometry of proteolytic fragments: Cleavage rate and accuracy. *Anal Chem 80*, 4154–4160.
7. Chaerkady, R., Harsha, H. C., Nalli, A., Gucek, M., Vivekanandan, P., Akhtar, J., Cole, R. N., Simmers, J., Schulick, R. D., Singh, S., Torbenson, M., Pandey, A., Thuluvath, P. J. (2008). A quantitative proteomic approach for identification of potential biomarkers in hepatocellular carcinoma. *J Proteome Res 7*, 4289–4298.
8. Gronborg, M., Kristiansen, T. Z., Iwahori, A., Chang, R., Reddy, R., Sato, N., Molina, H., Jensen, O. N., Hruban, R. H., Goggins, M. G., Maitra, A., Pandey, A. (2006). Biomarker discovery from pancreatic cancer secretome using a differential proteomic approach. *Mol Cell Proteomics 5*, 157–171.
9. Okamura, N., Masuda, T., Gotoh, A., Shirakawa, T., Terao, S., Kaneko, N., Suganuma, K., Watanabe, M., Matsubara, T., Seto, R., Matsumoto, J., Kawakami, M., Yamamori, M., Nakamura, T., Yagami, T., Sakaeda, T., Fujisawa, M., Nishimura, O., Okumura, K. (2008). Quantitative proteomic analysis to discover potential diagnostic markers and therapeutic targets in human renal cell carcinoma. *Proteomics 8*, 3194–3203.

10. Gygi, S. P., Rochon, Y., Franza, B. R., Aebersold, R. (1999). Correlation between protein and mRNA abundance in yeast. *Mol Cell Biol 19*, 1720–1730.

11. Anderson, L. (2005). Candidate-based proteomics in the search for biomarkers of cardiovascular disease. *J Physiol 563*(Pt 1) 23–60.

12. Ong, S. E., Blagoev, B., Kratchmarova, I., Kristensen, D. B., Steen, H., Pandey, A., Mann, M. (2002). Stable isotope labeling by amino acids in cell culture, SILAC, as a simple and accurate approach to expression proteomics. *Mol Cell Proteomics 1*, 376–386.

13. Ishihama, Y., Sato, T., Tabata, T., Miyamoto, N., Sagane, K., Nagasu, T., Oda, Y. (2005). Quantitative mouse brain proteomics using culture-derived isotope tags as internal standards. *Nat Biotechnol 23*, 617–621.

14. Oda, Y., Huang, K., Cross, F. R., Cowburn, D., Chait, B. T. (1999). Accurate quantitation of protein expression and site-specific phosphorylation. *Proc Natl Acad Sci USA 96*, 6591–6596.

15. Dieterich, D.C., A.J. Link, J. Graumann, D.A. Tirrell, and E.M. Schuman. (2006). Selective identification of newly synthesized proteins in mammalian cells using bioorthogonal noncanonical amino acid tagging (BONCAT). *Proc Natl Acad Sci U S A, 103*, 9482–9487.

16. Schaff, J. E., Mbeunkui, F., Blackburn, K., Bird, D. M., Goshe, M. B. (2008). SILIP: A novel stable isotope labeling method for in planta quantitative proteomic analysis. *Plant J 56*, 840–854.

17. Rose, K., M.G. Simona, R.E. Offord, C.P. Prior, B. Otto, and D.R. Thatcher. (1983). A new mass-spectrometric C-terminal sequencing technique finds a similarity between gamma-interferon and alpha 2-interferon and identifies a proteolytically clipped gamma-interferon that retains full antiviral activity. *Biochem J, 215*, 273–277.

18. Liu, T., W.J. Qian, E.F. Strittmatter, D.G. Camp, 2nd, G.A. Anderson, B.D. Thrall, and R. D. Smith. (2004). High-throughput comparative proteome analysis using a quantitative cysteinyl-peptide enrichment technology. *Anal Chem, 76*, 5345–5353.

19. Gygi, S. P., Rist, B., Gerber, S. A., Turecek, F., Gelb, M. H., Aebersold, R. (1999). Quantitative analysis of complex protein mixtures using isotope-coded affinity tags. *Nat Biotechnol 17*, 994–999.

20. Olsen, J.V., J.R. Andersen, P.A. Nielsen, M.L. Nielsen, D. Figeys, M. Mann, and J.R. Wisniewski. (2004). HysTag–a novel proteomic quantification tool applied to differential display analysis of membrane proteins from distinct areas of mouse brain. *Mol Cell Proteomics, 3*, 82–92.

21. Pasquarello, C., J.C. Sanchez, D.F. Hochstrasser, and G.L. Corthals. (2004). N-t-butyliodoacetamide and iodoacetanilide: two new cysteine alkylating reagents for relative quantitation of proteins. *Rapid Commun Mass Spectrom, 18*, 117–127.

22. Sebastiano, R., A. Citterio, M. Lapadula, and P.G. Righetti. (2003). A new deuterated alkylating agent for quantitative proteomics. *Rapid Commun Mass Spectrom, 17*, 2380–2386.

23. Gartner, C.A., J.E. Elias, C.E. Bakalarski, and S.P. Gygi. (2007). Catch-and-release reagents for broadscale quantitative proteomics analyses. *J Proteome Res, 6*, 1482–1491.

24. Guo, M., Galan, J., Tao, W. A. (2007). A novel quantitative proteomics reagent based on soluble nanopolymers. *Chem Commun (Camb) (12)* 1251–1253.

25. Ahrends, R., Pieper, S., Kuhn, A., Weisshoff, H., Hamester, M., Lindemann, T., Scheler, C., Lehmann, K., Taubner, K., Linscheid, M. W. (2007). A metal-coded affinity tag approach to quantitative proteomics. *Mol Cell Proteomics 6*, 1907–1916.

26. Rivera-Monroy, Z., Bonn, G. K., Guttman, A. (2008). Fluorescent isotope-coded affinity tag (FCAT). I: Design and synthesis. *Bioorg Chem 36*, 299–311.
27. Hoang, V.M., T.P. Conrads, T.D. Veenstra, J. Blonder, A. Terunuma, J.C. Vogel, and R.J. Fisher. (2003). Quantitative proteomics employing primary amine affinity tags. *J Biomol Tech*, *14*, 216–223.
28. Lu, Y., P. Bottari, F. Turecek, R. Aebersold, and M.H. Gelb. (2004). Absolute quantification of specific proteins in complex mixtures using visible isotope-coded affinity tags. *Anal Chem*, *76*, 4104–4111.
29. Asara, J.M., X. Zhang, B. Zheng, H.H. Christofk, N. Wu, and L.C. Cantley. (2006). In-Gel Stable-Isotope Labeling (ISIL): a strategy for mass spectrometry-based relative quantification. *J Proteome Res*, *5*, 155–163.
30. Panchaud, A., Hansson, J., Affolter, M., Bel Rhlid, R., Piu, S., Moreillon, P., Kussmann, M. (2008). ANIBAL, stable isotope-based quantitative proteomics by aniline and benzoic acid labeling of amino and carboxylic groups. *Mol Cell Proteomics 7*, 800–812.
31. Angel, P.M. and R. Orlando. (2007). Quantitative carbamylation as a stable isotopic labeling method for comparative proteomics. *Rapid Commun Mass Spectrom*, *21*, 1623–1634.
32. Brancia, F.L., M.E. Openshaw, and S. Kumashiro. (2002). Investigation of the electrospray response of lysine-, arginine-, and homoarginine-terminal peptide mixtures by liquid chromatography/mass spectrometry. *Rapid Commun Mass Spectrom*, *16*, 2255–2259.
33. Schmidt, A., Kellermann, J., Lottspeich, F. (2005). A novel strategy for quantitative proteomics using isotope-coded protein labels. *Proteomics 5*, 4–15.
34. Kuyama, H., M. Watanabe, C. Toda, E. Ando, K. Tanaka, and O. Nishimura. (2003). An approach to quantitative proteome analysis by labeling tryptophan residues. *Rapid Commun Mass Spectrom*, *17*, 1642–1650.
35. Qiu, Y., Sousa, E. A., Hewick, R. M., Wang, J. H. (2002). Acid-labile isotope-coded extractants: a class of reagents for quantitative mass spectrometric analysis of complex protein mixtures. *Anal. Chem. 74*, 4969–4979.
36. Shi, Y., R. Xiang, J.K. Crawford, C.M. Colangelo, C. Horvath, and J.A. Wilkins. (2004). A simple solid phase mass tagging approach for quantitative proteomics. *J Proteome Res*, *3*, 104–111.
37. Shen, M., L. Guo, A. Wallace, J. Fitzner, J. Eisenman, E. Jacobson, and R.S. Johnson. (2003). Isolation and isotope labeling of cysteine- and methionine-containing tryptic peptides: application to the study of cell surface proteolysis. *Mol Cell Proteomics*, *2*, 315–324.
38. Thompson, A., Schafer, J., Kuhn, K., Kienle, S., Schwarz, J., Schmidt, G., Neumann, T., Johnstone, R., Mohammed, A. K., Hamon, C. (2003). Tandem mass tags: A novel quantification strategy for comparative analysis of complex protein mixtures by MS/MS. *Anal Chem 75*, 1895–1904.
39. Wang, S. and F.E. Regnier. (2001). Proteomics based on selecting and quantifying cysteine containing peptides by covalent chromatography. *J Chromatogr A*, *924*, 345–357.
40. Ji, J., A. Chakraborty, M. Geng, X. Zhang, A. Amini, M. Bina, and F. Regnier. (2000). Strategy for qualitative and quantitative analysis in proteomics based on signature peptides. *J Chromatogr B Biomed Sci Appl*, *745*, 197–210.

41. Che, F.Y. and L.D. Fricker. (2002). Quantitation of neuropeptides in Cpe(fat)/Cpe(fat) mice using differential isotopic tags and mass spectrometry. *Anal Chem*, *74*, 3190–3198.
42. Zhang, X., Q.K. Jin, S.A. Carr, and R.S. Annan. (2002). N-Terminal peptide labeling strategy for incorporation of isotopic tags: a method for the determination of site-specific absolute phosphorylation stoichiometry. *Rapid Commun Mass Spectrom*, *16*, 2325–2332.
43. Munchbach, M., M. Quadroni, G. Miotto, and P. James. (2000). Quantitation and facilitated de novo sequencing of proteins by isotopic N-terminal labeling of peptides with a fragmentation-directing moiety. *Anal Chem*, *72*, 4047–4057.
44. Ross, P. L., Huang, Y. N., Marchese, J. N., Williamson, B., Parker, K., Hattan, S., Khainovski, N., Pillai, S., Dey, S., Daniels, S., Purkayastha, S., Juhasz, P., Martin, S., Bartlet-Jones, M., He, F., Jacobson, A., Pappin, D. J. (2004). Multiplexed protein quantitation in *saccharomyces cerevisiae* using amine-reactive isobaric tagging reagents. *Mol Cell Proteomics 3*, 1154–1169.
45. Cagney, G. and A. Emili. (2002). De novo peptide sequencing and quantitative profiling of complex protein mixtures using mass-coded abundance tagging. *Nat Biotechnol*, *20*, 163–170.
46. Mason, D.E. and D.C. Liebler. (2003). Quantitative analysis of modified proteins by LC-MS/MS of peptides labeled with phenyl isocyanate. *J Proteome Res*, *2*, 265–272.
47. Beardsley, R.L. and J.P. Reilly. (2003). Quantitation using enhanced signal tags: a technique for comparative proteomics. *J Proteome Res*, *2*, 15–21.
48. Lee, Y.H., H. Han, S.B. Chang, and S.W. Lee. (2004). Isotope-coded N-terminal sulfonation of peptides allows quantitative proteomic analysis with increased de novo peptide sequencing capability. *Rapid Commun Mass Spectrom*, *18*, 3019–3027.
49. Peters, E.C., D.M. Horn, D.C. Tully, and A. Brock. (2001). A novel multifunctional labeling reagent for enhanced protein characterization with mass spectrometry. *Rapid Commun Mass Spectrom*, *15*, 2387–2392.
50. Hsu, J.L., S.Y. Huang, N.H. Chow, and S.H. Chen. (2003). Stable-isotope dimethyl labeling for quantitative proteomics. *Anal Chem*, *75*, 6843–6852.
51. Fedjaev, M., S. Trudel, A.P.N. Tjon, A. Parmar, B.I. Posner, E. Levy, I. Nifant'ev, and A.V. Pshezhetsky. (2007). Quantitative analysis of a proteome by N-terminal stable-isotope labelling of tryptic peptides. *Rapid Commun Mass Spectrom*, *21*, 2671–2679.
52. Li, S., Zeng, D. (2007). *CILAT*—A new reagent for quantitative proteomics. *Chem Commun (Camb) (21)* 2181–2183.
53. Goodlett, D.R., A. Keller, J.D. Watts, R. Newitt, E.C. Yi, S. Purvine, J.K. Eng, P. von Haller, R. Aebersold, and E. Kolker. (2001). Differential stable isotope labeling of peptides for quantitation and de novo sequence derivation. *Rapid Commun Mass Spectrom*, *15*, 1214–1221.
54. Syka, J.E., J.A. Marto, D.L. Bai, S. Horning, M.W. Senko, J.C. Schwartz, B. Ueberheide, B. Garcia, S. Busby, T. Muratore, J. Shabanowitz, and D.F. Hunt. (2004). Novel linear quadrupole ion trap/FT mass spectrometer: performance characterization and use in the comparative analysis of histone H3 post-translational modifications. *J Proteome Res*, *3*, 621–626.
55. Atwood, J.A., 3rd, L. Cheng, G. Alvarez-Manilla, N.L. Warren, W.S. York, and R. Orlando. (2008). Quantitation by isobaric labeling: applications to glycomics. *J Proteome Res*, *7*, 367–374.

56. Alvarez-Manilla, G., N.L. Warren, T. Abney, J. Atwood, 3rd, P. Azadi, W.S. York, M. Pierce, and R. Orlando. (2007). Tools for glycomics: relative quantitation of glycans by isotopic permethylation using 13CH3I. *Glycobiology*, *17*, 677–687.

57. Ippel, J. H., Pouvreau, L., Kroef, T., Gruppen, H., Versteeg, G., van den Putten, P., Struik, P. C., van Mierlo, C. P. (2004). In vivo uniform (15)N-isotope labelling of plants: Using the greenhouse for structural proteomics. *Proteomics 4*, 226–234.

58. Nelson, C. J., Huttlin, E. L., Hegeman, A. D., Harms, A. C., Sussman, M. R. (2007). Implications of 15N-metabolic labeling for automated peptide identification in Arabidopsis thaliana. *Proteomics 7*, 1279–1292.

59. Veenstra, T. D., Martinovic, S., Anderson, G. A., Pasa-Tolic, L., Smith, R. D. (2000). Proteome analysis using selective incorporation of isotopically labeled amino acids. *J Am Soc Mass Spectrom 11*, 78–82.

60. Dong, M. Q., Venable, J. D., Au, N., Xu, T., Park, S. K., Cociorva, D., Johnson, J. R., Dillin, A., Yates, J. R., 3rd (2007). Quantitative mass spectrometry identifies insulin signaling targets in *C. elegans. Science 317*, 660–603.

61. Krijgsveld, J., Ketting, R. F., Mahmoudi, T., Johansen, J., Artal-Sanz, M., Verrijzer, C. P., Plasterk, R. H., Heck, A. J. (2003). Metabolic labeling of *C. elegans* and *D. melanogaster* for quantitative proteomics. *Nat Biotechnol 21*(8), 927–31.

62. Gouw, J. W., Tops, B. B., Mortensen, P., Heck, A. J., Krijgsveld, J. (2008). Optimizing identification and quantitation of 15N-labeled proteins in comparative proteomics. *Anal Chem 80*, 7796–7803.

63. Wu, C. C., MacCoss, M. J., Howell, K. E., Matthews, D. E., Yates, J. R., 3rd (2004). Metabolic labeling of mammalian organisms with stable isotopes for quantitative proteomic analysis. *Anal Chem 76*, 4951–4959.

64. Zhu, H., Pan, S., Gu, S., Bradbury, E.M., Chen, X. (2002). Amino acid residue specific stable isotope labeling for quantitative proteomics. *Rapid Commun Mass Spectrom 16*, 2115–2123.

65. Mann, M. (2006). Functional and quantitative proteomics using SILAC. *Nat Rev Mol Cell Biol 7*, 952–958.

66. Gu, S., Pan, S., Bradbury, E. M., Chen, X. (2003). Precise peptide sequencing and protein quantification in the human proteome through in vivo lysine-specific mass tagging. *J Am Soc Mass Spectrom 14*, 1–7.

67. Ong, S. E., Mittler, G., Mann, M. (2004). Identifying and quantifying in vivo methylation sites by heavy methyl SILAC. *Nat Meth 1*, 119–126.

68. Ong, S. E., Kratchmarova, I., Mann, M. (2003). Properties of 13C-substituted arginine in stable isotope labeling by amino acids in cell culture (SILAC). *J Proteome Res 2*, 173–181.

69. Scott, L., Lamb, J., Smith, S., Wheatley, D. N. (2000). Single amino acid (arginine) deprivation: rapid and selective death of cultured transformed and malignant cells. *Br J Cancer 83*, 800–810.

70. Wisniewski, J. R., Zougman, A., Kruger, S., Ziolkowski, P., Pudelko, M., Bebenek, M., Mann, M. (2008). Constitutive and dynamic phosphorylation and acetylation sites on NUCKS, a hypermodified nuclear protein, studied by quantitative proteomics. *Proteins 73*, 710–718.

71. Dobreva, I., Fielding, A., Foster, L. J., Dedhar, S. (2008). Mapping the integrin-linked kinase interactome using SILAC. *J Proteome Res 7*, 1740–1749.

72. Wang, Z., Pandey, A., Hart, G. W. (2007). Dynamic interplay between O-linked N-acetylglucosaminylation and glycogen synthase kinase-3-dependent phosphorylation. *Mol Cell Proteomics 6*, 1365–1379.

73. Bonenfant, D., Towbin, H., Coulot, M., Schindler, P., Mueller, D. R., van Oostrum, J. (2007). Analysis of dynamic changes in post-translational modifications of human histones during cell cycle by mass spectrometry. *Mol Cell Proteomics 6*, 1917–1932.

74. Meierhofer, D., Wang, X., Huang, L., Kaiser, P. (2008). Quantitative analysis of global ubiquitination in HeLa cells by mass spectrometry. *J Proteome Res 7*(10), 4566–76.

75. Liang, X., Zhao, J., Hajivandi, M., Wu, R., Tao, J., Amshey, J. W., Pope, R. M. (2006). Quantification of membrane and membrane-bound proteins in normal and malignant breast cancer cells isolated from the same patient with primary breast carcinoma. *J Proteome Res 5*, 2632–2641.

76. Spellman, D. S., Deinhardt, K., Darie, C. C., Chao, M. V., Neubert, T. A. (2008). Stable isotopic labeling by amino acids in cultured primary neurons: Application to brain-derived neurotrophic factor-dependent phosphotyrosine-associated signaling. *Mol Cell Proteomics 7*, 1067–1076.

77. Bendall, S. C., Hughes, C., Stewart, M. H., Doble, B., Bhatia, M., Lajoie, G. A. (2008). Prevention of amino acid conversion in SILAC experiments with embryonic stem cells. *Mol Cell Proteomics 9*, 1587–1597.

78. Van Hoof, D., Pinkse, M. W., Oostwaard, D. W., Mummery, C. L., Heck, A. J., Krijgsveld, J. (2007). An experimental correction for arginine-to-proline conversion artifacts in SILAC-based quantitative proteomics. *Nat Meth 4*, 677–678.

79. Gorg, A., Weiss, W., Dunn, M. J. (2004). Current two-dimensional electrophoresis technology for proteomics. *Proteomics 4*, 3665–3685.

80. Unlu, M., Morgan, M. E., Minden, J. S. (1997). Difference gel electrophoresis: a single gel method for detecting changes in protein extracts. *Electrophoresis 18*, 2071–2077.

81. Timms, J. F., Cramer, R. (2008). Difference gel electrophoresis. *Proteomics 23–24*, 4886–4897.

82. Reynolds, K. J., Yao, X., Fenselau, C. (2002). Proteolytic ^{18}O labeling for comparative proteomics: evaluation of endoprotease Glu-C as the catalytic agent. *J Proteome Res 1*, 27–33.

83. Schnolzer, M., Jedrzejewski, P., Lehmann, W. D. (1996). Protease-catalyzed incorporation of 18O into peptide fragments and its application for protein sequencing by electrospray and matrix-assisted laser desorption/ionization mass spectrometry. *Electrophoresis 17*, 945–953.

84. Angel, P. M., Orlando, R. (2006). Trypsin is the primary mechanism by which the (18)O isotopic label is lost in quantitative proteomic studies. *Anal Biochem 359*, 26–34.

85. Yao, X., Afonso, C., Fenselau, C. (2003). Dissection of proteolytic 18O labeling: endoprotease-catalyzed 16O-to-18O exchange of truncated peptide substrates. *J Proteome Res 2*, 147–152.

86. Rao, K. C., Carruth, R. T., Miyagi, M. (2005). Proteolytic 18O labeling by peptidyl-Lys metalloendopeptidase for comparative proteomics. *J Proteome Res 4*, 507–514.

87. Liu, H., Zhang, Y., Meng, L., Qin, P., Wei, J., Jia, W., Li, X., Cai, Y., Qian, X. (2007). Non–gel-based dual 18O labeling quantitative proteomics strategy. *Anal Chem 79*, 7700–7707.

88. Storms, H. F., van der Heijden, R., Tjaden, U. R., van der Greef, J. (2006). Considerations for proteolytic labeling-optimization of 18O incorporation and prohibition of back-exchange. *Rapid Commun Mass Spectrom 20*, 3491–3497.

89. Wang, Y. K., Ma, Z., Quinn, D. F., Fu, E. W. (2001). Inverse 18O labeling mass spectrometry for the rapid identification of marker/target proteins. *Anal Chem 73*, 3742–3750.

90. Li, J., Steen, H., Gygi, S. P. (2003). Protein profiling with cleavable isotope-coded affinity tag (cICAT) reagents: The yeast salinity stress response. *Mol Cell Proteomics 2*, 1198–1204.

91. Zhou, H., Ranish, J. A., Watts, J. D., Aebersold, R. (2002). Quantitative proteome analysis by solid-phase isotope tagging and mass spectrometry. *Nat Biotechnol 20*, 512–515.

92. Zhang, L., Guo, Y. L., Liu, H. Q. (2004). A novel class of chemically modified iodo-containing resins: design, synthesis and application to mass spectrometry-based proteome analysis. *J Mass Spectrom 39*, 447–457.

93. Hansen, K. C., Schmitt-Ulms, G., Chalkley, R. J., Hirsch, J., Baldwin, M. A., Burlingame, A. L. (2003). Mass spectrometric analysis of protein mixtures at low levels using cleavable 13C-isotope-coded affinity tag and multidimensional chromatography. *Mol Cell Proteomics 2*, 299–314.

94. Zhang, R., Regnier, F. E. (2002). Minimizing resolution of isotopically coded peptides in comparative proteomics. *J Proteome Res 1*, 139–147.

95. Tao, W. A., Wollscheid, B., O'Brien, R., Eng, J. K., Li, X. J., Bodenmiller, B., Watts, J. D., Hood, L., Aebersold, R. (2005). Quantitative phosphoproteome analysis using a dendrimer conjugation chemistry and tandem mass spectrometry. *Nat Meth 2*, 591–598.

96. Boas, U., Heegaard, P. M. (2004). Dendrimers in drug research. *Chem Soc Rev 33*, 43–63.

97. Galan, J. A., Guo, M., Sanchez, E. E., Cantu, E., Rodriguez-Acosta, A., Perez, J. C., Tao, W. A. (2008). Quantitative analysis of snake venoms using soluble polymer-based isotope labeling. *Mol Cell Proteomics 7*, 785–799.

98. Gerber, S. A., Rush, J., Stemman, O., Kirschner, M. W., Gygi, S. P. (2003). Absolute quantification of proteins and phosphoproteins from cell lysates by tandem MS. *Proc Nat Acad Sci USA 100*, 6940–6945.

99. Beynon, R. J., Doherty, M. K., Pratt, J. M., Gaskell, S. J. (2005). Multiplexed absolute quantification in proteomics using artificial QCAT proteins of concatenated signature peptides. *Nat Meth 2*, 587–589.

100. Pratt, J. M., Simpson, D. M., Doherty, M. K., Rivers, J., Gaskell, S. J., Beynon, R. J. (2006). Multiplexed absolute quantification for proteomics using concatenated signature peptides encoded by QconCAT genes. *Nat Protoc 1*, 1029–1043.

101. Anderson, N. L., Anderson, N. G., Haines, L. R., Hardie, D. B., Olafson, R. W., Pearson, T. W. (2004). Mass spectrometric quantitation of peptides and proteins using stable isotope standards and capture by anti-peptide antibodies (SISCAPA). *J Proteome Res 3*, 235–244.

102. Mallick, P., Schirle, M., Chen, S. S., Flory, M. R., Lee, H., Martin, D., Ranish, J., Raught, B., Schmitt, R., Werner, T., Kuster, B., Aebersold, R. (2007). Computational prediction of proteotypic peptides for quantitative proteomics. *Nat Biotechnol 25*, 125–131.

103. Panchaud, A., Affolter, M., Moreillon, P., Kussmann, M. (2008). Experimental and computational approaches to quantitative proteomics: Status quo and outlook. *J Proteomics 71*, 19–33.

104. Chelius, D., Zhang, T., Wang, G., Shen, R. F. (2003). Global protein identification and quantification technology using two-dimensional liquid chromatography nanospray mass spectrometry. *Anal Chem 75*, 6658–6665.

105. Liu, H., Sadygov, R. G., Yates, J. R., 3rd (2004). A model for random sampling and estimation of relative protein abundance in shotgun proteomics. *Anal Chem 76*, 4193–4201.

106. Rappsilber, J., Ryder, U., Lamond, A. I., Mann, M. (2002). Large-scale proteomic analysis of the human spliceosome. *Genome Res 12*, 1231–1245.

107. Zybailov, B. L., Florens, L., Washburn, M. P. (2007). Quantitative shotgun proteomics using a protease with broad specificity and normalized spectral abundance factors. *Mol Biosyst 3*, 354–360.

108. Chen, C. H. (2008). Review of a current role of mass spectrometry for proteome research. *Anal Chim Acta 624*, 16–36.

109. Colinge, J., Chiappe, D., Lagache, S., Moniatte, M., Bougueleret, L. (2005). Differential proteomics via probabilistic peptide identification scores. *Anal Chem 77*, 596–606.

110. Andersen, J. S., Wilkinson, C. J., Mayor, T., Mortensen, P., Nigg, E. A., Mann, M. (2003). Proteomic characterization of the human centrosome by protein correlation profiling. *Nature 426*, 570–574.

111. Chelius, D., Bondarenko, P. V. (2002). Quantitative profiling of proteins in complex mixtures using liquid chromatography and mass spectrometry. *J Proteome Res 1*, 317–323.

112. Lasonder, E., Ishihama, Y., Andersen, J. S., Vermunt, A. M., Pain, A., Sauerwein, R. W., Eling, W. M., Hall, N., Waters, A. P., Stunnenberg, H. G., Mann, M. (2002). Analysis of the *Plasmodium falciparum* proteome by high-accuracy mass spectrometry. *Nature 419*, 537–542.

113. Bondarenko, P. V., Chelius, D., Shaler, T. A. (2002). Identification and relative quantitation of protein mixtures by enzymatic digestion followed by capillary reversed-phase liquid chromatography-tandem mass spectrometry. *Anal Chem 74*, 4741–4749.

114. Old, W. M., Meyer-Arendt, K., Aveline-Wolf, L., Pierce, K. G., Mendoza, A., Sevinsky, J. R., Resing, K. A., Ahn, N. G. (2005). Comparison of label-free methods for quantifying human proteins by shotgun proteomics. *Mol Cell Proteomics 4*, 1487–502.

115. Ishihama, Y., Oda, Y., Tabata, T., Sato, T., Nagasu, T., Rappsilber, J., Mann, M. (2005). Exponentially modified protein abundance index (emPAI) for estimation of absolute protein amount in proteomics by the number of sequenced peptides per protein. *Mol Cell Proteomics 4*, 1265–1272.

116. Allet, N., Barrillat, N., Baussant, T., Boiteau, C., Botti, P., Bougueleret, L., Budin, N., Canet, D., Carraud, S., Chiappe, D., Christmann, N., Colinge, J., Cusin, I., Dafflon, N., Depresle, B., Fasso, I., Frauchiger, P., Gaertner, H., Gleizes, A., Gonzalez-Couto, E., Jeandenans, C., Karmime, A., Kowall, T., Lagache, S., Mahé, E., Masselot, A., Mattou, H., Moniatte, M., Niknejad, A., Paolini, M., Perret, F., Pinaud, N., Ranno, F., Raimondi, S., Reffas, S., Regamey, P. O., Rey, P. A., Rodriguez-Tomé, P., Rose, K., Rossellat. G., Saudrais, C., Schmidt, C., Villain, M., Zwahlen. C. (2004). In vitro and in silico processes to identify differentially expressed proteins. *Proteomics 4*, 2333–2351.

117. Anderle, M., Roy, S., Lin, H., Becker, C., Joho, K. (2004). Quantifying reproducibility for differential proteomics: Noise analysis for protein liquid chromatography-mass spectrometry of human serum. *Bioinformatics 20*, 3575–3582.

118. Strittmatter, E. F., Rodriguez, N., Smith, R. D. (2003). High mass measurement accuracy determination for proteomics using multivariate regression fitting: Application to electrospray ionization time-of-flight mass spectrometry. *Anal Chem 75*, 460–468.

119. Strittmatter, E. F., Ferguson, P. L., Tang, K., Smith, R. D. (2003). Proteome analyses using accurate mass and elution time peptide tags with capillary LC time-of-flight mass spectrometry. *J Am Soc Mass Spectrom 14*, 980–991.

120. Silva, J. C., Denny, R., Dorschel, C., Gorenstein, M. V., Li, G. Z., Richardson, K., Wall, D., Geromanos, S. J. (2006). Simultaneous qualitative and quantitative analysis of the *Escherichia coli* proteome: A sweet tale. *Mol Cell Proteomics 5*, 589–607.

121. Bantscheff, M., Schirle, M., Sweetman, G., Rick, J., Kuster, B. (2007). Quantitative mass spectrometry in proteomics: A critical review. *Anal Bioanal Chem 389*, 1017–1031.

122. Wang, W., Zhou, H., Lin, H., Roy, S., Shaler, T. A., Hill, L. R., Norton, S., Kumar, P., Anderle, M., Becker, C. H. (2003). Quantification of proteins and metabolites by mass spectrometry without isotopic labeling or spiked standards. *Anal Chem 75*, 4818–4826.

123. Wong, J. W., Sullivan, M. J., Cagney, G. (2008). Computational methods for the comparative quantification of proteins in label-free LCn-MS experiments. *Brief Bioinform 9*, 156–165.

124. Iliuk, A., J. Galan, and W.A. Tao. (2008). Playing tag with quantitative proteomics. *Anal Bioanal Chem.*

CHAPTER 5

Comparative Proteomics by Direct Tissue Analysis Using Imaging Mass Spectrometry

MICHELLE L. REYZER and RICHARD M. CAPRIOLI

5.1 INTRODUCTION

The modern study of proteomics has transformed the way new drug targets are identified, in large part owing to the technological advances in mass spectrometry (MS). Extensive comparative proteomics studies are routinely performed by using standard gel electrophoresis and high-performance liquid chromatography (HPLC) approaches coupled to electrospray (ESI) or matrix-assisted laser desorption/ionization (MALDI) mass spectrometric techniques for peptide and protein identification. These studies result in a catalog of potential biomarkers that are differentially expressed in disease, or in subjects treated with drugs, compared to healthy controls. The rationale behind these studies is that some of these changes are directly due to the disease pathology and thus present viable targets for drug discovery. Similar studies may also be used to assess toxicity, drug resistance, and drug efficacy, making MS-based proteomics an important tool for the development of new, safe, and effective pharmaceuticals.

With the development of imaging mass spectrometry (IMS) [1,2], these studies may now be performed directly on tissue sections, precluding the need for extensive sample preparation. MALDI mass spectrometry allows the use of thin sections of frozen tissues (biopsies, dissected animal organs, whole-body animal samples) to be analyzed. These tissue sections are directly thaw-mounted on conductive plates, coated or spotted with MALDI matrix, and mass spectra are obtained at discrete locations over the entire section. The mass spectra contain signals from proteins and peptides (as well as lipids, pharmaceuticals, endogenous small molecules, etc.) that are expressed in specific locations within the tissue. Spectra may be obtained with high spatial resolution (~50–200 μm, down to ~10 μm with special optics) and

Protein and Peptide Mass Spectrometry in Drug Discovery, Edited by Michael L. Gross, Guodong Chen, and Birendra N. Pramanik.

excellent mass accuracy (typically 5–10 ppm for most TOF-MS instruments) from a typical mass range of up to around 50 kDa.

Thus IMS may be used to perform routine comparative proteomics analyses, with signals of overexpressed and underexpressed materials being determined in the different sample groups. IMS, however, provides additional information compared to standard gel and HPLC-based analyses. For example, an assessment may be made as to *what* proteins are being expressed *where*. Because conventional approaches typically use homogenized tissues prior to gel and HPLC separation, any spatial information is lost. Furthermore signals obtained from a MALDI mass spectrum correspond to proteins in the form in which they are present in the tissues (i.e., including any modifications or cleavages). This highly specific output may provide important information for future drug discovery, as only proteins in a certain state (modified or unmodified) may be biologically relevant.

5.2 CONVENTIONAL COMPARATIVE PROTEOMICS

There are many examples in the literature that demonstrate the power of comparative proteomics as a discovery tool, with drug discovery as its ultimate goal. One example describes the use of comparative proteomics to identify patterns in cancer cells that become resistant to various chemotherapies [3]. The authors reviewed 29 studies employing conventional proteomics technologies to uncover differences in cell lines from cancer cells that had developed resistance to chemotherapy compared to cell lines that remained responsive. Thirty-eight different cell lines were examined, and differences were found in proteins involved in many relevant biochemical pathways. Half or more of the cell lines exhibited expression changes in proteins involved with calcium-binding, chaperone, cytoskeleton, and metabolism processes. The authors concluded that model cancer cell lines exhibit multiple mechanisms of drug resistance and that further study will help identify predictive factors to aid prognosis. In addition they state, "... these proteins may be used as targets for developing chemo-sensitizing therapeutics that can be used to enhance the chemo-sensitivity of cancers to currently available anticancer drugs in combination therapy."[3]

Another report describes the identification of proteomic markers for hepatic steatosis (fatty liver), which is often an early sign of drug toxicity [4]. The authors injected a compound (designated CDA) into rats and analyzed the protein expression differences in the livers after 2 and 5 days of treatment and compared the results with those from controls. Liver homogenates were subjected to 2D DIGE separation followed by spot excision, tryptic digestion, and MALDI mass fingerprinting for protein identification. Subsequently these experiments were repeated in an *in vitro* rat hepatocyte model. The spot pattern was similar between the liver and hepatocyte gels, and from the 14 spots identified from the hepatocyte gel, 6 proteins were identical to proteins identified from the liver gel, suggesting that similar pathways were being affected in the *in vitro* model as in the whole-animal. An important conclusion is that the *in vitro* model can be used as a surrogate to whole-animal studies. This can allow for rapid, high-throughput screening of drug toxicity earlier in the drug discovery process, minimizing costly, late-stage failures of drug candidates owing to toxicity issues [4].

5.3 COMPARATIVE PROTEOMICS USING IMAGING MS

Imaging mass spectrometry has been successfully applied to comparative proteomics studies involving human lung cancer [5–7], brain cancer [8,9], breast cancer [10], ovarian cancer [11], as well as other diseases. These studies are typically performed via histology-directed protein profiling [10] that is better suited for high-throughput analysis of many samples (tens to hundreds) with the ultimate goal of discovering statistically significant proteomic markers. This approach, however, is often complemented by a high-resolution image analysis of one or several representative tissue specimens. The images obtained can confirm the presence of the statistically significant biomarkers while highlighting their localization. Further visual inspection of the images can reveal other proteins that show similar localization patterns. Examples of these two modes of analysis are presented below.

5.3.1 Biomarker Discovery: Breast Cancer

In histology-directed protein profiling, optical images of tissue sections are examined by a pathologist who marks regions that are highly enriched in certain cell populations (typically $>80\%$ of one cell type; e.g., normal epithelial cells, tumor cells, or stroma cells). These optical images are overlayed onto images of serial sections that are mounted onto MALDI plates for mass spectral analysis. Precise x, y coordinates are generated for each marked region on the MALDI plate. Small volumes of matrix ($\sim$pL to nL) are then precisely deposited on the tissue by using robotic spotters [12] at the positions marked, affording discrete spots of matrix approximately 150 to 200 μm in diameter; the materials in those spots are subsequently analyzed. The resulting mass spectra thus correspond to unique positions on the tissue and contain profiles primarily from one cell type. These protein profiles are then evaluated by using biostatistical analysis, with the ultimate goal of determining differentially expressed proteins among the different cell types. This experiment can be thought of as directed low-resolution imaging, because although a high-resolution picture of protein distribution is not generated, each spectrum is linked to a specific location on the tissue.

Other approaches to molecular profiling of heterogeneous tissues utilize laser capture microdissection (LCM) [13,14]. LCM extracts small clusters of individual cells from a tissue section, which can then be analyzed by MALDI mass spectrometry to obtain molecular profiles. These analyses result in profiles that are unique to distinct cell populations, but the data may be compromised by the harsh washing and dehydration steps required for LCM. In addition LCM can be quite time-consuming, requiring hundreds of cells of a single type to be extracted manually onto a single region of a polymeric cap to produce good quality mass spectra.

An example of the histology-directed approach is the analysis of human breast cancer specimens [15]. Breast tissue is quite heterogeneous, containing discrete ductal and lobular functional units, surrounding stroma, and fatty tissue. In addition there are many types and stages of breast cancer, from localized ductal carcinoma *in situ* (DCIS) to invasive mammary cancer (IMC). These cancers may be present in the same sample in close proximity to each other. To derive meaningful data from these tissues, it is critical that the molecular profiles originate from, and thus represent, *primarily* the proteins present in the distinct cell types.

The process is illustrated in Figure 5.1 [15]. Figure 5.1*A* shows an image of an H&E stained section of a human breast cancer tissue that was annotated by a pathologist. In this case the marked circles are color-coded to differentiate the distinct cell types (red circles denote stroma cells, black circles are IMC, blue circles are DCIS, and green circles are nontumor epithelium). Figure 5.1*B* is a close-up view of one region of the tissue (Figure 5.1*A*), highlighting the close proximity of different cell types within the tissue section. The large circular shaded area represents the spot size that would result

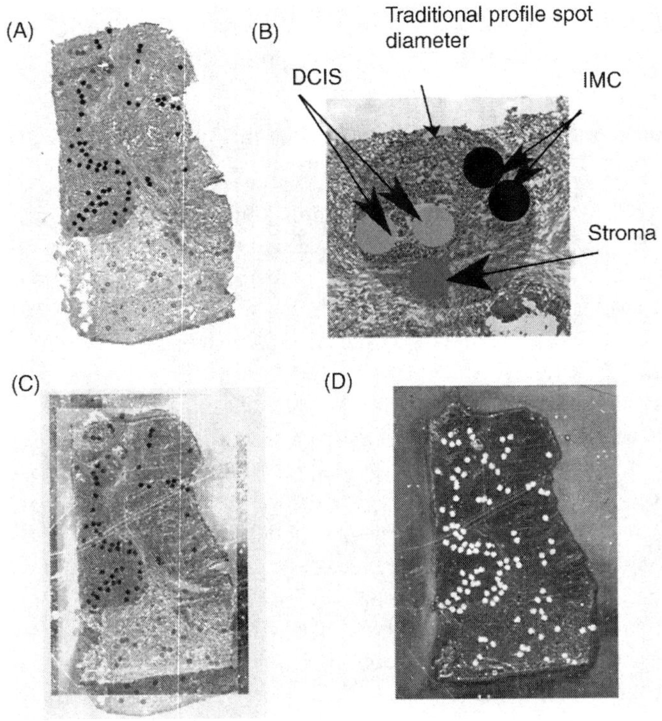

FIGURE 5.1 Histology-directed protein profiling for comparative proteomics. (*A*) H&E stained section of human breast cancer specimen annotated by a pathologist to locate regions of interest: *red,* peritumoral stroma; *black,* IMC; *blue,* DCIS; and *green,* non-tumor epithelium. (*B*) Illustration of the different surface areas profiled by the histology-directed strategy (colored spots) and traditional profiling (shaded area). (*C*) Overlay of the aligned H&E image with the section on the MALDI target plate for matrix spotting. (*D*) Optical image of the section on the MALDI target plate after robotic deposition of matrix onto the designated sites. Reproduced with permission from [15]. (*See the color version of this figure in Color Plates section.*)

after typical manual deposition of matrix with a mechanical pipette (~100 nL generates a spot ~1 mm in diameter). Protein profiles obtained from such a manually derived spot would contain proteins from all three cell types and would only provide an average protein distribution upon statistical analysis.

An overlay of the marked H&E image with the serial section on the MALDI plate (Figure 5.1*C*) shows that the images are aligned along both internal and external contours to maximize placement accuracy. Figure 5.1*D* shows an image of the tissue on the MALDI plate after spotting. A clear correlation is observed between the actual matrix spot locations in Figure 5.1*D* and the marked regions on the serial section in Figure 5.1*A*. Profiles acquired from this sample were analyzed by unsupervised classification followed by multidimensional scaling (MDS) of the distance in three dimensions. The MDS results show that normal epithelium, stroma, and cancer (DCIS and IMC) separate into three distinct groups. Further supervised analysis of the DCIS and IMC profiles reveal a small, but noticeable, separation, indicating that molecular differences can be observed even between similar pathologies.

This technology was also used to explore whether biomarkers may be found that correlate to drug response in a mouse model of breast cancer [16]. In this case tumors from transgenic mice expressing human HER2 (human epidermal growth factor receptor 2) were transplanted into wild-type mice, where they continued to grow and show HER2 overexpression. The antibody Herceptin binds to the HER2 receptor and shows antitumor activity in these tumors (notated as 1282 tumors). Another tumor line (Fo5), from the same transgenic founder, spontaneously developed resistance to Herceptin, while maintaining similar expression levels of HER2 and similar levels of Herceptin-HER2 binding. Fo5 and 1282 tumors were harvested after a single dose of Herceptin (30 mg/kg i.p.) and subjected to MS analysis. The mass spectra in the mass range of 9700 to 10,200 Da obtained for both sensitive (1282) and resistant (Fo5) tumor lines (Figure 5.2) show that treatment with Herceptin induces an increase of the species of *m/z* 9739, 9970, and 10,164 compared to the untreated tumor, whereas no change is evident for those signals in the resistant tumor after Herceptin treatment. Several other signals were found to exhibit a similar pattern, including those corresponding to ions of *m/z* 4795 and 9212 suggesting that these signals may be biomarkers of Herceptin resistance [16].

5.3.2 Biomarker Discovery: Toxicity

Identification of compounds that have adverse toxicity profiles is of critical importance in drug discovery. Similar studies to the one highlighted earlier involving hepatic steatosis have been conducted using MALDI IMS. For example, a group of monkeys was given the known nephrotoxicant gentamicin along with everninomicin to determine if the combination produced nephrotoxicity. After seven days of combination treatment (10 mg/kg gentamicin and 30 mg/kg everninomicin), the animals were sacrificed and their kidneys removed and flash frozen for MALDI analysis. Matrix was deposited on the kidney sections manually, and protein profiles were obtained to reveal an overall indication of proteomic effects. Several signals showed significant intensity differences compared to controls, including a signal for

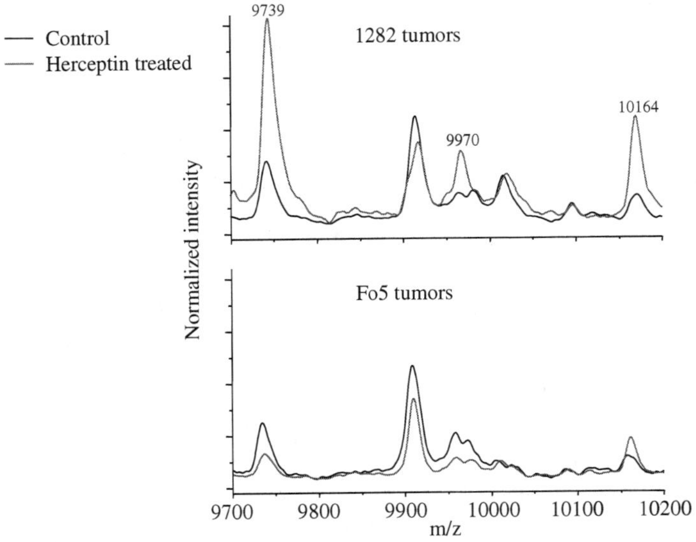

FIGURE 5.2 Drug-induced changes in the proteome predict for therapeutic resistance. Mice bearing Fo5 (Herceptin-resistant) and 1282 (Herceptin-sensitive) tumors were treated with a single dose of Herceptin (30 mg/kg i.p). Tumors were harvested after dosing and subjected to mass spectral proteomic analysis. Statistically significant changes observed after Herceptin-treatment in the 1282 tumors that are not observed in the Fo5 tumors are shown (an increase in *m/z* 9739 and 10,164). The sensitive tumor line traces consist of untreated tumors (average of 20 spectra from 6 tumors) and Herceptin-treated tumors (average of 13 spectra from 4 tumors). The resistant tumor line traces consist of untreated tumors (average of 11 spectra from 3 tumors) and Herceptin-treated tumors (average of 20 spectra from 4 tumors). (*See the color version of this figure in Color Plates section.*)

an ion of *m/z* 12,922 (shown in Figure 5.3*A*). This result was followed up with a high-resolution imaging analysis of kidneys from one control and one dosed monkey. As shown in Figure 5.3*B*, the treated monkey kidney shows an intense signal corresponding to the *m/z* 12,924 ion that is localized to the cortex. There is no noticeable signal in the medulla of the treated monkey, nor is there measureable signal anywhere in the control kidney. This protein was identified as transthyretin via HPLC-MS/MS and 2D gel electrophoresis. Subsequently these results were correlated with Western blot and immunohistochemistry experiments, which also showed a significant increase in transthyretin in the kidney cortex of treated animals. An analogous study undertaken in rats found similar results, with gentamicin-treated rats expressing an increase of transthyretin in the kidney cortex [17].

5.3.3 Correlating Drug and Protein Distributions

One unique advantage of imaging mass spectrometry is the ability to locate positionally a given compound with molecular specificity. This allows compound distributions to be obtained in a label-free manner, while differentiating signals originating from a parent compound and its metabolites. This has been demonstrated for olanzapine distribution

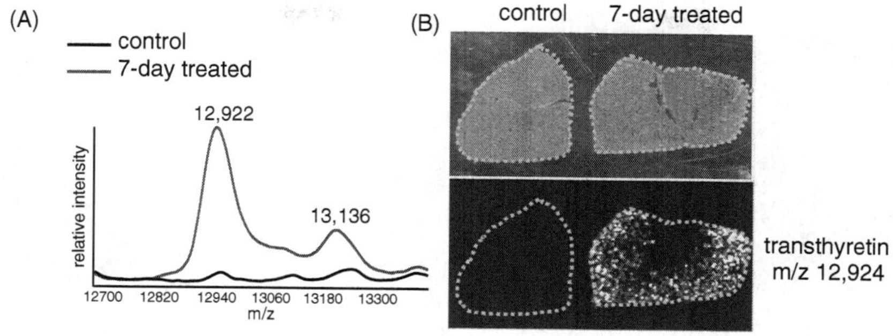

FIGURE 5.3 Drug-induced changes in the proteome correlate with drug-induced toxicity. Monkeys were dosed with a combination of the known nephrotoxicant gentamicin (10 mg/kg) and everninomicin (30 mg/kg) for 7 days. Kidneys were harvested and subjected to mass spectral proteomic analysis. (*A*) A signal at *m/z* 12,922 (subsequently identified as transthyretin) was found to be significantly increased in the dosed kidneys compared to controls. (*B*) High-resolution image analysis of kidneys from one control and one dosed monkey show the transthyretin ion is localized to the cortex of the dosed kidneys. (*See the color version of this figure in Color Plates section.*)

in whole-body rat sections [18]. In this case the distribution for olanzapine as well as for two first-pass metabolites, *N*-desmethyl-olanzapine and 2-hydroxymethyl-olanzapine, were detected from whole-body rat sections 2 h after dosing. Three unique fragmentation transitions were monitored (via MS/MS), one for each compound, thus generating three molecularly specific images from the same whole-body section.

This approach may be combined with protein distribution analyses to obtain a picture of proteomic changes associated with drug dosing. The information gleaned from such experiments may be useful for evaluating drug efficacy, elucidating biochemical pathways or drug mechanisms, discovering biomarkers for response or resistance, or establishing co-localization of a drug and its protein target.

One application of this technology being investigated in our laboratory is the distribution of antituberculosis drugs in infected rabbit lungs. Current chemotherapy regimens for people infected with *M. tuberculosis* last for six months or longer, but the factors that affect the slow rate of bacterial killing are largely unknown. It is known that tuberculosis infection results in the formation of a heterogeneous collection of pulmonary granulomas. One possibility for the different rates of bacterial killing may be related to the penetration of the drugs into the different granulomas. Proteins (or other endogenous compounds, e.g., lipids) present in the microenvironment surrounding the granulomas may influence drug accessibility and penetration. Imaging mass spectrometry allows the drugs themselves to be located and the surrounding proteins (and lipids) to be visualized, which may lead to a better understanding of the mechanism of action of the drugs *in vivo*.

An example is the imaging of a lung from a rabbit infected with *M. tuberculosis* and then orally dosed with a combination of antituberculosis drugs, including rifampin (RIF) at 30 mg/kg for 5 days, and sacrificed 1 h 5 min after the final dose (Figure 5.4). Figure 5.4*A* is an optical image of a thaw-mounted section of the lung on a gold-coated MALDI plate (which gives the tissue the overall yellow color). Several distinct

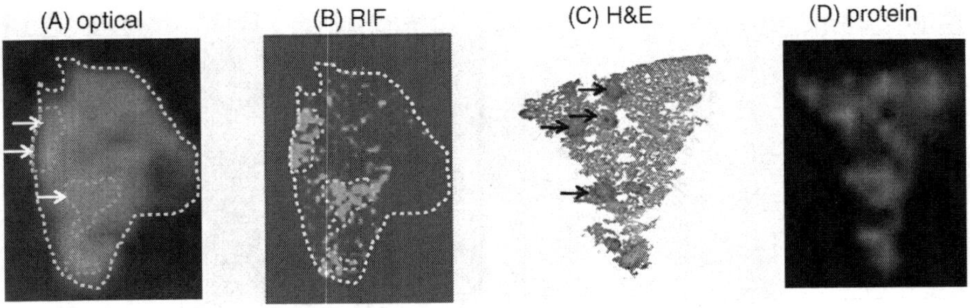

FIGURE 5.4 Examining drug distribution in the granuloma microenvironment in a rabbit model of tuberculosis infection. Rabbits were infected with *M. tuberculosis* and orally dosed with a combination of antituberculosis drugs, including rifampin at 30 mg/kg for 5 days. (*A*) Optical image of an infected rabbit lung section on a gold-coated MALDI target plate. This animal was sacrificed 1 h 5 min after the final dose. Granulomas are indicated with white arrows. (*B*) MALDI MS image of the distribution of rifampin (MS/MS *m/z* 821 → *m/z* 397 + *m/z* 722) in the lung section shown in *A*. Rifampin appears to localize to granulomas compared to surrounding lung. (*C*) H&E stained serial section of the lung tissue shown in *A*, with granulomas indicated by black arrows. (*D*) MALDI MS protein image showing the localization of *m/z* 11,345 (green) to the granuloma areas and *m/z* 15,787 (red) to adjacent uninvolved tissue. (*See the color version of this figure in Color Plates section.*)

granulomas can be observed on the optical section (light, whitish areas denoted by arrows). Figure 5.4*B* shows a reconstructed ion image for RIF, obtained via tandem mass spectrometry in negative-ion mode. The deprotonated RIF molecule of *m/z* 821 was dissociated and the primary fragments of *m/z* 397 and 722 were summed for the image. The drug can clearly be observed localizing in several of the granulomas compared to adjacent uninvolved lung tissue. Figure 5.4*C* shows an H&E stained slide of a separate section from the same tissue, where many small granulomas are clearly observed. Figure 5.4*D* is a reconstructed ion image for several protein signals obtained from a section cut serially to the H&E section. A signal for an *m/z* 11,345 ion (shown in green) shows distinct localization to the granuloma areas, and perhaps sub-localization to the outer ring of the granulomas. Another signal for an ion of *m/z* 15,787 is not present to a noticeable degree in the granulomas, but rather appears in the adjacent uninvolved tissue. Its distribution in the lung tissue, however, appears to be heterogeneous. Further study will be required to identify these proteins and to determine what role, if any, they may play in affecting the distribution of RIF or the other drugs in the lung tissue generally and granulomas specifically. Nevertheless, the information obtained from these experiments will certainly be beneficial for understanding how the existing antituberculosis therapies work and will help to facilitate the process of drug design for next-generation therapies.

5.4 CONCLUSIONS

Imaging mass spectrometry has tremendous potential for advancing and supporting the drug discovery process. Comparative proteomics experiments may be performed and the results used as a starting point for selecting promising drug targets, understanding the biological mechanisms of a pathology, or for probing the global effects of a drug on the proteome of an organism. Toxicity and ADME studies may also be performed by monitoring the drug and any known metabolites on individual organ or whole-body animal sections. The same (or serial) sections may also be analyzed for proteomic effects that may be indicators of drug toxicity or efficacy. Because these experiments are label-free and can make use of the same samples for multiple analyses, this technology has the potential to save money and resources now needed for the analyses performed. Furthermore the overall impact will be to save resources for the development because these experiments can be performed early in the drug discovery stage.

ACKNOWLEDGMENTS

The authors thank Walter Korfmacher, Ron Snyder, Eddie Yi-Zhong Gu, and Annette Erskine for the samples and analysis of transthyretin in monkey kidneys, and Clif Barry III, JoAnne Flynn, and Laura Via for the collaboration on tuberculosis imaging. Funding from NIH (5R01 GM058008), DOD (W81XWH-05-1-0179), and the Bill and Melinda Gates Foundation is acknowledged.

REFERENCES

1. Stoeckli, M., Chaurand, P., Hallahan, D. E., Caprioli, R. M. (2001). Imaging mass spectrometry: A new technology for the analysis of protein expression in mammalian tissues. *Nat Med 7*, 493–496.
2. Cornett, D. S., Reyzer, M. L., Chaurand, P., Caprioli, R. M. (2007). MALDI imaging mass spectrometry: molecular snapshots of biochemical systems. *Nat Meth 4*, 828–833.
3. Zhang, J.-T., Liu, Y. (2007). Use of comparative proteomics to identify potential resistance mechanisms in cancer treatment. *Can Treatm Revi 33*, 741–756.
4. Meneses-Lorente, G., Watt, A., Salim, K., Gaskell, S. J., Muniappa, N., Lawrence, J., Guest, P. C. (2006). Identification of early proteomic markers for hepatic steatosis. *Chem Res Toxicol 19*, 986–998.
5. Yanagisawa, K., Shyr, Y., Xu, B. J., Massion, P. P., Larsen, P. H., White, B. C., Roberts, J. R., Edgerton, M., Gonzalez, A., Nadaf, S., Moore, J. H., Caprioli, R. M., Carbone, D. P. (2003). Proteomic patterns of tumour subsets in non-small-cell lung cancer. *Lancet 362*, 433–439.
6. Taguchi, F., Solomon, B., Gregorc, V., Roder, H., Gray, R., Kasahara, K., Nishio, M., Brahmer, J., Spreafico, A., Ludovini, V., Massion, P. P., Dziadziuszko, R., Schiller, J., Grigorieva, J., Tsypin, M., Hunsucker, S. W., Caprioli, R., Duncan, M. W., Hirsch, F. R., Bunn, P. A., Jr., Carbone, D. P. (2007). Mass spectrometry to classify non–small-cell lung cancer patients for clinical outcome after treatment with epidermal growth factor receptor tyrosine kinase inhibitors: A multicohort cross-institutional study. *J Nat Can Inst 99*, 838–846.
7. Xu, B. J., Gonzalez, A. L., Kikuchi, T., Yanagisawa, K., Massion, P. P., Wu, H., Mason, S. E., Olson, S. J., Shyr, Y., Carbone, D. P., Caprioli, R. M. (2008). MALDI-MS derived

prognostic protein markers for resected non–small cell lung cancer. *Proteomics Clin Appl, 2*, 1508–1517.

8. Schwartz, S. A., Weil, R. J., Johnson, M. D., Toms, S. A., Caprioli, R. M. (2004). Protein profiling in brain tumors using mass spectrometry: Feasibility of a new technique for the analysis of protein expression. *Clin Can Res 10*, 981–987.
9. Schwartz, S. A., Weil, R. J., Thompson, R. C., Shyr, Y., Moore, J. H., Toms, S. A., Johnson, M. D., Caprioli, R. M. (2005). Proteomic-based prognosis of brain tumor patients using direct-tissue matrix-assisted laser desorption ionization mass spectrometry. *Can Res 65*, 7674–7681.
10. Cornett, D. S., Mobley, J. A., Dias, E. C., Andersson, M., Arteaga, C. L., Sanders, M. E., Caprioli, R. M. (2006). A novel histology-directed strategy for MALDI-MS tissue profiling that improves throughput and cellular specificity in human breast cancer. *Mol Cell Proteomics 5*, 1975–1983.
11. Lemaire, R., Menguellet, S. A., Stauber, J., Marchaudon, V., Lucot, J.-P., Collinet, P., Farine, M.-O., Vinatier, D., Day, R., Ducoroy, P., Salzet, M., Fournier, I. (2007). Specific MALDI imaging and profiling for biomarker hunting and validation: Fragment of the 11S proteasome activator complex, Reg Alpha fragment, is a new potential ovary cancer biomarker. *J Proteome Res 6*, 4127–4134.
12. Aerni, H.-R., Cornett, D. S., Caprioli, R. M. (2006). Automated acoustic matrix deposition for MALDI sample preparation. *Anal Chem 78*, 827–834.
13. Palmer-Toy, D. E., Sarracino, D. A., Sgroi, D., LeVangie, R., Leopold, P. E. (2000). Direct acquisition of matrix-assisted laser desorption/ionization time-of-flight mass spectra from laser capture microdissected tissues. *Clin Chem 46*, 1513–1516.
14. Xu, B. J., Caprioli, R. M., Sanders, M. E., Jensen, R. A. (2002). Direct analysis of laser capture microdissected cells by MALDI mass spectrometry. *J Am Soc Mass Spectrom 13*, 1292–1297.
15. Cornett, D. S., Mobley, J. A., Dias, E. C., Andersson, M., Arteaga, C. L., Sanders, M. E., Caprioli, R. M. (2006). A novel histology-directed strategy for MALDI-MS tissue profiling that improves throughput and cellular specificity in human breast cancer. *Mol Cell Proteomics 5*, 1975–1983.
16. Reyzer, M. L., Caldwell, R. L., Dugger, T. C., Forbes, J. T., Ritter, C. A., Guix, M., Arteaga, C. L., Caprioli, R. M. (2004). Early changes in protein expression detected by mass spectrometry predict tumor response to molecular therapeutics. *Can Res 64*, 9093–9100.
17. Meistermann, H., Norris, J. L., Aerni, H.-R., Cornett, D. S., Friedlein, A., Erskine, A. R., Augustin, A., De Vera Mudry, M. C., Ruepp, S., Suter, L., Langen, H., Caprioli, R. M., Ducret, A. (2006). Biomarker discovery by imaging mass spectrometry: Transthyretin is a biomarker for gentamicin-induced nephrotoxicity in rat. *Mol Cell Proteomics 5*, 1876–1886.
18. Khatib-Shahidi, S., Andersson, M., Herman, J. L., Gillespie, T. A., Caprioli, R. M. (2006). Direct molecular analysis of whole-body animal tissue sections by imaging MALDI mass spectrometry. *Anal Chem 78*, 6448–6456.

CHAPTER 6

Peptide and Protein Analysis Using Ion Mobility–Mass Spectrometry

JEFFREY R. ENDERS, MICHAL KLIMAN, SEVUGARAJAN SUNDARAPANDIAN, and JOHN A. MCLEAN

6.1 ION MOBILITY–MASS SPECTROMETRY: INSTRUMENTATION AND SEPARATION SELECTIVITY

Contemporary proteomics research generally involves either (1) structural/functional proteomics or (2) comprehensive protein identification from complex biological samples. The utility of rapid 2D separations on the basis of gas-phase ion mobility combined with mass spectrometry (IM-MS) in both structural proteomics and rapid protein identification are described in this chapter in the context of prevailing methodologies used in these areas. The selected examples are intended to underscore the advantages and limitations of peptide and protein analysis using IM-MS measurement strategies rather than being a comprehensive review of general IM-MS research. The reader is directed to several recent reviews for a more detailed description of IM-MS [1–5].

The native-state structures of large proteins and other biological molecules adopt conformations that regulate biological function. Detailed geometries of native state structures are typically obtained by using nuclear magnetic resonance [6–8] and X-ray crystallography [9–12]. Although these are considered the gold standards for structure elucidation they both have attendant limitations. Structural elucidation by NMR generally requires highly purified samples and is generally applicable only to proteins of moderate size. X-ray crystallography requires that the protein can be crystalized, which can be exceptionally challenging for membrane-bound and highly flexible proteins. The conformational changes associated with macromolecules in solution phase, or in gas phase, are due to transitions between native and denatured states, sometimes through thermodynamically stable intermediates. Some conformational

Protein and Peptide Mass Spectrometry in Drug Discovery, Edited by Michael L. Gross, Guodong Chen, and Birendra N. Pramanik.

changes associated with large biological molecules in solution are retained in the gas-phase [13–15]. Thus advances in gas-phase structural determinations may provide a unique means for performing structural proteomics directly from complex samples.

Rapid peptide and protein identification is facilitated by advances in liquid chromatography (LC) combined with electrospray ionization (ESI); this combination has resulted in reliable interfaces necessary for quantitative bioanalyses using LC-MS and LC-MS/MS. Although LC-MS/MS, combined with bioinformatics, is the work-horse in contemporary MS-based proteomics research, there remain several significant challenges, including: (1) long analysis times, (2) limited concentration dynamic range, (3) poor detection of low abundance species (e.g., post-translationally modified proteins), and (4) simultaneous detection of proteins with widely varying physical properties (e.g., hydrophobicity, pI).

Gas-phase IM-MS separations combined with LC-IM-MS and LC-IM-MS/MS can now address in part several of these challenges [16–24]. In contrast to liquid phase chromatography, IM separations performed in the gas phase can be achieved on the time scale of milliseconds. This speed is well suited for coupling with MS, whereby hundreds of mass spectra can be acquired across the IM elution profile and then subsequently stitched together to provide a 2D plot of IM versus mass spectra. For example, a single plasma profiling experiment using 2D LC coupled with IM-MS can be completed in about 3 h, which is nearly an order of magnitude faster than is required for conventional LC-MS/MS [25].

The primary aim of this chapter is to describe the instrumentation and information content that can be obtained by combining IM with MS for the study of peptides and proteins. Our hope is to encourage development and application of IM-MS in drug discovery involving peptides and proteins. In this context, the IM drift cell provides great experimental flexibility by acting as a compartmentalized tool that can be utilized in a variety of instrumental arrangements depending on the type of data desired. The recent commercial availability of IM-MS has significantly broadened its implementation in industrial and academic environments. The following sections provide a discussion of (1) the principles and operation of IM-MS instrumentation, (2) the unique separations selectivity of IM-MS, (3) the interpretation of peptide and protein structure, and (4) applications of IM-MS to the analysis of peptides and proteins.

6.1.1 Instrumentation

The primary components of an IM-MS instrument consist of an ion source, a mobility separation cell, a mass analyzer, and a detector as depicted in Figure 6.1*A*. Since the earliest reports of coupling IM with MS [26,27], a wide variety of instrumental arrangements have become available; they utilize a multitude of drift cells and mass analyzers. For example, IM drift cells can be interfaced with quadrupoles [15,27–30], quadrupole ion traps [31,32], double-focusing sector fields [33], Fourier-transform ion cyclotron resonance [34], and time-of flight [27,35,36] instruments. It is beyond the scope of this chapter to discuss the merits and pitfalls of each design, rather several illustrative and contemporary examples are highlighted specifically for IM-MS applications in the analysis of biological samples. Nevertheless, it is important to underscore the use of IM separation cells as a modular component that allows the order of various

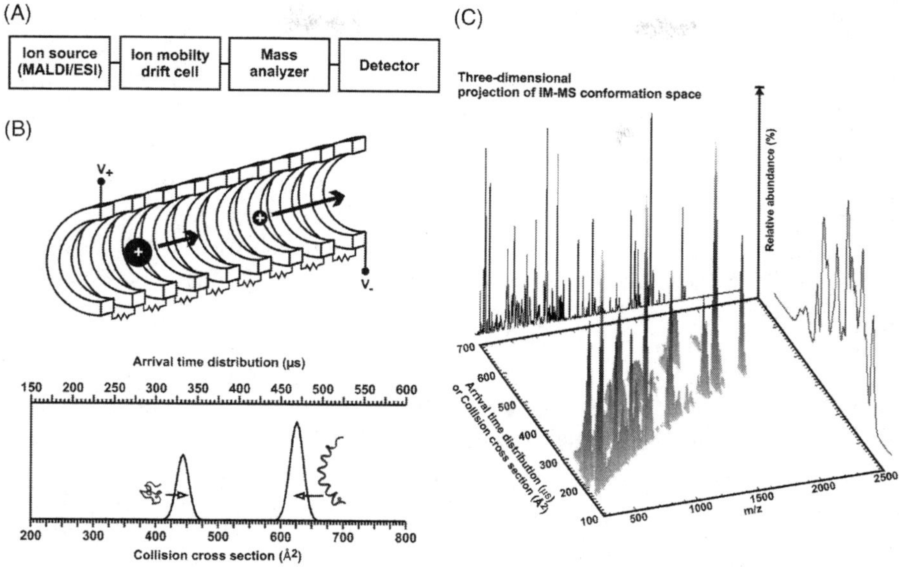

FIGURE 6.1 (*A*) A block diagram of the primary components of biological IM-MS instrumentation. (*B*, top) A conceptual depiction of an IM drift cell. A stack of ring electrodes are connected via resistors in series to form a voltage divider, which is typically designed to generate a relatively uniform electrostatic field along the axis of ion propagation. Ions of larger apparent surface area experience more collisions with the neutral drift gas and therefore elute slower than ions of smaller apparent surface area. (*B*, bottom) A hypothetical IM separation for peptide ions exhibiting two distinct structural subpopulations corresponding to globular (left) and to helical (right) conformations. The arrival time distribution data (top axis), or the observable, can be transformed to a collision cross-sectional profile (bottom axis) via equation [6]. (*C*) A 3D plot of conformation space obtained of a complex biological mixture using MALDI-IM-MS. The mass spectrum in back is what would be observed in the absence of IM, the electropherogram to the right is what would be observed by IM alone. By recording hundreds of TOFMS spectra across the elution profile of the IM, a 3D reconstruction of conformation space is obtained by stitching together the TOFMS spectra as a function of IM arrival time distribution.

components to change and creates new data acquisition modes. This discussion will focus mainly on the three forms of ion mobility that are in considerable use today: (1) drift tube ion mobility (DTIM), (2) traveling wave ion mobility (TWIM), and (3) high field asymmetric waveform ion-mobility spectrometry (FAIMS).

Fundamental Considerations for Ion Source Selection Both MALDI [37,38] and ESI [39] sources can be effectively coupled for IM-MS separations [4,34–47]. From an instrumental standpoint, however, there are clear fundamental differences that dictate the chosen source for particular experiments. To date, the majority of IM-MS designs have incorporated ESI [48–54] partly due to its compatibility with LC and capillary electrophoresis. Importantly, ESI produces an envelope of different charge state ions depending on the chemical composition (e.g., number of basic sites) and size of the analyte. In contrast, MALDI typically produces singly charged ions. The charge state is a critical factor for IM-MS because (1) the mobility of the ion is directly proportional to the charge state, (2) in complex samples multiple charge states yield an overabundance of signals that can result in congested spectra and reduced sensitivity by partitioning signals into multiple channels, and (3) in the low dielectric of the separation cell, multiple charges may alter the structure of the ion owing to electrostatic forces that are shielded in solution. Nevertheless, higher charge-state ions are advantageous when performing IM-MS/MS as they provide higher fragmentation efficiency and also to shift analyte ions to a lower *m/z* range, which may be necessary for certain mass analyzers (e.g., quadrupoles) and for achieving higher sensitivity depending on the detection scheme that is used (e.g., image current or microchannel plates).

Ion Migration and Data Dimensionality In contrast with the gas-phase collisions used in collision-induced dissociation (CID), ion-mobility separations utilize low-energy gas-phase collisions to separate ions predominantly on the basis of molecular surface area, also known as the ion-neutral collision cross section (Ω) with units of $Å^2$. Generally speaking, ions are injected into a separation cell filled with a neutral drift gas and migrate under the influence of a weak electric field (Figure 6.1*B*). The applied electric field is electrostatic for drift tube ion mobility (DTIM) and electrodynamic for both traveling wave ion mobility (TWIM) and field asymmetric waveform ion-mobility (FAIMS) separations. In the presence of the neutral drift gas larger ions have a lower mobility than smaller ions, resulting in longer drift times versus shorter drift times, respectively [48]. Thus IM separations can be thought of as a race between two skiers, one with his arms tucked at his sides and the other with his arms open wide. The skier, guided by the gradient of the hill (as ions are guided by the electric field), traveling with arms tucked will experience less drag and will reach the bottom of the hill faster. Following the elution profile of the ions from the IM-dimension, mass spectra are constantly being acquired and then subsequently stitched together. The result is a three-dimensional data set displaying mass to charge on the x-axis, drift time on the y-axis, and signal intensity on the z-axis (Figure 6.1*C*).

Time and Space Dispersive Ion Mobility Arrangements The common three types of ion-mobility separation cells are (1) drift tube ion mobility, (2) traveling wave ion mobility, and (3) field asymmetric waveform ion mobility. The original ion-mobility design was developed by using a drift tube composed of a series of concentric ring electrodes connected by resistors to create a uniform electrostatic field [48]. Under the influence of this electrostatic field and in the presence of a low molecular weight drift gas (e.g., helium, nitrogen, argon) the ions separate along the axis of motion based on their mobility (i.e., the number of low energy collisions between the ions and the drift gas molecules). The inherent simplicity of the drift tube design (Figure 6.1*B*) enables the transformation of measured drift time data into absolute collision cross section by using the kinetic theory of gases, as detailed in Section 6.2. Thus structural information can be inferred from the measurement of cross section when combined with computational approaches. The resolution afforded in typical DTIM arrangements such as that illustrated in Figure 6.2*A* ranges from 30 to 50 ($R = t/\Delta t$ at FWHM),

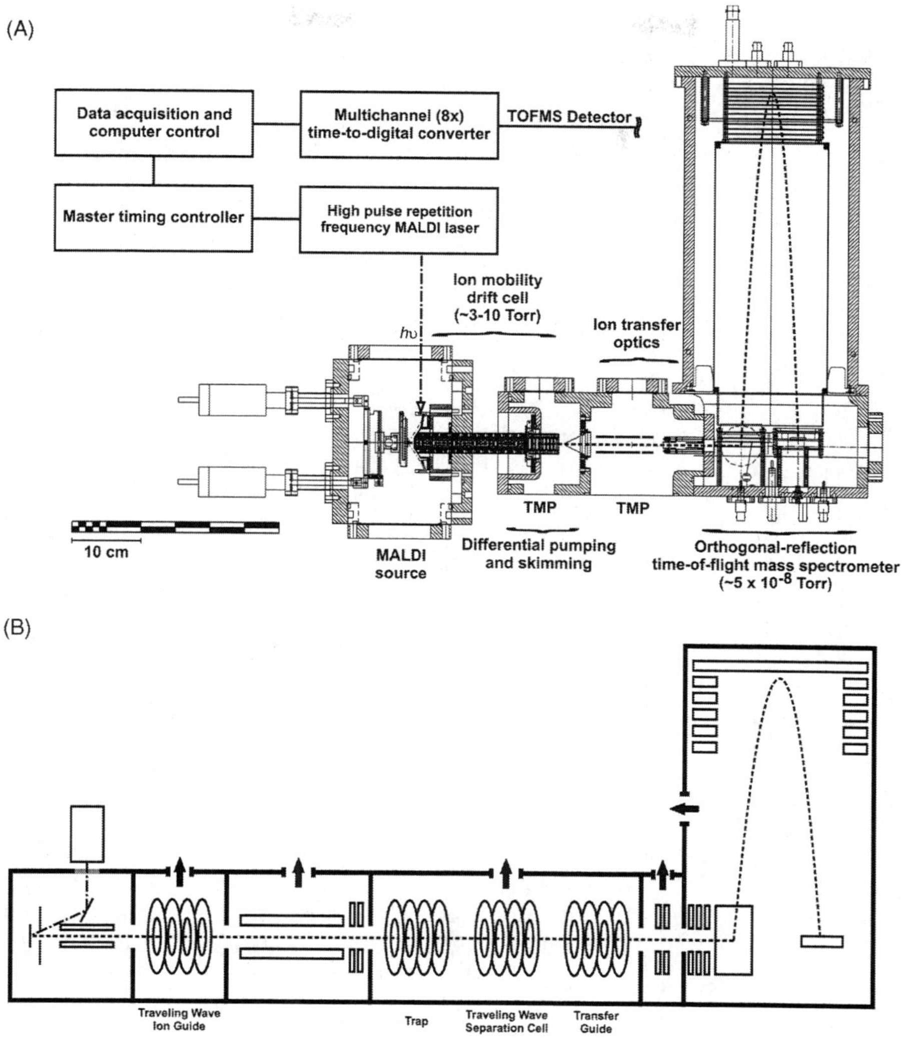

FIGURE 6.2 (*A*) An schematic diagram of a drift tube IM-MS. (*B*) A schematic diagram of a traveling wave IM-MS.

although longer, cryogenically cooled, or higher pressure drift tubes can be used to achieve resolutions exceeding 100 [50,55,56].

The commercial availability of traveling wave ion mobility instrumentation continues to make IM-MS technology accessible to a large number of users (Figure 6.2*B*). Similar to drift tube instruments, TWIM separates ions by time dispersion through collisions with a background drift gas, but by using electrodynamic fields rather than electrostatic fields [1,44]. This is accomplished by transmitting voltage pulses sequentially across a stack of ring electrodes; the voltage pulses create a travelling wave [57]. Conceptually, TWIM separations are performed based on the susceptibility of different ions to the influence of the specific wave characteristics and may be described as the ability of ions to "surf" on these waves [44]. Adjustable wave parameters include traveling wave pulse height, wave velocity, and ramping either of these variables. The commercial platforms are comprised of a MALDI (200-Hz pulse repetition rate) or ESI source, a mass resolving quadrupole, a trapping region for injecting pulses of ions into the TWIM, a TWIM drift cell, an ion transfer region, and an orthogonal TOFMS as depicted in Figure 6.2*B*. Importantly, CID can be performed in the regions before and after the TWIM drift cell. This allows either selection of a precursor ion on the basis of both IM and *m/z*, or for the structural separation of fragment ions, or both. Generally, IM resolution in the TWIM is similar to that of typical DTIM cells, around 30 to 50. Although there are protocols to obtain relative collision cross-sectional values by using TWIM experimental data, the calculations still rely on absolute values obtained by using drift tube instruments [58,59].

FAIMS devices are also commercially available. Ion separations utilizing FAIMS were first documented in the early 1990s by Buryakov and coworkers [60,61]. Separation selectivity in FAIMS occurs on the basis of the nonlinear dependence of the mobility coefficient in varying electric fields. Unlike time-dispersive DTIM and TWIM separations, FAIMS performs space-dispersive separations. In FAIMS devices, ions are subjected to positive and negative electric fields that oscillate perpendicular to the direction of ion propagation. As ions are injected between two parallel plates in a separation cell and this waveform directed perpendicularly across them, only selected ions will traverse the cell, and all other ions will strike either the lower or upper plate. The use of a compensation voltage roughly achieves in FAIMS a mobility-based selection similar to a radio frequency mass selection in a quadrupoles. For a more comprehensive overview of FAIMS, the reader is directed to several excellent reviews [62–71].

To contrast these three IM devices for structural interpretation, the ion-mobility data obtained from DTIM enables absolute collision cross-sectional calculations [72–75], whereas the TWIM and FAIMS provide estimated collision cross sections based on previously measured DTIM values [58,59]. Although efforts to understand the fundamental physical processes in TWIM and FAIMS separations are ongoing [62,69], gas-phase kinetics theory to describe these processes is still a major research endeavor [62–68].

Arrangements for Fragmentation Studies by Ion-Mobility–MS/MS An additional advantage of IM-MS separations is the ability to obtain mobility-correlated fragmentation information by placing a region of ion activation such as CID [76], or surface induced dissociation (SID) [77–79] between the drift tube and mass analyzer (i.e., IM-MS/MS). With ion selection performed in this manner, all fragment ions will possess the same drift time as the precursor ion and yield-correlated fragmentation patterns. An example of IM-MS/MS (Figure 6.3) indicates that with IM utilized to select precursor ions, multiple parent ions can be fragmented in the same analysis provided that they possess different IM drift times. This approach allows for multiplexing MS/MS experiments, rather than using a mass analyzer to select sequentially only a single precursor ion for fragmentation. It is advantageous in complex biological

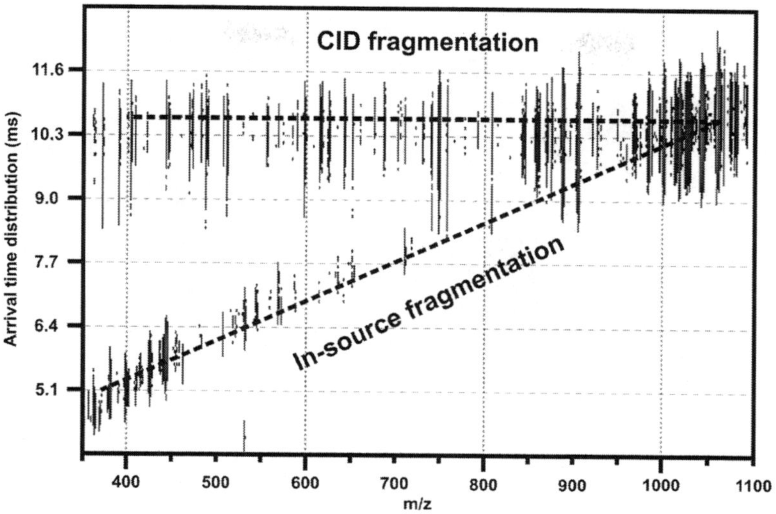

FIGURE 6.3 ESI-TWIM-MS/MS of the peptide, bradykinin (RPPGFSPFR), illustrating the two modes of IM-MS/MS. In-source decay fragmentation of bradykinin yields the diagonal line of peaks in the conformation space plot. Collision-induced dissociation following TWIM separation for this same analyte yields fragmentation ions that share the same arrival time distribution as the parent and so shows up as a horizontal line starting at the parent at the top right and extending to the left.

samples where analyte amount may be limited, or when interfering species can suppress or isobarically mask the ion of interest [80]. Furthermore, additional stages of IM, MS, MS/MS, or IM/IM can be performed depending on the specific goals of the experiment [81].

6.1.2 Separation Selectivity in Bioanalyses

An important aspect of the data dimensionality in IM-MS is the combination of these data forms into what is termed "conformation space," as it represents biomolecular structure, or conformation, as a function of *m/z* [82]. One of the main challenges in the analysis of complex biological samples is the large diversity of molecular species and the high probability for isobaric molecules. Although biomolecules are largely composed of a limited combination of elements (e.g., C, H, O, N, P, and S), different biomolecular classes (e.g., peptides, lipids, and carbohydrates) preferentially adopt structures at a given *m/z* correspondent to the prevailing gas-phase intermolecular folding forces for that class; these forces cause different characteristic molecular densities.

Structural separation on the basis of ion mobility allows isobaric species belonging to different biomolecular classes to be easily resolved, ultimately yielding more confident peak assignments. A hypothetical plot delineating regions of conformation space for which different biomolecular classes are predicted to occur is illustrated in Figure 6.4*A*, and for a suite of experimentally determined cross sections in Figure 6.4*B*.

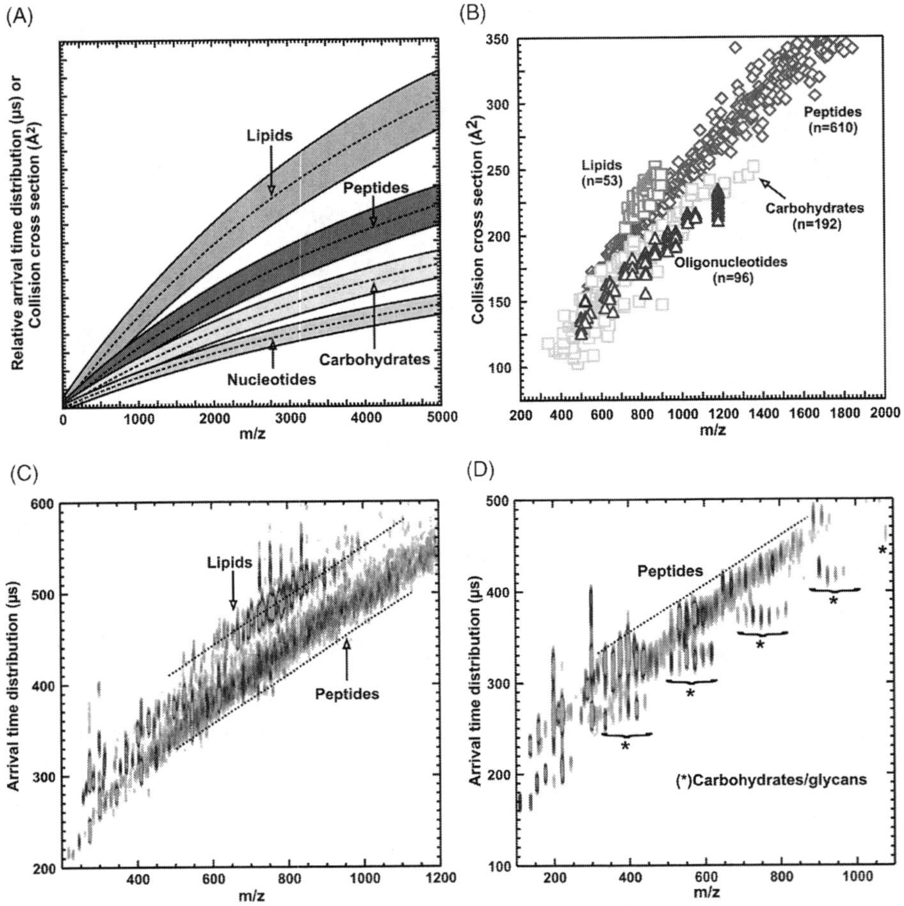

FIGURE 6.4 (*A*) A hypothetical plot highlighting where particular biomolecular classes are expected to appear in conformation space based on differing gas-phase packing efficiencies. (*B*) A plot showing the calculated collision cross-sectional data collected from these biomolecular classes, including oligonucleotides ($n = 96$), carbohydrates ($n = 192$), peptides ($n = 610$), and lipids ($n = 53$). All species correspond to singly charged ions generated using MALDI, where error $\pm 1\sigma$ is generally within the data point. Values for peptide species are from [73]. (*C*) A plot of conformation space illustrating the simultaneous separation of peptides and lipids. (*D*) A plot of conformation space illustrating the simultaneous separation of peptides and carbohydrates. Part (*a*) is adapted with kind permission from Springer Science + Business Media: *Anal. Bioanal. Chem.*, Biomolecular structural separations by ion mobility–mass spectrometry, 391, 2008, 906, L. S. Fenn and J. A. McLean, Fig. 2(a). Part (*b*) is adapted with kind permission from Springer Science + Business Media: *Anal. Bioanal. Chem.*, Characterizing ion mobility–mass spectrometry conformation space for the analysis of complex biological samples, 2009, in press, L. S. Fenn, M. Kliman, A. Mahsut, S. R. Zhao, and J. A. McLean, Fig. 1(a). (*See the color version of this figure in Color Plates section.*)

These plots show that the relative molecular densities increase in the order of nucleotides > carbohydrates > peptides > lipids at a given *m/z*. Although more pronounced for larger *m/z* values, at lower *m/z* values the correlations begin to overlap. Nevertheless, this molecular class selectivity provides a means for performing proteomics without the attendant need for extensive protein purification, because peptides and proteins are separated from what would otherwise be considered isobaric chemical noise.

This minimization of chemical noise overlap is illustrated for the simultaneous determination of peptides and lipids in Figure 6.4*C* and for peptides and carbohydrates in Figure 6.4*D*. Additional conformational differences within a particular molecular class (e.g., peptides/proteins) are also observed owing to conformational isomers arising from peptide/protein secondary and tertiary structure [83–87], intramolecular charge solvation [74,88–90], small-molecule peptide or noncovalent complexes (i.e., quaternary structure) [45,86,91–94], and time-dependent structural refolding of the analyte ion [95,96].

6.2 CHARACTERIZING AND INTERPRETING PEPTIDE AND PROTEIN STRUCTURES

Increasingly, research has focused on utilizing IM-MS for the determination and characterization of peptide and protein structures in support of structural proteomics, and these advances will be important in drug discovery. In this context absolute collision cross sections (CCS) are obtained using DTIM instruments and are then interpreted utilizing quantum mechanical and molecular dynamics calculations. The methods used to implement this process are discussed in three sections: (1) the theory of ion motion within a gas, (2) the practical considerations of calculating CCS values, and (3) the computational strategies for structural interpretation.

6.2.1 The Motion of Ions within Neutral Gases

The theory of ion movement through a neutral gas medium is rooted in the classical electrodynamics of directed diffusion and theoretical gas dynamics studies by Mason and McDaniel among others [97–101]. In principle, the size-dependent movement of ions in an ion-mobility cell is an intuitive process. After an ion is created in the source, it travels “downhill” through the mobility cell along a linearly decreasing electrostatic voltage gradient. As the ion traverses the mobility cell, it experiences numerous low-energy collisions (ca. 10^5–10^7) at moderate pressures and drift cell lengths. At sufficiently low electrostatic field strengths, where ion-neutral collisions are considered elastic, the ion is free to rotate and tumble through the drift cell. Owing to the combination of the large number of collisions and low field strength, the drift velocity (v_d) of the ion as it moves through the neutral gas is size dependent. Ions experiencing different numbers of collisions will have different drift velocities and therefore will spatially separate along the direction of movement.

6.2.2 Considerations for Calculating Collision Cross Sections

Several excellent works describe the mathematical foundation of ion mobility and the derivation of CCS from IM measurements, using the kinetic theory of gases [98,99,101]. At sufficiently low field strength, the drift velocity (v_d) is linearly dependent on the strength of the applied electrostatic field (E), and the linearity constant in this relationship is referred to as mobility (K):

$$v_d = KE. \tag{6.1}$$

Generally, mobility can be determined by measuring the drift time (t_d) of an ion across a drift cell of fixed length (L):

$$K = \frac{v_d}{E} = \frac{L}{t_d E}. \tag{6.2}$$

In practice, however, the time parameter obtained in an IM experiment is the arrival time distribution (t_{ATD}) of a packet of identical ions at the detector. The arrival time of an ion is the sum of the time the ion spends in the drift cell (t_d) and that it spends in the other parts of the instrument (i.e., the ion source, ion optics, and detector region). The time spent outside the drift cell can be determined by performing IM at multiple electrostatic field strengths by varying the potential (V) across the length of the drift cell. Through constructing a plot of t_{ATD} versus the inverse of the applied drift cell potential ($1/V$), a linear regression fit to these data can be seen to yield two important results. First, if the fit is linear, it indicates that low field conditions predominate. Second, the y-intercept corresponds to the time the ions spend outside the drift cell region, which is the limit where the applied potential across the drift cell is infinite and $t_d = 0$.

Ion-drift velocity changes with the number of collisions and therefore depends on gas number density (N), pressure (p), and gas velocity (v_{mean}); all these also depend on temperature (T). To derive CCS from K, the dependence of K on temperature and gas density must be further evaluated. By convention, K is reported as the standard or reduced mobility (K_0), which normalizes K to standard temperature and pressure (i.e., 0 °C/273 K and 760 Torr):

$$K_0 = K\frac{p}{760}\frac{273}{T} = \frac{L}{t_d E}\frac{p}{760}\frac{273}{T}. \tag{6.3}$$

We assume a Maxwellian distribution function of molecular and ion velocities at thermal equilibrium. So the mean thermal velocity of gas and ion molecules under zero electrostatic field conditions can be expressed as v_{mean}, where T is temperature (Kelvin), k is the Boltzman constant, and M_r is reduced mass:

$$v_{mean} = \left(\frac{8kT}{\pi M_r}\right)^{1/2}. \tag{6.4}$$

When a sufficiently weak electrostatic field is applied to a mixture of gas and charged molecules, the direction of movement (and velocity) of ions in the gas consists of two components, the random motion of ions at the temperature of the gas, on which a second small component in the direction of the electrostatic field is imposed. If the mean ion energy (and therefore velocity) as the ion traverses the mobility cell does not increase, the mobility separation is said to be performed under "low field" conditions, and interaction of the ion with the gas molecules via elastic collisions can be assumed. At higher electrostatic field strengths, the ion velocity depends less strongly on the gas temperature, and the mean ion energy increases as it traverses the drift region. Under these "high field" conditions, tumbling and elastic collisions can no longer be assumed, and K is no longer constant but depends on the ratio of electrostatic field to the gas-number density (E/N). Provided that the field strength is sufficiently weak to afford mobility separation at low field conditions (i.e., constant K), a closed equation for the dependence of K_0 on the ion-neutral CCS (Ω) of the ion-neutral pair can be expressed as follows:

$$K_0 = \frac{(18\pi)^{1/2}}{16} \frac{ze}{(k_B T)^{1/2}} \left[\frac{1}{m_i} + \frac{1}{m_n}\right]^{1/2} \frac{1}{N_0} \frac{1}{\Omega}, \tag{6.5}$$

where ze is the elementary charge of the ion, N_0 is the number density of the drift gas at STP (0 °C/273 K and 760 Torr), and k_B is the Boltzmann's constant. The middle term is the reduced mass of the ion-neutral collision pair (ion and neutral masses of m_i and m_n, respectively). Substituting for K_0 and rearranging to solve for Ω yields:

$$\Omega = \frac{(18\pi)^{1/2}}{16} \frac{ze}{(k_B T)^{1/2}} \left[\frac{1}{m_i} + \frac{1}{m_n}\right]^{1/2} \frac{t_d E}{L} \frac{760}{p} \frac{T}{273} \frac{1}{N_0}. \tag{6.6}$$

This is the functional form of the equation used to derive the collision cross section from mobility data. This equation holds when the collisions of ions with neutral gas molecules can be assumed to be elastic. The collision cross section thus derived is termed "hard sphere" (i.e., only momentum is transferred at the point of ion-neutral collision). Comparisons of experimentally measured and theoretically derived CCSs show that the hard-sphere, elastic-collision assumption holds for analytes larger than around 500 Da, which is a mass range well suited for peptide analysis [102].

6.2.3 Computational Approaches for Interpretation of Structure

Empirically determined collision cross-sectional values can be used to probe structural motifs of various biomolecular ions including peptides/proteins [103–105] and protein complexes [45,58]. This is accomplished by augmenting IM CCS measurements with computational modeling strategies. For the structural interpretation, the CCS is used as a size constraint and to discriminate and interpret a subset of structures from a large pool of computationally generated conformers. Structural populations can be expected to consist of structurally similar ions that either undergo thermally accessible structural/conformational isomerization/interconversion or, depending on intramolecular forces and atomic arrangement, remain in relatively fixed conformations on the time scale of the IM-MS experiment [106]. Therefore each IM-MS unique signal, ideally corresponding to identical ions with different conformations, has its own unique structural signature determined by the average shape of the populations of ions that create it. This structural signature is the basis for biomolecular separations observed in IM-MS (lipids from peptides, peptides from oligonucleotides and carbohydrates) [1,107,108]. It is also the explanation for the observed difference in average IM versus *m/z* trendline variability (spread, structural richness) of the different biomolecular classes observed by IM-MS [107,109]. For example, peptide mobility versus *m/z* IM-MS spectra can be expected to contain, and indeed to show, predictable variability due to PTMs [74,89,110] amino acid content, sequence [17,111,112], and type of cationizing agent.

Following the experimental determination of CCS, computational modeling is performed in the following general sequence: (1) the structure of the molecular ion is built by using one of the available structure-building software packages, (2) the atomic charge parameters for the built structure are determined by using quantum chemical calculations, and point charge derivation, and (3) molecular dynamics is performed to create a large number of structural conformers whose CCS can be computationally determined to allow experimental CCS based clustering analysis. A schematic of the work flow for combining the measured CCS with computational interpretation is illustrated in Figure 6.5.

Currently there are several software packages well suited for structure building of peptides and proteins, some proprietary (e.g., Molecular Operating Environment,

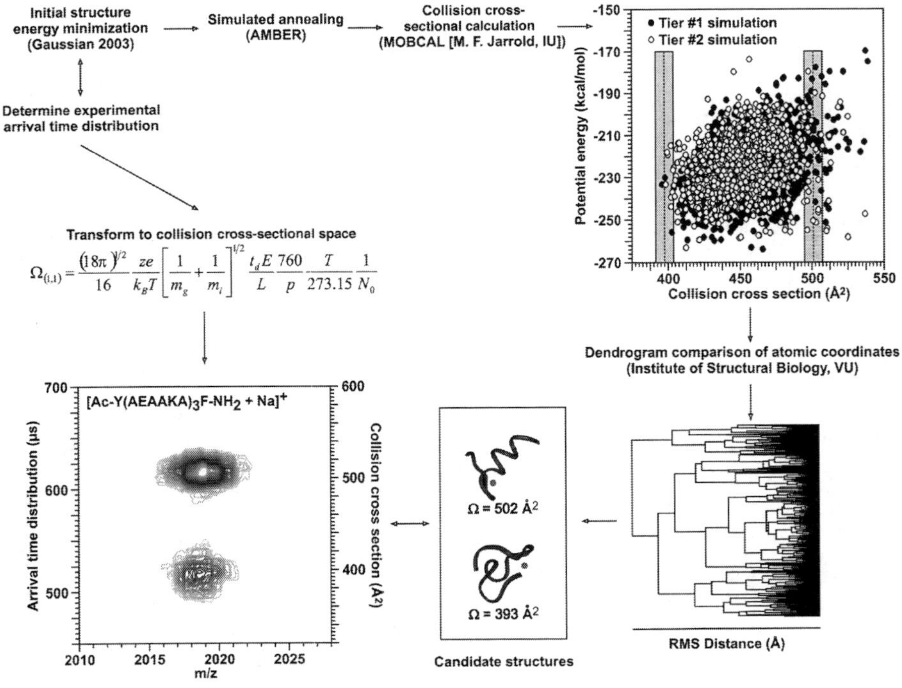

FIGURE 6.5 Modeling protocol used to interpret peptide and protein structure based on the absolute collision cross-sectional measurements acquired from IM-MS data. (*See the color version of this figure in Color Plates section.*)

Chemical Computing Group, Montreal, Canada) and some freely available (e.g., SIRIUS, University of California, San Diego). One should be aware of online structure databases and search for already available structures in publicly accessible databases such as the Research Collaboratory for Structural Bioinformatics Protein Data Bank (RCSB PDB, University of California, San Diego) and the Protein Databank (European Bioinformatics Institute, United Kingdom), which are both closely associated with the worldwide Protein Databank (wwPDB) project. Although computer processing speed has increased exponentially over several decades, computation still puts a limit on how large a system can be realistically studied by using the protocol described here. Quantum chemical minimization and electrostatic potential calculations in charge parameter development at intermediate level of quantum theory on the most powerful desktop workstations give reasonable 24 h results on systems of 20 atoms. Calculations, however, scale up exponentially not only with the level of theory and number of atoms, but also with mass of included elements. For example, studies of proteins containing metal ions may require additional specialized basis sets and time. Therefore quantum charge parameterization of larger systems has to be done on properly capped fragments, which are later connected for molecular dynamics sampling.

There are many free and proprietary quantum mechanical software packages capable of performing the needed minimization and electrostatic potential calculations (e.g., Cerius, Jaguar, Gaussian, and Spartan) at basic to intermediate levels of theory. Recently density functional theory (DFT) methods have made inroads into quantum minimization specifically for their ability to handle efficiently electron distribution of higher mass elements, but their accuracy is still being studied and molecular dynamics parameterization lags behind these in parameter development for higher mass elements.

The last step of the computational protocol is the conformational MD sampling coupled with computational clustering analysis based on the comparison of the computed CCS values for the generated structural conformers with the experimental CCS value. Once the charge parameters for each atom in the built structure are determined, one of the available molecular dynamics packages (e.g., Amber [113], CHARMM, ACCELRYS) can be used to perform conformational sampling. Conformational sampling is a molecular dynamics calculation that yields a large number of structural conformers of the studied peptide/protein ion. Two primary challenges of MD-based conformational sampling methods are the use and development of reliable MD parameters that ensure chemically and structurally relevant results and of temperature protocols that lead to a completely randomized pool of final structures. Aforementioned MD software packages typically contain reliable MD parameters for standard amino acids, carbohydrates, and nucleotides. Fully randomized sampling without any bias toward one or several conformations can normally be guaranteed by selecting a temperature scheme (heating and cooling algorithm) for the MD calculation in which the studied peptide/protein structure is heated to a temperature at which it can access all reasonably expected conformational states and then cooled slowly to relax into a low-energy structural conformer. An important output result of the MD calculation is the relative potential energy value of each conformer within the pool of generated structures. The computational collision cross-sectional value for each conformer is then calculated by using one of two currently available codes, Mobcal [114] or Sigma [115–117]. Once both the relative potential energy and collision cross section values are computed for each member of the computationally generated ensemble of structures, a plot of energy versus computed CCS is generated and visually inspected. Typically the lowest energy conformers with a computed CCS that matches the experimental value range are selected for further computational analysis (e.g., analysis of selected average distances, or clustering on the basis of atomic positions). Following cluster analysis, general structural motifs are obtained that are consistent with the measurement, while other structural motifs can be ruled out.

6.3 APPLICATIONS OF IM-MS TO PEPTIDE AND PROTEIN CHARACTERIZATIONS

IM-MS studies can now be used to structurally characterize peptide systems consisting of a few to many amino acids. The approach can be even extended to massive protein complexes in structural biology. IM-MS also plays an important role in medical research for developing a better understanding of Alzheimer's disease, Parkinson's disease, and Huntington's disease. When combined with LC approaches, IM-MS can be used to detect trace level proteins that are present in complex samples such as human urinary and plasma samples [16,21,22]. A survey of studies using IM-MS to investigate the structural biophysics of model peptides and proteins is presented in Table 6.1. A summary of studies aimed at instrumentation for the study of peptides and proteins, proteome profiling, and medicinal research is presented in Table 6.2. In the sections that follow, summaries of several studies in understanding peptide and protein biophysics and in understanding protein complexes are presented.

6.3.1 Fundamental Studies of Peptide and Protein Ion Structures

IM-MS can be used to examine the detailed structures of amino acids, as illustrated by the work of Bowers group in their gas-phase studies of cationized glycine; they showed that these molecular ions mostly exhibit charge-solvation structures [118]. Small alkali ions (e.g., sodium) bind to both the *N*- and the *C*-terminus whereas larger alkali ions (e.g., rubidium and cesium) bind preferentially to the *C*-terminus [118]. The stabilization of the charge-solvation structure compared to that of the salt bridge is larger for $GlyCs^+$ than for $GlyNa^+$. The approach is to use a high pressure drift cell with ESI introduction and augment with molecular dynamics calculations [119]. The outcome shows that solvating water molecules primarily bind to the ionic functional groups present in peptides. DFT calculations for dipeptide ions (Gly-Gly + H^+, Gly-Ala + H^+, Ala-Gly + H^+, Ala-Ala + H^+, Pro-Gly + H^+ and Gly-Trp + H^+) [120] show that the dominant water adsorption process involves a significant conformational change to accommodate the bridging water molecule. Molecular modeling and DFT calculations of deprotonated dimers of alanine and glycine [121] indicate that the charge is poorly shielded and readily available for noncovalent interaction with other molecules.

TABLE 6.1 Listing of Biophysical Studies of Model Peptides and Proteins

Model System	Type of Study	References
Alanine, glycine, arginine, mod AA	Gas-phase conformations noncovalent or cation/peptide interaction	118, 119, 161
Ala-, Gly-, Pro-, Trp-based dipeptides additional dipeptides poly (Ala) and modified poly (Ala)	Hydration and folding gas-phase conformations, helices and sheets	119–121, 162, 163
Modified pentapeptides	Hydration and folding	83, 104, 120, 122, 123, 164–176
Poly (Gly) and modified poly (Gly)	Gas-phase conformations, helices and sheets	122, 123, 166–168, 170, 177, 178
Poly (Pro)	Anhydrous helices/globules	124
Poly (TrpGly)	Gas-phase conformations	179
Modified poly (Leu), poly (Val)	Helices and sheets	123
di-AA clusters	Gas-phase conformations, noncovalent interactions	163
Serine clusters	Noncovalent or cation–peptide interaction	125, 180, 181
Proline clusters	Gas-phase conformations	182, 183
Amyloid beta protein	Gas-phase conformations, Noncovalent interactions	119, 161
Angiotensin II	Noncovalent or cation–peptide interaction	184
Apomyoglobin	Gas-phase conformations	126, 185
BPTI (bovine pancreatic trypsin inhibitor)	Hydration and folding	127, 186–188
Bradykinin	Gas-phase conformations, HD exchange, noncovalent or cation–peptide interaction	35, 88, 119, 176, 184, 189
Bradykinin fragment 1–5 (RPPGF)	H/D exchange	90
Cytochrome C	Gas-phase conformations	103, 127, 128, 185, 187, 190–192
Defensins (beta-def., P-defensin DEFB107)	Gas-phase conformations, hydration and folding	193, 194
Dynorphin fragments	Noncovalent or cation–peptide interaction	91, 184

(continued)

TABLE 6.1 (*Continued*)

Model System	Type of Study	References
Mini Gastrin I peptide	Noncovalent or cation–peptide interaction	91
Gonadotropin-releasing hormone	Gas-phase conformations, hydration and folding	193
Gramicidin S	Gas-phase conformations	184, 195
Hemoglobin	Gas-phase conformations, HD exchange	196
Oxidized Insulin chain A (ICA)	Noncovalent or cation–peptide interaction	176, 197
Kemptide	Noncovalent or cation–peptide interaction	184
LHRH (leutenizing hormone releasing hormone)	Hydration and folding	119
Lysozyme	Gas-phase conformations, hydration and folding	73
beta(2)-Microglobulin (beta(2)m)	Gas-phase conformations, folding, aggregates	198
Myoglobin	Gas-phase conformations, hydration and folding	119
Neurotensin 3+	Gas-phase conformations, hydration and folding	119, 161, 162
Oxytocin	Cation–peptide interaction	105, 199
alpha Synuclein	Noncovalent interaction	93, 200
Ubiquitin	Gas-phase conformations, hydration and folding	96, 106, 201–206
Mobility (676, 4000) and CCS (660) database of ESI (multiply charged) peptides	Gas-phase conformations	112, 130, 131
CCS database of MALDI (singly charged) peptides	Gas-phase conformations	73

TABLE 6.2 Topical Listing of IM-MS Studies Related to Instrumentation, Proteome Profiling, and Medicinal Research

Topic	References
Collision-induced activation	81, 159, 207, 208
Combinatorial studies	209
Conserved solution phase structure	84, 129, 203, 210
H/D exchange/labeling structure study	85, 171, 194, 211–213
IM-(CID, SID)/MS	16, 20, 21, 23, 24, 52, 78, 87, 208, 211, 214–228
IM-MS resolution/separation power	2, 17, 18, 20, 21, 52, 74, 78, 81, 84, 85, 87, 89, 108–112, 130, 163, 207, 208, 211, 215, 216, 224, 226–235
IM-MS enhanced detection	130, 232, 233
Instrumental advances	20, 21, 23, 52, 78, 130, 159, 208, 214–216
Ionization Type	
ESI studies	16–18, 20–25, 52, 81, 87, 110–112, 130, 159, 163, 194, 207, 209, 211, 214–231
MALDI studies	2, 74, 78, 84, 85, 89, 108, 109, 232–235
Peak capacity of IM-MS approach	228, 234, 235
Peptide libraries	18, 52, 87, 89, 109, 110, 163, 226, 229–232, 235
Peptide aggregation in gas phase	94, 163, 176, 194, 236, 237
Peptide mobility/CCS prediction algorithm	131, 238–244
Phosphorylated peptides	74, 89, 110
Proteome profiling	
Drosophila	20, 23, 24, 219–223
Human urinary proteome	21
Human plasma proteome	16, 22
Medicinal research	
Alzheimer's disease	94, 161, 236, 237, 245–248
Parkinson's disease	93, 200, 220–223
Huntington's disease	218
Other	16, 21, 105, 111, 194, 249, 250
Shift reagents	231

High-resolution IM measurements and molecular dynamics simulations of larger systems, namely protonated polyglycine and polyalanine (Gly_nH^+ and Ala_nH^+, where $n = 3$–20) in the gas-phase can also be done, as illustrated by the Jarrold group [122]. They found that the measured CCS for both the polyglycine and the polyalanine peptides are consistent with a self-solvated globule conformation. Here the change in conformation is due to the peptide chain wrapping around the charge located on the terminal amine. The conformations of unsolvated leucine-based peptides form helices more readily than alanine, but less readily than valine (Val > Leu > Ala) [123]. Interestingly, the side-chain entropy is not the determining factor in helix formation when dealing with unsolvated peptides.

IM-MS methods combined with molecular modeling calculations can be used to examine the conformations and charge states of polyproline peptides, $[Pro_n + zH]^{z+}$ (where $n = 3$–56; and $z = 1$–6), as was demonstrated by Clemmer and coworkers [124]. Their investigations show that all proline residues are in the cis configuration, and protonation at the *N*-terminus allows hydrogen bonds to be formed with backbone carbonyl groups of the second and third proline residues in each polyproline chain. Protonation at the *N*-terminus also stabilizes the helix macrodipole. Singly charged ions formed from aqueous solutions favor globular and hairpin-like conformers that contain both *cis*- and *trans*-proline residues, whereas higher charge state ions ($z = 3$–6) formed from aqueous solutions favor relatively extended conformations. With the increase in the number of proline residues, however, higher charge state ions became more compact. Similar studies can be performed on serine clusters [125] by using a sonic-spray ionization method in an IM-MS instrument. Experimentally derived cross sections with values calculated for trial geometries indicate that large clusters favored tightly packed, roughly spherical geometries.

The collision cross-sectional values of several cationized forms of bradykinin ($(BK + H)^+$, $(BK + Na)^+$, and $(BK - H + 2Na)^+$), introduced by MALDI, were calculated by Bowers and his coworkers. All three species had very similar cross sections of 245 ± 3 Å^2, and these cross sections were independent of temperature from 300 to 600 K. This particular property of bradykinin makes it the standard calibrant peptide of choice for many IM studies.

Ion mobility measurements can be used to characterize the conformations of various charge states of proteins. An example is the +4 to +22 charge states of apomyoglobin [126]. When collisionally heated, the higher (> +7) charge states undergo proton stripping to produce +4 to +7 charge states. This results in spontaneous collapse into partially folded conformations. Only extended conformations exist for the high (> +10) charge states in the gas phase, and they become more extended as the number of charges increases, whereas the lower charge states (< +7) adopt a slightly more compact structure than the native protein in solution. This study also shows that the CCS per residue for the conformations of apomyoglobin and cytochrome c is similar [126,127]. An important result from these studies is that different proteins share common structural motifs in the gas phase, as they do in solution [103,127,128]. One can use desorption electrospray ionization (DESI) assisted by IM-MS analysis as shown for solid-phase horse heart cytochrome c and lysozyme from chicken egg; the outcome shows that charge state distributions are comparable to those produced by ESI [129]. Elongated structures of these two proteins are formed when the DESI solvent is mixed with a small amount (~2%) of acetic acid [129].

A peptide library containing 4000 different peptides having the general form NH_2-Xxx-Xxx-Xxx-CO_2H, NH_2-Ala-Xxx-Xxx-Xxx-CO_2H, NH_2-Ser-Ala-Xxx-Xxx-Xxx-CO_2H, and NH_2-Leu-Ser-Ala-Xxx-Xxx-Xxx-CO_2H (where Xxx represents Ala, Arg, Asp, Glu, Gly, Leu, Lys, Phe, Ser, or Val) is now available from LC measurements combined with IM-MS [130]. By LC-IM-MS approximately 82% of the 4000 library components can be resolved [130]. There is also available a CCS database of 420 singly charged and 240 doubly charged peptide ions introduced by utilizing an ESI source followed by IM-MS [131]. A similar database of ion-neutral collision cross sections for 607 singly charged peptide ions introduced with a MALDI source is available from Russell and coworkers [73].

6.3.2 Studies in Structural Biology—Protein Complex Characterization

Protein complexes are large aggregates of often two or more proteins that are noncovalently bound. These intact complexes are able to perform intricate chemical processes and are often the biological machinery at work behind many of the most important physiological functions in the body. Understanding their properties will likely be important in drug discovery and development. Electrospray ionization–ion-mobility measurements of these large protein complexes is now at the forefront of structural biology research following a number of studies throughout the 1990s and 2000s that showed the predominant secondary, tertiary, and quaternary structures to be largely conserved during the transition from liquid to gas phase [132]. The ability of IM-MS for characterizing molecular structure makes it well suited for gaining insights in such structural biology studies.

TABLE 6.3 A Listing of IM Studies of Protein Complexes

Protein Complex	Type of Study	References
20S proteasome from Methanosarcina thermophila	Analysis of the binding and stoichiometry, conservation of structure in the gas phase	133
Trp RNA binding protein (TRAP)	Stability of complex, effect of Trp, RNA binding	45
Plasma protein complex 56 kDa tetrameric transthyretin (TTR)	Study of stable collision activated protein unfolded structures over ms time scales	137
Wheat heat shock protein TaHSP16.9	Stability of dodecamers in solution and gas phases	58
Protein kinase G (PKG)	Investigation of global conformational changes of intact 152 kDa dimeric PKG; structural changes upon cGMP-binding	251, 252
Hepatitis B Virus (HBV) capsid protein	Preservation of solution phase structure in vacuo; study of viral capsid assembly	253
Bacteriophage P22 portal protein	Structure of P22 portal protein; conformational changes upon its association with tail factor gp4	254
Acr1 mycobacterium tuberculosis heat shock protein	Collision induced activation of Acr 1, shock protein a 12-subunit 197-kDa dodecameric protein	255

Most IM protein complex studies are aimed at probing gas-phase structure as a reliable representation of the solution counterpart [133,134]. Many such studies attempt to justify the results elicited from ion mobility by comparing them to electron microscopy, X-ray crystallography, and NMR data through the use of sources such as the protein data bank [133,135]. Work has progressed to larger and larger species, culminating in the successful ionization and structural investigation of the MVP vault protein complex at around 9.4 MDa [136]. A summary of studies using IM-MS to characterize protein complexes is provided in Table 6.3.

Computational methods have also been used to study these systems [45,58,137]. Although current computational techniques are not able to handle the large complexes as of yet, approximations and coarse-grained models can now be used in these studies, and this work does provide significant insight into the experimental ion-mobility measurements observed.

6.4 FUTURE DIRECTIONS

6.4.1 Applications

Any drug-based treatment (both prophylactic and therapeutic) has its own limitations due to the efficacy of the drug and induced side effects that vary from one individual to another. Identification of biomarker signatures of individual patients allows for diagnosis of the disease at a very early stage and improved treatment strategies with fewer side effects. Biomarkers can be identified by a variety of standard laboratory techniques including common blood tests, immunohistochemistry [138,139], and flow cytometry [140], or more recent omics methods such as genomics [141], transcriptomics [141], proteomics [2,142,143], peptidomics [144], glycomics [18,31,145], lipidomics [146] and metabolomics [72]. Linking the individual patients' condition with the correct drug based on clinical biomarkers opens the door for personalized medicine. Even though the instrumentation is now expensive and complex, personalized medicine may become widely accessible with advances in high-throughput and routine instrumentation. Advances in IM-MS technology over the past few decades have enhanced its ability to analyze samples collected from a large number of individuals in a reasonable amount of time; these discoveries may contribute to the discovery of biomarkers and play an important role in the field of personalized medicine.

In a related way, the process of drug screening is greatly dependent on the use of modern technologies to determine in vitro therapeutic efficacy. In vitro cellular response has traditionally been characterized by using live/dead stains, simple optical and electrochemical analysis, or tagging [147] techniques that monitor only a limited number of specified analytes. Given that the cellular secretion profile is complex in nature (encompassing many different bimolecular classes), the use of traditional methods fails to capture the richness of the cellular response. Work is underway to implement novel drug screening methods through the use of combining microfluidics with IM-MS. By using IM-MS after a type of microfluidic cell trapping device, termed a multi-trap nanophysiometer [148–153], one can direct medium over the trapped cells and subsequently analyze their secreted biomolecules [154,155]. IM-MS is uniquely suited for expediently delineating contributions from individual biomolecular classes (i.e., carbohydrates, nucleotides, proteins, and lipids) that may be present in this exudate without the limitation of isobaric interferences. This way cellular response can be gauged on the basis of a broad biomolecular output profile [156]. Ineffective drug candidates would be eliminated early in the testing process based on cost-efficient personalized screening assays.

6.4.2 Instrumentation

Progress in the field of IM-MS is governed not only by the areas to which it is applied but also by the development of new technologies. An example is the work of Russell's group. Advancing upon initial studies [4,33,157], they recently developed a variable temperature IM-TOFMS [158] for studying the structural differences between isomers. At low temperatures a number of favorable changes occur: (1) ion diffusion within the buffer gas decreases, leading to less diffusional broadening, (2) the internal energy of the ion is lowered, and (3) long-range ion-neutral gas interactions become more pronounced. Therefore low temperature IM-MS offers high mobility resolution, an important advantage for structural separation of isomers having very similar cross-sectional values. Future directions of variable temperature IM-MS will likely focus on further eliciting the folding kinetics of peptides, proteins, protein complexes, and drug/protein receptor interactions.

A second goal in instrumentation development is to increase the peak capacity of the measurement to facilitate even more comprehensive characterization of complex samples. There exist two paths to increasing peak capacity: (1) through combining separation techniques, so-called hyphenated methods (*e.g.* LC-IM-MS, GC-IM-MS, and IM/IM-MS) with IM-MS and (2) improvements to current designs that lead to increased mobility resolution. The introduction of the "combing" technique by Clemmer and colleagues [159] is an example of hyphenated strategy to increase peak capacity. Recent experiments on a circular drift tube by Clemmer and colleagues [55,160] are expected to lead to increased resolution and peak capacity in IM separations.

In summary, the ability of IM to function as a compartmentalized tool that combines well with different types of mass spectrometers and different ionization sources has led to a wide variety of applications in peptide and protein science. Ion mobility has been demonstrated to be highly useful in the characterization of samples ranging from small peptides to large protein complexes, both as a structural characterization tool and as a means for characterizing complex biological samples.

ACKNOWLEDGMENTS

Financial assistance for this work was provided by the National Institutes of Health (RC2DA028981), the US Defense Threat Reduction Agency (HDTRA-09-1-0013), the Vanderbilt University College of Arts and Sciences, the Vanderbilt Institute for Chemical Biology, and the Vanderbilt Institute for Integrative Biosystems Research and Education.

REFERENCES

1. Fenn, L. S., McLean, J. A. (2008). Biomolecular structural separations by ion mobility-mass spectrometry. *Anal Bioanal Chem 391*, 905–909.
2. McLean, J. A., Ruotolo, B. T., Gillig, K. J., Russell, D. H. (2005). Ion mobility-mass spectrometry: A new paradigm for proteomics. *Int J Mass Spectrom 240*, 301–315.
3. Wyttenbach, T., Bowers, M. T. (2003). Gas-phase conformations: The ion mobility/ion chromatography method. *Mod Mass Spectrom 225*, 207–232.
4. Clemmer, D. E., Jarrold, M. F. (1997). Ion mobility measurements and their applications to clusters and biomolecules. *J Mass Spectrom 32*, 577–592.
5. Jarrold, M. F. (2000). Peptides and proteins in the vapor phase. *An Rev Phys Chem 51*, 179–207.
6. Wuthrich, K. (1986). *NMR of Proteins and Nucleic Acids*. Wiley, New York, p 320.
7. Bax, A. (1989). Two-dimensional NMR and protein-structure. *An Rev Biochem 58*, 223–256.
8. Wuthrich, K. (1989). The development of nuclear magnetic-resonance spectroscopy as a technique for protein-structure determination. *Accounts Chem Res 22*, 36–44.
9. Lipscomb, W. N. (1953). Experimental crystallography. *An Rev Biochem 58*, 223–256.
10. Perutz, M. F., Rossmann, M. G., Cullis, A. F., Muirhead, H., Will, G., North, A. C. T. (1960). Structure of haemoglobin: a three-dimensional Fourier synthesis at 5.5 Å resolution, obtained by X-ray analysis. *Nature 185*, 416–422.
11. Kendrew, J. C., Dickerson, R. E., Strandberg, B. E., Hart, R. G., Davies, D. R. (1960). Structure of myoglobin: A three-dimensional Fourier synthesis at 2 Å resolution. *Nature 185*, 422–427.
12. Blundell, T. L., Johnson, L. N. (1976). *Protein Crystallography*, Academic Press, New York, p. 565.
13. Loo, R. R. O., Smith, R. D. (1994). Investigation of the gas-phase structure of electrosprayed proteins using ion-molecule reactions. *J Am Soc Mass Spectrom 5*, 207–220.
14. Covey, T., Douglas, D. J. (1993). Collision cross-sections for protein ions. *J Am Soc Mass Spectrom 4*, 616–623.
15. Hudgins, R. R., Woenckhaus, J., Jarrold, M. F. (1997). High resolution ion mobility measurements for gas phase proteins: Correlation between solution phase and gas phase conformations. *Int. J. Mass Spectrom. Ion Processes 165*, 497–507.
16. Liu, X. Y., Valentine, S. J., Plasencia, M. D., Trimpin, S., Naylor, S., Clemmer, D. E. (2007). Mapping the human plasma proteome by SCX-LC-IMS-MS. *J Am Soc Mass Spectrom 18*, 1249–1264.
17. Valentine, S. J., Counterman, A. E., Hoaglund, C. S., Reilly, J. P., Clemmer, D. E. (1998). Gas-phase separations of protease digests. *J Am Soc Mass Spectrom 9*, 1213–1216.
18. Taraszka, J. A., Counterman, A. E., Clemmer, D. E. (2001). Gas-phase separations of complex tryptic peptide mixtures. *Fresenius J Anal Chem 369*, 234–245.
19. Valentine, S. J., Kulchania, M., Srebalus Barnes, C. A., Clemmer, D. E. (2001). Multidimensional separations of complex peptide mixtures: A combined high-performance liquid chromatography/ion mobility/time-of-flight mass spectrometry approach. *Int J Mass Spectrom 212*, 97–109.

20. Myung, S., Lee, Y. J., Moon, M. H., Taraszka, J., Sowell, R., Koeniger, S., Hilderbrand, A. E., Valentine, S. J., Cherbas, L., Cherbas, P., Kaufmann, T. C., Miller, D. F., Mechref, Y., Novotny, M. V., Ewing, M. A., Sporleder, C. R., Clemmer, D. E. (2003). Development of high-sensitivity ion trap ion mobility spectrometry time-of-flight techniques: A high-throughput nano-LC-IMS-TOF separation of peptides arising from a Drosophila protein extract. *Anal Chem 75*, 5137–5145.

21. Moon, M. H., Myung, S., Plasencia, M., Hilderbrand, A. E., Clemmer, D. E. (2003). Nanoflow LC/ion mobility/CID/TOF for proteomics: Analysis of a human urinary proteome. *J Proteome Res 2*, 589–597.

22. Liu, X., Plasencia, M., Ragg, S., Valentine, S. J., Clemmer, D. E. (2004). Development of high throughput dispersive LC-ion mobility-TOFMS techniques for analysing the human plasma proteome. *Briefings Funct Genomics Proteomics 3*, 177–186.

23. Taraszka, J. A., Kurulugama, R., Sowell, R. A., Valentine, S. J., Koeniger, S. L., Arnold, R. J., Miller, D. F., Kaufman, T. C., Clemmer, D. E. (2005). Mapping the proteome of *Drosophila melanogaster*: Analysis of embryos and adult heads by LC-IMS-MS methods. *J Proteome Res 4*, 1223–1237.

24. Taraszka, J. A., Gao, X. F., Valentine, S. J., Sowell, R. A., Koeniger, S. L., Miller, D. F., Kaufman, T. C., Clemmer, D. E. (2005). Proteome profiling for assessing diversity: Analysis of individual heads of *Drosophila melanogaster* using LC-ion mobility-MS. *J Proteome Res 4*, 1238–1247.

25. Valentine, S. J., Liu, X. Y., Plasencia, M. D., Hilderbrand, A. E., Kurulugama, R. T., Koeniger, S. L., Clemmer, D. E. (2005). Developing liquid chromatography ion mobility mass spectometry techniques. *Expert Rev Proteomics, 2*, 553–565.

26. Barnes, W. S., Martin, D. W., McDaniel, E. W. (1961). Mass spectrographic identification of the ion observed in hydrogen mobility experiments. *Phy Rev Lett 6*, 110–111.

27. McAfee, K. B. J., Edelson, D. (1963). Identification and mobility of ions in a Townsend discharge by time-resolved mass spectrometry. *Proc Phys Soc London, 81*, 382–384.

28. Karasek, F. W., Hill, H.H., Kim, S.H. (1976). Plasma chromatography of heroin and cocaine with mass-identified mobility spectra. *J Chromatogr 117*, 327–336.

29. Liu, Y. S., Valentine, S. J., Counterman, A. E., Hoaglund, C. S., Clemmer, D. E. (1997). Injected-ion mobility analysis of biomolecules. *Anal Chem 69*, A728–A735.

30. Wu, C., Siems, W. F., Asbury, G. R., Hill, H. H. (1998). Electrospray ionization high-resolution ion mobility spectrometry - Mass spectrometry. *Anal Chem 70*, 4929–4938.

31. Hoaglund, C. S., Valentine, S. J., Clemmer, D. E. (1997). An ion trap interface for ESI-ion mobility experiments. *Anal Chem 69*, 4156–4161.

32. Creaser, C. S., Benyezzar, M., Griffiths, J. R., Stygall, J. W. (2000). A tandem ion trap/ion mobility spectrometer. *Anal Chem 72*, 2724–2729.

33. Vonhelden, G., Wyttenbach, T., Bowers, M. T. (1995). Inclusion of a Maldi ion-source in the ion chromatography technique—Conformational information on polymer and biomolecular ions. *Int J Mass Spectrom Ion Processes 146*, 349–364.

34. Bluhm, B. K., Gillig, K. J., Russell, D. H. (2000). Development of a Fourier-transform ion cyclotron resonance mass spectrometer-ion mobility spectrometer. *Rev Sci Instr 71*, 4078–4086.

35. Hoaglund, C. S., Valentine, S. J., Sporleder, C. R., Reilly, J. P., Clemmer, D. E. (1998). Three-dimensional ion mobility TOFMS analysis of electrosprayed biomolecules. *Anal Chem 70*, 2236–2242.

36. Gillig, K. J., Ruotolo, B., Stone, E. G., Russell, D. H., Fuhrer, K., Gonin, M., Schultz, A. J. (2000). Coupling high-pressure MALDI with ion mobility/orthogonal time-of flight mass spectrometry. *Anal Chem 72*, 3965–3971.
37. Tanaka, K., Waki, H., Ido, Y., Akita, S., Yoshida, Y., Yoshida, T. (1988). Protein and polymer analyses up to *m/z* 100 000 by laser ionization time-of-flight mass spectrometry. *Rapid Commun Mass Spectrom 2*, 151–153.
38. Karas, M., Hillenkamp, F. (1988). Laser Desoprtion Ionization of proteins with molecular masses exceeding 10000 Daltons. *Anal Chem 60*, 2299–2301.
39. Fenn, J. B., Mann, M., Meng, C. K., Wong, S. F., Whitehouse, C. M. (1989). Electrospray ionization for mass-spectrometry of large biomolecules. *Science 246*, 64–71.
40. Sundarapandian, S., May, J. C., McLean, J. A. (2010). Dual source ion mobility-mass spectrometer for direct comparison of electrospray ionization and MALDI collision cross section measurements. *Anal Chem, 82*, 3247–3254.
41. Javahery, G., Thomson, B. (1997). A segmented radiofrequency-only quadrupole collision cell for measurements of ion collision cross section on a triple quadrupole mass spectrometer. *J Am Soc Mass Spectrom, 8*, 697–702.
42. Guo, Y. Z., Wang, J. X., Javahery, G., Thomson, B. A., Siu, K. W. M. (2005). Ion mobility spectrometer with radial collisional focusing. *Anal Chem 77*, 266–275.
43. Guo, Y. Z., Ling, Y., Thomson, B. A., Siu, K. W. M. (2005). Combined ion-mobility and mass-spectrometry investigations of metallothionein complexes using a tandem mass spectrometer with a segmented second quadrupole. *J Am Soc Mass Spectrom, 16*, 1787–1794.
44. Giles, K., Pringle, S. D., Worthington, K. R., Little, D., Wildgoose, J. L., Bateman, R. H. (2004). Applications of a travelling wave-based radio-frequencyonly stacked ring ion guide. *Rapid Commun Mass Spectrom, 18*, 2401–2414.
45. Ruotolo, B. T., Giles, K., Campuzano, I., Sandercock, A. M., Bateman, R. H., Robinson, C. V. (2005). Evidence for macromolecular protein rings in the absence of bulk water. *Science 310*, 1658–1661.
46. Steiner, W. E., Clowers, B. H., English, W. A., Hill, H. H. (2004). Atmospheric pressure matrix-assisted laser desorption/ionization with analysis by ion mobility time-of-flight mass spectrometry. *Rapid Commun Mass Spectrom 18*, 882–888.
47. Vonhelden, G., Hsu, M. T., Kemper, P. R., Bowers, M. T. (1991). Structures of carbon cluster ions from 3 to 60 atoms—linears to rings to fullerenes. *J Chem Phys 95*, 3835–3837.
48. Fenn, L. S., McLean, J. A. (2009). Simultaneous glycoproteomics on the basis of structure using ion mobility-mass spectrometry. *Molecular BioSystems, 5*, 1298–1302.
49. Shumate, C. B., Hill, H. H. (1989). Coronaspray nebulization and ionization of liquid samples for ion mobility spectrometry. *Anal Chem 61*, 601–606.
50. Wyttenbach, T., Kemper, P. R., Bowers, M. T. (2001). Design of a new electrospray ion mobility mass spectrometer. *Int J Mass Spectrom 212*, 13–23.
51. Kinnear, B. S., Hartings, M. R., Jarrold, M. F. (2001). Helix unfolding in unsolvated peptides. *J Am Chem Soc 123*, 5660–5667.
52. Hoaglund-Hyzer, C. S., Lee, Y. J., Counterman, A. E., Clemmer, D. E. (2002). Coupling ion mobility separations, collisional activation techniques, and multiple stages of MS for analysis of complex peptide mixtures. *Anal Chem 74*, 992–1006.
53. Valentine, S. J., Koeniger, S. L., Clemmer, D. E. (2003). A split-field drift tube for separation and efficient fragmentation of biomolecular ions. *Anal Chem 75*, 6202–6208.

54. Clowers, B. H., Hill, H. H. (2005). Mass analysis of mobility-selected ion populations using dual gate, ion mobility, quadrupole ion trap mass spectrometry. *Anal Chem 77*, 5877–5885.

55. Merenbloom, S. I., Glaskin, R. S., Henson, Z. B., Clemmer, D. E. (2009). High-resolution ion cyclotron mobility spectrometry. *Anal Chem 81*, 1482–1487.

56. Dugourd, P., Hudgins, R. R., Clemmer, D. E., Jarrold, M. F. (1997). High-resolution ion mobility measurements. *Rev Sci Instr 68*, 1122–1129.

57. Pringle, S. D., Giles, K., Wildgoose, J. L., Williams, J. P., Slade, S. E., Thalassinos, K., Bateman, R. H., Bowers, M. T., Scrivens, J. H. (2007). An investigation of the mobility separation of some peptide and protein ions using a new hybrid quadrupole/travelling wave IMS/oa-ToF instrument. *Int J Mass Spectrom, 261*, 1–12.

58. Ruotolo, B. T., Benesch, J. L. P., Sandercock, A. M., Hyung, S. J., Robinson, C. V. (2008). Ion mobility–mass spectrometry analysis of large protein complexes. *Nat Protoc, 3*, 1139–1152.

59. Williams, J. P., Scrivens, J. H. (2008). Coupling desorption electrospray ionisation and neutral desorption/extractive electrospray ionisation with a travelling-wave based ion mobility mass spectrometer for the analysis of drugs. *Rapid Commun Mass Spectrom 22*, 187–196.

60. Buryakov, I. A., Krylov, E. V., Makas, A. L., Nazarov, E. G., Pervukhin, V. V., Rasulev, U. K. (1991). Ion division by their mobility in high tension alternating electric field. *Z Tekhnich Fiz 17*, 60–65.

61. Buryakov, I. A., Krylov, E. V., Nazarov, E. G., Rasulev, U. K. (1993). A new method of separation of mult-atomic ions by mobility at atmospheric-pressure using a high-frequency amplitude-asymmetric strong electric-field. *Int J Mass Spectrom Ion Processes 128*, 143–148.

62. Krylov, E. V., Nazarov, E. G., Miller, R. A. (2007). Differential mobility spectrometer: Model of operation. *Int J Mass Spectrom 266*, 76–85.

63. Krylov, E. V. (2003). Comparison of the planar and coaxial field asymmetrical waveform ion mobility spectrometer (FAIMS). *Int J Mass Spectrom 225*, PII S1387-3806(02) 01037-0.

64. Shvartsburg, A. A., Tang, K., Smith, R. D. (2005). Optimization of the design and operation of FAIMS analyzers. *J Am Soc Mass Spectrom 16*, 2–12.

65. Shvartsburg, A. A., Smith, R. D. (2007). Scaling of the resolving power and sensitivity for planar FAIMS and mobility-based discrimination in flow- and field-driven analyzers. *J Am Soc Mass Spectrom 18*, 1672–1681.

66. Shvartsburg, A. A., Li, F. M., Tang, K. Q., Smith, R. D. (2007). Distortion of ion structures by field asymmetric waveform ion mobility spectrometry. *Anal Chem 79*, 1523–1528.

67. Shvartsburg, A. A., Smith, R. D. (2008). Optimum waveforms for differential ion mobility spectrometry (FAIMS). *J Am Soc Mass Spectrom 19*, 1286–1295.

68. Krylov, E. V., Coy, S. L., Nazarov, E. G. (2009). Temperature effects in differential mobility spectrometry. *Int J Mass Spectrom 279*, 119–125.

69. Shvartsburg, A. A., Smith, R. D. (2008). Fundamentals of traveling wave ion mobility spectrometry. *Anal Chem 80*, 9689–9699.

70. Shvartsburg, A. A., Li, F. M., Tang, K. Q., Smith, R. D. (2006). High-resolution field asymmetric waveform ion mobility spectrometry using new planar geometry analyzers. *Anal Chem 78*, 3706–3714.

71. Shvartsburg, A. A. (2009). *Differential Ion Mobility Spectrometry: Nonlinear Ion Transport and Fundamentals of FAIMS*. CRC Press, Boca Raton.

72. Dwivedi, P., Wu, P., Klopsch, S. J., Puzon, G. J., Xun, L., Hill, H. H. (2008). Metabolic profiling by ion mobility mass spectrometry (IMMS). *Metabolomics 4*, 63–80.

73. Tao, L., McLean, J. R., McLean, J. A., Russell, D. H. (2007). A collision cross-section database of singly-charged peptide ions. *J Am Soc Mass Spectrom 18*, 1232–1238.

74. Ruotolo, B. T., Verbeck, G. F., Thomson, L. M., Woods, A. S., Gillig, K. J., Russell, D. H. (2002). Distinguishing between phosphorylated and nonphosphorylated peptides with ion mobility-mass spectrometry. *J Proteome Res 1*, 303–306.

75. Furche, F., Ahlrichs, R., Weis, P., Jacob, C., Gilb, S., Bierweiler, T., Kappes, M. M. (2002). The structures of small gold cluster anions as determined by a combination of ion mobility measurements and density functional calculations. *J Chem Phys 117*, 6982–6990.

76. Reyzer, M. L., Hsieh, Y. S., Ng, K., Korfmacher, W. A., Caprioli, R. M. (2003). Direct analysis of drug candidates in tissue by matrix-assisted laser desorption/ionization mass spectrometry. *J Mass Spectrom 38*, 1081–1092.

77. Stone, E. G., Gillig, K. J., Ruotolo, B. T., Russell, D. H. (2001). Optimization of a matrix-assisted laser desorption ionization-ion mobility-surface-induced dissociation-orthogonal-time-of-flight mass spectrometer: Simultaneous acquisition of multiple correlated MS1 and MS2 spectra. *Int J Mass Spectrom 212*, 519–533.

78. Stone, E., Gillig, K. J., Ruotolo, B., Fuhrer, K., Gonin, M., Schultz, A., Russell, D. H. (2001). Surface-induced dissociation on a MALDI-ion mobility-orthogonal time-of-flight mass spectrometer: Sequencing peptides from an "in-solution" protein digest. *Anal Chem 73*, 2233–2238.

79. Sun, W. J., May, J. C., Russell, D. H. (2007). A novel surface-induced dissociation instrument for ion mobility-time-of-flight mass spectrometry. *Int J Mass Spectrom 259*, 79–86.

80. Domon, B., Costello, C. E. (1988). A systematic nomenclature for carbohydrate fragmentations in Fab-Ms Ms spectra of glycoconjugates. *Glycoconjugate J5*, 397–409.

81. Merenbloom, S. I., Koeniger, S. L., Valentine, S. J., Plasencia, M. D., Clemmer, D. E. (2006). IMS-IMS and IMS-IMS-IMS/MS for separating peptide and protein fragment ions. *Anal Chem 78*, 2802–2809.

82. McLean, J. A. (2009). The mass-mobility correlation redux: The conformational landscape of anhydrous biomolecules. *J Am Soc Mass Spectrom 20*, 1775–1781.

83. Hudgins, R. R., Ratner, M. A., Jarrold, M. F. (1998). Design of helices that are stable in vacuo. *J Am Chem Soc 120*, 12974–12975.

84. Ruotolo, B. T., Verbeck, G. F., Thomson, L. M., Gillig, K. J., Russell, D. H. (2002). Observation of conserved solution-phase secondary structure in gas-phase tryptic peptides. *J Am Chem Soc 124*, 4214–4215.

85. Ruotolo, B. T., Russell, D. H. (2004). Gas-phase conformations of proteolytically derived protein fragments: Influence of solvent on peptide conformation. *J Phys Chem B108*, 15321–15331.

86. McLean, J. R., McLean, J. A., Wu, Z. X., Becker, C., Perez, L. M., Pace, C. N., Scholtz, J. M., Russell, D. H. (2010). Factors that influence helical preferences for singly charged gas-phase peptide ions: The effects of multiple potential charge-carrying sites. *J Phys Chem B, 114*, 809–816.

87. Counterman, A. E., Clemmer, D. E. (2002). Cis-trans signatures of proline-containing tryptic peptides in the gas phase. *Analy Chem 74*, 1946–1951.

88. Wyttenbach, T., vonHelden, G., Bowers, M. T. (1996). Gas-phase conformation of biological molecules: Bradykinin. *J Am Chem Soc 118*, 8355–8364.

89. Ruotolo, B. T., Gillig, K. J., Woods, A. S., Egan, T. F., Ugarov, M. V., Schultz, J. A., Russell, D. H. (2004). Analysis of phosphorylated peptides by ion mobility-mass spectrometry. *Anal Chem 76*, 6727–6733.

90. Sawyer, H. A., Marini, J. T., Stone, E. G., Ruotolo, B. T., Gillig, K. J., Russell, D. H. (2005). The structure of gas-phase bradykinin fragment 1–5 (RPPGF) ions: An ion mobility spectrometry and H/D exchange ion-molecule reaction chemistry study. *J Am Soc Mass Spectrom 16*, 893–905.

91. Woods, A. S., Koomen, J. M., Ruotolo, B. T., Gillig, K. J., Russel, D. H., Fuhrer, K., Gonin, M., Egan, T. F., Schultz, J. A. (2002). A study of peptide–peptide using MALDI ion mobility o-TOF and ESI mass spectrometry. *J Am Soc Mass Spectrom 13*, 166–169.

92. Kaleta, D. T., Jarrold, M. F. (2002). Noncovalent interactions between unsolvated peptides. *J Phys Chem A106*, 9655–9664.

93. Bernstein, S. L., Liu, D. F., Wyttenbach, T., Bowers, M. T., Lee, J. C., Gray, H. B., Winkler, J. R. (2004). α-Synuclein: Stable compact and extended monomeric structures and pH dependence of dimer formation. *J Am Soc Mass Spectrom 15*, 1435–1443.

94. Bernstein, S. L., Wyttenbach, T., Baumketner, A., Shea, J. E., Bitan, G., Teplow, D. B., Bowers, M. T. (2005). Amyloid β-protein: Monomer structure and early aggregation states of A beta 42 and its Pro(19) alloform. *J Am Chem Soc 127*, 2075–2084.

95. Badman, E. R., Myung, S., Clemmer, D. E. (2005). Evidence for unfolding and refolding of gas-phase cytochrome c ions in a Paul trap. *J Am Soc Mass Spectrom 16*, 1493–1497.

96. Myung, S., Badman, E. R., Lee, Y. J., Clemmer, D. E. (2002). Structural transitions of electrosprayed ubiquitin ions stored in an ion trap over similar to 10 ms to 30 s. *J Phys Chem A106*, 9976–9982.

97. Koutselos, A. D., Mason, E. A. (1991). Generalized Einstein relations for ions in molecular gases. *Chem Phys 153*, 351–370.

98. McDaniel, E. W., Viehland, L. A. (1984). The transport of slow ions in gases—Experiment, theory, and applications. *Phys Rep/Rev Sec Phys Lett 110*, 333–367.

99. Mason, E. A., McDaniel, E. W. (1988). *Transport Properties of Ions in Gases*. Wiley, New York.

100. Carr, T. W. (1984). *Plasma Chromatography*. Springer-Verlag, New York, p. 274.

101. Revercomb, H. E., Mason, E. A. (1975). Theory of plasma chromatography gaseous electrophoresis—Review. *Anal Chem 47*, 970–983.

102. Shvartsburg, A. A., Jarrold, M. F. (1996). An exact hard-spheres scattering model for the mobilities of polyatomic ions. *Chem Phys Lett 261*, 86–91.

103. Clemmer, D. E., Hudgins, R. R., Jarrold, M. F. (1995). Naked protein conformations—Cytochrome-C in the gas-phase. *J Am Chem Soc*, *117*, 10141–10142.

104. Kohtani, M., Jones, T. C., Schneider, J. E., Jarrold, M. F. (2004). Extreme stability of an unsolvated alpha-helix. *J Am Chem Soc 126*, 7420–7421.

105. Wyttenbach, T., Liu, D. F., Bowers, M. T. (2008). Interactions of the hormone oxytocin with divalent metal ions. *J Am Chem Soc 130*, 5993–6000.

106. Koeniger, S. L., Merenbloom, S. I., Clemmer, D. E. (2006). Evidence for many resolvable structures within conformation types of electrosprayed ubiquitin ions. *J Phys Chem B110*, 7017–7021.

107. Fenn, L. S., Kliman, M., Mahsut, A., Zhao, S. R., McLean, J. A. (2009). Characterizing ion mobility–mass spectrometry conformation space for the analysis of complex biological samples. *Anal Bioanal Chem 394*, 235–244.

108. Woods, A. S., Ugarov, M., Egan, T., Koomen, J., Gillig, K. J., Fuhrer, K., Gonin, M., Schultz, J. A. (2004). Lipid/peptide/nucleotide separation with MALDI-ion mobility-TOF MS. *Anal Chem 76*, 2187–2195.

109. Ruotolo, B. T., Gillig, K. J., Stone, E. G., Russell, D. H. (2002). Peak capacity of ion mobility mass spectrometry: Separation of peptides in helium buffer gas. *J Chromatogr B-Anal Technol Biomed Life Sci 782*, 385–392.

110. Thalassinos, K., Grabenauer, M., Slade, S. E., Hilton, G. R., Bowers, M. T., Scrivens, J. H. (2009). Characterization of phosphorylated peptides using traveling wave-based and drift cell ion mobility mass spectrometry. *Anal Chem 81*, 248–254.

111. Williams, J. P., Giles, K., Green, B. N., Scrivens, J. H., Bateman, R. H. (2008). Ion mobility augments the utility of mass spectrometry in the identification of human hemoglobin variants. *Rapid Commun Mass Spectrom 22*, 3179–3186.

112. Srebalus, C. A., Li, J. W., Marshall, W. S., Clemmer, D. E. (1999). Gas phase separations of electrosprayed peptide libraries. *Anal Chem 71*, 3918–3927.

113. Case, D. A., Cheatham, T. E., Darden, T., Gohlke, H., Luo, R., Merz, K. M., Onufriev, A., Simmerling, C., Wang, B., Woods, R. J. (2005). The Amber biomolecular simulation programs. *J Computat Chem 26*, 1668–1688.

114. Mesleh, M. F., Hunter, J. M., Shvartsburg, A. A., Schatz, G. C., Jarrold, M. F. (1996). Structural information from ion mobility measurements: Effects of the long-range potential. *J Phys Chem 100*, 16082–16086.

115. von Helden, G., Hsu, M. T., Gotts, N., Bowers, M. T. (1993). Carbon cluster cations with up to 84 atoms—Structures, formation mechanism, and reactivity. *J Phys Chem 97*, 8182–8192.

116. Wyttenbach, T., von Helden, G., Batka, J. J., Carlat, D., Bowers, M. T. (1997). Effect of the long-range potential on ion mobility measurements. *J Am Soc Mass Spectrom 8*, 275–282.

117. Wyttenbach, T., Witt, M., Bowers, M. T. (2000). On the stability of amino acid zwitterions in the gas phase: The influence of derivatization, proton affinity, and alkali ion addition. *J Am Chem Soc 122*, 3458–3464.

118. Wyttenbach, T., Witt, M., Bowers, M. T. (1999). On the question of salt bridges of cationized amino acids in the gas phase: glycine and arginine. *Int J Mass Spectrom 183*, 243–252.

119. Wyttenbach, T., Liu, D. F., Bowers, M. T. (2005). Hydration of small peptides. *Int J Mass Spectrom 240*, 221–232.

120. Kohtani, M., Breaux, G. A., Jarrold, M. F. (2004). Water molecule adsorption on protonated dipeptides. *J Am Chem Soc 126*, 1206–1213.

121. Liu, D. F., Wyttenbach, T., Carpenter, C. J., Bowers, M. T. (2004). Investigation of noncovalent interactions in deprotonated peptides: Structural and energetic competition between aggregation and hydration. *J Am Chem Soc 126*, 3261–3270.

122. Hudgins, R. R., Mao, Y., Ratner, M. A., Jarrold, M. F. (1999). Conformations of Gly(n)H(+) and Ala(n)H(+) peptides in the gas phase. *Biophys J, 76*, 1591–1597.

123. Kinnear, B. S., Jarrold, M. F. (2001). Helix formation in unsolvated peptides: Side chain entropy is not the determining factor. *J Am Chem Soc 123*, 7907–7908.

124. Counterman, A. E., Clemmer, D. E. (2004). Anhydrous polyproline helices and globules. *J Phys Chem B108*, 4885–4898.

125. Myung, S., Julian, R. R., Nanita, S. C., Cooks, R. G., Clemmer, D. E. (2004). Formation of nanometer-scale serine clusters by sonic spray. *J Phys Chem B108*, 6105–6111.

126. Shelimov, K. B., Jarrold, M. F. (1997). Conformations, unfolding, and refolding of apomyoglobin in vacuum: An activation barrier for gas-phase protein folding. *J Am Chem Soc 119*, 2987–2994.

127. Shelimov, K. B., Clemmer, D. E., Hudgins, R. R., Jarrold, M. F. (1997). Protein structure in vacuo: Gas-phase confirmations of BPTI and cytochrome c. *J Am Chem Soc 119*, 2240–2248.

128. Shelimov, K. B., Jarrold, M. F. (1996). "Denaturation" and refolding of cytochrome c in vacuo. *J Am Chem Soc 118*, 10313–10314.

129. Myung, S., Wiseman, J. M., Valentine, S. J., Takats, Z., Cooks, R. G., Clemmer, D. E. (2006). Coupling desorption electrospray ionization with ion mobility/mass spectrometry for analysis of protein structure: Evidence for desorption of folded and denatured states. *J Phys Chem B110*, 5045–5051.

130. Barnes, C. A. S., Hilderbrand, A. E., Valentine, S. J., Clemmer, D. E. (2002). Resolving isomeric peptide mixtures: A combined HPLC/ion mobility-TOFMS analysis of a 4000-component combinatorial library. *Anal Chem 74*, 26–36.

131. Valentine, S. J., Counterman, A. E., Clemmer, D. E. (1999). A database of 660 peptide ion cross sections: Use of intrinsic size parameters for bona fide predictions of cross sections. *J Am Soc Mass Spectrom 10*, 1188–1211.

132. Loo, J. A. (1997). Studying noncovalent protein complexes by electrospray ionization mass spectrometry. *Mass Spectrom Rev 16*, 1–23.

133. Loo, J. A., Berhane, B., Kaddis, C. S., Wooding, K. M., Xie, Y. M., Kaufman, S. L., Chernushevich, I. V. (2005). Electrospray ionization mass spectrometry and ion mobility analysis of the 20S proteasome complex. *J Am Soc Mass Spectrom 16*, 998–1008.

134. Leary, J. A., Schenauer, M. R., Stefanescu, R., Andaya, A., Ruotolo, B. T., Robinson, C. V., Thalassinos, K., Scrivens, J. H., Sokabe, M., Hershey, J. W. B. (2009). Methodology for measuring conformation of solvent-disrupted protein subunits using T-WAVE ion mobility MS: An investigation into eukaryotic initiation factors. *J Am Soc Mass Spectrom 20*, 1699–1706.

135. Heck, A. J. R., van den Heuvel, R. H. H. (2004). Investigation of intact protein complexes by mass spectrometry. *Mass Spectrom Rev 23*, 368–389.

136. Kaddis, C. S., Lomeli, S. H., Yin, S., Berhane, B., Apostol, M. I., Kickhoefer, V. A., Rome, L. H., Loo, J. A. (2007). Sizing large proteins and protein complexes by electrospray ionization mass spectrometry and ion mobility. *J Am Soc Mass Spectrom 18*, 1206–1216.

137. Ruotolo, B. T., Hyung, S. J., Robinson, P. M., Giles, K., Bateman, R. H., Robinson, C. V. (2007). Ion mobility–mass spectrometry reveals long-lived, unfolded intermediates in the dissociation of protein complexes. *Angewandte Chem Int Ed 46*, 8001–8004.

138. Gown, A. M. (2002). Genogenic immunohistochemistry: A new era in diagnostic immunohistochemistry. *Curr Diagnost Pathol 8*, 193–200.

139. Ramos-Vara, J. A. (2005). Technical aspects of immunohistochemistry. *Vet Pathol 42*, 405–426.

140. Steinkamp, J. A. (1984). Flow-cytometry. *Rev Sci Instr*, *55*, 1375–1400.

141. Seda, O., Tremblay, J., Sedova, L., Hamet, P. (2005). Integrating genomics and transcriptomics with geo-ethnicity and the environment for the resolution of complex cardiovascular diseases. *Curr Opin Mol Therapeut 7*, 583–587.

142. Gabryelski, W., Froese, K. L. (2003). Rapid and sensitive differentiation of anomers, linkage, and position isomers of disaccharides using high-field asymmetric waveform ion mobility spectrometry (FAIMS). *J Am Soc Mass Spectrom 14*, 265–277.

143. Aebersold, R., Mann, M. (2003). Mass spectrometry-based proteomics. *Nature 422*, 198–207.

144. Wei, X., Li, L. (2009). Mass spectrometry-based proteomics and peptidomics for biomarker discovery in neurodegenerative diseases. *International J Clin Exper Pathol 2*, 132–148.

145. Jin, L., Barran, P. E., Deakin, J. A., Lyon, M., Uhrin, D. (2005). Conformation of glycosaminoglycans by ion mobility mass spectrometry and molecular modelling. *Phys Chem Chem Phys 7*, 3464–3471.

146. Watson, A. D. (2006). Lipidomics: A global approach to lipid analysis in biological systems. *J Lipid Res 47*, 2101–2111.

147. Warnement, M. R., Faley, S. L., Wikswo, J. P., Rosenthal, S. J. (2006). Quantum dot probes for monitoring dynamic cellular response: Reporters of T cell activation. *IEEE Trans Nanobiosci 5*, 268–272.

148. Faley, S., Seale, K., Hughey, J., Schaffer, D. K., VanCornpernolle, S., McKinney, B., Baudenbacher, F., Unutmaz, D., Wikswo, J. P. (2008). Microfluidic platform for real-time signaling analysis of multiple single T cells in parallel. *Lab Chip 8*, 1700–1712.

149. Werdich, A. A., Lima, E. A., Ivanov, B., Ges, I., Anderson, M. E., Wikswo, J. P., Baudenbacher, F. J. (2004). A microfluidic device to confine a single cardiac myocyte in a sub-nanoliter volume on planar microelectrodes for extracellular potential recordings. *Lab Chip 4*, 357–362.

150. Eklund, S. E., Snider, R. M., Wikswo, J., Baudenbacher, F., Prokop, A., Cliffel, D. E. (2006). Multianalyte microphysiometry as a tool in metabolomics and systems biology. *J ElectroAnal Chem 587*, 333–339.

151. Toh, Y. C., Zhang, C., Zhang, J., Khong, Y. M., Chang, S., Samper, V. D., van Noort, D., Hutmacher, D. W., Yu, H. R. (2007). A novel 3D mammalian cell perfusion-culture system in microfluidic channels. *Lab Chip 7*, 302–309.

152. Prokop, A., Prokop, Z., Schaffer, D., Kozlov, E., Wikswo, J., Cliffel, D., Baudenbacher, F. (2004). NanoLiterBioReactor: Long-term mammalian cell culture at nanofabricated scale. *Biomed Microdev 6*, 325–339.

153. Addae-Mensah, K. A., Wikswo, J. P. (2008). Measurement techniques for cellular biomechanics in vitro. *Exper Biol Med, 233*, 792–809.

154. Beebe, D. J., Mensing, G. A., Walker, G. M. (2002). Physics and applications of microfluidics in biology. *An Rev Biomed Eng 4*, 261–286.

155. Andersson, H., van den Berg, A. (2003). Microfluidic devices for cellomics: A review. *Sensors Actuators B-Chem 92*, 315–325.

156. Zguris, J. C., Itle, L. J., Hayes, D., Pishko, M. V. (2005). Microreactor microfluidic systems with human microsomes and hepatocytes for use in metabolite studies. *Biomed Microdev 7*, 117–125.

157. Mao, Y., Woenckhaus, J., Kolafa, J., Ratner, M. A., Jarrold, M. F. (1999). Thermal unfolding of unsolvated cytochrome c: Experiment and molecular dynamics simulations. *J Am Chem Soc, 121*, 2712–2721.

158. May, J. C., Russell D. H. (2011). A Mass-selective variable-temperature drift tube ion mobility-mass spectrometer for temperature dependent ion mobility studies. *J Am Soc Mass Spectrom, 22*, 1134–1145.

159. Merenbloom, S. I., Koeniger, S. L., Bohrer, B. C., Valentine, S. J., Clemmer, D. E. (2008). Improving the efficiency of IMS-IMS by a combing technique. *Anal Chem 80*, 1918–1927.

160. Bohrer, B. C., Mererbloom, S. I., Koeniger, S. L., Hilderbrand, A. E., Clemmer, D. E. (2008). Biomolecule analysis by ion mobility spectrometry. *Anl. Rev Anal Chem 1*, 293–327.

161. Wyttenbach, T., Bowers, M. T. (2007). Intermolecular interactions in biomolecular systems examined by mass spectrometry. *An Rev Phys Chem 58*, 511–533.

162. Liu, D. F., Wyttenbach, T., Barran, P. E., Bowers, M. T. (2003). Sequential hydration of small protonated peptides. *J Am Chem Soc 125*, 8458–8464.

163. Counterman, A. E., Hilderbrand, A. E., Barnes, C. A. S., Clemmer, D. E. (2001). Formation of peptide aggregates during ESI: Size, charge, composition, and contributions to noise. *J Am Soc Mass Spectrom 12*, 1020–1035.

164. Kohtani, M., Jarrold, M. F. (2002). The initial steps in the hydration of unsolvated peptides: Water molecule adsorption on alanine-based helices and globules. *J Am Chem Soc 124*, 11148–11158.

165. Hudgins, R. R., Jarrold, M. F. (1999). Helix formation in unsolvated alanine-based peptides: Helical monomers and helical dimers. *J Am Chem Soc 121*, 3494–3501.

166. Breaux, G. A., Jarrold, M. F. (2003). Probing helix formation in unsolvated peptides. *J Am Chem Soc 125*, 10740–10747.

167. Hartings, M. R., Kinnear, B. S., Jarrold, M. F. (2003). The energy landscape of unsolvated peptides: The role of context in the stability of alanine/glycine helices. *J Am Chem Soc 125*, 3941–3947.

168. Sudha, R., Kohtani, M., Breaux, G. A., Jarrold, M. F. (2004). Pi-helix preference in unsolvated peptides. *J Am Chem Soc 126*, 2777–2784.

169. Zilch, L. W., Kaleta, D. T., Kohtani, M., Krishnan, R., Jarrold, M. F. (2007). Folding and unfolding of helix-turn-helix motifs in the gas phase. *J Am Soc Mass Spectrom 18*, 1239–1248.

170. Jarrold, M. F. (2007). Helices and sheets in vacuo. *Phys Chem Chem Phys 9*, 1659–1671.

171. Wyttenbach, T., Paizs, B., Barran, P., Breci, L., Liu, D. F., Suhai, S., Wysocki, V. H., Bowers, M. T. (2003). The effect of the initial water of hydration on the energetics, structures, and H/D exchange mechanism of a family of pentapeptides: An experimental and theoretical study. *J Am Chem Soc 125*, 13768–13775.

172. Counterman, A. E., Clemmer, D. E. (2001). Large anhydrous polyalanine ions: Evidence for extended helices and onset of a more compact state. *J Am Chem Soc 123*, 1490–1498.

173. Taraszka, J. A., Counterman, A. E., Clemmer, D. E. (2001). Large anhydrous polyalanine ions: Substitution of Na+ for H+ destabilizes folded states. *Int J Mass Spectrom 204*, 87–100.

174. Counterman, A. E., Clemmer, D. E. (2002). Gas phase polyalanine: Assessment of $i \rightarrow i + 3$ and $i \rightarrow i + 4$ helical turns in [Ala(n) + 4H](4+) ($n = 29$–49) ion. *J Phys Chem B106*, 12045–12051.

175. Counterman, A. E., Clemmer, D. E. (2003). Compact → extended helix transitions of polyalanine in vacuo. *J Phys Chem B107*, 2111–2117.

176. Counterman, A. E., Valentine, S. J., Srebalus, C. A., Henderson, S. C., Hoaglund, C. S., Clemmer, D. E. (1998). High-order structure and dissociation of gaseous peptide aggregates that are hidden in mass spectra. *J Am Soc Mass Spectrom 9*, 743–759.
177. Hudgins, R. R., Jarrold, M. F. (2000). Conformations of unsolvated glycine-based peptides. *J Phys Chem B104*, 2154–2158.
178. Wyttenbach, T., Bushnell, J. E., Bowers, M. T. (1998). Salt bridge structures in the absence of solvent? The case for the oligoglycines. *J Am Chem Soc 120*, 5098–5103.
179. Antoine, R., Compagnon, I., Rayane, D., Broyer, M., Dugourd, P., Breaux, G., Hagemeister, F. C., Pippen, D., Hudgins, R. R., Jarrold, M. F. (2002). Electric dipole moments and conformations of isolated peptides. *Eur Phys J D20*, 583–587.
180. Counterman, A. E., Clemmer, D. E. (2001). Magic number clusters of serine in the gas phase. *J Phys Chem B105*, 8092–8096.
181. Julian, R. R., Myung, S., Clemmer, D. E. (2004). Spontaneous anti-resolution in heterochiral clusters of serine. *J Am Chem Soc 126*, 4110–4111.
182. Myung, S., Lorton, K. P., Merenbloom, S. I., Fioroni, M., Koeniger, S. L., Julian, R. R., Baik, M. H., Clemmer, D. E. (2006). Evidence for spontaneous resolution of icosahedral proline. *J Am Chem Soc 128*, 15988–15989.
183. Myung, S., Fioroni, M., Julian, R. R., Koeniger, S. L., Baik, M. H., Clemmer, D. E. (2006). Chirally directed formation of nanometer-scale proline clusters. *J Am Chem Soc 128*, 10833–10839.
184. Wu, C., Klasmeier, J., Hill, H. H. (1999). Atmospheric pressure ion mobility spectrometry of protonated and sodiated peptides. *Rapid Commun Mass Spectrom 13*, 1138–1142.
185. Fye, J. L., Woenckhaus, J., Jarrold, M. F. (1998). Hydration of folded and unfolded gas-phase proteins: Saturation of cytochrome c and apomyoglobin. *J Am Chem Soc 120*, 1327–1328.
186. Woenckhaus, J., Hudgins, R. R., Jarrold, M. F. (1997). Hydration of gas-phase proteins: A special hydration site on gas-phase BPTI. *J Am Chem Soc 119*, 9586–9587.
187. Jarrold, M. F. (1999). Unfolding, refolding, and hydration of proteins in the gas phase. *Accounts Chem Res 32*, 360–367.
188. Mao, Y., Ratner, M. A., Jarrold, M. F. (2000). One water molecule stiffens a protein. *J Am Chem Soc 122*, 2950–2951.
189. Wyttenbach, T., Bowers, M. T. (1999). Gas phase conformations of biological molecules: The hydrogen/deuterium exchange mechanism. *J Am Soc Mass Spectrom 10*, 9–14.
190. Woenckhaus, J., Mao, Y., Jarrold, M. F. (1997). Hydration of gas phase proteins: Folded +5 and unfolded +7 charge states of cytochrome c. *J Phys Chem B101*, 847–851.
191. Mao, Y., Ratner, M. A., Jarrold, M. F. (1999). Molecular dynamics simulations of the charge-induced unfolding and refolding of unsolvated cytochrome c. *J Phys Chem B103*, 10017–10021.
192. Mao, Y., Woenckhaus, J., Kolafa, J., Ratner, M. A., Jarrold, M. F. (1999). Thermal unfolding of unsolvated cytochrome c: Experiment and molecular dynamics simulations. *J Am Chem Soc 121*, 2712–2721.
193. Thalassinos, K., Slade, S. E., Jennings, K. R., Scrivens, J. H., Giles, K., Wildgoose, J., Hoyes, J., Bateman, R. H., Bowers, M. T. (2004). Ion mobility mass spectrometry of proteins in a modified commercial mass spectrometer. *Int J Mass Spectrom 236*, 55–63.
194. McCullough, B. J., Eastwood, H., Clark, D. J., Polfer, N. C., Campopiano, D. J., Dorin, J. A., Maxwell, A., Langley, R. J., Govan, J. R. W., Bernstein, S. L., Bowers, M. T., Barran,

P. E. (2006). Characterisation of DEFB107 by mass spectrometry: Lessons from an anti-antimicrobial defensin. *Int J Mass Spectrom 252*, 180–188.

195. Ruotolo, B. T., Tate, C. C., Russell, D. H. (2004). Ion mobility-mass spectrometry applied to cyclic peptide analysis: Conformational preferences of gramicidin S and linear analogs in the gas phase. *J Am Soc Mass Spectrom 15*, 870–878.

196. Valentine, S. J., Sevugarajan, S., Kurulugama, R. T., Koeniger, S. L., Merenbloom, S. I., Bohrer, B. C., Clemmer, D. E. (2006). Split-field drift tube/mass spectrometry and isotopic labeling techniques for determination of single amino acid polymorphisms. *J Proteome Res 5*, 1879–1887.

197. Taraszka, J. A., Li, J. W., Clemmer, D. E. (2000). Metal-mediated peptide ion conformations in the gas phase. *J Phys Chem B104*, 4545–4551.

198. Smith, D. P., Giles, K., Bateman, R. H., Radford, S. E., Ashcroft, A. E. (2007). Monitoring copopulated conformational states during protein folding events using Electrospray ionization-ion mobility spectrometry-mass spectrometry. *J Am Soc Mass Spectrom 18*, 2180–2190.

199. Badman, E. R., Hoaglund-Hyzer, C. S., Clemmer, D. E. (2001). Monitoring structural changes of proteins in an ion trap over similar to 10–200 ms: Unfolding transitions in cytochrome c ions. *Anal Chem 73*, 6000–6007.

200. Grabenauer, M., Bernstein, S. L., Lee, J. C., Wyttenbach, T., Dupuis, N. F., Gray, H. B., Winkler, J. R., Bowers, M. T. (2008). Spermine binding to Parkinson's protein alpha-synuclein and its disease-related A30P and A53T mutants. *J Phys Chem B112*, 11147–11154.

201. Segev, E., Wyttenbach, T., Bowers, M. T., Gerber, R. B. (2008). Conformational evolution of ubiquitin ions in electrospray mass spectrometry: Molecular dynamics simulations at gradually increasing temperatures. *Phys Chem Chem Phys 10*, 3077–3082.

202. Valentine, S. J., Counterman, A. E., Clemmer, D. E. (1997). Conformer-dependent proton-transfer reactions of ubiquitin ions. *J Am Soc Mass Spectrom 8*, 954–961.

203. Li, J. W., Taraszka, J. A., Counterman, A. E., Clemmer, D. E. (1999). Influence of solvent composition and capillary temperature on the conformations of electrosprayed ions: Unfolding of compact ubiquitin conformers from pseudonative and denatured solutions. *Int J Mass Spectrom 187*, 37–47.

204. Badman, E. R., Hoaglund-Hyzer, C. S., Clemmer, D. E. (2002). Dissociation of different conformations of ubiquitin ions. *J Am Soc Mass Spectrom 13*, 719–723.

205. Koeniger, S. L., Merenbloom, S. I., Sevugarajan, S., Clemmer, D. E. (2006). Transfer of structural elements from compact to extended states in unsolvated ubiquitin. *J Am Chem Soc 128*, 11713–11719.

206. Koeniger, S. L., Clemmer, D. E. (2007). Resolution and structural transitions of elongated states of ubiquitin. *J Am Soc Mass Spectrom 18*, 322–331.

207. Koeniger, S. L., Merenbloom, S. I., Valentine, S. J., Jarrold, M. F., Udseth, H. R., Smith, R. D., Clemmer, D. E. (2006). An IMS-IMS analogue of MS-MS. *Anal Chem 78*, 4161–4174.

208. Fernandez-Lima, F. A., Becker, C., Gillig, K. J., Russell, W. K., Tichy, S. E., Russell, D. H. (2009). Ion mobility-mass spectrometer interface for collisional activation of mobility separated ions. *Anal Chem 81*, 618–624.

209. Barnes, C. A. S., Clemmer, D. E. (2001). Assessment of purity and screening of peptide libraries by nested ion mobility TOFMS: Identification of RNase S-protein binders. *Anal Chem 73*, 424–433.

210. Barran, P. E., Polfer, N. C., Campopiano, D. J., Clarke, D. J., Langridge-Smith, P. R. R., Langley, R. J., Govan, J. R. W., Maxwell, A., Dorin, J. R., Millar, R. P., Bowers, M. T. (2005). Is it biologically relevant to measure the structures of small peptides in the gas-phase? *Int J Mass Spectrom 240*, 273–284.
211. Kindy, J. M., Taraszka, J. A., Regnier, F. E., Clemmer, D. E. (2002). Quantifying peptides in isotopically labeled protease digests by ion mobility/time-of-flight mass spectrometry. *Anal Chem 74*, 950–958.
212. Valentine, S. J., Clemmer, D. E. (2002). Temperature-dependent H/D exchange of compact and elongated cytochrome c ions in the gas phase. *J Am Soc Mass Spectrom 13*, 506–517.
213. Valentine, S. J., Clemmer, D. E. (1997). H/D exchange levels of shape-resolved cytochrome c conformers in the gas phase. *J Am Chem Soc 119*, 3558–3566.
214. Lee, Y. J., Hoaglund-Hyzer, C. S., Taraszka, J. A., Zientara, G. A., Counterman, A. E., Clemmer, D. E. (2001). Collision-induced dissociation of mobility-separated ions using an orifice-skimmer cone at the back of a drift tube. *Anal Chem 73*, 3549–3555.
215. Hoaglund-Hyzer, C. S., Clemmer, D. E. (2001). Ion trap/ion mobility/quadrupole/time of flight mass spectrometry for peptide mixture analysis. *Anal Chem 73*, 177–184.
216. Badman, E. R., Myung, S., Clemmer, D. E. (2002). Gas-phase separations of protein and peptide ion fragments generated by collision-induced dissociation in an ion trap. *Anal Chem 74*, 4889–4894.
217. Valentine, S. J., Plasencia, M. D., Liu, X. Y., Krishnan, M., Naylor, S., Udseth, H. R., Smith, R. D., Clemmer, D. E. (2006). Toward plasma proteome profiling with ion mobility-mass spectrometry. *J Proteome Res 5*, 2977–2984.
218. Liu, X. Y., Miller, B. R., Rebec, G. V., Clemmer, D. E. (2007). Protein expression in the striatum and cortex regions of the brain for a mouse model of Huntington's disease. *J Proteome Res 6*, 3134–3142.
219. Sowell, R. A., Hersberger, K. E., Kaufman, T. C., Clemmer, D. E. (2007). Examining the proteome of *Drosophila* across organism lifespan. *J Proteome Res 6*, 3637–3647.
220. Xun, Z., Sowell, R. A., Kaufman, T. C., Clemmer, D. E. (2007). Lifetime proteomic profiling of an A30P alpha-synuclein *Drosophila* model of Parkinson's disease. *J Proteome Res 6*, 3729–3738.
221. Xun, Z. Y., Sowell, R. A., Kaufman, T. C., Clemmer, D. E. (2007). Protein expression in a *Drosophila* model of Parkinson's disease. *J Proteome Res 6*, 348–357.
222. Xun, Z. Y., Sowell, R. A., Kaufman, T. C., Clemmer, D. E. (2008). Quantitative proteomics of a presymptomatic A53T alpha-synuclein *Drosophila* model of Parkinson disease. *Mol Cell Proteomics 7*, 1191–1203.
223. Xun, Z. Y., Kaufman, T. C., Clemmer, D. E. (2008). Proteome response to the panneural expression of human wild-type alpha-synuclein: A *Drosophila* model of Parkinson's disease. *J Proteome Res 7*, 3911–3921.
224. Polfer, N. C., Bohrer, B. C., Plasencia, M. D., Paizs, B., Clemmer, D. E. (2008). On the dynamics of fragment isomerization in collision-induced dissociation of peptides. *J Phys Chem A112*, 1286–1293.
225. Alves, P., Arnold, R. J., Clemmer, D. E., Li, Y. X., Reilly, J. P., Sheng, Q. H., Tang, H. X., Xun, Z. Y., Zeng, R., Radivojac, P. (2008). Fast and accurate identification of semi-tryptic peptides in shotgun proteomics. *Bioinformatics 24*, 102–109.

226. Kurulugama, R. T., Valentine, S. J., Sowell, R. A., Clemmer, D. E. (2008). Development of a high-throughput IMS-IMS-MS approach for analyzing mixtures of biomolecules. *J Proteomics 71*, 318–331.

227. Riba-Garcia, I., Giles, K., Bateman, R. H., Gaskell, S. J. (2008). Evidence for structural variants of a- and b-type peptide fragment ions using combined ion mobility/mass spectrometry. *J Am Soc Mass Spectrom 19*, 609–613.

228. Merenbloom, S. I., Bohrer, B. C., Koeniger, S. L., Clemmer, D. E. (2007). Assessing the peak capacity of IMS-IMS separations of tryptic peptide ions in He at 300 K. *Anal Chem 79*, 515–522.

229. Henderson, S. C., Valentine, S. J., Counterman, A. E., Clemmer, D. E. (1999). ESI/ion trap/ion mobility/time-of-flight mass spectrometry for rapid and sensitive analysis of biomolecular mixtures. *Anal Chem 71*, 291–301.

230. Valentine, S. J., Kulchania, M., Barnes, C. A. S., Clemmer, D. E. (2001). Multidimensional separations of complex peptide mixtures: A combined high-performance liquid chromatography/ion mobility/time-of-flight mass spectrometry approach. *Int J Mass Spectrom 212*, 97–109.

231. Hilderbrand, A. E., Myung, S., Clemmer, D. E. (2006). Exploring crown ethers as shift reagents for ion mobility spectrometry. *Anal Chem 78*, 6792–6800.

232. Ruotolo, B. T., Gillig, K. J., Stone, E. G., Russell, D. H., Fuhrer, K., Gonin, M., Schultz, J. A. (2002). Analysis of protein mixtures by matrix-assisted laser desorption ionization–ion mobility–orthogonal-time-of-flight mass spectrometry. *Int J Mass Spectrom 219*, 253–267.

233. McLean, J. A., Russell, D. H. (2003). Sub-femtomole peptide detection in ion mobility-time-of-flight mass spectrometry measurements. *J Proteome Res 2*, 427–430.

234. Ruotolo, B. T., McLean, J. A., Gillig, K. J., Russell, D. H. (2004). Peak capacity of ion mobility mass spectrometry: The utility of varying drift gas polarizability for the separation of tryptic peptides. *J Mass Spectrom 39*, 361–367.

235. Ruotolo, B. T., McLean, J. A., Gillig, K. J., Russell, D. H. (2005). The influence and utility of varying field strength for the separation of tryptic peptides by ion mobility–mass spectrometry. *J Am Soc Mass Spectrom 16*, 158–165.

236. Monien, B. H., Fradinger, E. A., Spring, S. M., Bernstein, S. L., Bowers, M. T., Bitan, G. (2006). A novel approach to Alzheimer's disease therapy: Inhibition of A beta 42 oligomerization by *C*-terminal A beta 42 fragments. *J Peptide Sci 12*, 147–147.

237. Spring, S. M., Bernstein, S. L., Lazo, N. D., Urbanc, B., Stanley, H. E., Bowers, M. T., Teplow, D. B., Bitan, G. (2006). Towards inhibition of amyloid beta-protein oligomerization. *Underst Biol Using Peptides 19*, 515–516.

238. Henderson, S. C., Li, J. W., Counterman, A. E., Clemmer, D. E. (1999). Intrinsic size parameters for Val, Ile, Leu, Gln, Thr, Phe, and Trp residues from ion mobility measurements of polyamino acid ions. *J Phys Chem B103*, 8780–8785.

239. Valentine, S. J., Counterman, A. E., Hoaglund-Hyzer, C. S., Clemmer, D. E. (1999). Intrinsic amino acid size parameters from a series of 113 lysine-terminated tryptic digest peptide ions. *J Phys Chem B103*, 1203–1207.

240. Counterman, A. E., Clemmer, D. E. (1999). Volumes of individual amino acid residues in gas-phase peptide ions. *J Am Chem Soc 121*, 4031–4039.

241. Shvartsburg, A. A., Siu, K. W. M., Clemmer, D. E. (2001). Prediction of peptide ion mobilities via a priori calculations from intrinsic size parameters of amino acid residues. *J Am Soc Mass Spectrom 12*, 885–888.

242. Mosier, P. D., Counterman, A. E., Jurs, P. C., Clemmer, D. E. (2002). Prediction of peptide ion collision cross sections from topological molecular structure and amino acid parameters. *Anal Chem 74*, 1360–1370.

243. Barnes, C. A. S., Clemmer, D. E. (2003). Assessing intrinsic side chain interactions between i and i + 4 residues in solvent-free peptides: A combinatorial gas-phase approach. *J Phys Chem A107*, 10566–10579.

244. Hilderbrand, A. E., Clemmer, D. E. (2005). Determination of sequence-specific intrinsic size parameters from cross sections for 162 tripeptides. *J Phys Chem B109*, 11802–11809.

245. Baumketner, A., Bernstein, S. L., Wyttenbach, T., Bitan, G., Teplow, D. B., Bowers, M. T., Shea, J. E. (2006). Amyloid beta-protein monomer structure: A computational and experimental study. *Protein Sci 15*, 420–428.

246. Baumketner, A., Bernstein, S. L., Wyttenbach, T., Lazo, N. D., Teplow, D. B., Bowers, M. T., Shea, J. E. (2006). Structure of the 21–30 fragment of amyloid beta-protein. *Protein Sci 15*, 1239–1247.

247. Teplow, D. B., Lazo, N. D., Bitan, G., Bernstein, S., Wyttenbach, T., Bowers, M. T., Baumketner, A., Shea, J. E., Urbanc, B., Cruz, L., Borreguero, J., Stanley, H. E. (2006). Elucidating amyloid beta-protein folding and assembly: A multidisciplinary approach. *Accounts Chem Res 39*, 635–645.

248. Krone, M. G., Baumketner, A., Bernstein, S. L., Wyttenbach, T., Lazo, N. D., Teplow, D. B., Bowers, M. T., Shea, J. E. (2008). Effects of familial Alzheimer's disease mutations on the folding nucleation of the amyloid beta-protein. *J Mol Biol 381*, 221–228.

249. Barran, P. E., Roeseke, R. W., Pawson, A. J., Sellar, R., Bowers, M. T., Morgan, K., Lu, Z. L., Tsuda, M., Kusakabe, T., Millar, R. P. (2005). Evolution of constrained gonadotropin-releasing hormone ligand conformation and receptor selectivity. *J Biol Chem 280*, 38569–38575.

250. Liu, D. F., Seuthe, A. B., Ehrler, O. T., Zhang, X. H., Wyttenbach, T., Hsu, J. F., Bowers, M. T. (2005). Oxytocin-receptor binding: Why divalent metals are essential. *J Am Chem Soc 127*, 2024–2025.

251. Alverdi, V., Mazon, H., Versluis, C., Hemrika, W., Esposito, G., van den Heuvel, R., Scholten, A., Heck, A. J. R. (2008). cGMP-binding prepares PKG for substrate binding by disclosing the *C*-terminal domain. *J Mol Biol 375*, 1380–1393.

252. Scholten, A., Aye, T. T., Heck, A. J. R. (2008). A multi-angular mass spectrometric view at cyclic nucleotide dependent protein kinases: In vivo characterization and structure/function relationships. *Mass Spectrom Rev 27*, 331–353.

253. Uetrecht, C., Versluis, C., Watts, N. R., Wingfield, P. T., Steven, A. C., Heck, A. J. R. (2008). Stability and shape of hepatitis B virus capsids in vacuo. *Angewandte Chem Int Ed 47*, 6247–6251.

254. Lorenzen, K., Olia, A. S., Uetrecht, C., Cingolani, G., Heck, A. J. R. (2008). Determination of stoichiometry and conformational changes in the first step of the P22 tail assembly. *J Mol Biol 379*, 385–396.

255. Benesch, J. L. P. (2009). Collisional activation of protein complexes: Picking up the pieces. *J Am Soc Mass Spectrom 20*, 341–348.

CHAPTER 7

Chemical Footprinting for Determining Protein Properties and Interactions

SANDRA A. KERFOOT and MICHAEL L. GROSS

7.1 INTRODUCTION TO HYDROGEN–DEUTERIUM EXCHANGE

Understanding protein-ligand binding requires that an interacting system be thoroughly characterized to afford not only the structure of the protein but also its interaction with ligands of interest. Insight is often desired into solvent accessibility, affinity for the ligand, interfaces with the ligand (and with other proteins), and location of binding sites. These have been explored in a variety of ways including site-directed mutagenesis, crosslinking, and chemical footprinting or mapping, and with a variety of instrumental approaches including NMR, X-ray crystallography, circular dichroism, fluorescence, and FRET. Many of these approaches are costly and consume time and material. Although NMR and X-ray crystallography offer excellent resolution of protein structure, the former is limited to smaller proteins (MW $<$ 20 kDa), and the latter, of course, requires that the protein crystallize. Both require approximately 1 mg or more of material.

The combination of chemical footprinting and mass spectrometry is an emerging approach that offers insight into structure, binding, and stability. This approach, however, gives lower resolution results than X-ray crystallography and NMR. Nevertheless, what chemical footprinting coupled to mass spectrometry lacks in detailed information output, it more than makes up for in speed, sensitivity, and cost with respect to NMR and X-ray cystallography. As mass spectrometry laboratories proliferate for analytical proteomics applications, the accessibility for measurement in structural proteomics will also increase.

Chemical footprinting methods come in two main categories: reversible and irreversible modifications. Irreversible modifications include oxidation, iodination,

Protein and Peptide Mass Spectrometry in Drug Discovery, Edited by Michael L. Gross, Guodong Chen, and Birendra N. Pramanik.

alkylation, and crosslinking. The classic reaction that produces reversible modifications is hydrogen–deuterium amide exchange (HDX). Owing to the more extensive background for HDX experiments compared to irreversible footprinting, the first part of this chapter will be dedicated to the basic experimental approach and applications of HDX coupled to mass spectrometry.

7.1.1 Fundamentals of Hydrogen–Deuterium Amide Exchange in Proteins

HDX has been used for over 50 years to study proteins and peptides; most of this work has been in the NMR community [1,2]. Although there were earlier applications with fast atom bombardment mass spectrometry, the development of electrospray ionization (ESI) promoted the demonstrations of HDX coupled with mass spectrometry to study protein conformation [3] and permitted the field to evolve rapidly. ESI has been used in over 1600 publications (as of late 2009), in which the application of HDX and MS answers various questions in protein science. Before describing some examples of HDX MS, it is important that we review the fundamentals of the experiment.

There are three categories of hydrogens in a peptide or protein. The side-chain functional group hydrogens and those at the *N*-terminus and *C*-terminus are considered fast exchangers. These hydrogens exchange with deuterium too quickly to be measured by most techniques. Hydrogens bound to carbon are considered slow exchangers owing to their highly covalent nature. Hydrogens on the peptide bonds, also referred to as backbone hydrogens, exchange on a time scale that is easily measurable. Backbone hydrogens in a folded protein exchange spontaneously with aqueous solution, and the times of exchange, range from ms for solvent-accessible amides to years for highly H-bonded amides [4]. In a denatured peptide, the time for exchange of amide hydrogens is in the range of 10 to 1000 ms [5]. Neighboring side chains affect the rates in a linear peptide and are responsible for the range [6]. In a folded protein, however, the rates can vary by 10^8 [7]. Backbone hydrogens that are not involved in hydrogen bonding and are solvent accessible, such as in a “floppy loop” region of a protein, exchange quickly, whereas those that are buried in the protein or are involved in a hydrogen-bonding network (α helix or β sheet) exchange more slowly. It is the difference in exchange rates between two protein conformations that is exploited in most HDX experiments.

7.1.2 EX1 and EX2 Rates of HDX

Many complex kinetic models are available to explain the factors responsible for the varying rates of H/D exchange in a folded protein [8–10]. A simple, two-step model describes most exchanges in a folded protein, as demonstrated by equations 7.1 and 7.2:

$$F(H) \xrightarrow{k_f} F(D) \tag{7.1}$$

$$F(H) \underset{k_{-1}}{\overset{k_1}{\Leftrightarrow}} U(H) \underset{D_2O}{\overset{k_2}{\rightarrow}} U(D) \underset{k_1}{\overset{k_{-1}}{\Leftrightarrow}} F(D) \tag{7.2}$$

where F and U represent the folded and unfolded protein and H and D symbolize hydrogen and deuterium. Many HDX experiments are done in a solvent containing $<90\%$ deuterium; therefore, once a hydrogen is exchanged for a deuterium, it stays put, making the reaction unidirectional. If the HDX goes to completion, the protein will have approximately the same deuterium composition as the solvent. Equation (7.1) describes exchange in the folded form, likely of solvent-accessible amide hydrogens that are not involved in hydrogen bonding. The rate constant for this type of exchange is k_f. The protein structures, however, are often in equilibrium between the folded and unfolded states. Equation (7.2) describes exchange in unfolded protein. To exchange, amides buried in the protein require unfolding, which may be localized (i. e., for a specific loop) or involve the entire protein as in a global unfolding event [11]. Rate constants for unfolding and refolding are k_1 and k_{-1}, respectively. In this model, exchange occurs after unfolding has "opened" new sites in the protein for exchange by either releasing hydrogen bonds or increasing solvent accessibility.

The rate constants for exchange at each individual amide have contributions from the folded (k_f) and unfolded (k_u) exchange (equation 7.3). The equilibrium constant for unfolding (K_{unf}), along with a probability factor (β) and the rate of exchange from a totally denatured peptide (k_2) define the rate constant for exchange:

$$k_{ex} = k_f + k_u = (\beta + K_{unf})k_2 \tag{7.3}$$

The variable β is a probability factor for exchange from folded forms [9]. β values range from 0 to 1 based on solvent accessibility and hydrogen bonding. As β increases, so does the probability that an amide hydrogen will be accessible to D_2O; thus the rate of exchange increases.

For a stable protein, $k_1 \ll k_{-1}$. For each amide site to exchange, $k_2 \ll k_{-1}$, and the protein must unfold and refold many times. The rate constant for exchange (k_u) under these conditions is directly related to the folding equilibrium (equation 7.4):

$$k_u = \frac{k_1}{k_{-1}} k_2 = K_{unf} k_2 \tag{7.4}$$

The situation described by equation (7.4) is termed EX2 kinetics or uncorrelated exchange. This can be thought of as a sum of many quick, indiscriminate movements within a protein, each of which allows exchange. Most proteins undergo HDX by EX2 kinetics under physiological conditions.

If $k_2 \gg k_{-1}$, the rate constant for exchange from the unfolded form is equal to the rate constant for unfolding (equation 7.5); this is called EX1 kinetics and the exchange is sometimes termed "correlated":

$$k_u = k_1 \tag{7.5}$$

EX1 is often explained as a cooperative unfolding event where all of the unfolded amide sites exchange before refolding. Given that the rate constant for exchange in the unfolded form, k_2, is greatly affected by the pH, temperature, presence of denaturants, and the nature of the neighboring amino acids, one can induce EX1 kinetics by exploiting these properties [5,12,13].

7.2 EXPERIMENTAL PROCEDURES

7.2.1 Global Hydrogen–Deuterium Exchange

To label, or exchange hydrogen for deuterium, one simply admits the protein to a deuterated solvent. This step typically involves starting with the protein in H_2O and diluting by 10- to 20-fold with a deuterated buffer. Other methods involving buffer exchange, however, have been explored [14], and the protein can be fully deuterated by soaking in D_2O, and the back-exchange can be followed. When the protein is admitted to a deuterated solution, spontaneous exchange occurs, and when the concentration of deuterium is greater than 90%, the reaction is essentially unidirectional until the extent of deuteration of the protein is equal to the level of D in the solvent. In a typical experiment, the protein is incubated with its ligand in a biologically relevant buffer until equilibrium is reached. At that point an identical buffer in D_2O is added to give a 90% to 99% deuterated solution, although solvents of lower deuterium content can be used. The amount of time the protein complex is left in the D_2O buffer is determined by the type of experiment. To examine the kinetics of HDX, the extent of exchange is measured as the length of D_2O incubation time is varied; the outcome is a kinetic curve for deuterium uptake. If the goal is to compare a protein bound to its ligand versus the unbound protein, then only one time point may be needed. Many experiments are done by comparing the total amount of deuterium uptake at a "steady state," a time at which the increase in deuterium uptake slows considerably and becomes nearly constant with time. It is best to determine the exchange incubation time by comparing kinetics curves of both bound and unbound protein. In most cases the maximum difference in exchange, ΔD, occurs once the uptake reaches "steady state." There are occasions where the ΔD is small at the steady state, though at earlier time points it is large; in such cases it is better to compare bound versus unbound protein exchange at the incubation time with the largest ΔD. Once the protein has been incubated in the deuterated buffer for the desired time, the exchange reaction is nearly stopped by lowering the pH by adding strong acid and lowering the temperature by submerging in ice. This process is called quenching. Quenching takes advantage of the fact that the rate of HDX is both acid and base-catalyzed; thus its rates are sensitive to pH (Figure 7.1) [7,15]. By carefully controlling the pH, one can decrease the exchange rate constant from approximately $10\,s^{-1}$ at pH 7.5 to $0.001\,s^{-1}$. The exchange rate constant can be lowered by another order of magnitude by decreasing the temperature from 25°C to nearly 0°C. This step is done by simply placing the sample and the tubing used to handle it into an ice bath. At pH 2.5 and 0°C, the half life of exchange is nearly 2 h. The resulting decrease in rates is important because it provides time to work with the sample; for example, to desalt on a short column and to conduct an enzymatic digest.

In a typical work flow for an HDX experiment (illustrated in Figure 7.2), one quenches the exchange and either measures the intact protein to obtain a global HDX

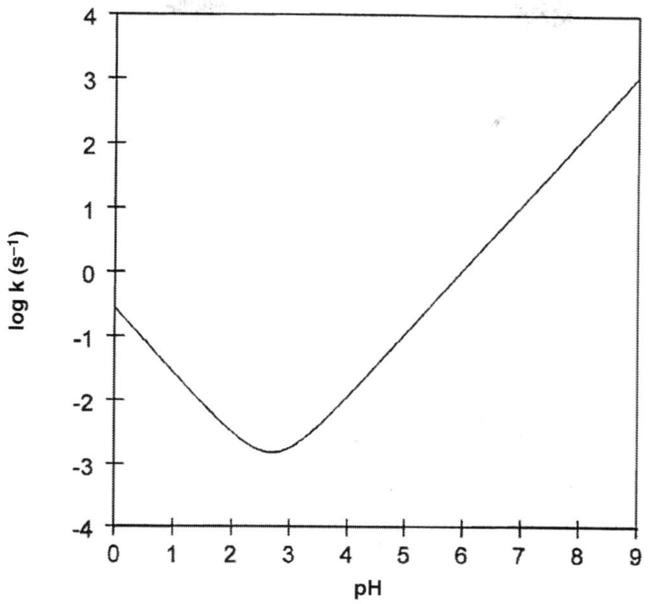

FIGURE 7.1 Variation of the intrinsic exchange rate constant of peptide amide hydrogens as a function of pH. From, D. L. Smith, Y. Deng, Z. Zhang, Probing the non-covalent structure of proteins by amide hydrogen exchange and mass spectrometry. *J Mass Spectrom* 32(2),135–146, Copyright © 1997 by John Wiley & Sons, Ltd. Reprinted by permission of John Wiley & Sons, Inc.

profile or digests the protein to find information at the peptide level (sometimes called "local level" or "local exchange"). Because the protein-ligand complex is typically diluted 10- to 20-fold for exchange, and starting concentrations are normally low (ideally they are close to the K_d), the sample must often be concentrated before analysis. Samples in biologically relevant buffer also must be desalted to improve signal quality. In an ESI experiment, both steps are done by using a short column (typically containing C-8 resin). To remove salts, chilled water at pH 2.5 is applied to the column; this H_2O also back exchanges the fast-exchanging side chains to the H form so that they are not counted in the HDX experiment. The protein is eluted by using solvents that are 20% to 50% organic depending on the specific protein. Although most researchers use ESI for ionization in HDX, matrix-assisted laser desorption ionization (MALDI) can also be chosen. In MALDI experiments these steps are accomplished by using zip-tips or spin columns to concentrate quickly and desalt the sample. Often, using MALDI shows a greater extent of back exchange than does using ESI.

7.2.2 HDX at the Peptide Level

When information at the peptide level is desired, it is necessary to digest the protein first, then concentrate, desalt, and elute [16]. Enzymes like trypsin are commonly used in proteomics for digestion because they cut the peptide at specific amino acids

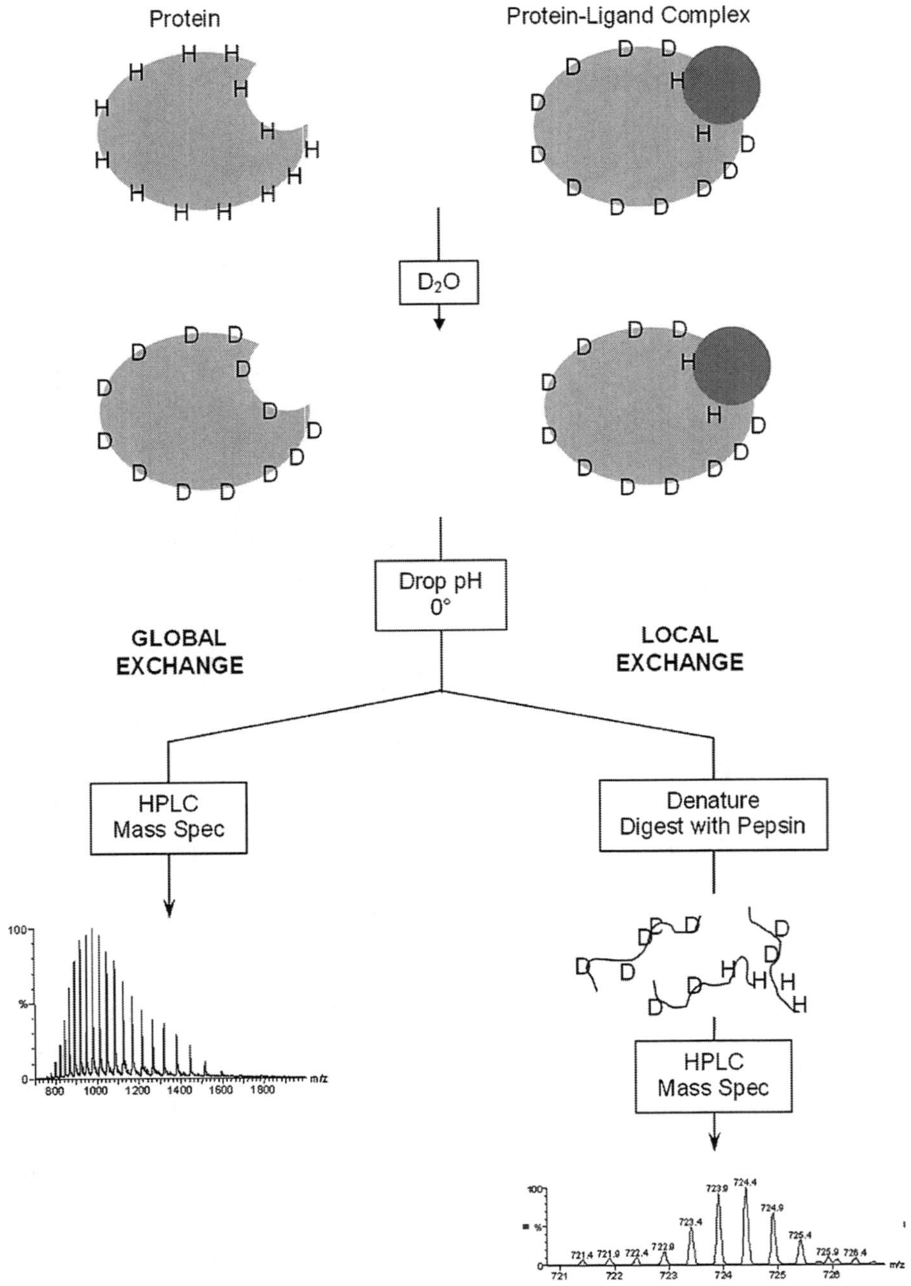

FIGURE 7.2 General procedure for determining the HDX of protein–ligand complexes. Protein, either bound or unbound, is added to D_2O; after the desired length of exchange, the process is quenched by decreasing the pH and lowering the temperature of the sample. The sample can then either be analyzed for global exchange by submitting intact protein to the mass spectrometer or to the exchange at the peptide level, which requires enzymatically digesting the protein and separating the resultant labeled peptides by HPLC.

(Arg, Lys), thus allowing the user to predict the peptide sequences if the protein is known, or to constrain database searching if the sequence is unknown. Unfortunately, the low pH requirements for quenching HDX render trypsin and most other enzymes useless. One exception is pepsin, an enzyme found in the stomach of mammals. This enzyme works only at acidic pH and, while reproducible under constant conditions, it is not highly predictable. To identify accurately peptic peptides, it is necessary to sequence them by using MS/MS followed by database searching. In addition to being unpredictable, pepsin produces many overlapping and/or large peptides. This can be viewed as a complication (dispersion of information) or as an advantage (information in overlapping regions). Despite its drawbacks, pepsin does work quickly; most digestions take less than 5 min, even at low temperatures. Furthermore, overlapping peptides can prove useful because they improve the resolution for locating specific residues responsible for deuterium uptake or protection. Although other acid proteases are available, pepsin is still the most efficient [17]. Pepsin is available immobilized on either beads or column supports, and both forms allow for rapid removal of the enzyme to ensure consistent digestion times.

After digestion, peptides are loaded onto a column, typically C-18, the column washed to remove salts, and a gradient applied to separate peptides as well as possible contaminants before entering the mass spectrometer for analysis. Currently most research involves monitoring the amount of deuterium on a peptide as measured by a full mass spectrum. Programs like HX-Express measure the centroid of the deuterated peptide and compare it to that of the undeuterated peptide (Figure 7.3) [18].

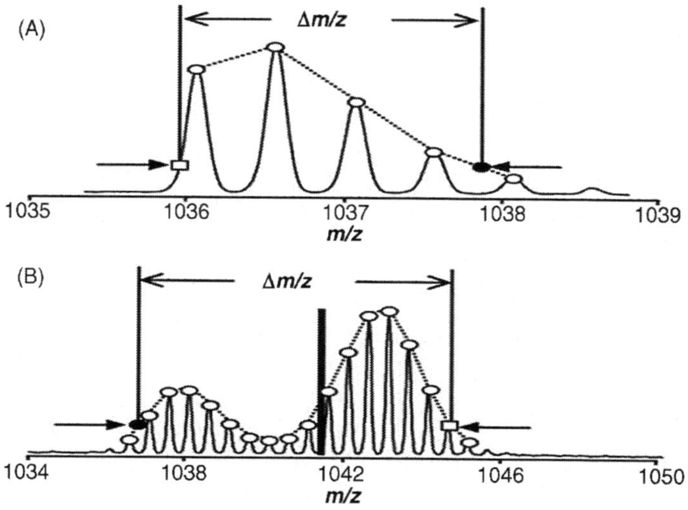

FIGURE 7.3 Illustrations of HX-Express processing of (*A*) undeuterated peptide and (*B*) deuterated peptide with a bimodal distribution. The centroid of the envelope is shown in spectrum (*B*) as the vertical solid line for an EX1 system (the width of envelop is also reported). From D. D. Weis, J. R. Engen, I. J. Kass, Semi-automated data processing of hydrogen exchange mass spectra using HX-Express. *J Am Soc Mass Spectrom* 17(12), 1700–1703. Copyright © 2006 by Elsevier. Reprinted by permission of Elsevier.

It would be ideal to monitor not only the deuterium uptake of peptic peptides, but to locate D sites on the individual amino acids. Typically monitoring changes at specific amino acids (such as post-translation modifications) is accomplished by performing tandem mass spectrometry experiments on the peptides, revealing the precise residue that has been altered. Collision induced dissociation (CID) is traditionally the method used to fragment peptides, and it has been evaluated for fragmenting deuterated peptides [19]. Unfortunately, CID of deuterated peptides is generally not useful because "scrambling" of D's and H's occurs in a partially labeled protein. When a fragment is scrambled the original information concerning the precise location of deuterium is lost. More instantaneous methods of activation (e.g., electron capture dissociation [ECD] and electron transfer dissociation [ETD]) offer hope that they can be used to fragment peptides with little D scrambling. In both ECD and ETD, radical cations are produced. These cations appear to decompose to c- and z-type fragment ions with little scrambling [20]. Early experiments suggest they may prove useful in deciphering HDX at the amino acid level [21–23].

7.3 MASS SPECTROMETRY-BASED HDX IN PRACTICE

7.3.1 Protein–Ligand Interactions by Automated HDX

One of the main drawbacks of HDX is that executing the experiment is labor intensive and has low throughput. In an attempt to alleviate this, the Griffin Lab at Scripps introduced an automated system for HDX [24]. The automated system is constructed around an autosampler with cooled sample stacks. HDX experiments are initiated by the autosampler in one of two modes. The first mode measures kinetics from 130 ms to 30 s, the second mode handles the longer time points, up to 24 h. Long exchange time samples are performed in parallel to reduce the total time of the experiment. The autosampler is connected to a three-valve unit that is contained in a custom-built cooling chamber that keeps the mobile phases at around 1°C. The quenched samples are submitted to online pepsin digestion, peptide desalting, and reverse-phase HPLC separation prior to ESI MS analysis.

The automated HDX platform can analyze ligand interactions; an example is the ligand-dependent nuclear receptor PPARγ. Full agonist drugs that target PPARγ have had some success in clinical trials; however, they are plagued by side effects. Partial agonists may be associated with fewer side effects while preserving efficacy. Both full and partial agonists cause a conformation change in PPARγ. Figure 7.4 illustrates the HDX patterns found for PPARγ bound to a full agonist (GW1929) and to a partial agonist (nTZDpa). The patterns are shown by color-shading regions of the X-ray crystal structure using the HDX data. Both ligands decrease HDX in helix 3, a known binding site, but only the full agonist decreases deuterium uptake in helices 11 and 13. This clear distinction between a full and partial agonist, combined with the speed and reproducibility enabled by automation, demonstrate the utility of HDX to screen ligands as potential drugs.

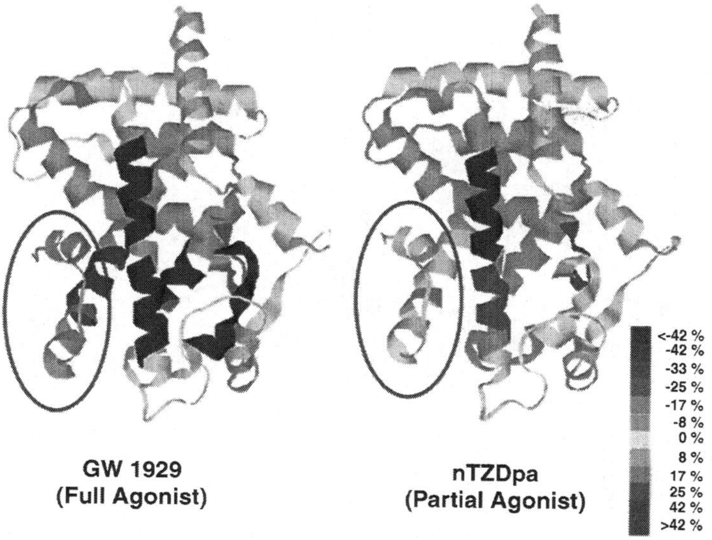

FIGURE 7.4 Change in HDX rate constants as a result of binding a full agonist (*left*) and a partial agonist (*right*) to PPARγ. From M. J. Chalmers et al., Probing protein ligand interactions by automated hydrogen/deuterium exchange mass spectrometry. *Anal Chem* 78(4), 1005–1014. Copyright © 2006 by American Chemical Society. Reprinted by permission of American Chemical Society. (*See the color version of this figure in Color Plates section.*)

In addition to automating deuterium exchange, advances in data analysis have contributed to making HDX a more efficient experiment. Software packages, including HXExpress [18], mentioned previously, HD Desktop [25], and Hydra [26], reduce the data processing time from weeks to days and in some cases hours.

7.3.2 Solvent Accessibility by HDX and MALDI-TOF Mass Spectrometry

Komives and coworkers were the first to conduct HDX and use MALDI-TOF mass spectrometry for mass analysis. An example is their use of this approach to reveal differences between globular and nonglobular proteins [27]. In a globular protein, amide exchange is typically very slow in the core where amides are involved in hydrogen bonding and typically inaccessible to solvent. The surface amides exchange much more quickly but at a rate slower than that of a denatured peptide. To probe the surface amides, it is necessary to determine the HDX kinetics at early time points, requiring the use of a stop-flow device. To do this, the researchers diluted the protein solution into a deuterated buffer and exchanged for 0.05 to 120 s by varying the length of the delay line that carries the exchanged protein solution into the quench solution (Figure 7.5). The quenched protein was immediately digested with immobilized pepsin, aliquoted, frozen in liquid N_2, and stored at –80°C. The samples were prepared and analyzed by MALDI-MS, one at a time, by rapidly thawing, mixing with chilled matrix, spotting on a chilled plate, and drying in vacuum. The mass spectra

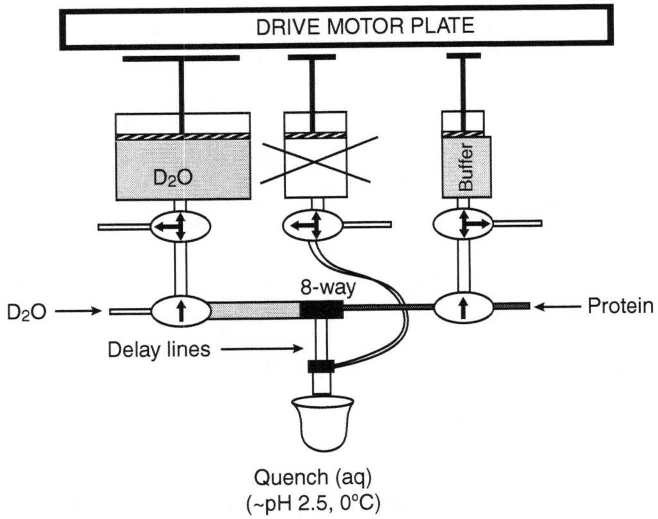

FIGURE 7.5 A schematic diagram for the Kintek QF-3 instrument used for quench-flow hydrogen deuterium exchange. From S. M. Truhlar et al., Solvent accessibility of protein surfaces by amide H/2H exchange MALDI-TOF mass spectrometry. *J Am Soc Mass Spectrom* 17(11), 1490–1497. Copyright © 2006 by Elsevier. Reprinted by permission of Elsevier.

obtained show an increase in deuterium uptake over time, and the reproducibility with stop-flow is excellent; the error bars are too small to be seen on the graph (Figure 7.6).

The authors were able to reach several conclusions on the basis of this experiment. First, although some surface amides exchange at rates comparable to those of unstructured peptides, most exchange was slowed significantly. Second, the effect of hydrogen bonding on the rate of amide exchange varies for globular and nonglobular proteins. Third, HDX correlates well with solvent accessible surface area, but only after one accounts for exchange on side chains. This implies that amino-acid side chains are protected in part from HDX.

7.3.3 High-Throughput Screening of Protein Ligands by SUPREX

SUPREX (stability of unpurified proteins from rates of H/D exchange) is an HDX MS approach developed in the Michael Fitzgerald lab at Duke University. The binding free energies determined by SUPREX match well those determined by spectroscopic and calorimetric methods [28–30]. SUPREX can be applied to many ligands (small molecule, peptide, oligonucleotides, and proteins) over a range of K_d values. This experiment requires only picomoles of protein and is adaptable to automation and relatively high throughput.

In a typical SUPREX experiment, the binding free energy is determined by incubating a protein (with or without ligand) with a denaturant. At least 10 increasing concentrations of denaturant are used. Each sample undergoes HDX for a set period of time (e.g., 1 h). Figure 7.7 shows that as the concentration of denaturant increases, the extent of HDX also increases [31]. The curve is a best fit to the data, affording the

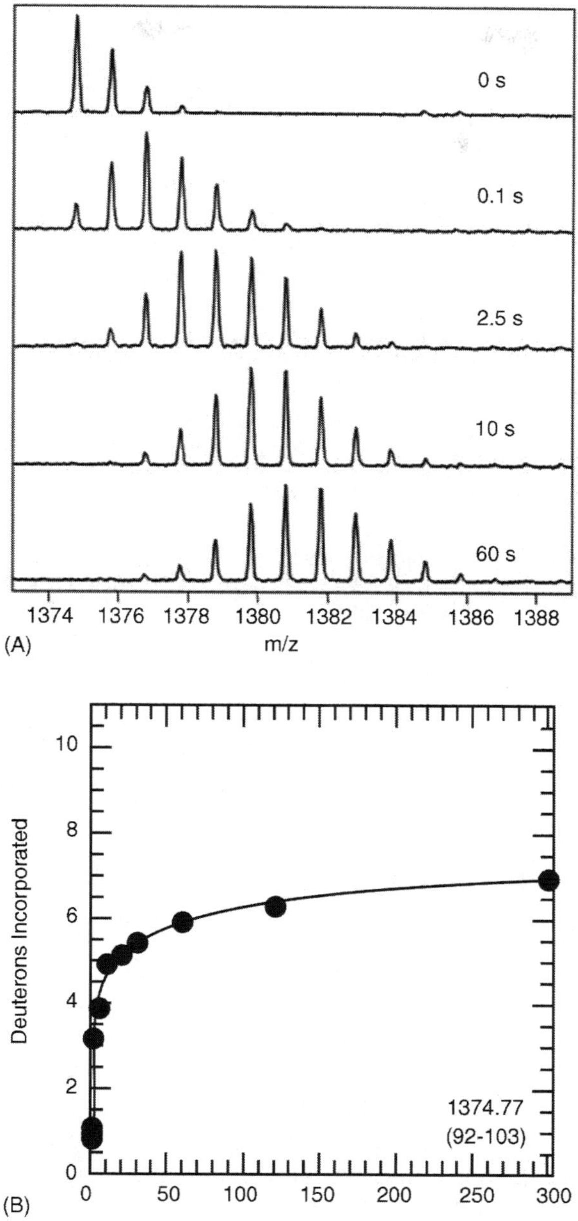

FIGURE 7.6 (*A*) Stacked plot of the isotopic distribution of quench-flow labeled peptic peptide. (*B*) The kinetic plot of deuterium uptake as a function of time. Error bars indicating the precision of the measurements are too small to be seen on this plot. From S. M. Truhlar et al., Solvent accessibility of protein surfaces by amide H/2H exchange MALDI-TOF mass spectrometry. *J Am Soc Mass Spectrom* 17(11), 1490–1497. Copyright © 2006 by Elsevier. Reprinted by permission of Elsevier.

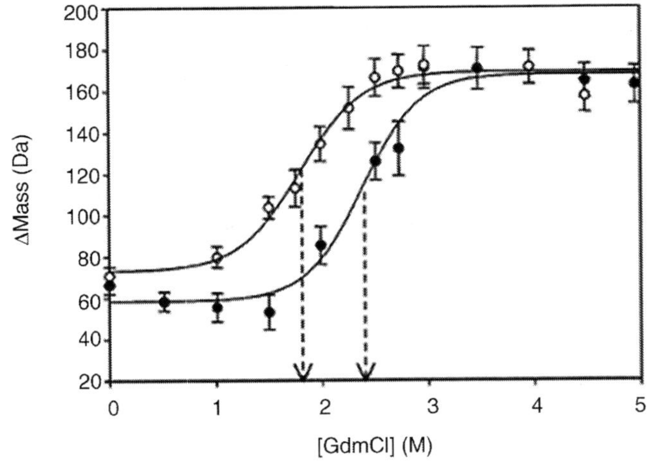

FIGURE 7.7 Typical SUPREX curve demonstrating the effect of denaturant on protein structure. From P. L. Roulhac et al., SUPREX (Stability of Unpurified Proteins from Rates of H/D Exchange) analysis of the thermodynamics of synergistic anion binding by ferric-binding protein (FbpA), a bacterial transferrin. *Biochemistry* 43(50), 15767–15774. Copyright © 2004 American Chemical Society. Reprinted by permission of American Chemical Society.

denaturant concentration at the transition midpoint ($C^{1/2}{}_{\text{SUPREX}}$). The $C^{1/2}{}_{\text{SUPREX}}$ is used to calculate ΔG.

Recently SUPREX was applied to a high-throughput protocol and demonstrated efficient screening for ligand peptides in a combinatorial library to a model protein [32]. The high-throughput screening (HTS) experiment is done is two parts: (1) preparing the ligand library and (2) using a single point SUPREX analysis (Figure 7.8). A single-point SUPREX analysis is the Δ*mass* of the target protein obtained from the H/D exchange experiment. For this single point to be meaningful, a SUPREX analysis of the protein must be obtained in which a curve consisting of at least 10 points at increasing concentrations of denaturant (e.g., urea) while the protein is in the presence of a ligand with a known binding affinity is acquired. From the reference curve one only needs to compare the single-point analysis: if the Δ*mass* is greater than that at the midpoint of the reference curve, the ligand binds more tightly; if it is lower, the ligand is a weaker binder. The experiment can be calibrated to suit different needs by changing the concentration of urea for the single-point experiment. Increasing the urea concentration allows one to screen ligands that bind more tightly (bind with a lower K_d); lower concentrations of urea are better for weaker binding systems (Figure 7.9).

The Fitzgerald lab demonstrated that applying HTS to SUPREX allows fast and accurate screening of ligands. In this experiment, they show that they are able to select ligands of varying K_d values by using different urea concentrations to screen the library. Of particular interest is their demonstration that a tight-binding ligand can be screened in a library composed mostly of weak binders, as is often the case in

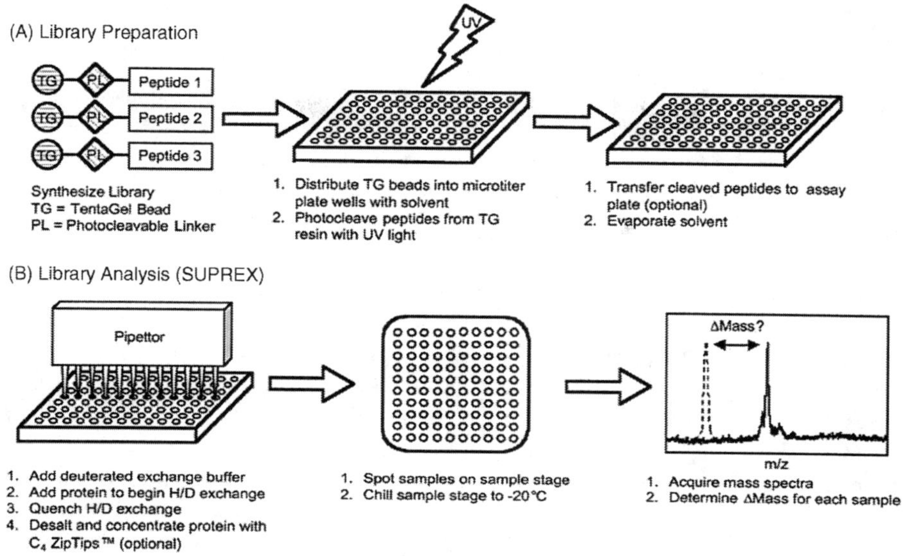

FIGURE 7.8 High throughput screening protocol employed in SUPREX experiments. From K. D. Powell, M. D. Fitzgerald, High-throughput screening assay for the tunable selection of protein ligands. *J Comb Chem* 6(2), 262–269. Copyright © 2004 American Chemical Society. Reprinted by permission of American Chemical Society.

screening for drug targets. This experiment was repeated with three different libraries; in each case they identified the tight-binding ligands without false positives. Single-point SUPREX shows promise in improving the throughput of SUPREX, making it more useful to pharmaceutical researchers who work in drug discovery.

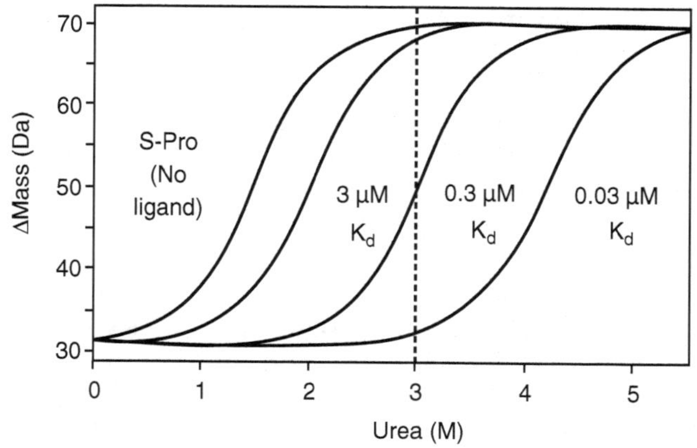

FIGURE 7.9 Idealized SUPREX curves for the S-Pro system, without ligand, in the presence of a 3 μM ligand, a 0.3 μM ligand, and a 0.03 μM ligand. From K. D. Powell, M. C. Fitzgerald, High-throughput screening assay for the tunable selection of protein ligands. *J Comb Chem* 6(2), 262–269. Copyright © 2004 American Chemical Society. Reprinted by permission of American Chemical Society.

7.3.4 Functional Labeling and Multiple Proteases

Functional labeling is a HDX technique that serves to characterize only those amide sites that change when a protein changes from one functional structure to another [33]. This method differs from traditional HDX experiments because it does not consider the change in deuterium uptake in the bound versus unbound forms of the protein. Instead, the approach is to forward-label the protein in the absence of ligand, add ligand, then back-exchange the protein. The deuteriums that remain are functionally labeled.

In a collaborative effort, Englander and Woods coupled functional labeling to multiple protease digestion in an automated MS approach. In this example, they used hemoglobin (Hb) to demonstrate how functional labeling can reveal allosterically sensitive amides. The fast-exchanging oxy-Hb was exchanged in on a short time scale, labeling both allosterically sensitive and insensitive sites. The protein was then rapidly switched to deoxy state and exchanged out for a longer period of time. At this point only the sites that became protected as a function of the switch from oxy-Hb to deoxy-Hb were labeled. This methodology eliminates the background, or functionally insensitive sites, thus reducing the uncertainty that can arise from traditional HDX experiments.

To locate the position of the functionally labeled sites within the protein, one can proteolytically digest it and analyze the fragments by HPLC/ESI MS. Studying the peptide fragments gives information on the region in which the label is located; however, locating the specific amino acid that is labeled is often difficult. Success can be achieved if there is suitable overlap of the peptides formed in the digestion. For example, by comparing the extent of HDX for a peptide with that of a peptide containing one flanking amino acid reveals the exchange occurring at that amino acid. Reliable results from tandem MS experiments are not yet convincingly demonstrated for deuterated samples. The strategy of invoking overlapping peptides is better served by having multiple overlapping fragments achieved by using both pepsin and fungal protease XIII rather than by using a single enzyme. Online digestion of Hb with these two proteases followed by HPLC-MS yields 100% sequence coverage with many overlapping fragments (Figure 7.10). Figure 7.11 illustrates how overlapping fragments can give localized exchange information. In this example β130-146 has an overall exchange of four deuteriums; however, three are located in the *C*-terminus of the peptide.

7.3.5 PLIMSTEX: Application in Protein–DNA Interactions

Protein–DNA interactions are a common and important target of drug therapies [34,35]. These interactions can be difficult to characterize owing to the unstructured nature of many DNA-binding proteins. Studying the changes in the solvent accessibility of a protein–DNA complex is not prejudiced by the nature of the protein. HDX and mass spectrometry are a convenient way to reveal solvent accessibility and to characterize further protein–DNA interactions. We will use this application to illustrate another approach to determining protein/ligand affinity,

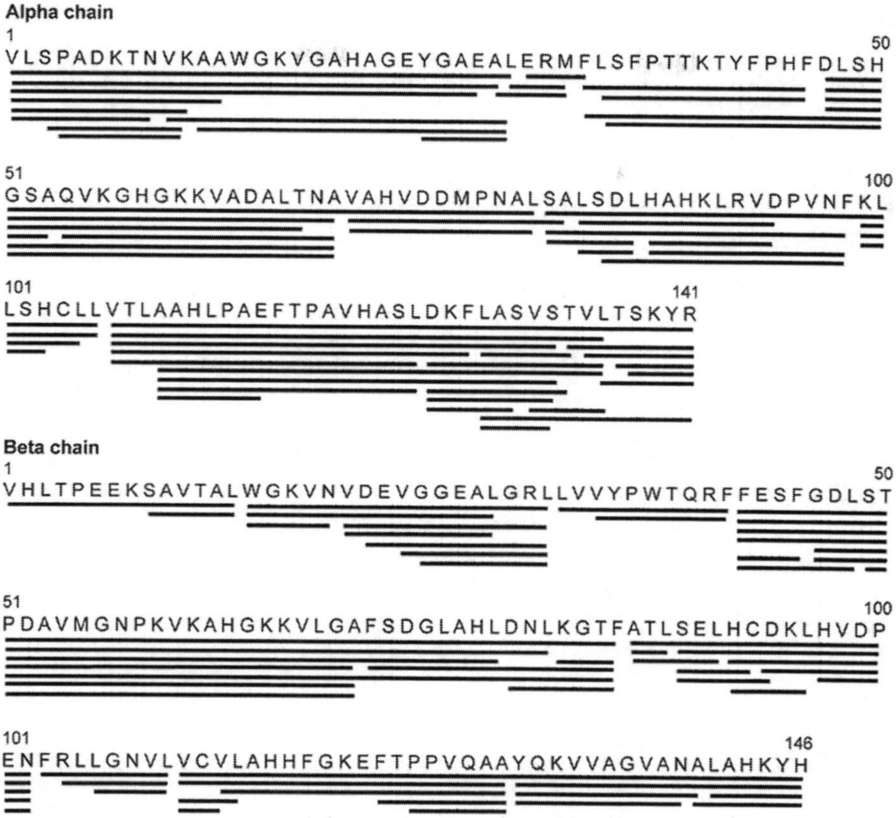

FIGURE 7.10 Peptides from Hemoglobin identified by HPLC/MS after tandem digestion with pepsin and fungal protease XII. From J. J. Englander et al., Protein structure change studied by hydrogen-deuterium exchange, functional labeling, and mass spectrometry. *Proc Natl Acad Sci USA* 100(12),7057–7062. Copyright © 2003 National Academy of Science. Reprinted with permission of the National Academy of Science.

namely "protein-ligand interactions by mass spectrometry, titration, and HD exchange (PLIMSTEX) [36,37].

M. Gross and his group at Washington University developed PLIMSTEX [36,37] and then applied it to the interaction of human telomeric repeat binding factor 2 (hTRF2) with telomeric DNA [38]. The approach is unique because is measures not only the HDX of bound and unbound protein, but also that of intermediate mixtures of apo-protein with ligand. The approach is to conduct a titration, generating a PLIMSTEX curve. This curve can be fit with a 1:*n* (protein:*n* ligands) sequential binding model to extract the binding constants. In addition to analyzing the global HDX, pepsin digestion can be used to produce constituent peptides whose extents of HDX serve to localize regions where exchange occurs.

In general, the first step in a PLIMSTEX experiment is to acquire the global kinetics; that is, the rates at which the apo- and holo-forms of the protein gain

			Percent recovery		Mass increment			Zero time
Sequence	Mass	Z	Obs	$Theo_{10min}$	Obs	Bkgd	(Obs-bkgd)/ recov	Corrected ×1.25
β130–146	1868.01	1;3	86 (3)	78	2.6 (0.7)	0.0 (0.1)	3.2	4.0
β140–146	838.44	1;2	62 (4)	78	2.0 (0.4)	0.1 (0.2)	2.8	3.5
β141–146	767.41	1;2	60 (10)	72	1.7 (0.4)	0.1 (0.1)	2.4	3.0
β130–140	1118.61	1	87 (3)	82	0.3 (0.1)	0.0 (0.1)	0.3	0.4
β130–139	1047.57	1	86 (6)	85	0.3 (0.1)	0.1 (0.1)	0.3	0.4

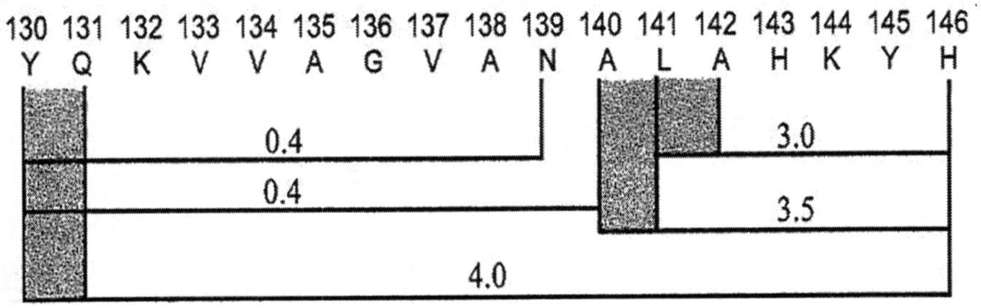

FIGURE 7.11 Deuterium uptake of overlapping peptides can be used to increase the specificity in the determination of protection in HDX. From J. J. Englander et al., Protein structure change studied by hydrogen-deuterium exchange, functional labeling, and mass spectrometry. *Proc Natl Acad Sci USA* 100(12),7057–7062. Copyright © 2003 National Academy of Science. Reprinted with permission of the National Academy of Science.

deuterium. For the hTRF2-DNA example, the apoprotein exchanges roughly the same number of amides at long time points, but the rate of exchange is much faster than that of the holoprotein (Figure 7.12*A*). The time at which the difference in deuterium (ΔD) is the greatest and approximately at steady state is used for the titration step. In this step, the ligand is added to the protein in increasing concentrations from zero to excess and allowed to equilibrate. After the solution has reached equilibrium, a deuterated buffer is added so that the final solution has a reasonable concentration of D_2O (e.g., 95% deuterium although lower percentages can be used depending on the extent of dilution that one can tolerate). The exchange is "quenched" at the time specified by the global kinetics experiment, in this case, 3 min. After measuring the extent of HDX with a mass spectrometer, one obtains a curve showing the HDX extent in going from the apo to holò form (Figure 7.12*B*). Mathematical modeling [39] reveals the binding constant; in this example, the K_D for hTRF2-DNA is 580 ± 140 nM, in good agreement with a value previously determined by surface plasmon resonance.

Although the global exchange data are useful for characterizing the protein–DNA complex and determining affinity, exchange at regions of the protein or even at the amino-acid level can reveal binding regions and perhaps report on allosteary. As is the case with other HDX approaches, enzymatic digestion of the exchanged protein followed by MS analysis of the resulting peptides provides this specificity. Owing to the small size of hTRF2, only five peptic fragments are needed to give complete coverage. Peptides located in the unstructured region are highly deuterated, and the

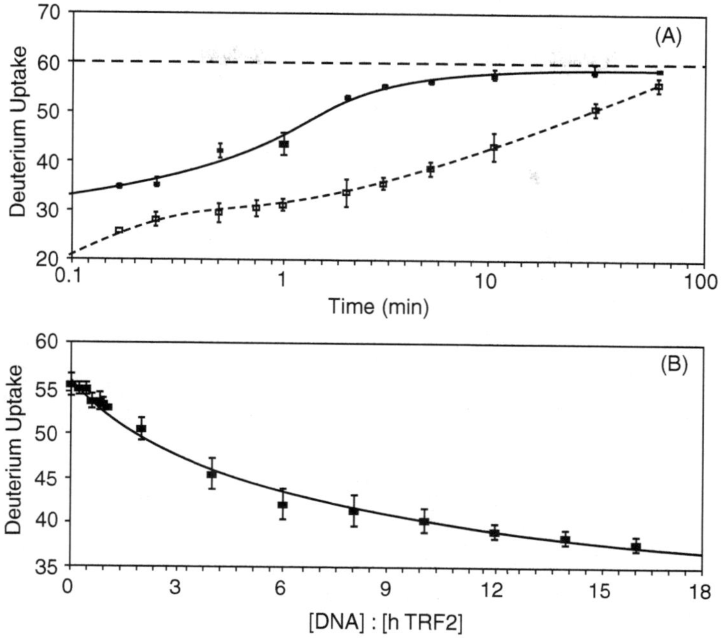

FIGURE 7.12 (*A*) Global deuterium uptake kinetics over 60 min for apo (■) and holo (□) hTRF2 (1 M). The maximum deuterium uptake (60 amides) was achieved at 37 °C after 5 h. (B) PLIMSTEX titration curve obtained by monitoring the deuterium uptake of hTRF2 (100 nM) upon adding increasing amounts of the ODN ligand. The HDX was quenched at 3 min for each point of the titration. A K_D of 580 ± 140 nM, a ΔD of 25.1 ± 1.5 Da, and a D_0 of 56.1 ± 0.4 Da were determined from the fit of the PLIMSTEX data. From J.B. Sperry, X. Shi, D.L. Rempel, Y. Nishimura, S. Akashi, and M.L. Gross (2008), A mass spectrometric approach to the study of DNA-binding proteins: interaction of human TRF2 with telomeric DNA. *Biochemistry* 47(6), 1797–1807. Copyright © 2008 American Chemical Society. Reprinted by permission of American Chemical Society.

ΔD is small. Those in the binding region show a large ΔD; the holo exchanged far less than the apo. This is one example showing that PLIMSTEX provides a reliable means of determining affinity at both the global and the peptide level of a protein.

In the course of this application, a rapid means of separating the protein and peptides from oligodeoxynucucleotide was developed to ensure that the protein/peptides could be effectively measured by MS [40].

7.3.6 HDX and Tandem Mass Spectrometry Analysis

As noted, the most desirable information to be gained from HDX would be localized at the single amino-acid residue. This type of information appears to be available using electron-capture or electron-transfer tandem mass spectrometry. Traditionally, peptide tandem MS is performed by colliding the charged molecule with an inert gas, causing fragmentation into the well-known b- and y-ions. This type of tandem MS utilizes collision-induced dissociation (CID), but it is not sufficient for localizing

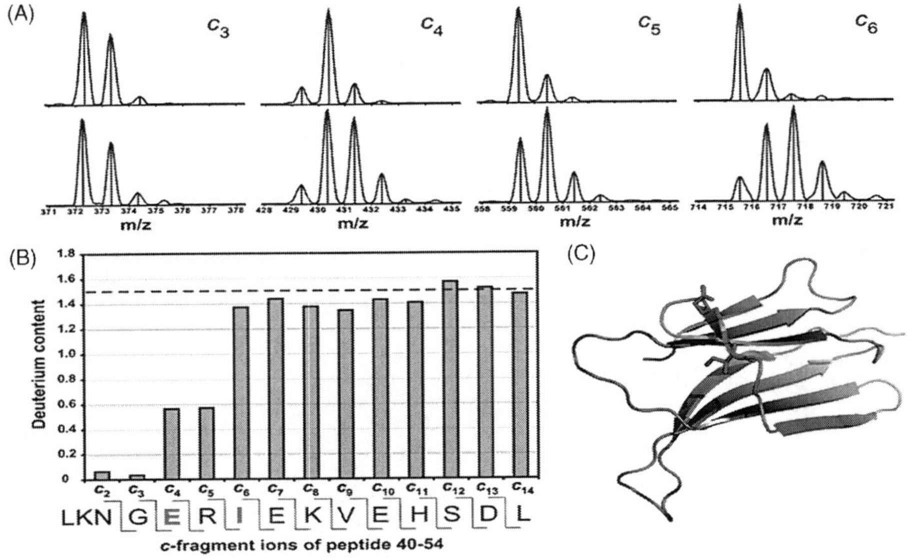

FIGURE 7.13 A. Tandem MS of deuterated peptide. B. Deuterium content of c-ion series. From J. Villanueva et al., Increase in the conformational flexibility of beta 2-microglobulin upon copper binding: A possible role for copper in dialysis-related amyloidosis. *Protein Sci* 13(3),797–809. Copyright © 2004 by John Wiley & Sons, Inc. Reprinted by permission of John Wiley & Sons, Inc.

HDX information because the deuteriums "scramble" along the backbone after activation but before the fragmentation process [41,42]. Electron transfer dissociation (ETD) is an alternative to CID and does not fall victim to scrambling. The Jorgensen Group at the University of Southern Denmark recently used HDX and ETD to explore the backbone of β_2-microglobulin [43]. The HDX and the pepsin digestion portions of the experiment are straight forward; therefore our focus will be the use of ETD tandem MS to analyze the labeled peptides.

β_2-Microglobulin was exchanged and digested with pepsin under normal HDX conditions. The resultant peptides were separated by fast HPLC and analyzed by using ESI MS/MS (see Figure 7.13 for the mass spectra in the regions of a series of c-ions). The bar graph represents the average number of deuteriums in each fragment from which one can easily identify the protected residues. The locations of the protected residues identified by HDX MS/MS agree well with those identified by NMR, where scrambling of D and H is not an issue [44].

7.3.7 Optimizing HDX with High Pressure

The first ultra performance liquid chromatography (UPLC) mass spectrometry paper was published in 2004 [45]. Since then, UPLC has been gaining popularity owing to its ability to separate rapidly analytes. While initially used for metabolomics [46–48], UPLC for peptide separation has further improved HDX [49,50]. UPLC instruments use high pressure (10,000–20,000 psi) to achieve a better separation (peaks widths of 1 s can be achieved) in short times. In a local HDX experiment, the isotopic envelope of each peptide expands as it gains deuterium, and these envelopes may overlap, making it difficult, if not impossible, to analyze the data. This is especially true with large proteins that produce many peptides upon digestion. Chromatographic separation of these peptides eliminates overlap and increases protein coverage. Engen [49] recently demonstrated a chromatographic separation of model peptides using UPLC that could not be achieved on a HDX time scale with traditional HPLC. The extent of deuterium loss using UPLC was greater than that achieved using HPLC; however, much of this loss is stems from sample injection time and temperature of solvent. The utility of using UPLC separations for HDX studies of antibodies was documented [51].

High pressure is also being used to improve the digestion of proteins. Incomplete digestion of proteins in deuterated solutions is common. As a result, peptide coverage is not as complete as when digestion is performed in an aqueous solution. Proteins that are difficult to denature and those that are members of a protein complex, such as antibodies, are especially difficult to digest fully in the short time frame of a HDX experiment. The Gross group addressed this by using an online high-pressure digestion system that takes advantage of UPLC [52]. It was established that protein digestion is accelerated under high pressure [53,54]. The Gross group demonstrated that high pressure also increases digestion efficiency for proteins that aggregate and are generally harder to digest. They were able to increase digestion coverage of an aggregate protein from 53% to 80% within the HDX time scale. Additionally the high-pressure digestion did not increase back-exchange in comparison with a protein digested in the absence of high pressure. All these advances in the technology of HDX have increased the impact of the approach, and new advances will continue to make the method more widely available and useful.

7.4 PROTEIN FOOTPRINTING VIA FREE-RADICAL OXIDATION

Although HDX is a valuable approach for footprinting proteins and investigating their interactions and interfaces, its disadvantages are that it is a relatively slow process, taking place from seconds to hours, and it introduces a reversible covalent change in the protein. A complementary approach that is fast and introduces an irreversible change is OH radical oxidation. This process, although less general than HDX, does modify a large fraction of the amino acids in a short time. In fact, the rates of reaction of a number of amino-acid residues (e.g., those bearing aromatic, sulfur-containing, and aliphatic with a tertiary H side chains) are sufficiently fast that they approach the limit of diffusion control. Like HDX, oxidation can be used to probe protein–protein interfaces, map solvent accessibility, and study protein folding and unfolding. A benefit of oxidation is that the label is introduced irreversibly; once a site is oxidized, it does not readily return to its reduced form. This contrasts significantly with HDX, where there are technical challenges of controlling back exchange (HDX is a reversible change). This aspect of the modification allows for more deliberate manipulation of the protein after the label has been introduced. For example, trypsin digestion, long HPLC gradients, and tandem mass spectrometry analysis can be performed without perturbing the information gained by labeling.

There is a risk of perturbation during the oxidation step. Too much oxidation can denature the protein, therefore, changing the solvent accessibility or protein–protein interface and introducing oxidation sites that would not be accessible in the folded protein. For this reason, oxidation must be conducted under more stringent conditions than HDX by controlling the concentration of hydroxyl radicals, the duration of radical exposure, or the number of "hits" the oxidation produces.

There are multiple methods for producing hydroxyl radicals. Chemical generation, using hydrogen peroxide with transition metals, is inexpensive but can be difficult to control because it produces a steady and relatively constant source of radicals and is conducted over a relatively long time frame. Electrochemical generation of hydroxyl radicals is possible by turning up the electrical voltage applied to the emitter of an ESI source. This is an inexpensive approach, but it is limited to labeling molecules in the gas phase. Radiolytic generation uses either radioactive material or a high-intensity radiation source such as a synchrotron. This method offers better experimental control but working with radioactive material is not ideal, and access to a synchrotron is not convenient or immediate. Mark Chance at Case Western has used radiolytic generation to study many protein interactions, as will be discussed below [55–57]. A recent development described by Hambly and Gross [58] at Washington University uses laser photolysis for pulsed and fast hydroxyl generation and labeling. This work takes as its foundation earlier work where oxidation was demonstrated to be an appropriate tool for footprinting protein/DNA interactions and RNA folding [59], protein interfaces [60], and protein folding [61]. Before discussing their development, we will consider Fenton reaction for generating radicals and Chance's work in radiolytic generation.

7.4.1 Fenton Chemistry Oxidation

Generating hydroxyl radicals by converting Fe(II) to Fe(III) and reducing hydrogen peroxide is simple and readily applied. The reaction is named after its inventor, H. J. H Fenton (Fenton, 1894). The mechanism for the generation of hydroxyl radicals and regeneration of Fe(II) is described by equations (7.6) to (7.7) [62,63]. When using iron to promote oxidation of proteins, it is important to use chelated iron to prevent iron from binding to the protein and giving specific oxidation [64].

$$Fe^{2+} + H_2O_2 \rightarrow Fe^{3+} + HO^{\bullet} + OH^- \quad (7.6)$$

$$Fe^{2+} + HO^{\bullet} + H^+ \rightarrow Fe^{3+} + H_2O \quad (7.7)$$

In 1994 Heyduk and Heyduk [64] demonstrated that Fenton chemistry-induced oxidation and analysis by gel electrophoresis can map protein domains. Given that gel electrophoresis is a low resolution technique, this study was limited to following protein backbone cleavage caused by oxidation. With the relatively new advances of

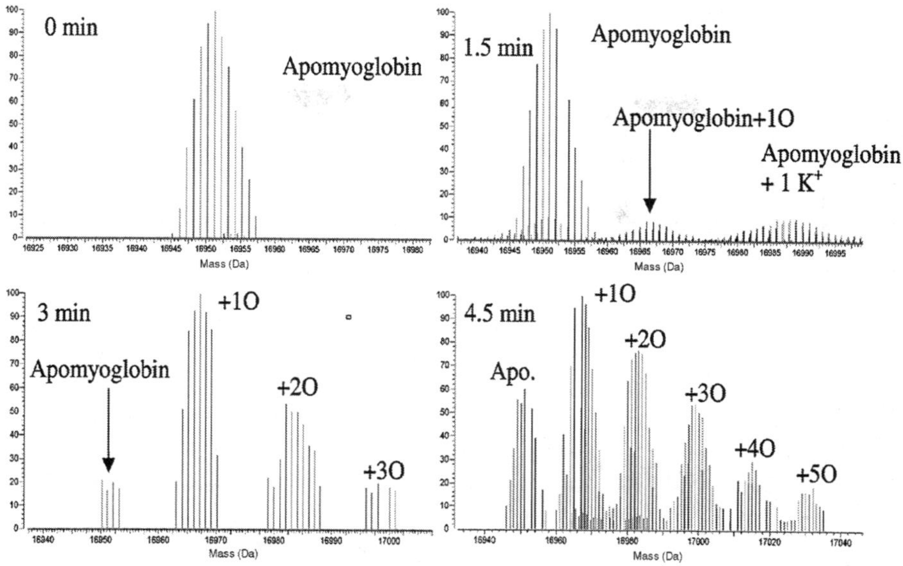

FIGURE 7.14 Mass spectra showing that the amount of oxidation resulting from solution-phase hydroxyl radicals generated by the Fenton reaction increases with time. From J. S. Sharp, J. M. Becker, B. L. Hettich, Protein surface mapping by chemical oxidation: structural analysis by mass spectrometry. *Anal Biochem* 313(2), 216–225. Copyright © 2003 by Elsevier Science (USA). Reprinted by permission of Elsevier.

HPLC and electrospray mass spectrometry, especially with high-performance FT instruments, oxidation located on side chains can be identified by digesting proteins and analyzing the peptides. With this technology, Sharp and Hettich [65] pursued mapping the solvent accessibility of apomyoglobin, a well-characterized protein, using chelated iron and hydrogen peroxide to provide the reagent radicals.

The apomyoglobin was oxidized by solution-phase hydroxyl radicals that were generated by Fenton chemistry. The oxidation of apomyoglobin was conducted in a time-dependent manner; mass spectra show that with time the protein becomes more oxidized (Figure 7.14). Analysis of the protein during the experiment is important because it offers a way to limit the extent of oxidation, reducing, but not eliminating, the prospect that the protein structure was perturbed by excessive oxidation. To determine specifically where the oxidations were located, the protein was digested with trypsin and the resulting peptides were analyzed by tandem MS. Product-ion spectra show that an oxidation site can be identified by assigning y- and b-ions. In Figure 7.15, one can see that b_6 and b_7 contain an oxidized residue, but not b_5. One can also see that y_5 and y_6 contain an oxidized residue, indicating that the oxidized amino acid must be *C*-terminal of the leucine. With this information, one can deduce that the oxidation is located on the phenylalanine. A comparison of the oxidation results with solvent-accessibility data from an NMR structure reveals that all residues oxidized are highly solvent-accessible. This is evidence that oxidation provides an accurate map of a protein's surface.

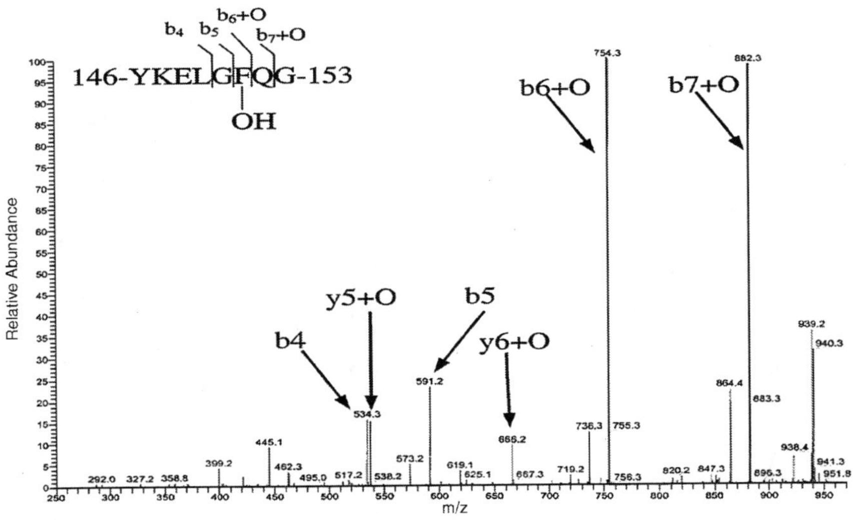

FIGURE 7.15 Tandem MS spectra of a solution-phase oxidized peptide. The presence of $b_6 + O$ but no $b_5 + O$ indicates that it is the Phe that is oxidized. From J. S. Sharp, J. M. Becker, R. L. Hettich, Protein surface mapping by chemical oxidation: structural analysis by mass spectrometry. *Anal Biochem* 313(2), 216–225. Copyright © 2003 by Elsevier Science (USA). Reprinted by permission of Elsevier.

7.4.2 Radiolytic Generation of Hydroxyl Radicals

Hydroxyl radicals can be radolytically derived using X rays, γ rays, or electron beams [60,66]. Equation (7.8) describes how a synchrotron beam, delivering between 10^{14} and 10^{15} photons/s, generates hydroxyl radicals.

$$2H_2O + h\nu \rightarrow H_2O + H_2O^{\bullet +} + e^-_{\text{ejected}} \rightarrow H_3O^+ + HO^\bullet + e^-_{\text{aqueous}} \qquad (7.8)$$

The hydroxyl radicals react with amino-acid side chains that are located on the surface of protein complexes. Amino acids react with the radicals at different rates, ranging from diffusion control to orders of magnitude slower. The order of reactivity is cysteine, methionine $\gg$ phenylalanine, tyrosine, tryptophan $>$ proline $>$ histidine, leucine [57]. Using an amino-acid scavenger can decrease the lifetime to the low microsecond time scale, in some cases faster than the tertiary structure of a protein changes and on the order of secondary perturbations.

Like other footprinting methods, synchrotron-generated oxidation can be used in many different types of experiments. Chance and coworkers [55] employed time-resolved synchrotron footprinting to reveal dynamic changes in the protein, gelsolin, upon calcium binding. In this experiment the X-ray exposure time was kept constant while varying the gelsolin-Ca^{2+} reaction time. After the labeling reaction, the protein complex was digested with trypsin, and peptides from three of the six gelsolin domains were analyzed. The amount of labeling was quantified and normalized to construct progress curves for each peptide. The data clearly illustrate conformational changes associated with ligand binding. This technique has also been utilized in examining RNA folding [67].

7.4.3 Fast Photochemical Oxidation of Proteins (FPOP)

An alternative approach to generating hydroxyl radicals utilizes laser photolysis of hydrogen peroxide [58]. The aim of this development, termed "fast photochemical oxidation of proteins (FPOP)," is to form, in a nanosecond pulse, the radicals and then limit their lifetime in solution by using a scavenger to ensure that the protein does not have time to deviate from its native state while being oxidized. Small proteins can unfold in a few microseconds, and localized unfolding of α-helices can be as fast as tens of nanoseconds.

A KrF excimer laser beam at 248 nm is absorbed by hydrogen peroxide and cleaves it into hydroxyl radicals. Although the quantum yield of this reaction is high at 248 nm, the protein itself absorbs only weakly at this wavelength. Hydroxyl radicals need approximately 100 μs to self-quench; in that time span, a protein can partially unfold, allowing buried residues to be labeled. Using a scavenger (e.g., an amino acid whose rate of reaction with OH radicals is known), the life span of the hydroxyl radicals is lowered to around 1 μs. In other types of oxidation experiments, the radical can react with the protein for ms or longer, ultimately reacting with sites that are not solvent accessible in the folded protein. A recent effort showed that the distribution of products at +16, 32, 48, . . . , constitute a Poisson distribution, indicating that the labeling is indeed for a single state of the protein and therefore, faster than protein unfolding [68].

An early test of this method utilized myoglobin as a model protein. In contrast to the continuous labeling, the protein underwent multiple oxidations. The issue of unfolding induced by oxidation was resolved by keeping the reaction time short (in the μs range). At these short times, not only does sufficient labeling takes place, but also the hydroxyl radicals are quenched before the protein can unfold. This permits the use of higher radical concentrations, thus labeling more sites on the protein surface. In previous studies with other oxidation techniques, 9 sites were found to be oxidized. In this experiment, 23 of the 50 possible oxidation sites were observed.

FPOP has been further developed to include sulfate radical anions as reagents [69]. This approach is similar to that for OH radicals, but it takes advantage of the characteristics of ionic sulfate radical anions, specifically $SO_4^{-\bullet}$, which has a higher reduction potential than $^{\bullet}OH$ and can encourage protein crosslinks [70–72]. Apomyoglobin and calmodulin were used as model proteins, and results indicate that $SO_4^{-\bullet}$ labels nonspecifically. Direct comparisons between $^{\bullet}OH$ and $SO_4^{-\bullet}$ show that aliphatic residues and Phe, Thr, Gln, and Lys are more readily labeled by peroxide FPOP.

Recently FPOP was employed as a structural probe to investigate protein folding [73]. This experiment uses two lasers, one to supply a temperature jump (T-jump), the second to generate hydroxyl radicals. This is a standard "pump/probe" experiment, but is unique in its utilization of chemical reactions as the structural probe as compared to spectroscopic methods [74,75].

7.4.4 SPROX: Stability of Proteins from Rates of Oxidation

A new technique, termed SPROX (stability of proteins from rates of oxidation), takes advantage of oxidation to unveil the folding free energy of proteins and evaluate protein-ligand affinities [76]. Similar to SUPREX (described in Section 7.3.3), SPROX probes protein stability in the presence of increasing concentrations of a chemical denaturant. Instead of using deuterium as the label, SPROX employs hydrogen peroxide to oxidize proteins (specifically Met). Ultimately, the denaturant concentration dependence of the oxidation reaction rate is used to evaluate a folding free energy (ΔG_f). Protein affinity can be extrapolated from the $\Delta\Delta G_f$ for a protein and the protein bound to its ligand.

The SRPOX methodology has been applied to a complex mixture to identify protein drug interactions [77]. In this experiment, yeast cell lysate was incubated with a drug, CsA, and a SPROX curve was generated using 12 denaturant concentration points. The cell lysate was digested and peptides containing methionine were monitored by LC-MS/MS. The methionine-containing peptides, 886 in total, represented 327 unique proteins. As in any biological system, there is a large range of protein expression; however, this does not influence the extent of oxidation of each protein as all are exposed to the same oxidation conditions. Proteins that are CsA targets were identified by the significant differences between the SPROX curves in the presence and absence of CsA. This type of experiment can be applied to many systems, and its impact is significant.

7.5 CHEMICAL CROSSLINKING

Crosslinks in the context of this chapter are bonds that are introduced to link one polymer chain to another for the purpose of probing those regions of a biomolecule that are adjoining. In biological systems, the systems are usually proteins, although they can be peptides, DNA, or small molecule ligands [78]. The goal of chemical crosslinking is to map a biopolymer's structure to reveal its interaction with other molecules. This can be especially effective for large proteins, DNA, or RNA complexes that cannot be crystallized. Crosslinking gives information on tertiary and quaternary structure, as well as intermolecular interactions. This is achieved by chemically tethering two regions. For example, a crosslinker may join two cysteine residues, either intramolecular or intermolecular. The distance, through space, is controlled by the length of the linker; it should be less than or equal to that length. Identification of the crosslinks can be used to infer the interaction of the two polymers (intermolecular) or the structure of the polymer itself (intramolecular). Following crosslinking with MS analysis is particularly advantageous. The size of the protein complex is not a constraint because the complex can be digested, and the peptides analyzed. MS analysis is fast and requires little sample, as discussed earlier.

To exploit fully the benefits of cross linking, one must use several types of crosslinkers and look for many cross-linked products. The more cross-linked products identified, the more restraints available for protein modeling. It may even be possible to assign structure at the atomic level if enough restraints are given [79], much like is done in NMR protein structure determination. Achieving sufficient crosslinks requires using multiple crosslinkers.

Chemical crosslinkers are typically classified in terms of chemical specificity and length of "spacer arm." The chemical specificity determines which functional groups within the biomolecule will react with the crosslinker. Only 9 of the 20 amino acids have reactive side chains (Arg, Lys, Asp, Glu, Cys, His, Met, Trp, and Tyr). Crosslinkers can be monofunctional, directly linking two reactive groups to a single reagent without a spacer, or bifunctional, whereby the reactant ends are tethered by a spacer arm. Monofunctional crosslinkers can only be used to map direct contact within a protein, whereas the reach of a bifunctional crosslinker is defined by the length of the spacer arm. Bifunctional crosslinkers can be either homobifunctional or heterobifunctional. Another type of crosslinker is photosensitive; one end binds specifically, the other, upon activation by radiation, binds nonspecifically. Sinz provided a good review on the specific types of crosslinkers and their reaction mechanisms [80], and Chapter 4 of Kaltashov's *Mass Spectrometry in Biophysics* [81] further details using crosslinking and MS to gain three-imensional structural information in protein complexes.

As with HDX and oxidative footprinting, analysis of crosslinked proteins can be done either with bottom-up or top-down approaches (Figure 7.16) [82]. Both approaches begin with the protein or protein complex in its native state; the protein is then subjected to the crosslinking reaction. Once the sample is crosslinked, the two approaches diverge. In bottom-up analysis, the crosslinked species is proteolytically digested, purified, and the peptides analyzed by MS. The mass spectrometer determines the masses of the peptides and, if possible, sequences the peptides by using tandem MS approaches. In a top-down experiment, the crosslinked protein complex is sent directly to the mass spectrometer where it is fragmented in the gas phase. Both approaches have pluses and minuses. Bottom-up is not restricted by complex size, but requires significant sample handling. Conversely, the top-down approach has very little sample preparation time, but currently does not give good coverage for proteins with MWs greater than approximately 20,000, and it is not as sensitive as the bottom-up approach.

Fragmenting crosslinked proteins, like sequencing any post-translationally labeled protein, affords many product ions, whether being produced proteolytically or in the gas phase. Because crosslinking introduces a new species (two peptides joined together by the crosslinker), interpretation of product-ion spectra can be difficult. Intramolecular crosslinks can be identified easily through in silico digestion by accounting for modification by the crosslinker. Intermolecular crosslinks, however, are more difficult to identify and often require manual interpretation. Several software packages are now available to assist in interpreting spectra of peptides that are crosslinked by special chemical reagents [79,83–86] or by disulfide linkages [87].

7.5.1 Drawbacks of Crosslinking

Chemical crosslinking has several drawbacks. The principal one is the additional complexity introduced by crosslinks. Considering that many crosslinking experiments

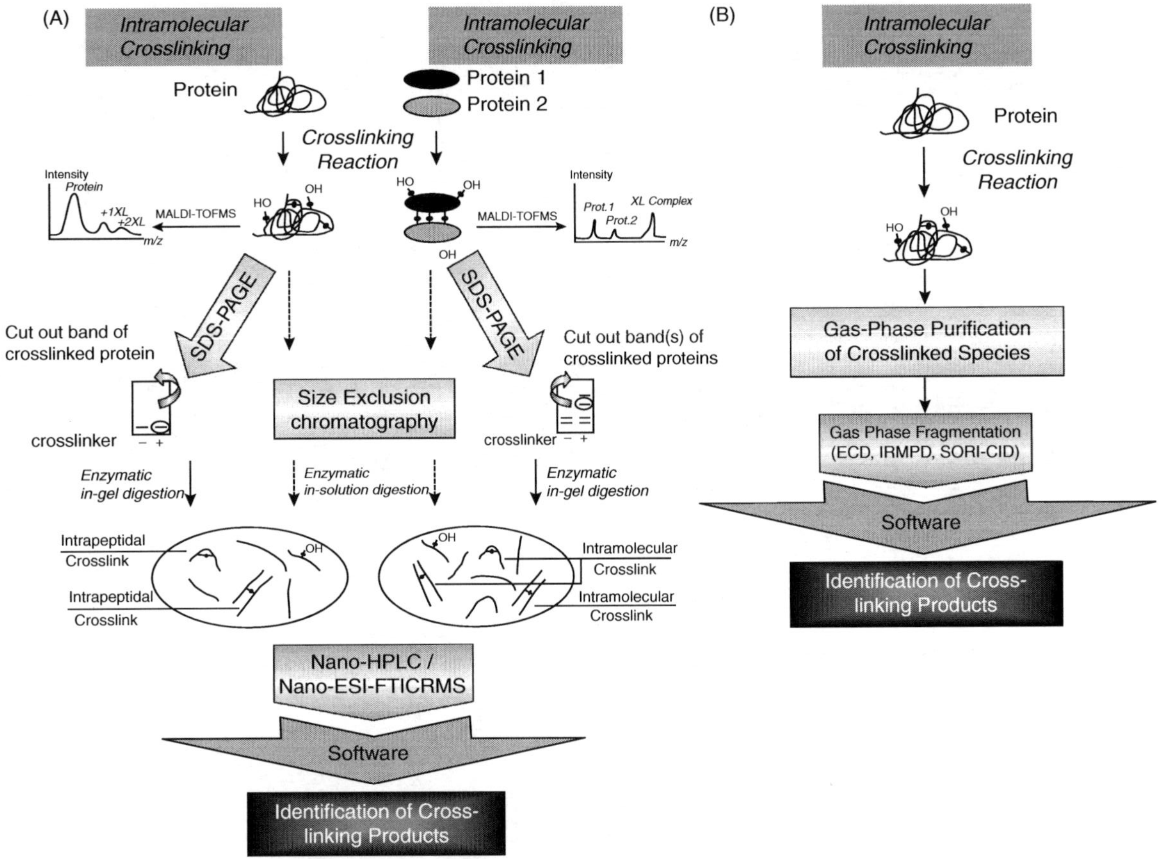

FIGURE 7.16 General strategies for protein structure characterization by chemical crosslinking and mass spectrometry. A. Bottom-up approach. B. Top-down approach. A. Sinz, Chemical cross-linking and FTICR mass spectrometry for protein structure characterization. *Anal Bioanal Chem* 381(1), 44–47. Copyright © 2004 by Spring-Verlag. Reprinted by permission of Springer-Verlag.

involve two or more large proteins, we realize that the peptides resulting from a tryptic digest of the unmodified complex constitute a complex mixture. Crosslinking introduces even more complexity by joining some of those peptides. One should realize that some of the protein molecules (sometimes a major fraction) do not undergo the reactions, adding unmodified peptides to those that are modified. Depending on the specificity and length of the crosslinker, there may be many combinations of peptide crosslinks. This increased complexity may lead to ion suppression when the separation by HPLC shows high overlap. Crosslinked peptides are, in general, also larger than normal peptides, making them more difficult to fragment.

When dealing with a complex mixture, it can be difficult to distinguish easily crosslinked products from normal peptides. Several advances in methodology are available to correct this problem. Of course, improvement in chromatographic and mass resolving power and mass accuracy continue to make identification of cross-linked species more tractable, but there are some options that do not require using a high-performance mass spectrometer. One approach is the use of isotopically labeled crosslinkers [88]. A unique isotopic pattern is created by crosslinking with both unlabeled and stable isotope crosslinkers. Although this pattern is more recognizable than when there is no labeling, the protocol dilutes the sample, decreasing signal to noise. Another option is to use ^{18}O enriched water during the trypsin digestion. Intermolecular crosslinks will have an 8 μ shift, as compared to normal peptides with a 4 μ shift [89]. Yet another solution is to use a trifunctional crosslinker that has an attached biotin molecule [90]. Once the protein is digested, the crosslinked peptides can be enhanced by using a nickel affinity column, decreasing the sample complexity significantly.

Chemical crosslinking, despite its limitations, is a powerful tool for investigating the three-dimensional structure of large protein complexes. Recent advances in MS, including increased sensitivity, resolving power, and mass accuracy, are increasing the power of this approach. As technology continues to advance, chemical cross-linking coupled with MS may continue to improve and become an effective complement to NMR and X-ray crystallography, and possibly a viable alternative, for studying large proteins and protein complexes.

7.6 SELECTIVE AND IRREVERSIBLE CHEMICAL MODIFICATION

Selective chemical modification provides another avenue to explore protein structure, interactions, and solvent accessibility; this classic approach is now experiencing a renaissance given the analytical capabilities of modern MS. It has been used extensively in enzymology since the early 1970s [91,92]. Selective chemical modification is very similar to crosslinking in that only specific amino acids are targeted by the modifier. Like other modification techniques, selective modification can be used to obtain low-resolution protein structure information and to probe interfaces and protein orientation in complexes. There are many different chemical modifiers, each labeling a specific amino acid [93]. The specificity can be both an advantage and a disadvantage. If large portions of a protein interface are to be probed, for example, the specificity is a disadvantage. If, however, the question is focused on a given residue (e.g., which of the cysteines in a protein are solvent-accessible?), a specific labeling approach is advantageous as the information is also highly focused, and the analysis of the modified peptides is relatively simple. Specific chemical modification can be used effectively in combination with other approaches, as is discussed in the first example that follows.

7.6.1 Acetylation of Lysine

An example that illustrates the utility of selectively modifying amino acids is a probe of protein–DNA interactions. In many cases the negatively charged DNA molecule interacts with the positively charged lysine side chains of a protein. Using acetic anhydride to probe these side chains, the Gross lab not only mapped the interface but, in conjunction with HDX data, determined the binding constant of the protein–DNA complex [38]. The experiment allowed the number of acetylations of the apo-protein and the protein–DNA complex to be compared. The results clearly pinpoint those lysines located near the binding site that directly interact with the DNA backbone (Figure 7.17). The complementary information gained by combining HDX and acetic anhydride labeling provides a clearer picture of the protein–DNA interface than would be obtained by either technique alone. This approach is likely to be important in drug development because these interfaces are common drug targets.

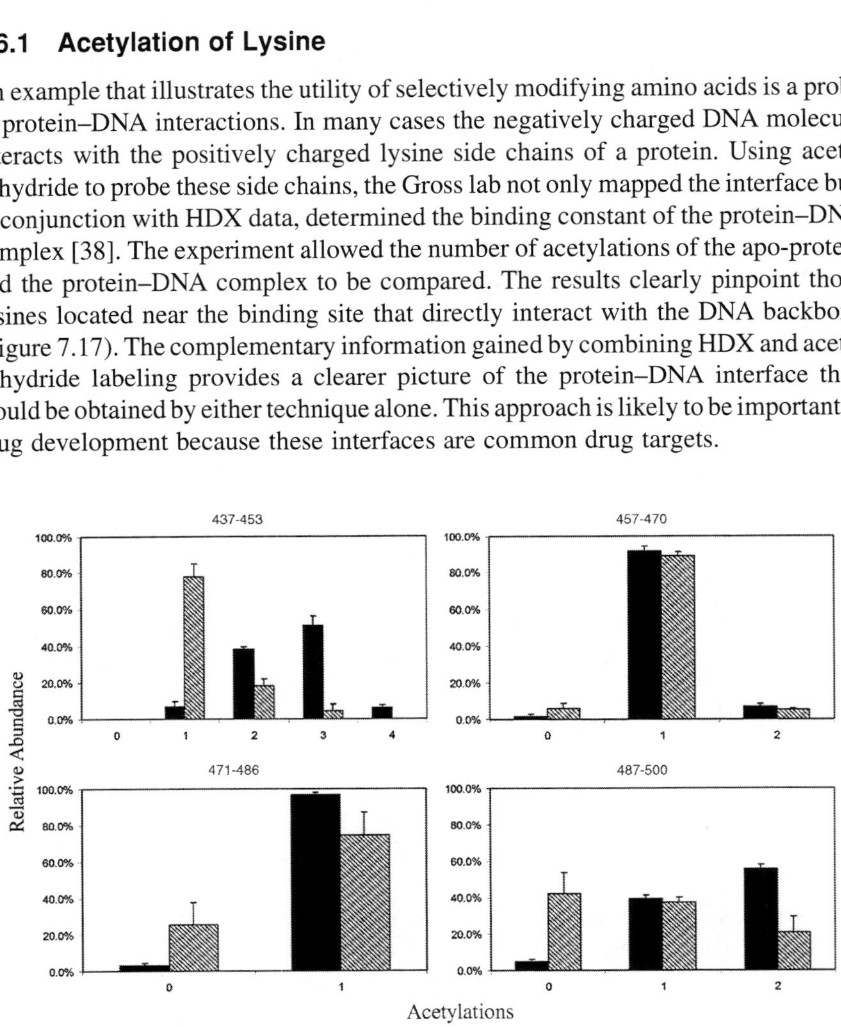

FIGURE 7.17 Extent of acetylation for four peptides of hTRF2 for the apo (black) and holo (gray) forms. The holo form is prepared by binding with an oligodeoxynucleotide. From J. B. Sperry et al., A mass spectrometric approach to the study of DNA-binding proteins: interaction of human TRF2 with telomeric DNA. *Biochemistry* 47(6), 1797–1807. Copyright © 2008 American Chemical Society. Reprinted by permission of American Chemical Society.

7.6.2 Thiol Derivatization of Cysteines

Another example of selective chemical modifications provides a view of a zinc finger protein. Zinc fingers typically bind DNA and are rich in cysteines. Oxidation of cysteines can result in impaired protein–DNA binding [94]. In the case of human estrogen receptor (ER), oxidation of the zinc fingers results in impaired gene expression, possibly linked to breast cancer [95]. The zinc finger of ER contains two cysteines, and it is of interest to characterize the reactivity of each. To probe them, the thiols were reduced with iodoacetic acid (IAA), a thiol alkylating fluorophore (BrB), and menadione, a vitamin K analogue (K3) [96]. The length of time each reducing agent was reacted with ER was varied over a 60-min time course, and the appropriate data processing yielded pseudo-first-order rate constants for the disappearance of unreacted peptide. As expected, IAA reduces ER more completely than BrB and K3, which are sterically bulkier reactants. The rate constants were also measured at acidic, neutral, and basic pH. IAA and BrB react more quickly in basic conditions, consistent with the nucleophilic attack being dependent on thiolate anions. K3 reacts via a Michael addition, which is much slower under basic conditions. IAA reacts with both cysteines; however, BrB and K3 yield only a single adduct. MS/MS indicates that BrB and K3 specifically react with Cys-240. This bias is a result of the flanking amino acids that influence the thiol pK_a.

7.6.3 Footprinting FMO Protein in Photosynthetic Bacteria

Another demonstration of the power of specific footprinting is mapping the surface of the membrane-attached Fenna–Matthews–Olson (FMO) antenna protein in green-sulfur bacteria [97]. This protein has important implications for the high–energy-transfer efficiency found in photosynthetic organisms. Achieving high efficiency relies on the optimal pigment–protein binding geometry in the protein complexes and also on the overall architecture of the photo system. In green-sulfur bacteria, the FMO protein serves as a "wire" to connect the large peripheral chlorosome antenna complex, which gathers in light, with a reaction center (RC), and is embedded in a cytoplasmic membrane (CM). Energy collected by the chlorosome is funneled through the FMO to the RC. Although there has been considerable effort to understand the relationships between structure and function of the individual isolated complexes, the specific architecture for in vivo interactions of the FMO protein, the CM, and the chlorosome, ensuring highly efficient energy transfer, was still not completely established experimentally at the time of the footprinting study. Put more simply, it was whether the protein trimeric complex is "up" as in Figure 7.18*A* (left) or "down" as in Figure 7.18*A* (right).

Answering this question is beyond the abilities of solution NMR and X-ray crystallography. Even sophisticated linear dichroism and STEM imaging have not given an answer. HDX would be difficult to apply to this complex because considerable time and effort for isolation are needed after the footprinting, leading to extensive and variable back-exchange. FPOP, like HDX, may also be inappropriate because the footprinting is too extensive, and the photosystem is inherently sensitive to laser-light

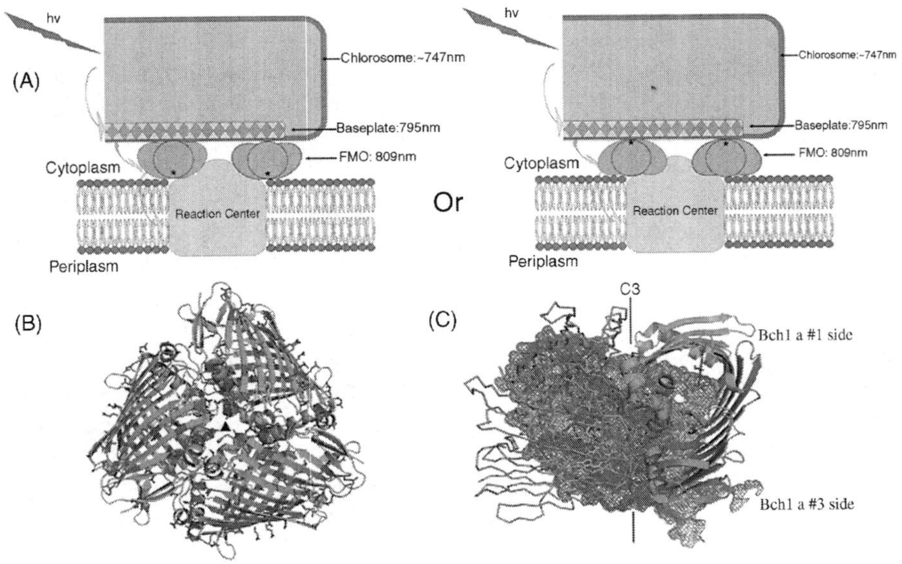

FIGURE 7.18 Photosystem from *C. tepidum* and structure of FMO. (*A*) Model architecture of photosystem from *C. tepidum*. The two possible orientations of FMO on the CM are presented. Bchl a #3 is shown as a star. (*B*) Top view of the FMO trimer with the Bchl a #3 side shown. All the pigments are omitted except Bchl a #3 which is colored cyan. (*C*) Side view of the FMO trimer shown as cartoon, ribbon, and mesh for clarity. Positions of Bchl a #3 (cyan) and Bchl a #1 (red) are labeled in the monomer. From J. Wen et al., Membrane orientation of the FMO antenna protein from Chlorobaculum tepidum as determined by mass spectrometry-based footprinting. *Proc Nat Acad Sci USA* 106(15), 6134–6139. (*See the color version of this figure in Color Plates section.*)

exposure. Therefore the decision was to utilize a specific mapping approach, labeling solvent-exposed aspartic and glutamic acid residues on the FMO protein.

The locations and extents of labeling of FMO on the native membrane in comparison with labeling FMO alone and labeling on a chlorosome-depleted membrane afforded an answer. Modification of the D/E residues on the protein by glycine ethyl ester (GEE), under physiological conditions, by using the zero-length crosslinker, 1-ethyl-3-(3-dimethylaminopropyl)carbodiimide hydrochloride (EDC) (equation 7.9).

$$-CH_2COOH + GEE + EDC \rightarrow -CH_2CONHCH_2COOEt \rightarrow -CH_2CONHCH_2COOH \quad (7.9)$$

Once the samples were modified, they were submitted to SDS-PAGE. Separate samples of the FMO protein, visualized as appropriate gel bands, were in-gel digested, and the peptides were loaded to LC-MS/MS and analyzed with accurate mass measurement to identify, locate, and quantify the modified Asp and Glu. An advantage of accurate mass capability is that locating the modified peptides was made relatively simple and secure by searching for specific increases in the masses of unmodified peptides, corresponding to the accurate mass of [$NHCH_2COOH$ – OH] and [$NHCH_2COOEt$ – OH].

The large differences in the modification of certain peptides show that the Bchl on side 3 of the FMO trimer interacts with the CM (shown in Figure 7.18*A*, left), which is consistent with recent theoretical predictions [98–100]. The results provide direct experimental evidence for the overall architecture of the photosystem from *Chlorobaculum tepidum* and give information on the packing of the FMO protein in its native environment, answering a question that had eluded answering by many other approaches.

7.6.4 Potential Pitfalls

There is a word of caution when considering specific chemical modification. Unlike HDX, the presence or absence of label is not necessarily directly related to protein structure. Of course, if the reagent cannot physically reach a site, labeling will not occur, but other factors also influence the rate of reaction. The chemical microenvironment, determined by the amino acids in close proximity, can either promote or inhibit binding. The reactivity of labeling molecules can also benefit catalytically from neighboring residues. The hydrophobicity of the label can also determine whether it will interact with the surface or the interior of a protein. Although these factors must be taken into consideration when interpreting data, the general outcome of specific labeling or footprinting can be highly informative as the examples discussed above show.

7.7 CONCLUSION

Although there are powerful and highly informative structural tools in protein chemistry, notably X-ray crystallography, multidimensional NMR, and molecular modeling, NMR and X-ray crystallography, as experimental approaches, require time and typically mg quantities of proteins. MS can play a role in the analysis of structure and, more specifically, in locating interfaces and binding sites, and determining affinity because its throughput and sensitivity are higher than those of NMR and X-ray crystallography. The MS approach is similar to that used in NMR; that is, some factor (i.e., chemical reactivity or footprinting strategies) provides constraints to structure, and therefore, information on regions that are in contact with or part of an interface. The constraints are relatively crude with respect to the distance constraints of NMR, however, and the structural inferences are correspondingly of low resolution. Nevertheless, when used appropriately, detailed information can be obtained. For example, in drug discovery the structure of a protein–ligand complex may be characterized at the expense of time and material by using NMR or X-ray crystallography. Structure with other ligands can be more readily determined to be similar or different on the basis of chemical footprinting rather than carrying out a full structure study. Another example is the determination of the binding site. Although the

3D structure of the site cannot be realized by footprinting and MS, the location of the site can be determined.

MS now plays a vital role in protein science. It is extensively used today in proteomics (1) to identify proteins based on sequence analysis, (2) to quantify proteins on the basis of isotope labeling strategies or even by simple spectral counting, and (3) to determine the number and locations of post-translational modifications. We believe that MS-based structural proteomics tools, such as those discussed in this chapter, will become as well an important asset in structural protein science (structural biology) and drug discovery. In time, as its full potential is realized, MS-based footprinting can be expected to join the other three roles of MS mentioned above to become a "fourth pillar of proteomics"

REFERENCES

1. Hvidt, A., Linderstrom-Lang, K. (1954). Exchange of hydrogen atoms in insulin with deuterium atoms in aqueous solutions. *Biochim Biophys Acta 14*(4), 574–5.
2. Dyson, H. J., Wright, P. E. (2004). Unfolded proteins and protein folding studied by NMR. *Chem Rev 104*, 3607–22.
3. Chowdhury, S. K., Katta, V., Chait, B. T. (1990). Probing conformational changes in proteins by mass spectrometry. *J Am Chem Soc 112*, 9012:9013.
4. Voet, D., Voet, J. G. (2004). *Biochemistry*, 3rd ed., J. Wiley & Sons, Hoboken, NJ, 2 v. (xv, 1560).
5. Bai, Y., et al. (1993). Primary structure effects on peptide group hydrogen exchange. *Proteins 17* (1), 75–86.
6. Molday, R. S., Englander, S. W., Kallen, R. G. (1972). Primary structure effects on peptide group hydrogen exchange. *Biochemistry 11* (2), 150–8.
7. Smith, D. L., Deng, Y., Zhang, Z. (1997). Probing the non-covalent structure of proteins by amide hydrogen exchange and mass spectrometry. *J Mass Spectrom 32* (2), 135–46.
8. Englander, S. W. Kallenbach, N. R. (1983). Hydrogen exchange and structural dynamics of proteins and nucleic acids. *Q Rev Biophys 16* (4), 521–655.
9. Kim, K. S., Woodward, C. (1993). Protein internal flexibility and global stability: Effect of urea on hydrogen exchange rates of bovine pancreatic trypsin inhibitor. *Biochemistry 32* (37), 9609–13.
10. Miller, D. W., Dill, K. A. (1995). A statistical mechanical model for hydrogen exchange in globular proteins. *Protein Sci 4* (9), 1860–73.
11. Wales, T. E., Engen, J. R. (2006). Hydrogen exchange mass spectrometry for the analysis of protein dynamics. *Mass Spectrom Rev 25* (1), 158–70.
12. Swint-Kruse, L., Robertson, A. D. (1996). Temperature and pH dependences of hydrogen exchange and global stability for ovomucoid third domain. *Biochemistry 35* (1), 171–80.
13. Deng, Y., Smith, D. L. (1998). Identification of unfolding domains in large proteins by their unfolding rates. *Biochemistry 37* (18), 6256–62.
14. Engen, J. R., Smith, D. L. (2000). Investigating the higher order structure of proteins. Hydrogen exchange, proteolytic fragmentation, and mass spectrometry. *Methods Mol Biol 146*, 95–112.

15. Dempsey, C. E., (2001). Hydrogen exchange in peptides and proteins using NMR spectroscopy. *Progress in Nuclear Magnetic Resonance Spectroscopy 39* (2), 135–170.
16. Zhang, Z., Smith, D. L. (1993). Determination of amide hydrogen exchange by mass spectrometry: A new tool for protein structure elucidation. *Protein Sci 2* (4), 522–31.
17. Cravello, L., Lascoux, D., Forest, E. (2003). Use of different proteases working in acidic conditions to improve sequence coverage and resolution in hydrogen/deuterium exchange of large proteins. *Rapid Commun Mass Spectrom 17* (21), 2387–93.
18. Weis, D. D., Engen, J. R., Kass, I. J. (2006). Semi-automated data processing of hydrogen exchange mass spectra using HX-Express. *J Am Soc Mass Spectrom 17* (12), 1700–3.
19. Deng, Y., Pan, H., Smith, D.L., (1999). Selective isotope labeling demonstrates that hydrogen exchange at individual peptide amide linkages can be determined by collision-induced dissociation mass spectrometry. *J. Am. Chem. Soc. 121* (9), 1966–1967.
20. Zubarev, R. A., Kelleher, N. L., McLafferty, F. W. (1998). Electron capture dissociation of multiply charged protein cations. a nonergodic process. *Journal of the American Chemical Society 120* (13), 3265–3266.
21. Charlebois, J. P., Patrie, S. M., Kelleher, N. L. (2003). Electron capture dissociation and 13C, 15N depletion for deuterium localization in intact proteins after solution-phase exchange. *Anal Chem 75* (13), 3263–6.
22. Zehl, M., et al. (2008). Electron transfer dissociation facilitates the measurement of deuterium incorporation into selectively labeled peptides with single residue resolution. *J Am Chem Soc 130* (51), 17453–9.
23. Pan, J., et al. (2009). Hydrogen/deuterium exchange mass spectrometry with top-down electron capture dissociation for characterizing structural transitions of a 17 kDa protein. *J Am Chem Soc 131* (35), 12801–8.
24. Chalmers, M. J., et al. (2006). Probing protein ligand interactions by automated hydrogen/deuterium exchange mass spectrometry. *Anal Chem 78* (4), 1005–14.
25. Pascal, B. D., et al. (2009). HD desktop: An integrated platform for the analysis and visualization of H/D exchange data. *J Am Soc Mass Spectrom 20* (4), 601–10.
26. Slysz, G. W., et al. (2009). Hydra: Software for tailored processing of H/D exchange data from MS or tandem MS analyses. *BMC Bioinformatics 10*, 162.
27. Truhlar, S. M., et al. (2006). Solvent accessibility of protein surfaces by amide H/2H exchange MALDI-TOF mass spectrometry. *J Am Soc Mass Spectrom 17* (11), 1490–7.
28. Ghaemmaghami, S., Fitzgerald, M. C., Oas, T. G. (2000). A quantitative, high-throughput screen for protein stability. *Proc Natl Acad Sci U S A 97* (15), 8296–301.
29. Ma, L., Fitzgerald, M. C. (2003). A new H/D exchange- and mass spectrometry-based method for thermodynamic analysis of protein-DNA interactions. *Chem Biol 10* (12), 1205–13.
30. Powell, K. D., Fitzgerald, M. C. (2003). Accuracy and precision of a new H/D exchange- and mass spectrometry-based technique for measuring the thermodynamic properties of protein-peptide complexes. *Biochemistry 42* (17), 4962–70.
31. Roulhac, P. L., et al. (2004). SUPREX (stability of unpurified proteins from rates of H/D exchange) analysis of the thermodynamics of synergistic anion binding by ferric-binding protein (FbpA), a bacterial transferrin. *Biochemistry 43* (50), 15767–74.
32. Powell, K. D., Fitzgerald, M. C. (2004). High-throughput screening assay for the tunable selection of protein ligands. *J Comb Chem 6* (2), 262–9.

33. Englander, J. J., et al. (2003). Protein structure change studied by hydrogen-deuterium exchange, functional labeling, and mass spectrometry. *Proc Natl Acad Sci U S A 100* (12), 7057–62.

34. Chan, L. L., et al. (2008). A general method for discovering inhibitors of protein-DNA interactions using photonic crystal biosensors. *ACS Chem Biol 3* (7), 437–48.

35. Sobell, H. M., et al. (1977). Visualization of drug-nucleic acid interactions at atomic resolution. III. Unifying structural concepts in understanding drug-DNA interactions and their broader implications in understanding protein-DNA interactions. *J Mol Biol 114* (3), 333–65.

36. Zhu, M. M., et al. (2003). Quantification of protein-ligand interactions by mass spectrometry, titration, and H/D exchange: PLIMSTEX. *J Am Chem Soc 125* (18), 5252–3.

37. Zhu, M. M., Rempel, D. L., Gross, M. L. (2004). Modeling data from titration, amide H/D exchange, and mass spectrometry to obtain protein-ligand binding constants. *J Am Soc Mass Spectrom 15* (3), 388–97.

38. Sperry, J. B., et al. (2008). A mass spectrometric approach to the study of DNA-binding proteins: Interaction of human TRF2 with telomeric DNA. *Biochemistry 47* (6), 1797–807.

39. Zhu, M. M., Rempel, D. L., Gross, M. L. (2004). Modeling data from titration, amide H/D exchange and mass spectrometry to obtain protein-ligand binding constants. *J. Am. Soc. Mass Spectrom 15*, 388–397.

40. Sperry, J. B., Wilcox, J. M., Gross, M. L. (2008). Strong anion exchange for studying protein-DNA interactions by H/D exchange mass spectrometry. *J Am Soc Mass Spectrom 19* (6), 887–90.

41. Ferguson, P. L., et al. (2007). Hydrogen/deuterium scrambling during quadrupole time-of-flight MS/MS analysis of a zinc-binding protein domain. *Anal Chem 79* (1), 153–60.

42. Demmers, J. A., et al. (2002). Factors affecting gas-phase deuterium scrambling in peptide ions and their implications for protein structure determination. *J Am Chem Soc 124* (37), 11191–8.

43. Rand, K. D., et al. (2009). Protein hydrogen exchange measured at single-residue resolution by electron transfer dissociation mass spectrometry. *Anal Chem 81* (14), 5577–84.

44. Villanueva, J., et al. (2004). Increase in the conformational flexibility of beta 2-microglobulin upon copper binding: A possible role for copper in dialysis-related amyloidosis. *Protein Sci 13* (3), 797–809.

45. Plumb, R., et al. (2004). Ultra-performance liquid chromatography coupled to quadrupole-orthogonal time-of-flight mass spectrometry. *Rapid Commun Mass Spectrom 18* (19), 2331–7.

46. Castro-Perez, J., et al. (2005). Increasing throughput and information content for in vitro drug metabolism experiments using ultra-performance liquid chromatography coupled to a quadrupole time-of-flight mass spectrometer. *Rapid Commun Mass Spectrom 19* (6), 843–8.

47. Wilson, I. D., et al. (2005). High resolution "ultra performance" liquid chromatography coupled to oa-TOF mass spectrometry as a tool for differential metabolic pathway profiling in functional genomic studies. *J Proteome Res 4* (2), 591–8.

48. Wilson, I. D., et al. (2005). HPLC-MS-based methods for the study of metabonomics. *J Chromatogr B Analyt Technol Biomed Life Sci 817* (1), 67–76.

49. Wu, Y., Engen, J. R., Hobbins, W. B. (2006). Ultra performance liquid chromatography (UPLC) further improves hydrogen/deuterium exchange mass spectrometry. *J Am Soc Mass Spectrom 17* (2), 163–7.

50. Wales, T. E., et al. (2008). High-speed and high-resolution UPLC separation at zero degrees Celsius. *Anal Chem 80* (17), 6815–20.

51. Houde, D., et al. (2009). Characterization of IgG1 conformation and conformational dynamics by hydrogen/deuterium exchange mass spectrometry. *Anal Chem 81* (14), 5966.

52. Jones, L. M., et al. (2010). Online, high-pressure digestion system for protein characterization by hydrogen/deuterium exchange and mass spectrometry. *Anal Chem. 82* (4), 1171–4.

53. Lopez-Ferrer, D., et al. (2008). On-line digestion system for protein characterization and proteome analysis. *Anal Chem 80*, 1871–1882.

54. Lopez-Ferrer, D., et al. (2008). Application of pressurized solvents for ultrafast trypsin hydrolysis in proteomics: Proteomics on the fly. *J Proteome Res 7* (8), 3276–81.

55. Kiselar, J. G., Chance, M. R. (2010). Future directions of structural mass spectrometry using hydroxyl radical footprinting. *J Mass Spectrom 45*, 1373–1382.

56. Sclavi, B., et al. (1998). RNA folding at millisecond intervals by synchrotron hydroxyl radical footprinting. *Science 279* (5358), 1940–3.

57. Maleknia, S. D., Brenowitz, M., Chance, M. R. (1999). Millisecond radiolytic modification of peptides by synchrotron X-rays identified by mass spectrometry. *Anal Chem 71* (18), 3965–73.

58. Hambly, D. M., Gross, M. L. (2005). Laser flash photolysis of hydrogen peroxide to oxidize protein solvent-accessible residues on the microsecond timescale. *J Am Soc Mass Spectrom 16* (12), 2057–63.

59. Maleknia, S. D., et al. (2001). Determination of macromolecular folding and structure by synchrotron x-ray radiolysis techniques. *Anal Biochem 289* (2), 103–15.

60. Xu, G., Takamoto, K., Chance, M. R. (2003). Radiolytic modification of basic amino acid residues in peptides: Probes for examining protein-protein interactions. *Anal Chem 75* (24), 6995–7007.

61. Stocks, B. B., Konermann, L. (2009). Structural characterization of short-lived protein unfolding intermediates by laser-induced oxidative labeling and mass spectrometry. *Anal Chem 81* (1), 20–7.

62. Barb, W. G., Baxendale, J. H., George, P., Hargrave, K. R. (1951). Reactions of ferrous and ferric ions with hydrogen peroxide. Part II.—The ferric ion reaction. *Transactions of the Faraday Society 47*, 591–616.

63. Barb, W. G., Baxendale, J. H., George, P., Hargrave, K. R. (1951). Reactions of ferrous and ferric ions with hydrogen peroxide. Part I.—The ferrous ion reaction. *Transactions of the Faraday Society 47*, 462–500.

64. Heyduk, E., Heyduk, T. (1994). Mapping protein domains involved in macromolecular interactions: A novel protein footprinting approach. *Biochemistry 33* (32), 9643–50.

65. Sharp, J. S., Becker, J. M., Hettich, R. L. (2003). Protein surface mapping by chemical oxidation: Structural analysis by mass spectrometry. *Anal Biochem 313* (2), 216–25.

66. Greiner, D. P., Hughes, K. A., Meares, C. F. (1996). Radiolytic protein surface mapping. *Biochem Biophys Res Commun 225* (3), 1006–8.

67. Ralston, C. Y., et al. (2000). Time-resolved synchrotron X-ray footprinting and its application to RNA folding. *Methods Enzymol 317*, 353–68.

68. Gau, B. C., et al. (2009). Fast photochemical oxidation of protein footprints faster than protein unfolding. *Anal Chem 81* (16), 6563–6571.

69. Gau, B. C., et al. (2010). Sulfate radical anion as a new reagent for fast photochemical oxidation of proteins. *Anal Chem 82*, 7821–7827.

70. Davies, M. J., Dean, R. T. (1997). *Radical-Mediated Protein Oxidation: From Chemistry to Medicine* , Oxford university Press, Oxford, U.K.

71. Fancy, D. A., Kodadek, T. (1999). Chemistry for the analysis of protein-protein interactions: Rapid and efficient cross-linking triggered by long wavelength light. *Proc Natl Acad Sci U S A 96* (11), 6020–4.

72. Rayshell, M., Ross, J., Werbin, H. (1983). Evidence that N-acetoxy-N-acetyl-2-aminofluorene crosslinks DNA to protein by a free radical mechanism. *Carcinogenesis 4* (5), 501–7.

73. Chen, J., Rempel, D. L., Gross, M. L. (2010). Temperature jump and fast photochemical oxidation probe submillisecond protein folding. *J Am Chem Soc 132*, 15502–15504 (2010).

74. Gulotta, M., et al. (2003). Primary folding dynamics of sperm whale apomyoglobin: Core formation. *Biophys J 84* (3), 1909–18.

75. Kubelka, J. (2009). Time-resolved methods in biophysics. 9. Laser temperature-jump methods for investigating biomolecular dynamics. *Photochem Photobiol Sci 8* (4), 499–512.

76. West, G. M., Tang, L., Fitzgerald, M. C. (2008). Thermodynamic analysis of protein stability and ligand binding using a chemical modification- and mass spectrometry-based strategy. *Anal Chem 80* (11), 4175–85.

77. West, G. M., et al. (2010). Quantitative proteomics approach for identifying protein-drug interactions in complex mixtures using protein stability measurements. *Proc Natl Acad Sci U S A 107* (20), 9078–82.

78. Wong, S. S., (1991). *Chemistry of Protein Conjugation and Cross-Linking*, CRC Press, Boca Raton, 340.

79. Young, M. M., et al. (2000). High throughput protein fold identification by using experimental constraints derived from intramolecular cross-links and mass spectrometry. *Proc Natl Acad Sci U S A 97* (11), 5802–6.

80. Sinz, A. (2006). Chemical cross-linking and mass spectrometry to map three-dimensional protein structures and protein-protein interactions. *Mass Spectrom Rev 25* (4), 663–82.

81. Kaltashov, I. A., Eyles, S. J. (2005). *Mass Spectrometry in Biophysics: Conformation and Dynamics of Biomolecules*, Wiley-Interscience, Hoboken, N. J., xvi, 458, [4] of plates.

82. Sinz, A. (2005). Chemical cross-linking and FTICR mass spectrometry for protein structure characterization. *Anal Bioanal Chem 381* (1), 44–7.

83. Kruppa, G. H., Schoeniger, J., Young, M. M. (2003). A top down approach to protein structural studies using chemical cross-linking and Fourier transform mass spectrometry. *Rapid Commun Mass Spectrom 17* (2), 155–62.

84. Peri, S., Steen, H., Pandey, A. (2001). GPMAW--a software tool for analyzing proteins and peptides. *Trends Biochem Sci 26* (11), 687–9.

85. Schnaible, V., et al. (2002). Screening for disulfide bonds in proteins by MALDI in-source decay and LIFT-TOF/TOF-MS. *Anal Chem 74* (19), 4980–8.
86. Wefing, S., Schnaible, V., Hoffmann, D. (2006). SearchXLinks. A program for the identification of disulfide bonds in proteins from mass spectra, *Anal Chem, 78,* 1235–1241.
87. Xu, H., Zhang, L., Freitas, M. A. (2008). Identification and characterization of disulfide bonds in proteins and peptides from tandem MS data by use of the MassMatrix MS/MS search engine. *J Proteome Res 7* (1), 138–44.
88. Back, J. W., et al. (2001). A new crosslinker for mass spectrometric analysis of the quaternary structure of protein complexes. *J Am Soc Mass Spectrom 12* (2), 222–7.
89. Back, J. W., et al. (2002). Identification of cross-linked peptides for protein interaction studies using mass spectrometry and 18O labeling. *Anal Chem 74* (17), 4417–22.
90. Trester-Zedlitz, M., et al. (2003). A modular cross-linking approach for exploring protein interactions. *J Am Chem Soc 125* (9), 2416–25.
91. Motorin, Y., et al. (2007). Identification of modified residues in RNAs by reverse transcription-based methods. *Methods Enzymol 425*, 21–53.
92. Busenlehner, L. S., Armstrong, R. N. (2005). Insights into enzyme structure and dynamics elucidated by amide H/D exchange mass spectrometry. *Arch Biochem Biophys 433* (1), 34–46.
93. Shaw, E. (1970). Selective chemical modification of proteins. *Physiol Rev 50* (2), 244–96.
94. Baldwin, M. A., Benz, C. C. (2002). Redox control of zinc finger proteins. *Methods Enzymol 353*, 54–69.
95. Liang, X., et al. (1998). Oxidant stress impaired DNA-binding of estrogen receptor from human breast cancer. *Mol Cell Endocrinol 146* (1–2), 151–61.
96. Atsriku, C., et al. (2005). Reactivity of zinc finger cysteines: Chemical modifications within labile zinc fingers in estrogen receptor. *J Am Soc Mass Spectrom 16* (12), 2017–26.
97. Wen, J., et al. (2009). Membrane orientation of the FMO antenna protein from Chlorobaculum tepidum as determined by mass spectrometry-based footprinting. *Proc. Nat'l Acad. Sci (USA) 106 (15)*, 6134–6139.
98. Engel, G. S., et al. (2007). Evidence for wavelike energy transfer through quantum coherence in photosynthetic systems. *Nature 446* (7137), 782–6.
99. Muh, F., et al. (2007). Alpha-helices direct excitation energy flow in the Fenna Matthews Olson protein. *Proc Natl Acad Sci U S A 104* (43), 16862–7.
100. Read, E. L., et al. (2008). Visualization of excitonic structure in the Fenna-Matthews-Olson photosynthetic complex by polarization-dependent two-dimensional electronic spectroscopy. *Biophys J 95* (2), 847–56.

CHAPTER 8

Microwave Technology to Accelerate Protein Analysis

UROOJ A. MIRZA, BIRENDRA N. PRAMANIK, and AJAY K. BOSE

8.1 INTRODUCTION

Proteins and peptides are of central interest in biological studies and in pharmaceutical research for the discovery of novel drugs. The identification of proteins and their modifications can provide important clues to the understanding of biological activities/processes and will lead to better understanding of the biological pathways of the target proteins, thereby providing a means of selecting new targets for therapeutic intervention [1,2].

The identification of protein biomarkers as measurable signals for normal and diseased states has been one of the major goals of proteomics research. One important goal of modern drug discovery, however, is to develop highly potent, small-molecule compounds that are chemically and metabolically stable, sustain good serum concentration, and selectively bind noncovalently with target proteins to produce a desired therapeutic response with minimal side effects. The process generally includes identifying a disease target (usually a protein), screening of synthetic compounds or compound libraries for a lead compound, and optimizing the lead compound for activity, selectivity, and pharmacokinetics for a potential clinical candidate [3,4].

Advances in recombinant DNA technology have provided reliable means of producing therapeutic proteins/protein products in bulk quantities. Facile preparation methods promote the discovery of complementary biologic drugs including antibodies. One of the first recombinant protein drugs is interferon (IFN) α-2b, a synthetic *E. coli* recombinant DNA-derived protein [5] that is effective in treating hepatitis C virus and a variety of tumors. Other recombinant protein drugs are

This chapter is dedicated to the memory of Prof. Ajay K. Bose, deceased February 12, 2010.

Protein and Peptide Mass Spectrometry in Drug Discovery, Edited by Michael L. Gross, Guodong Chen, and Birendra N. Pramanik.

used for the treatment of rheumatoid arthritis; they function by binding to and blocking the action of the protein tumor necrosis factor alpha (TNF-alpha) in the human body before TNF-alpha can trigger inflammation, thus reducing inflammatory symptoms [6].

Biological and pharmaceutical research is a competitive field aimed at developing novel therapeutics through the production of biologically important proteins by recombinant DNA techniques and the generation of their chemically or genetically modified variants. Time is an essential factor for winning the competition to develop new drugs just as is accurate structural information on the new compounds produced. Rapid structure characterization of proteins has become an integral part of the success of the total process.

Mass spectrometry (MS) in combination with electrospray ionization and matrix-assisted laser desorption ionization (MALDI) [7,8] has emerged as an effective analytical tool in protein analysis. To obtain detailed structural information, proteins are selectively cleaved into smaller polypeptide fragments by controlled chemical or enzymatic reactions [9]. The resulting mixture of peptides is then analyzed by MALDI-MS or liquid chromatography ESI-MS (LC-ESI-MS) [10–12]. This approach is known as peptide mapping [13–15]. Each polypeptide fragment can be further studied by tandem mass spectrometry (MS-MS) to determine the amino-acid sequence and to identify post-translational modifications of the original protein. A broad understanding of the application of LC-ESI-MS to the analysis of peptides and proteins is described in the book by Pramanik et al. (2002) [16]. MS-based proteomics is covered in Chapter 3 by Boyne and Bose in this volume.

New technologies are emerging that enable faster characterization of biomolecules with increased sensitivity [17]. MS has benefited, for example, from developments in higher flow rate LC and improved software. Moreover development continues of mass spectrometers with faster scanning, higher sensitivity, greater mass-measurement accuracy [18], and increased bioinformatics tools for data base searching [19]. Nevertheless, there is a need to improve sample handling to enhance the speed of protein characterization.

Although protein and peptide separation and identification are now highly automated, sample preparation and protein digestion are considerably slower. To deal with time-consuming sample preparation, immobilized enzymes [20,21] and acid-labile surfactants [22] have become available as have protocols to speed up the proteolytic and/or chemical cleavage of proteins and to remove post-translational modifications (PTMs) [23].

Another approach to speed protein digestion is microwave irradiation [24]. Two 1986 papers demonstrated that a variety of organic reactions can be completed in minutes instead of hours when conducted in sealed glass or Teflon vessels and irradiated with microwave radiation [25,26]. Although there were some explosions in the early studies, owing to the rise of pressure in the vessels, these problems can be avoided by using MORE (Microwave Induced Organic Reaction Enhancement) chemistry [27,28]. Microwave irradiation has also been applied by synthetic chemists to improve chemical processes and to modify chemo-, regio, or stereo-selectivity [29–31]. Interested readers are referred to an extensive review by Caddick et al. [32]. Recently Comer and Organ [33] designed a continuous flow, microwave-assisted, micro-reactor and used it in chemical synthesis to afford mg quantities in excellent yield and purity. Microwave technology is also available for bioanalytical chemistry. One example is in tissue-culture studies for histology [34–36].

In proteomics, analysis of complex mixtures of proteins is carried out by bottom-up or top-down methods (see Chapter 3 by Boyne and Bose in this volume). In the bottom-up approach, thousands of peptides are generated from enzymatic digestions of proteins and commonly analyzed by MS [37,38]. Speed and ease of digestion of mixtures of proteins are critical for the success of protein identification.

The goal of this chapter is to provide an overview on current microwave methodology to accelerate the structural analysis of proteins and peptides for MS analysis. We discuss identification of *C*-terminal amino acids based on Akabori reaction, characterization of cyclic polypeptides, increased enzymatic and chemical cleavages of proteins for peptide mapping, and various biological sample preparation procedures. Last we address the effect of microwave irradiation on conformational state of proteins.

8.2 MICROWAVE TECHNOLOGY

Microwaves are electromagnetic waves with wave lengths longer than those of infrared light but shorter than those of radio waves; the wavelengths fall in the range of 1 mm to 30 cm, corresponding to a frequency range of 1 to 30 GHz. The theory of microwave irradiation was predicted in 1864 and first physically demonstrated to exist in 1888 [39,40]. A microwave oven as a kitchen appliance employs a beam of microwaves to cook food; it consists of a magnetron control circuit, a magnetron, a wave guide, and a cooking chamber. In 1946, Percy Spencer, a Raytheon Corp. self-taught engineer, recognized that exposure to low-density microwave energy could cook food quickly, and he patented the microwave cooking process. In 1947, Raytheon built the first commercial microwave oven; in 1978, and CEM Corporation introduced the first commercial microwave for laboratory use.

8.2.1 Application of Microwave Iirradiation to Akabori Reaction

The Akabori reaction [41], devised more than 50 years ago for the identification of the *C*-terminus of peptides, utilizes the cleavage of amide bonds in peptides by hydrazine in a heated sealed tube at 125°C for several hours. The carboxy terminus is released as a free amino acid and identified. Bose et al. [42] used the Akabori reaction under microwave irradiation and observed that the time required to complete the reaction was reduced from hours to minutes. This approach allows rapid identification of the *C*-terminus amino-acid in a polypeptide including its amino-acid sequence information at both *C*-terminus and *N*-terminus. Additionally this method provides quantitative data on the presence of certain amino-acid residues with functional groups that are altered by the reaction with hydrazine (Arg, Asn, Asp, Cys, Glu, Gln, etc.). This was one of the early experiments of microwave-assisted peptide reaction using an open-vessel system.

We extended these efforts in the application of microwave irradiation to sequencing of cyclopeptide analysis by microwave-enhanced Akabori reaction [43]. Sequence determination of cyclic peptides is very challenging by using MS because the ring openings are largely indiscriminate to give a set of acylium (or rearranged acylium) ions of the same *m/z* as the closed ring. Tandem MS can also be used for the sequence determination of cyclic peptides [44–46]. Microwave irradiation can be used to obtain rapid sequence information of a cyclic peptide in minutes by the Akabori reaction. In the case of several glycine containing cyclic oligopeptides, microwave-assisted hydrazinolysis produced the corresponding open chain hydrazide (s) at glycine residues. The approach affords the sequence information of intractable cyclic peptides by using a combination of the microwave-enhanced Akabori reaction followed by LC-MS and LC-MS-MS experiments.

8.2.2 Protein Characterization by Microwave Irradiation and MS

Microwave irradiation can be used to accelerate the hydrolysis of peptides and proteins with 6 M HCl in a sealed tube [47]. It can also be used for the rapid enzymatic digestion of proteins. This was first demonstrated by Pramanik et al. [48], and since then several groups have utilized the power of microwave in mapping protein structures by chemical cleavage [49,50]. Microwave can also be exploited for characterizing the oligosaccharide moieties of intact glycoproteins [51,52], and it can be applied to protein mixtures [53], and used with immobilized trypsin [54]. A review article by Lill et al. [55] is available on microwave-assisted proteomics.

An important strategy in protein characterization and identification is confirming protein sequence. With the completion of the human-genome project and with advances made in recombinant DNA technology for the production of therapeutic proteins, a fast approach to sequence proteins for their characterization and identification is needed. Application of a bottom-up approach is the most reliable and readily applied with MS to identify protein sequences. In the bottom-up approach, proteins are subjected to proteolysis, and the resulting peptides are analyzed by MS [9,56,57]. In the standard enzymatic digestion of a protein at 37°C, 16 to 18 h or more are needed to give nearly complete digestion of a protein. Needed are approaches to reduce incubation times without compromising sequence coverage and specificity. One can, for example, add methanol and/or acetonitrile, detergents, urea, to facilitate the digestion [58,59]. The use of immobilized enzymes and immobilization of proteins to PVDF membranes for subsequent incubation with non-ionic surfactants and proteolytic enzymes also can decrease proteolysis times [60,61].

The acceleration of protein digestion under controlled microwave irradiation can be achieved by using endoproteases trypsin or lysine C enzymes [48]. Example proteins whose accelerated digestion has been established are cytochrome C, bovine ubiquitin, horse heart myoglobin, chicken egg lysozyme, and recombinant human interferon α-2b (rh-IFN- α-2b). Cytochrome C, after treating with 1: 25 protease:protein (by weight) and 10 min of microwave irradiation, gives a digest that is ready for MALDI-MS analysis. The protein coverage can be more than 90%,

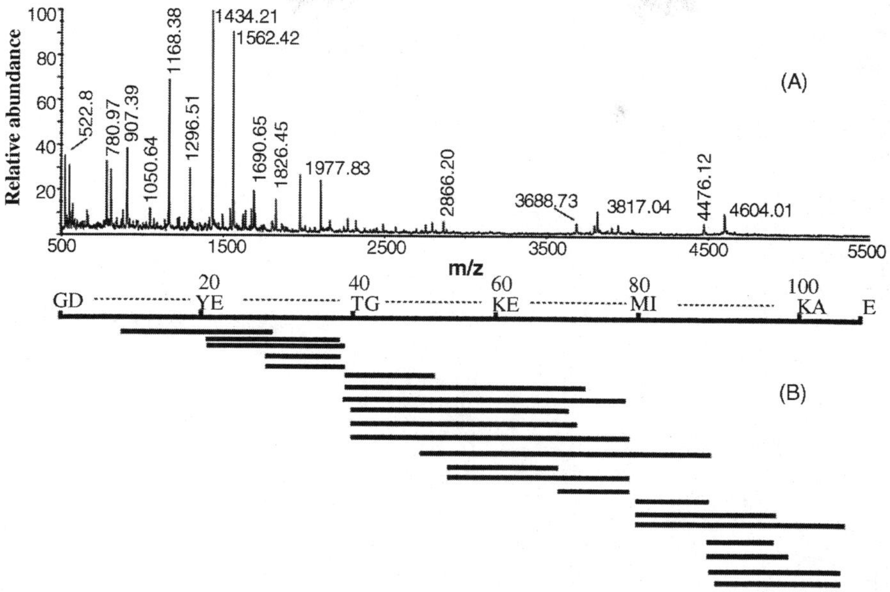

FIGURE 8.1 (*A*) MALDI mass spectrum of tryptic fragments of cytochrome C after 10 min of microwave irradiation at 1: 25 protease to protein ratio by weight and (*B*) tryptic fragments in (*A*) are represented in the form of horizontal lines, showing the total sequence coverage of cytochrome C.

and importantly, no nonspecific peptides result (Figure 8.1, Table 8.1). Enhancement in the digestion process occurs at an incubation temperature of 60°C, as can be provided by a multi-inlet Star-6 microwave apparatus from CEM Corp using 30% of the available power or 144 W [48]. Microwave irradiation of proteins even at 60°C does not produce any autolysis of the enzymes, trypsin or lysine C. Although the enzyme remains active for the first 10 min of irradiation, its activity is completely reduced after 20 min. In the absence of an enzyme, no digestion of the protein can be detected. Whereas the conventional approach of incubating a protein with enzyme can require 6 h at 37°C (Figure 8.2), the microwave-enhanced approach takes only 10 min at 37 °C and affords high yields (nearly complete protein coverage of protein).

8.2.3 Temperature and Microwave Irradiation Effects on the Enzyme in Protein Digestion

Microwave-assisted enzymatic digestion of glycated hemoglobin HbA1C can be completed by using trypsin for 20 min at 50°C [62]. The enzyme Glu-C, however, is inactivated at temperatures above 50°C. The inefficient digestion may be due to heat-induced inactivation of the enzyme. Clearly, the two different enzymes behave differently under microwave irradiation, and one can maximize the activity of the enzyme by optimizing the temperature during the proteolysis.

TABLE 8.1 Peptides with Mass Values Observed in MALDI Mass Spectrum of Cytochrome C after 10 min of Trypsin Digestion with Microwave Irradiation

Peptide Sequence	Expected Mass Values	Observed Peptide with Mass Values
T1 GDVEK	546.6	
T2 GK	203.2	
T3 K	146.2	$T_{3-7} = 2273$
T4 IFVQK	633.8	
T5 CAQCHTVEK	1018.2	
T6 GGK	260.3	$T_{6-8} = 1674.6$ $T_{6-9} = 1802.7$
T7 HK	283.3	
T8 TGPNLHGLFGR	1168.3	$T_8 = 1168.3$, $T_{8,9} = 1296.3$
T9 K	146.2	$T_{9-11} = 1825.6$ $T_{9-13} = 3947.7$
T10 TGQAPGFSYTDANK	1456.5	$T_{10-12} = 3690.4$ $T_{10-13} = 3816.0$
T11 NK	260.3	$T_{11-15} = 3798.5$
T12 GITWGEETLMEYLENPK	2010.2	$T_{12} = 2010$ $T_{12-14} = 2796$
T13 K	146.2	$T_{13} = 805.4$
T14 YIPGTK	677.8	$T_{10-14} = 44.75$ $T_{9-14} = 4603.0$
T15 MIFAGIK	779.0	$T_{15} = 788.9$, $T_{15,16} = 906.4$,
T16 K	146.2	$T_{16-19} = 1561.2$
T17 K	146.2	$T_{17-21} = 1976.8$
T18 GER	360.4	$T_{18,19} = 1305.3$
T19 EDLIAYLK	964.1	$T_{15-19} = 2322$
T20 K	146.2	$T_{16-20} = 1689.6$
T21 ATNE	433.4	$T_{15-21} = 2865.2$, $T_{16-21} = 2104.9$

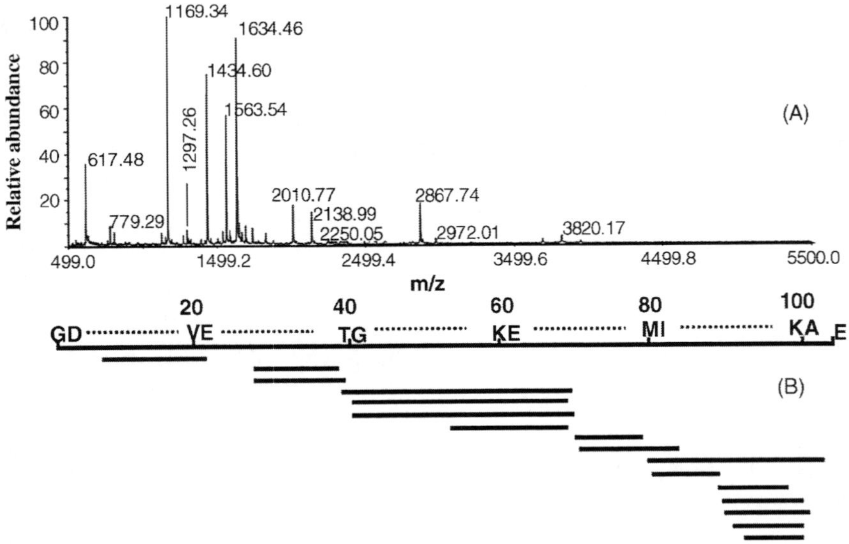

FIGURE 8.2 (*A*) MALDI mass spectrum of tryptic peptides of cytochrome C generated after 6 h of digestion (1: 25, w/w) at 37°C (classic approach) and (*B*) tryptic fragments in a is represented in the form of horizontal lines showing the total sequence coverage of cytochrome C.

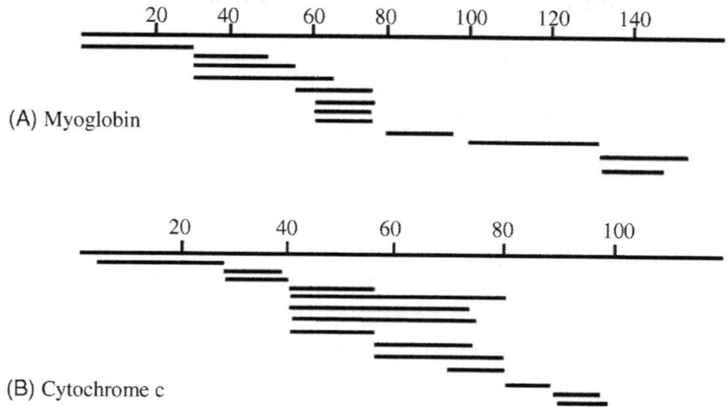

FIGURE 8.3 Coverage of tryptic fragments (represented in the form of lines) obtained after 20 min of microwave-assisted ingel digestion and MALDI-MS analysis. (*A*) Myoglobin sequence coverage and (*B*) cytochrome C sequence coverage.

8.2.4 Use of Microwave Digestion of Proteins from SDS-PAGE Gels

Microwave-assisted digestion can be applied to proteins separated on SDS-PAGE gels (unpublished results from our laboratory). One uses microwave to de-stain a coomassie-stained gel band in 15 min and to complete the digestion in 15 min with high peptide recovery and protein coverage (Figure 8.3). Juan et al. [63] demonstrated the efficacy of this approach by in-gel digestion of several proteins including, lysozyme, albumin, conalbumin, and ribonuclease A. Utilizing microwave technology, they could decrease the in-gel digestion time from 16 h to 5 min.

Recently Marchetti-Deschmann et al. [64] applied microwave irradiation for gel staining and in-gel digestion of glycoprotein. Results suggest that band intensities and sensitivities are comparable to original staining protocols; however, the staining procedure was dramatically reduced to 30 min for silver nitrate and 1.5 h for CBB staining. They also found an increase in the total number of tryptic peptides with the application of microwave irradiation.

8.2.5 Extraction of Intact Proteins from SDS-PAGE Using Microwave Irradiation

We [65] developed a methodology to extract intact proteins from Cu and commassie-stained SDS-PAGE gels and implemented the approach for the characterization of adenovirus proteins. The important steps in the destaining and extraction process involved washing the excised band with a combination of solvents, requiring 7 to 8 h. This long time for destaining and extraction can be reduced to 30 min with the use of microwave irradiation, affording a sample ready for MALDI MS analysis, as demonstrated for cytochrome C and myoglobin (Figure 8.4). Microwave irradiation can be used not only for digesting proteins but also for efficiently extracting intact proteins from SDS-PAGE gels.

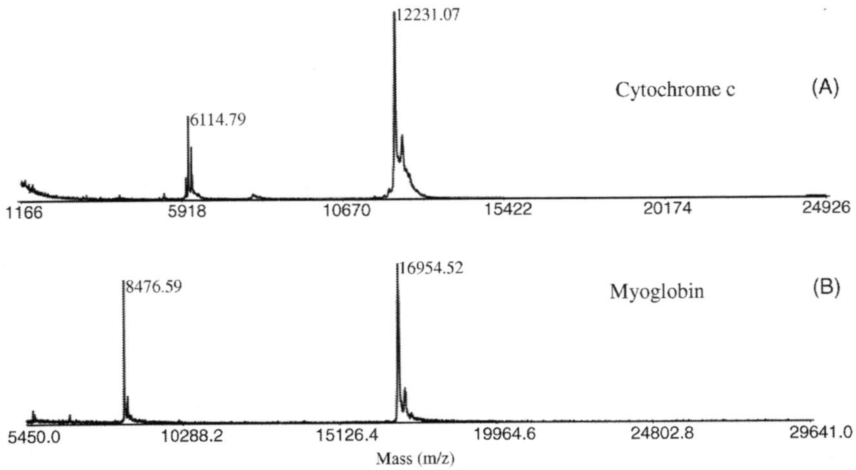

FIGURE 8.4 MALDI mass spectra of (*A*) cytochrome C and (*B*) myoglobin after directly extracted from SDS-PAGE using 20 min of microwave irradiation.

8.2.6 Application of Microwave-Assisted Proteolysis Using Trypsin-Immobilized Magnetic Silica Microspheres

Shuang et al. [66] applied immobilized trypsin in the microwave-assisted enzymatic digestion of proteins. A significant increase in the proteolyisis efficiency can be achieved using microwave irradiation with immobilized trypsin; peptides can be generated in 15 s of microwave irradiation of BSA and myoglobin and confidently identified by MALDI-TOF MS. The enhanced speed of digestion suggests that the approach may be useful for a complex sample. Combination of microwave-assisted proteolysis of a rat-liver sample with immobilized trypsin on magnetic silica microspheres promotes the successful identification of two proteins in 15 s of microwave irradiation.

Microwave-assisted enzymatic digestion can be accelerated further by using multifunctional magnetic beads. With these beads, trypsin digestion of cytochrome C can be achieved in 30 s and myoglobin in 1 min [67]. The decrease in the digestion time may be due to the denaturation of adsorbed protein on the beads, which in turn assists the enzyme in the digestion. Other functions of the magnetic beads are to concentrate low levels of protein and possibly denature the adsorbed protein.

An interesting application of trypsin-immobilized magnetic nanoparticles (TIMNs) and microwave-assisted protein digestion is the study of proteins of the human lens. The digestion of these proteins can be done in approximately 1 min with a microwave power of 400 W, using a trypsin-to-protein ratio of 1:5 [68]. Twenty-six proteins can be identified compared to only 11 with a traditional, 12 h in-solution digestion. The microwave-assisted approach can also be used for in-tip digestion of proteins [69]. In 2 min, with full automation by a robotic system, one can digest and analyze 96 samples in a total of 80 min with this highly effective proteolysis.

Organic solvents enhance the rate of protein digestion compared to aqueous buffers [70,71]. One can also use a combination of organic solvents (acetonitrile, methanol, and chloroform) and microwave irradiation to accelerate the digestion of proteins, as demonstrated for cytochrome C, myoglobin, and lysozyme. Although the accelerated digestion in the presence of excess acetonitrile indicates that this solvent does not inhibit the enzyme activity, excess methanol decreases the enzyme activity, and its use does not accelerate the digestion process.

8.2.7 Acid Hydrolysis of Proteins with Microwave Irradiation

One can also use microwave irradiation to obtain protein structural information by chemical proteolysis, in particular acid hydrolysis [72]. One of the benefits of using acid hydrolysis is to obtain structural information of proteins that are otherwise resistant to enzymatic digestion. Using acid hydrolysis in 6 M HCl, with a short exposure to microwave irradiation for about 1 min, one can digest cytochrome C and proteins from *E. coli.* K12 cells can be completely hydrolyzed, giving full protein coverage. One of the limitations is that this approach requires fairly pure protein [73].

This approach of acid hydrolysis has advantages for sequencing transmembrane proteins. An example is bacteriorhodopsin, a very hydrophobic protein, which can be hydrolyzed with 25% TFA in aqueous solution [74]. In using microwave-assisted acid hydrolysis (MAAH), one can keep membrane proteins free from aggregation, which otherwise would be problematic with conventional methods, and achieve efficient cleavage of the membrane domain with good sequence coverage. This approach can also be applied to the mass spectrometric profiling of the human-heart tissue samples [75]. Even those extracted proteins that are insoluble in SDS solution can be analyzed by LC-MS-MS using MAAH.

MAAH can also be applied to myoglobin, BSA, and proteins isolated from *E. coli* K12 cells, allowing cleavage at the *C*-terminal side of aspartyl (Asp) residue, within 10 min of irradiation [76]. An advantage, in addition to speed of digestion, is that one can control the extent of chemical digestion by varying the exposure time of microwave irradiation [77].

Another application of microwave-assisted acid hydrolysis is of small acid-soluble spore proteins from *B. anthracis* [78]. Formic acid (12%) cleaves at both *N*-terminal and *C*-terminal sides of aspartic acid residues [79]. One can take advantage of the incomplete acid digestion to observe both the digested products and the undigested precursor protein, increasing the confidence of identification.

One can use 50% acetic acid in microwave-assisted solubilization and subsequent digestion of the bacteriophage MS2 virus by using a CEM Discover Benchmate Microwave system (CEM corp., Matthews, NC) [80]. Various microwave powers can be used (130 to 220 W) to optimize the extent of chemical digestion efficiency. The most practical microwave power is 190 W, and the best signal-to-noise ratios (*S/N*) for detecting the digestion products is with 108°C. MAAH caused cleavage of all the Asp sites in less than 30 s with nearly 100% sequence coverage. This application was evaluated with yeast ribosome for high-throughput automated workflows [81]. Using 12.5% acetic acid with benchmate laboratory CEM microwave for 20 min at 140°C, one can identify nearly 250 peptides to facilitate identification of 58 of the 79 ribosomal proteins. This work demonstrates another aspect of the power and speed of using MAAH for protein identification.

The microwave accelerated proteolysis for proteins and peptides is also useful for investigating post-translational modifications [82]. When using acetic acid for hydrolysis at a high temperature of 140°C for 2 to 20 min, one observes no oxidation of methionine or cysteine. Application of a hot-acid treatment to the phosphopeptide, *TRDIYETDYphosYRK*, for 45 s followed by 5 min with a Benchmate Proteome Discover microwave (CEM) can hydrolyze the phosphate group. The hydrolysis, however, is incomplete, allowing both the modified and unmodified peptide to be observed together. Overall, the use of MAAH is a rapid and practical approach in proteomic research, and its utility is nicely illustrated by ongoing work in the Fenselau group [83].

In most applications microwave-assisted enzymatic or chemical digestion of proteins is carried out offline; the protein samples are placed in 0.35-mL Eppendorf tubes [48], a 5-mL loosely capped vials [66], or in a septum sealed 10-mL glass test tube [78] for microwave irradiation. The sample is then transferred manually for MALDI-MS analysis. One can utilize an online, nonenzymatic digestion with a microwave-heated flow cell for acid hydrolysis at aspartic acid (D) sites [83]. The rapid hydrolysis of aspartic acid in a fused silica flow cell with continuous microwave irradiation is termed "microwave D-cleavage." The total volume of the reaction loop can be as low as 5 μL, and the reaction rate can be controlled by changing the flow rate. The maximum digestion time used in the process can be less than 5 min. This setup conveniently allows direct connection to either MALDI-MS or LC-ESI-MS. Disulphide reduction can also be carried out online, and no alkylation of the reduced protein is required. This approach was tested with insulin, α-lactalbumin, lysozyme, myoglobin, carbonic anhydrase, albumin, and *E. coli* lysate.

The application of microwave-assisted acid proteolysis can be extended to sequencing of peptides by electron transfer dissociation (ETD) tandem mass spectrometry [84]. One can take advantage of the size of the peptide (15–25 amino acids) generated by the *C*-terminal cleavage of the aspartic acid residues to give peptides that often contain multiple arginine and lysine residues. As a result the peptides are highly charged, which is ideal for ETD experiments. In using the c- and z-type fragment ions produced by ETD [85], one can achieve better sequence coverage than that from CID fragmentation of the same precursor ion.

8.2.8 Do Protein Denature During Microwave Irradiation?

The dramatic improvement in digestion efficiency of proteins by microwave irradiation may be due to the unfolding of the protein upon interaction with the irradiation. An important question is, Does microwave irradiation of proteins under controlled conditions cause unfolding or denaturation? To answer this question, we infused a 5-mM ammonium acetate solution of cytochrome C at pH 6.8, before and after 10-min exposure to microwave irradiation at 37°C, and used ESI-MS to view the charge-state

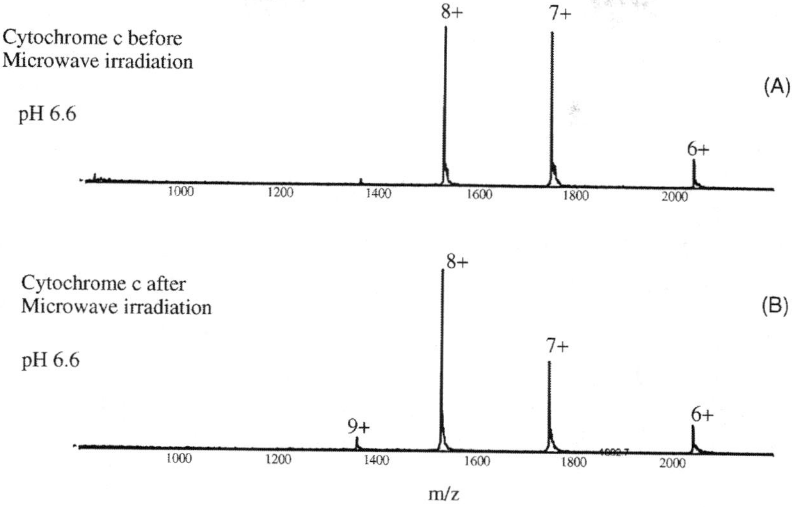

FIGURE 8.5 ESI spectra of cytochrome C obtained from 5-mM ammonium acetate solution, at pH 6.8, before (*A*) and after (*B*) 10 min of microwave irradiation.

distribution. Protein unfolding induced by acid, solvent and heat can be monitored [86–88] using ESI MS to observe the charge-state distribution because unfolding usually leads to a higher and broader charge-state distribution. In fact there is no change-in the charge state distribution (Figure 8.5), providing preliminary evidence that proteins may be generally stable as a result of microwave irradiation.

Another approach establishing whether conformational changes occur in proteins is hydrogen–deuterium amide exchange [89]. We carried out two sets of experiments. In one, cytochrome C (calculated molecular mass of 12,231 Da) was incubated in deuterated water for a set time and directly infused into a triple quadrupole instrument. In another, cytochrome C was incubated with deuterated water and exposed to microwave irradiation for about 10 min at 37°C before the MS analysis. A comparison of the mass spectral data indicates that the charge-state distribution remains unchanged before and after microwave irradiation, reinforcing the conclusion stated above. The measured molecular weight, however, is 12,378 Da (±3 Da, 81% deuterium incorporation) after microwave treatment, and 12,324 Da (±3 Da, 81% deuterium incorporation) for the untreated protein. The difference (54 Da) in the measured masses of the same protein strongly suggests that microwave treatment causes some significant unfolding of the protein [90,91]. The lack of change in the charge-state distribution is supported by CD experiments before and after the microwave irradiation. One explanation for these observations is that microwave irradiation increases protein breathing or other subtle motions of the protein that can be probed by H/D exchange but not by CD or charge-state distributions. The HD exchange result may bear on the observation that enzymatic digestion rates increase but trypsin remained active at least for the first 5 to 10 min.

8.3 SUMMARY

The combination of microwave technology and MS provides some unique analytical capabilities to accelerate structural characterization of proteins and peptides in general and in drug discovery. One advantage is that chemical and enzymatic digestion of proteins to peptide fragments occurs in minutes rather than the hours required for conventional methods. Microwave irradiation is also important for sequence determination of cyclic peptides in using an enhanced Akabori reaction followed by MS/MS analysis of the linear polypeptide hydrazides. Microwave irradiation should also be considered in complex biological matrices for improved solvent extraction of tissue samples to enrich low-level proteins. Improvements of microwave instruments, particularly in temperature control and compatibility with small sample volumes, will make the approach more amenable for proteomics research. More research, however, is needed to understand fully the effect of microwave irradiation in enhancing the reaction rates. Radiation has an obvious effect on protein but in ways that are only apparent thus far with H/D amide exchange.

ACKNOWLEDGMENTS

The authors would like to thank Dr. Yao-Hain Ing for technical support, Dr. William Greenlee for his support of this project, and Prof. Michael Gross for editing this manuscript.

REFERENCES

1. Pandey, A., Mann, M. (2000). Proteomics to study genes and genomes. *Nature 405*, 837–846.
2. Aebersold, R., Mann, M. (2003). Mass spectrometery-based proteomics. *Nature 422*, 198–207.
3. Kola, I., Landis, J. (2004). Can the pharmaceutical industry reduce attrition rates? *Nat Rev Drug Discov 3*, 711–716.
4. MacCoss, M., Baillie, T. A. (2004). Organic chemistry in drug discovery. *Science 303*, 1810–1813.
5. Bordens, R., Grossberg, S. E., Trotta, P. P., Nagabhushan, T. L. (1997). Molecular and biological characterization of recombinant interferon α-2b. *Semi Oncology 24*, S9–S51.
6. Olsen, N. J., Stein, C. M. (2004). New drugs for rheumatoid arthritis. *N Eng J Med 350*, 2167–2179.
7. Fenn, J. B., Mann, M., Meng, C. K., Wong, S. F., Whitehouse, C. M. (1989). Electrospray ionization of large molecules. *Science 246*, 64–71.
8. Karas, M., Hillenkamp, F. (1988). Laser desorption ionization of proteins with molecular mass exceeding 10000 Daltons. *Anal Chem 60*, 2299–2301.

9. Chait, B. T., Kent, S. B. H. (1991). Weighing naked proteins: Practical, high accuracy mass measurement of peptides and proteins. *Science 257*, 1885–1894.

10. Gibson, B. W., Biemann, K. (1984). Strategy for the mass spectrometric verification and correction of the primary structures of proteins deduced form their DNA sequence. *Proc Nat Acad Sci USA 81*, 1956–1960.

11. Griffin, P. R., Coffman, J. A., Hood, L. E., Yates, J.R. III, (1991). Structural analysis of proteins by capillary HPLC electrospray tandem mass spectrometry. *Int J Mass Spectrom Ion Process 11*, 131–149.

12. Davis, M. T., Terry, D. L. (1992). Analysis of peptide mixtures by capillary high performance liquid chromatography: A practical guide of small-scale separations. *Protein Sci 1*, 935–944.

13. Chowdhury, S. K., Katta, V., Chait, B. T. (1990). Electrospray ionization mass spectrometric peptide analysis. *Biochem Biophys Res Commun 167*, 686–692.

14. Yates, J. R. III, Speicher, S., Griffin, P., Hunkapiller, R.T. (1993). *Anal Biochem 214*, 497–408.

15. Nguyen D. N., Becker, G. W. Riggin, R. N. (1995). Protein mass spectrometry: Application to analytical biotechnology. *J Chromatogr A705*, 21–45.

16. Pramanik, B. N., Ganguly, A. K., Gross, M. L. (2002). Applied electrospray mass spectrometry. Dekker, New York.

17. Plumb, R., Castro-Perez, J., Granger, J., Beatte, I., Joncour, K., Wright, A. (2004). Ultra-Performance liquid chromatography coupled to quadrupole-Orthogonal time-of-flight mass spectrometry. *Rapid Commun Mass Spectrom 18*, 2331–2337.

18. Schwartz, J. C., Senko, M.W., Syka, J. E. (2002). A two dimensional quadrupole ion trap mass spectrometry. *J Am Soc Mass Spectrom 13*, 659–669.

19. Frewen, B. E., Merrihew, G. E., Wu, C. C., Noble, W. S., MacCoss, M. J. (2006) Analysis of peptide MS/MS spectra from large-scale proteomics experiments using spectrum libraries. *Anal Chem 78*, 5678–5684.

20. Massolini, G., Calleri, E. (2005). Immobilized trypsin systems coupled on-line to separation methods: Recent developments and analytical applications. *J Separation Sci 28*, 7–21.

21. Duan, J., Liang, Z., Yang, C., Zhang, J., Zhang, L., Zhang, W., Zhang, Y. (2006). Rapid protein identification using monolithic enzymatic microreactor and LC-ESI-MS/MS. *Proteomics 6*, 412–419.

22. Umar, A., Dalebout, J. C., Timmerman, A. M., Foekens, J. A., Luider, T. M. (2005). Method optimization for peptide profiling of microdissected breast carcinoma tissue by matrix-assisted laser desorption/ionization-time of flight mass spectrometry. *Rapid Commun Mass Spectrom 19*, 2865–2870.

23. Papac, D. I., Chin, E. T., Jones, A. J. (1998). A high-throughput microscale method to release *N*-linked oligosaccharides from glycoproteins for matrix-assisted laser desorption/ionization time-of-flight mass spectrometric analysis. *Glygcobiology 8*, 445–454.

24. Abu-Samara, A., Morris, J.S., Koirtyohann, S. R. (1975). Wet Ashing of some biological samples in a microwave oven. *Anal Chem 47*, 1475–1477.

25. Gedye, R., Smith, F., Westaway, K., Ali, H., Baldisera, L., Rousell, J. (1986). The use of microwave ovens for rapid organic synthesis. *Tetrahed Lett 27*, 279–282.

26. Giguere, R. J., Bray, T. L., Duncan, S. C., Majetich, G. (1986). Application of commercial microwave ovens to organic synthesis. *Tetrahed Lett 27*, 4945–4948.

27. Bose, A. K., Manhas, M. S., Ghosh, M., Raju, V. S., Tabei, K., Urbanczyk, L. Z., (1990). Highly accelerated reactions in microwave oven: Synthesis of heterocyles. *Heterocyles 20*, 741–744.
28. Bose, A. K., Banik, B. K., Lavlinskaia, N., Jayaraman, M., Manhas, M. S. (1997) More chemistry in a microwave. *Chemtech 27*, 18–24.
29. Erickson, B. (1998). Standardizing the world with microwave. *Anal Chem 70*, 467A–471A.
30. Reichert, R. C., Link, D., Kingston, H. M. S. (2001). Microwave-enhanced chemistry. *Analy Chem 73*, no. 1, 30A–37A.
31. Lidstrom, P., Tierney, J., Wthey, B., Westman, J. (2001). Microwave-assisted organic synthesis. *Terahedron 57*, 9225–9283.
32. Caddick, S. (1995). Microwave-assisted organic reactions. *Tetrahedron 51*, 10403–10432.
33. Comer, E., Organ, M. G. (2005). A micro reactor for microwave-assisted capillary (continuous flow) organic synthesis. *J Am Chem Soc 127*, 8160–8167.
34. Hafajee, Z.A., Leong, A. S. (2004). Ultra-rapid microwave-stimulated tissue processing with a modified protocol incorporating microwave fixation. *Pathology 36*, 325–329.
35. Emerson, L. L., Tripp, S. R., Baird, B. C., Layfield, L.J., Rohr, L. R. (2006). A comparison of immunohistochemical stain quality in conventional and rapid microwave processed tissues. *Am J Clin Pathol 125*, 176–183.
36. Chaudhari, K., Chattopadhyay, A., Dutta, S. K. (2000). Microwave technique in histopathology and its comparison with the conventional technique. *Ind J Pathol Microbiol 43*, 387–394.
37. Washburn, M. Walters, P., Yates, J. R. (2001). Large-scale analysis of the yeast proteome by multidimensional protein identification technology. *Nat Biotechnol 19*, 242–247.
38. Walters, D. A., Washburn, M., Yates, J. R. (2001). An automated multidimensional protein identification technology for shotgun proteomics. *Anal Chem 73*, 5683–5690.
39. Maxwell, J. C. (1888). *An elementary Treatise on Electricity*. 2nd ed. Clarendon Press, Oxford, UK.
40. Buchwald, J. Z. (1994). *The Creation of Scientific Effects: Heinrich Hertz and Electric Waves*. University of Chicago Press, Chicago.
41. Akabori, S., Ohno, K., Narita, K. (1952). On the hydrolysis of proteins and peptide a method for the characterization of carboxyl-terminal. Amino acids in Proteins. *Bull Chem Soc Jap 25*, 214–218.
42. Bose, A. K., Ying, Y.-H., Pramanik, B. N., Bartner, P.L., Liu, Y.-H., Heimark, L., Lavlinskaia, N., Sareen, C. (2002). Microwave enhanced Akabori reaction for peptide analysis. *J Am Soc Mass Spectrom 13*, 839–850.
43. Pramanik B.N., Ing, Y. H., Bose, A. K., Zhang, L. K., Liu, Y. H., Ganguly, S. N., Bartner, P. (2003). Rapid cyclopeptide analysis by microwave enhanced Akabori reaction. *Tetrahed Lett 44*, 2565–2568.
44. Eckart, K., Schwart, H., Tomer, K. B., Gross, M. L. (1985). Tandem mass spectrometry methodology for the sequence determination of cyclic peptides. *J Am Chem Soc 107*, 6765–6769.
45. Eckart, K. (1994). Mass spectrometry of cyclic polypeptides. *Mass Spectrom Rev 13*, 23–35.
46. Ngoka, L., Gross, M. L. (1999). Multistep tandem mass spectrometry for sequencing cyclic peptides in an ion-trap mass spectrometer. *J Am Soc Mass Spectrom 10*, 732–746.

47. Chen, S. T., Chiou, S. H., Wang, K. T. (1991). Enhancement of chemical reactions by microwave irradiation. *J Ch Chem Soc 38*, 85–91.
48. Pramanik, B. N., Mirza, U. A., Ing, Y. H., Liu, Y. H., Bartner, P. L., Weber, P. C., Bose, A. K. (2002). Microwave-enhanced enzyme reaction for protein mapping by mass spectrometry: A new approach to protein digestion in minutes. *Protein Sci 11*, 2676–2687.
49. Zhong, H., Marcus, S. L., Li, L. (2005). Microwave-assisted acid hydrolysis of proteins combined with liquid chromatography MALDI MS/MS for protein identification. *J Am Soc Mass Spectrom 16*, 471–481.
50. Hua, L., Low, T. Y., Sze, S. K. (2005). Microwave assisted specific chemical digestion for rapid protein identification. *Proteomics 6*, 586–591.
51. Lee, B. S., Krishnanchettiar, S., Lateef, S. S., Gupta, S. (2005). Characterization of oligosaccharide moieties of glycopeptides by microwave-assisted partial acid hydrolysis and mass spectrometry. *Rapid Commun Mass Spectrom 19*, 1545–1550.
52. Lee, B. S., Krishnanchettiar, S., Lateef, S. S., Lateef, N.S., Gupta, S. (2005) Characterization of oligosaccharide moieties of glycopeptides by microwave- assisted partial acid hydrolysis and mass spectrometry. *Rapid Commun Mass Spectrom 19*, 2629–2635.
53. Sun, W., Gao, S., Wang, L., Chen, Y., Wu, S., Wang, X., Zheng, D., Gao, Y. (2006). Microwave-assisted protein preparation and enzymatic digestion in proteomic. *Mol Cell Proteomics 5*, 769–776.
54. Lin S., Yao, G., Qi, D., Li, Y., Deng, C., Yang, P., Zhang, X. (2008). Fast and efficient proteolysis by microwave-assisted protein digestion using trypsin- immobilized magnetic silica microspheres. *Anal Chem 80*, 3655–3665.
55. Lill, J. R., Ingle, E. S., Liu, P. S., Pham, V., Sandoval, W. N. (2007). Microwave- assisted proteomics. *Mass Spectrom Rev 26*, 657–671.
56. Henzel, W. J., Watanabe, C., Stults, J. T. (2003). Protein identification: The origins of peptide mass fingerprinting. *J Am Soc Mass Spectrom 14*, 931–942.
57. Hunt, D. F., Yates, J. R., Shabanowitz, J., Winston, S., Hauer, C. (1986). Protein sequencing by tandem mass spectrometry. *Proc Nat Acad Sci USA 83*, 6233–6237.
58. Houen, G., Sando, T. (1991). Peptide mapping in sodium dodecyl containing buffers control of proteolytic cleavage by organic solvent. *Anal Biochem 193*, 186–190.
59. Russell, W. K., Park, Z. Y., Russell, D. H. (2001). Proteolysis in mixed organic aqueous solvent systems: Applications for peptide mass mapping using mass spectrometry. *Anal Chem 73*, 2682–2685.
60. Massolini, G., Calleri, E. (2005). Immobilized trypsin systems coupled on-line to separation methods: Recent developments and analytical applications. *J Separation Sci 28*, 7–21.
61. Pham, V. C., Henzel, W. J., Lill, J. R. (2005). Rapid on-membrane proteolytic cleavage for Edman sequencing and mass spectrometric identification of proteins. *Electrophoresis 26*, 4243–4251.
62. Vesper, H. W., Mi, L., Enada, A., Myers, G. L. (2005). Assessment of microwave- assisted enzymatic digestion by measuring glycated hemoglobin A1C by mass spectrometry. *Rapid Commun Mass Spectrom 19*, 2865–2870.
63. Juan, H.-F., Chang, S.-C., Huang, H.-C., Chen, S.-T. (2005). A new application of microwave technology to proteomics. *Proteomics 5*, 840–842.
64. Marchetti-Deschmann, M., Kempter, J., Reichel, C., Allamaier, G. (2009). Comparing standard and microwave assisted staining protocols for SDS- PAGE of glycoproteins followed by subsequent PMF with MALDI-MS. *J Proteomics 72*, 628–639.

65. Mirza, U. A., Liu, Y.-H., Tang, J. T., Porter, F., Bondoc, L., Chen, G., Pramanik, B., Nagabhushan, N. T. L. (2000). Extraction and characterization of adenovirus proteins from sodium dodecyl sulphate poly acrylamide gel electrophoresis by matrix assisted laser desorption/ionization mass spectrometry. *J Am Soc Mass Spectrom 11*, 356–361.

66. Shuang, L., Guoping, Y., Dawei, Q., Yan, L., Chunhui, D., Pengyuan, Y., Xiangmin, Z. (2008). Fast and efficient proteolysis by microwave-assisted protein digestion using trypsin-immobilized magnetic silica microspheres. *Anal Chem 80*, 3655–3665.

67. Chen, W.-Y., Chen, Y.-C. (2007). Acceleration of microwave-assisted enzymatic digestion reactions by magnetite beads. *Anal Chem 79*, 2394–2401.

68. Miao, A., Dai, Y., Ji, Y., Jiang, Y., Lu, Y. (2009). Liquid-chromatogarphic and mass-spectrometric identification of lens proteins using microwave-assisted digestion with trypsin-immobilized magnetic nanoparticles. *Biochem Biophys Res Commun 380*, 603–608.

69. Hahn, H. W., Rainer, M., Ringer, T., Huck, C. W., Bonn, G. K. (2009). Ultrafast microwave-assisted in-tip digestion of proteins. *J Proteome Res 8*, 4225–4230.

70. Lin, S.-S., Wu, C.-H., Sun, S. M.-C., Sun, C.-M., Ho, Y.-P. (2005). Microwave- assisted enzyme-catalyzed reactions in various solvent systems. *J Am Soc Mass Spectrom 16*, 581–588.

71. Russell, W. K., Park, Z.-Y., Russell, D. H. (2001). Proteolysis in mixed organic- aqueous solvent systems: Application for peptide mass mapping using mass spectrometry. *Anal Chem 73*, 2682–2685.

72. Kishiyama, A., Zhang, Z., Henzel, W. J. (2000). Cleavage and identification of proteins: A modified aspartyl-prolyl cleavage. *Anal Chem 72*, 5431–5436.

73. Zhong, H., Zhang, Wen, Y. Z., Li, L. (2004). Protein sequencing by mass analysis of polypeptide ladders after controlled protein hydrolysis. *Nat Biotechnol 22*, 1291–1296.

74. Zhong H., Marcus, S. L., Li, L. (2005). Microwave-assisted hydrolysis of proteins combined with liquid chromatography MALDI MS/MS for protein identification. *J Am Soc Mass Spectrom 16*, 471–481.

75. Gebremedhin, M., Zhong, H., Wang, S., Weinfeld, M., Li, L. (2007). Benefit of microwave-assisted acid hydrolysis of proteins for mass spectrometric profiling of the human heart tissue proteome. *Rapid Commun Mass Spectrom 21*, 2779–2783.

76. Hua, L., Low, T. Y., Sze, S. K. (2006). Microwave-assisted specific chemical digestion for rapid protein identification. *Proteomics 6*, 586–591.

77. Li, A., Sowder, R. C., Henderson, L. E., Moore, S. P. (2001). Chemical cleavage at aspartyl residues for protein identification. *Anal Chem 73*, 5395–5402.

78. Swatkoski, S., Russell, S. C., Edwards, N., Fenselau, C. (2006). Rapid chemical digestion of small acid-soluble spore proteins for analysis of bacillus spores. *Anal Chem 78*, 181–188.

79. Schultz J. (1967). Cleavage at aspartic acid. *Meth Enzymol 11*, 255–267.

80. Swatkoski, S., Russell, S. C., Edwards, N., Fenselau, C. (2007). Analysis of a model virus using residue-specific chemical cleavage and MALDI-TOF mass spectrometry. *Anal Chem 79*, 654–658.

81. Swatkoski, S., Gutierrez, P., Ginter. J., Petrov, A., Dinman, J. D., Edwards, N., Fenselau, C. (2007). Integration of residue-specific acid cleavage into proteomic workflows. *J Proteome Res 6*, 4525–4527.

82. Swatkoski, S., Gutierrez, P., Wynne, C, Petrov, A., Dinman, J. D., Edwards, N., Fenselau, C. (2008). Evaluation of Microwave-acclerated residue-specific acid cleavage for proteomic applications. *J Proteome Res 7*, 579–586.

83. Hauser, N. J., Basile, F. (2008). Online microwave D-cleavage LC-ESI-MS/MS of proteins: Site-specific cleavage at aspartic acid residues and disulphide bonds. *J Proteome Res 7*, 1012–1026.

84. Hauser, N. J., Han, H., McLuckey, S. A., Basile, F. (2008). Electron transfer dissociation of peptides generated by microwave D-cleavage digestion of proteins. *J Proteome Res 7*, 1867–1872.

85. Syka, J. E. P., Coon, J. J., Schroeder, M. J., Shabanowitz, J., Hunt, D. F. (2004) Peptide and protein sequence analysis by electron capture dissociation mass spectrometry. *Proc Nat Acad Sci USA 101*, 9528–9533.

86. Loo, J. A., Oogorzalek, R. R., Udseth, H. R., Edmons, C. G., Smith, D. (1991) Solvent-induced conformational changes of polypeptide probed by electrospray-ionization mass spectrometry. *Rapid Commun Mass Spectrom 5*, 101–105.

87. Mirza, U. A., Cohen, S. L., Chait, B. T. (1993). Heat-induced conformational changes in proteins studied by electrospray ionization mass spectrometry. *Anal Chem 61*, 1–6.

88. Chaudhury, S. K., Katta, V., Chait, B. T. (1990). Probing conformational changes in proteins by mass spectrometry. *J Am Chem Soc 112*, 9012–9013.

89. Mirza, U. A., Bose, A. K., Pramanik, B. N. (2008). Hydrogen-deuterium exchange: Electrospray ionization mass spectrometry for probing structural changes in proteins upon microwave irradiation. Presented at the 56th ASMS conference on mass spectrometry and allied topics. June 1–5, Denver, CO.

90. Katta, V., Chait, B. T. (1993). Hydrogen-deuterium exchange electrospray ionization mass spectrometry: A method for probing protein conformational changes in solution. *J Am Chem Soc 115*, 6317–6321.

91. Zhu, M. M., Rempel, L., Zhao, D. E., Giblin, J., Gross, M. L. (2003). Probing Ca^{2+} -induced conformational changes in porcine calmodulin by H/D exchange and ESI-MS: Effect of cations and ionic strength. *Biochemistry 42*, 15388–15397.

CHAPTER 9

Bioinformatics and Database Searching

SURENDRA DASARI and DAVID L. TABB

9.1 OVERVIEW

Shotgun proteomics has become the method of choice for identifying proteins in biological samples. This strategy relies on digesting proteins into peptides and sequencing them by using tandem mass spectrometry (MS/MS). Peptides for which product-ion spectra are captured are then identified using a database search engine. Resulting peptide identifications are filtered to retain the most reliable candidates. To complete this process, a list of protein identifications is inferred from filtered peptide identifications. In this chapter, we introduce basic concepts of peptide sequencing using MS/MS. We provide a detailed picture of automated peptide identification using the MyriMatch database search engine. Finally, we present the process of identification filtering and protein inference using IDPicker software.

9.2 INTRODUCTION TO TANDEM MASS SPECTROMETRY

9.2.1 Protein Sequencing

Protein identification is the principal success story for proteomics. Several methods exist to sequence proteins. Classical methods like Edman degradation can isolate, purify, and sequence only one protein at a time. The introduction of tandem mass spectrometry (MS/MS) to proteomics, however, has radically changed the way we sequence proteins. This new technique has enabled the simultaneous identification of thousands of proteins.

Most proteins assume higher order structures in their native states. Such intact proteins are not amenable to thorough sequence analysis. Hence the process of

Protein and Peptide Mass Spectrometry in Drug Discovery, Edited by Michael L. Gross, Guodong Chen, and Birendra N. Pramanik.

protein sequencing using MS/MS starts with denaturation of proteins. All denaturing methods disrupt higher order structures of proteins while leaving their backbones intact. A denatured protein molecule contains from tens to a few thousands of amino acids, which are linearly arranged like beads on a string. Current MS/MS techniques (12 T FTICR mass spectrometer with electron-capture dissociation) can effectively sequence approximately 60% of the amino acids of myoglobin in a few minutes via a process called "top-down" sequencing or "top-down" proteomics. Shotgun techniques digest proteins into smaller peptides by using proteolytic enzymes like trypsin. Trypsin is highly preferred for this step because it consistently cuts proteins at very specific amino acids (Lys and Arg) and produces peptides that are amenable to identification through runs of contiguous fragment ions.

Tandem mass spectrometry of peptides proceeds in three stages. In the first stage, peptides are ionized and introduced into the mass spectrometer by electrospray ionization (ESI) or matrix assisted laser desorption ionization (MALDI). In the second stage, mass-to-charge (*m/z*) ratios of peptide ions are measured as they pass through a mass analyzer or as they undergo resonant motions in a Fourier transform instrument. A detector measures the ion current of the mass-analyzed peptides. The *m/z* and intensity information ("fingerprint") of peptides is recorded and stored as a mass spectrum. In the third stage, a peptide ion of interest is selected and fragmented using collision induced dissociation (CID). A tandem mass spectrum (product-ion spectrum) records the *m/z* ratios and intensities of fragment ions. This product-ion mass spectrum can then be used to determine the amino-acid sequence of a peptide. A basic knowledge of peptide fragmentation principles is necessary to understand peptide sequencing with MS/MS. Readers are encouraged to consult Chapter 2 by Lin and O'Connor of this volume for a detailed discussion of ion activation and MS/MS.

9.2.2 Peptide Fragmentation

The process of peptide fragmentation can be explained with a model peptide "GEMFILEKGEYPR." A simplified product-ion spectrum from MS/MS for this model peptide is shown in Figure 9.1. The CID process energizes isolated peptide ions via collisions with an inert gas. As the peptide gains energy, its ionizing protons begin to migrate throughout its structure, destabilizing peptide bonds [1]. Because multiple peptide bonds are fragmenting simultaneously, many different fragments from the peptide will be produced. Low-energy CID conditions typically yield both *N*- and *C*-terminal fragments, named b- and y-ions, respectively (Figure 9.2). For example, the model peptide cleaves between the Leu and Glu to produce b6- and y7-ions (Figure 9.1). The number in the fragment names describes the number of amino acids they encompass. For example, a b6-ion contains six amino acids from the *N*-terminal of a peptide; the y7-ion incorporates seven amino acids from the *C*-terminal of a peptide (Figure 9.1). B-ions are formed when protons ionize the fragment on the *N*-terminal side of the broken bond; y-ions are formed when the *C*-terminal side adopts a charge.

Fragments produced at different peptide bonds of the same peptide constitute a series of fragments (Figure 9.1). Successive fragment ions in a series differ by the

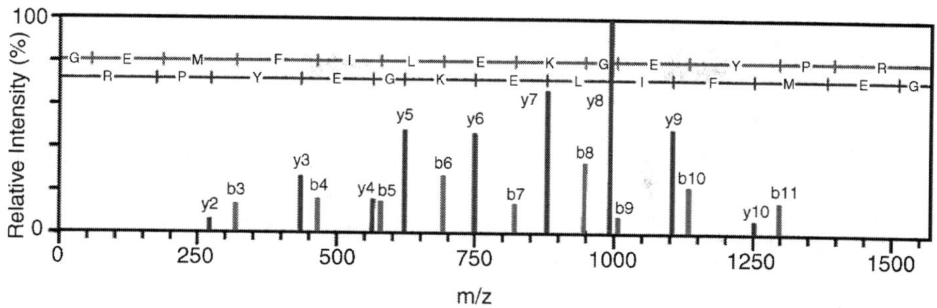

FIGURE 9.1 Simplified product-ion spectrum of the model peptide "GEMFILEKGEYPR." The b-ion series is shown in lighter gray; the y-ion series is darker gray. Low and high mass fragment ions generally fall outside the mass range of the mass spectrometer and hence are not shown in the product-ion spectrum.

mass of an amino acid. For example, the mass difference between the b6 and b7 fragment ions of the model peptide is equal to the mass of glutamate, indicating that this residue is at position seven. Sequentially chaining these mass differences together decodes the entire peptide sequence provided, of course, that all the peptide bonds undergo fragmentation to produce product ions for the MS/MS experiment (Figure 9.1). This technique is called de novo sequencing.

Of course, the CID process produces spectra far more complex than this simple model. *N*-terminal b-ions can lose carbon monoxide to produce ions of type a (Figure 9.2). Both b- and y-ion types can lose NH_3 or H_2O without loss of charge. Several alternative dissociation strategies have also been developed over the years, such as electron capture dissociation (ECD), electron transfer dissociation (ETD), and surface induced dissociation (SID) [2]. Different dissociation techniques produce different distributions of fragment ions. These alternative dissociation strategies, however, have been slow to achieve widespread adoption in tandem mass spectrometry.

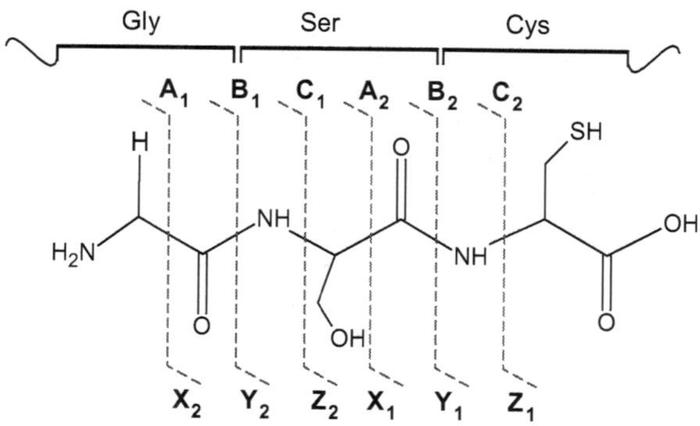

FIGURE 9.2 Fragment ions resulting from dissociation of a tri-peptide. Low-energy CID produces b and y fragment ions. Some b-ions undergo neutral loss of CO to form ions of type a. ETD and ECD produces breakages between the backbone nitrogen and the alpha carbon instead, giving rise to c- and z-ions.

9.3 OVERVIEW OF PEPTIDE IDENTIFICATION WITH DATABASE SEARCHING

How can we infer a peptide sequence from its product-ion spectrum? Before the advent of database searching, peptide identification was performed manually, and it was a very cumbersome process. Researchers can use de novo sequencing to derive a list of potential sequences from a product-ion spectrum (MS/MS). These sequences can then be homology searched against a protein database using BLAST [3]. Results should be critically evaluated to retain the most likely peptide match for each spectrum. This type of manual analysis may require many hours to analyze a modest dataset containing a few hundred spectra. Modern mass spectrometers, however, routinely generate tens of thousands of spectra per hour, making this type of manual analysis unsustainable.

Several database search algorithms that can automatically assign peptide sequences to product-ion spectra are available. Table 9.1 lists the most commonly employed commercial and open source database search tools. All of these database search software operate in a similar manner, as outlined in Figure 9.3. The software accepts two types of inputs: a product-ion spectrum and a protein sequence database. The specificity of the protease (usually trypsin) is used to derive a list of potential peptides from protein sequences. This list is filtered to retain candidates whose masses match the mass of the peptide that produced the product-ion spectrum. Spectra are predicted for these candidates by using basic fragmentation principles. Predicted fragments of each candidate are matched to observed peaks in the experimental spectrum, and a match quality score is computed. Candidates are ranked based on their scores, and the best scoring peptide is usually assigned to the experimental spectrum.

TABLE 9.1 List of Database Search Engines

Search Engine	Website
Sequest	http://www.thermo.com
Mascot	http://www.matrixscience.com
SpectrumMill	http://www.chem.agilent.com
ProteinLynx	http://www.waters.com
Phenyx	http://www.genebio.com
Paragon	http://www3.appliedbiosystems.com
X! Tandem	http://www.thegpm.org/TANDEM
MyriMatch	http://fenchurch.mc.vanderbilt.edu/software.php
OMSSA	http://pubchem.ncbi.nlm.nih.gov/omssa
PepProbe	http://bart.scripps.edu/public/search/pep_probe/search.jsp
Crux	http://noble.gs.washington.edu/proj/crux
MS-Tag	http://prospector.ucsf.edu/prospector
Greylag	http://greylag.org

Note: The bottom seven search engines in the list are open source software.

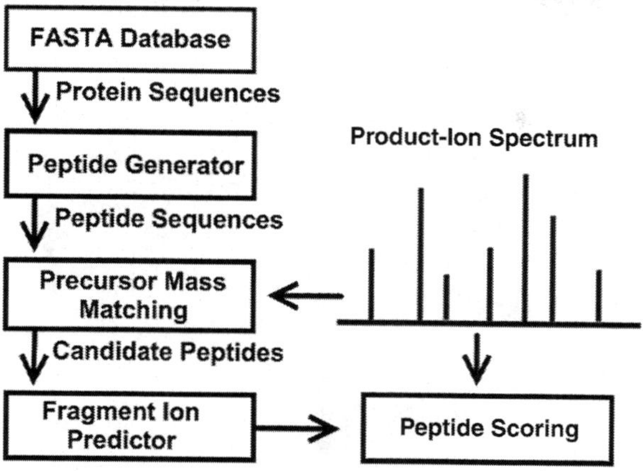

FIGURE 9.3 Peptide identification with database searching. The candidate peptide producing the best score is reported as the match for that spectrum.

Compared to manual analysis, peptide identification with database searching is a very efficient process. Typically a user spends a few minutes to define various search parameters and launches the database search software with appropriate input files. The software proceeds without any additional user intervention and produces a corresponding output file containing peptide identifications.

9.4 MyriMatch-IDPicker PROTEIN IDENTIFICATION PIPELINE

The previous section oversimplified the identification process for didactic purposes. Automated protein identification via database searching, however, is a fairly complex process encompassing several steps. We will now describe this process in greater detail by using pipeline shown in Figure 9.4. The key components of the identification pipeline are the MyriMatch [4] database search engine and IDPicker [5,6] protein assembly software. MyriMatch starts the database search by matching the raw product-ion spectra to peptide sequences derived from a protein sequence database. The software writes the resulting raw peptide identifications to a pepXML file [10]. IDPicker reads these raw peptide identifications, filters them to a set of reliable identifications, and assembles protein identifications from these peptides. IDPicker finishes the protein identification process by creating user-oriented reports.

9.4.1 Raw Data File Formats

Mass spectrometers store product-ion spectra inside binary encoded RAW files. Formats of these RAW files are specific to the instrument's vendor. Typically vendor-purchased software has exclusive access to binary data in RAW files. This becomes problematic when sharing data between laboratories. Thus binary data in RAW files are transcribed to universally accessible text files. A wide variety of open-source

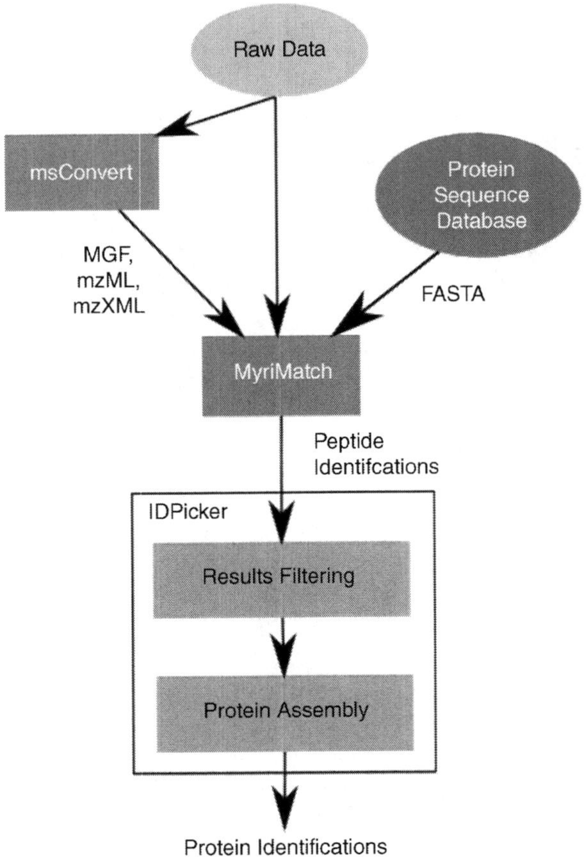

FIGURE 9.4 MyriMatch-IDPicker protein identification pipeline. MyriMatch is a database search engine. IDPicker is a protein assembler.

formats and software exist for this purpose. For example, the msConvert [7] software can transcode raw data from a variety of instruments into open formats like mzML, mzXML, and MGF [8].

The basic structure of a product-ion spectrum in MGF format is composed of a header followed by a list of observed fragment ions (Figure 9.5). The header contains three basic pieces of information: the mass of peptide (precursor) that produced the product-ion spectrum, the charge state of the peptide (if known), and a unique scan event identifier for the spectrum. The header may also contain additional metadata, such as the retention time at which the product-ion spectrum was collected and the intensity of the peptide ion that produced the spectrum. Following the header, the fragment ions observed for the peptide are listed as (*m/z*, intensity) pairs.

Over the years many different file formats have been developed for efficient storage and representation of data extracted from RAW data files. Some file formats can accommodate additional experimental metadata like activation type and energy used

```
BEGIN IONS
PEPMASS=801.41
CHARGE=1+            <- Header
SCANS=129
162.13    1701.50
305.21    1582.11
306.17    4692.75 <- Fragment
307.18    4817.92 <- Ion Peaks
320.19    1507.60 <-
321.18    1724.69
   .         .
   .         .
   .         .
699.43    8710.89
701.43    7019.75
746.34    1348.68
END IONS
```

FIGURE 9.5 Product-ion spectrum representation. An example product-ion spectrum represented in Mascot Generic File (MGF) format.

for peptide fragmentation and number of fragment ions observed for a peptide. The lowest common denominator for these file formats, however, is precursor mass, precursor charge, *m/z* and intensities of observed fragment peaks, and corresponding product-ion scan event identifier. All database search engines use each of these data types in the peptide identification process.

9.4.2 Protein Sequence Databases

Database search engines match the product-ion spectra directly to peptide sequences in protein databases. If we obtain a product-ion spectrum of a peptide for which the sequence is in the protein database, then algorithms will be able to match sequence to the spectrum. If instead we obtain a spectrum of a peptide for which the sequence is not in the protein database, then database search would not be able to obtain the correct match. Hence completeness and accuracy of the protein sequence databases have a large bearing on the quality of the database search results.

Publicly available protein sequence databases (Table 9.2) differ in terms of quality of sequence annotations, completeness, degree of redundancy, and number of supported species. Choice of sequence database depends on the experiment and final goals of the search. For example, curated databases like Swiss-Prot, RefSeq, and PDB contain sequence entries that were previously verified by other biological experiments. Hence searching against these curated databases generally produce highly reliable protein identifications.

Curated databases, however, are far from complete, and they often contain only data for the major isoforms of proteins. Furthermore these curated databases include only "wild-type" proteins and ignore the natural sequence diversity that exists in a population. More comprehensive databases like TrEMBL and Entrez exist for

TABLE 9.2 Protein Sequence Databases

Database Name	Website
UniProtKB/Swiss-Prot	ca.expasy.org
Protein Databank (PDB)	www.rcsb.org
Reference Sequence (RefSeq)	www.ncbi.nlm.nih.gov/RefSeq
UniProtKB/TrEMBL	ca.expasy.org
Ensembl	www.ensembl.org
UniRef	www.uniprot.org/help/uniref
UniParc	www.uniprot.org/help/uniparc
Internation Protein Index (IPI)	www.ebi.ac.uk/IPI/IPIhelp.html
Entrez	www.ncbi.nlm.nih.gov
Protein Information Resource (PIR)	pir.georgetown.edu

Note: All databases are available via the Internet for free use. More information about databases can be found at the corresponding websites.

these purposes. Typically these comprehensive databases contain both verified and unverified genomic translations of an organism. Entrez also contains mutant proteins detected in the population. Hence these comprehensive databases are useful when sequencing proteomes of a new organisms or detecting mutant proteins in an individual.

Comprehensive databases like Entrez should be used with caution because they often contain multiple entries for a single protein (wild type, mutants, sequence variants, etc.). Such sequence redundancy can complicate protein-list inference. Comprehensive databases are also several orders of magnitude larger than curated databases, requiring longer search times. The peptide diversity of these databases also increases the likelihood of obtaining an incorrect peptide match by random chance. Despite their disadvantages, sometimes comprehensive databases are the only available choice for researchers, because not all organisms have high-quality curated databases.

As of late 2010 the UniProt provides several protein sequence databases that strike a good balance between comprehensiveness, redundancy, and size for a large number of species. The UniProtKB sequence database is a hybrid of all curated and predicted sequence databases. Additionally this database includes known alternative isoforms of proteins. Nonredundant sequences are extracted from the UniProtKB and included in the UniRef sequence database in order to speed up sequence similarity searches. UniProtKB also contains "complete proteome sets" for 1579 organisms (set of proteins thought to be expressed by an organism whose genome has been completely sequenced). These databases are highly recommended for initial exploration of proteomes, whenever possible.

All databases listed in Table 9.2 are available as FASTA formatted plain text files, often a requirement for database search tools. Each protein entry in the FASTA file is composed of a header followed by a sequence. The header primarily contains the name and a unique identifier of the protein. Often it may also contain other useful information like the name of the gene that encodes the protein and the species to which the protein belongs. Following the header, the sequence of each protein is given as a linear string of standard amino-acid codes. Additional information about these databases can be found at website links provided in Table 9.2.

9.4.3 MyriMatch Database Search Engine

All database search engines (see Table 9.1 for list) take a similar approach to the database searching as outlined in Figure 9.3. Here we provide a detailed picture of the database searching using MyriMatch as a representative tool (other search engines follow the same model as MyriMatch). The MyriMatch search engine requires two types of inputs: raw MS/MS data and protein sequence database (Figure 9.4). The software accepts the raw MS/MS data in a variety of file formats like RAW, WIFF, YEP, MGF, mzXML, and mzML. Protein-sequence databases are accepted in FASTA format (Figure 9.4). After reading all input files, MyriMatch proceeds through several stages. First, experimental product-ion spectra are preprocessed to make them amenable for comparison. Second, peptides are generated from the database for comparing against the product-ion spectra. Finally, theoretical product-ion spectra are generated for database candidates and compared against experimental product-ion spectra.

Spectrum Preprocessing Product-ion spectra often contain peaks corresponding to fragment ions mixed with noise peaks. These extraneous peaks should be removed from the spectra. Otherwise, they interfere with peptide scoring and produce false-positive matches to candidate fragments. Removing peaks indiscriminately, however, may remove true fragment ion peaks from the spectra, whereas a relaxed filter would leave noise peaks behind. MyriMatch employs a tunable noise filter for this task. First, the software computes total ion current (TIC) in the spectrum and lets the user choose a proportion of the TIC to be retained. Next, MyriMatch sorts peaks in the spectrum by decreasing order of intensity and retains the minimum number of peaks to satisfy the target TIC. This approach is comparable to peak screening in X! Tandem, which routinely retains the top 50 peaks for each spectrum. Likewise Sequest filters out all but the top 200 peaks for each spectrum and normalizes intensities in 10 zones spanning the product-ion spectrum. Preprocessing takes a slightly different form for each search engine.

Selecting Candidates from Databases MyriMatch selects peptides from a protein database by using knowledge of the digestion enzyme. For example, if trypsin were used for digestion, MyriMatch may generate only tryptic peptides from the sequence database. The software can also be instructed to ignore enzyme specificity to select from the database all possible peptide sequences that correspond to the mass of the peptide that produced the spectrum. The use of enzyme specificity in MyriMatch is very similar to that of other algorithms, though the software enables users to specify the removal of initial methionine residues from proteins or disallow digestion events *N*-terminal to proline residues.

After this step MyriMatch filters the selected peptides to retain candidates whose masses match the precursor mass of peptide that produced the spectrum (within a user-specified tolerance). The appropriate choice of the precursor mass tolerance (PMT) parameter depends on the mass accuracy of the mass spectrometer used to acquire the data. Lower resolving power instruments like ion traps and quadrupoles provide precursor masses that are accurate to a part per thousand or slightly better. Hence PMT for these instruments is typically set to 2 to 3 Da. Q-TOF instruments provide better mass accuracy of 10 parts per million (ppm). High resolving power FTICR and Orbitraps are capable of providing sub ppm mass accuracy [9]. Setting PMT too low can prevent the predicted product-ion spectrum of correct database peptides from being compared to the experimental spectrum, because their calculated mass is "out of range" from observed mass. A wide definition of this parameter allows predicted spectra of irrelevant peptides to be compared to experimental spectrum because their calculated mass happens to be within the range of observed mass. Both these scenarios could reduce the numbers of identified peptides. Ultimately the precursor mass tolerance determines which database peptides' predicted product-ion spectra get compared to each experimental product-ion spectrum. Like X!Tandem and Mascot, MyriMatch allows users to configure PMT in both *m/z* and ppm units. Sequest, however, restricts this option to Dalton units.

Comparing Candidate Spectra with Experimental Spectra and Evaluating Matches The problem of comparing two product-ion spectra for similarity is well understood. Numerous computational methods exist to address this issue. Comparing a candidate sequence to an experimental product-ion spectrum is a slightly different problem that requires a bit of an abstraction. First, the candidate sequence is converted into a theoretical spectrum, which is the list of *m/z* locations where one may expect fragment ions for a given sequence. Next, peaks in the theoretical spectrum are compared to peaks in the experimental spectrum. Finally, a peptide score is computed for the match, which measures the similarity between expected and observed fragmentation patterns.

MyriMatch uses basic peptide fragmentation rules to predict theoretical spectra for the candidates. The software predicts the b- and y-ions that would be produced from each peptide bond breakage in a peptide. If a peptide is singly or doubly charged, the software will predict only singly charged fragments. For more highly charged peptides, the software will predict multiply charged fragments. The software attempts to determine which side of a peptide bond for a $+3$ peptide will take on a double charge, leaving the other terminus as singly charged. This model is not very different from other search engines, though X!Tandem and Sequest both predict peak lists with projected intensity values, as suits their scoring strategies.

Many scoring algorithms have been developed for evaluating matches between theoretical and experimental spectra. For example, a rudimentary "shared peak count" (SPC) reports the total number of peaks matched between a theoretical and an experimental spectrum. MyriMatch uses a sophisticated intensity-based scoring system for matching theoretical and experimental spectra. First, filtered peaks in the experimental spectrum are separated into three intensity classes. These classes differ in the number of peaks they hold so that the highest intensity class holds the fewest peaks, whereas remaining classes each double the number of peaks of the next, more intense, class. Matching a peak from a sparsely populated, high-intensity class contributes more to the peptide score than matching a peak from a more populous, low-intensity class. For each predicted peak in a theoretical spectrum, MyriMatch looks at the corresponding location in the experimental spectrum to determine if a peak is observed, and if so, the class of the matching peak. A peak match is valid if and only if a predicted peak falls within a certain range of the observed peak (defined by fragment mass tolerance). The score for the peptide candidate is computed as the probability of observing this particular distribution of peak classes by random chance. MyriMatch employs a multivariate hypergeometric (MVH) distribution to compute this probability and reports the negative of the logarithmic probability. Matching predominantly the peaks of the most intense class is highly unlikely to occur solely by random chance, thus producing a high score. The best-scoring five candidate sequences are reported for each experimental product-ion spectrum.

The statistically based MVH score is the principal scorer for MyriMatch. In Sequest this role is filled by the cross-correlation-based XCorr score. Sequest differs by applying this scoring strategy to only the best 500 candidates, as judged by a preliminary scoring algorithm. The MVH score is interpretable by its statistical derivation, while XCorr scores do not bear inherent statistical meanings. The same might be said of the dot-product-based hyperscore in X!Tandem. X!Tandem solves this problem, however, by using the hyperscore distribution for each spectrum to derive the expectation value associated with the highest scoring sequence. Principal scoring is generally the feature that most differentiates database search algorithms.

The appropriate specification of fragment mass tolerance is crucial for accurate peptide scoring. This parameter defines how far away a predicted peak may be from its expected *m/z* location in the product-ion spectrum. If fragment mass tolerance is defined too narrowly, predicted peaks may fail to match because they differ from their expected positions in the spectrum. If this parameter is defined too widely, predicted peaks will match to random peaks in the spectrum. Both scenarios produce peptide identification scores that lack discrimination. As with precursor mass tolerance, the appropriate choice of this parameter depends on the mass accuracy of the instrument. In MyriMatch the fragment mass tolerance parameter is typically set to 0.5 *m/z* for lower mass resolving power ion trap instruments. For higher resolving power Orbitrap and Q-TOF instruments this parameter could be set at 50 to 75 ppm. Although Sequest and Mascot limit fragment mass tolerance to be defined in Dalton or *m/z* units, other search engines like X!Tandem allow for this to be configured in ppm units.

A brief summary of the database searching with MyriMatch includes the following steps: MyriMatch selects peptide candidates from a protein database for which the masses match the observed precursor mass of spectrum. The software creates theoretical spectra for each candidate and compares them to the experimental spectrum. Peptide scores are computed to evaluate each match. Peptide matches to the experimental spectrum are ranked from best to worst by their scores.

```
<msms_pipeline_analysis name="mam_20090403x_Yeast_60nguL_1" date="2009-04-09T13:59:10" ... >
 <msms_run_summary base_name="mam_20090403x_Yeast_60nguL_1" raw_data_type="mzXML" raw_data="unknown">
  .
  .
  .
  <search_summary ...
     .
     .
     .
  </search_summary>
  <spectrum_query spectrum="mam_20090403x_Yeast_60nguL_1.9848.9848.2" start_scan="9848" end_scan="9848"
     spectrumNativeID="scan=9848" spectrumIndex="9847" precursor_neutral_mass="1504.10" assumed_charge="2"
     retention_time_sec="4260.78">
     <search_result num_target_comparisons="73630" num_decoy_comparisons="73655">
     <search_hit hit_rank="1" peptide="SKVNERDIESSIK" peptide_prev_aa="K" peptide_next_aa="P" protein="YDR227W"
        peptide_offset="732" num_tot_proteins="1" num_matched_ions="13" tot_num_ions="19"
        calc_neutral_pep_mass="1502.6" massdiff="-1.45" num_tol_term="2" num_missed_cleavages="2">
             <search_score name="mvh" value="20.04" />
     </search_hit>
     .
     .
     .
     <search_hit hit_rank="5" peptide="TSIRGDSPSNPKDK" peptide_prev_aa="I" peptide_next_aa="Q" protein="YKL071W"
        peptide_offset="35" num_tot_proteins="1" num_matched_ions="8" tot_num_ions="23"
        calc_neutral_pep_mass="1501.53" massdiff="-2.60" num_tol_term="1" num_missed_cleavages="2">
             <search_score name="mvh" value="10.2" />
     </search_hit>
  </spectrum_query>
  .
  .
  .
 </msms_run_summary>
</msms_pipeline_analysis>
```


FIGURE 9.6 A snapshot of MyriMatch results file. The software produces peptide identifications in an industry standard pepXML format.

9.4.4 Peptide Identification Reporting

Modern mass spectrometers generate thousands of spectra per LC-MS/MS experiment. MyriMatch processes all spectra in an experiment as a single batch. Resulting raw peptide identifications are written to an output file in pepXML format [10]. This data format was developed at Institute of Systems Biology (ISB) to provide a common platform for reporting database search results. Figure 9.6 shows a snapshot of an example results file generated by MyriMatch. The file starts by describing the raw data used to obtain the search results. Next the format summarizes the configuration used for database search. For example, the summary details the database search engine employed, search parameters specified for analysis, name of protein database used for search, database statistics, date and time of search execution, and total search time. Following the search summary, the format describes input product-ion mass spectra, including spectrum names, unique spectrum identifiers assigned by the instrument, mass and charge of the peptide that produced the spectrum, and retention time at which the spectrum was acquired. After each spectrum description, peptide sequences assigned to that spectrum are listed. A typical peptide identification description contains the amino-acid sequence of the peptide, calculated mass of the peptide, the protein from which the peptide was generated, the position at which the peptide starts in the corresponding protein, the number of missed enzymatic cleavage sites in the peptide, and peptide identification score(s). Peptide identifications of each spectrum are sorted and ranked from best to worst by their identification score. This pepXML results file is used for downstream processing and reporting of peptide–protein identifications.

Judging Quality of Peptide Matches By default, MyriMatch reports the top five peptide matches for every spectrum. The quality of these matches can be judged

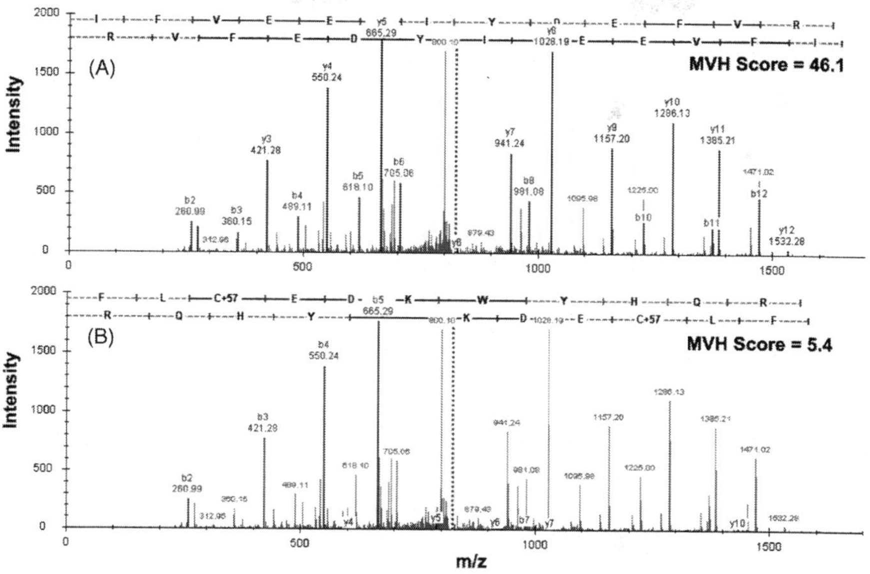

FIGURE 9.7 MVH score separates reliable from unreliable peptide matches. Peptide interpretations highlighting predicted b and y ions matches in an experimental spectrum. (*A*) The alignment shows a good match to predicted ions. (*B*) A different alignment to the same spectrum shows a poor match to predicted ions. Peptide scores (MVH) are shown for both interpretations.

on the basis of reported MVH scores or by simply overlaying the predicted fragment peaks over the spectrum and visually inspecting each interpretation for the number of matched peaks [11]. For example, a candidate sequence whose calculated spectrum matches the most intense peaks in an experimental spectrum receives a higher MVH score (Figure 9.7*A*). When predicted peaks do not match major peaks in experimental spectrum, a lower MVH score results (Figure 9.7*B*). Thus the MVH score can be used to separate reliable from unreliable peptide matches. It is important to note, however, that MyriMatch does not attempt to judge posterior error probabilities of the matches assigned to the spectra. The software will assign peptides to every spectrum in a dataset, even though the matches are of poor quality. Thus raw identifications must be filtered to retain the most reliable peptide identifications.

9.4.5 Post-processing of Search Results Using IDPicker

Modern proteomic datasets contain tens of thousands of identifications per experiment, requiring automated filtration. Furthermore most researchers prefer to see a list of proteins rather than of peptides. Hence filtered peptide identifications should be grouped according to their parent proteins. A variety of software exists for postprocessing search results and generating user-oriented protein identifications lists. Here we discuss IDPicker, which uses False Discovery Rates (FDR) for peptide identification filtering and transparent parsimony for generating protein identification lists.

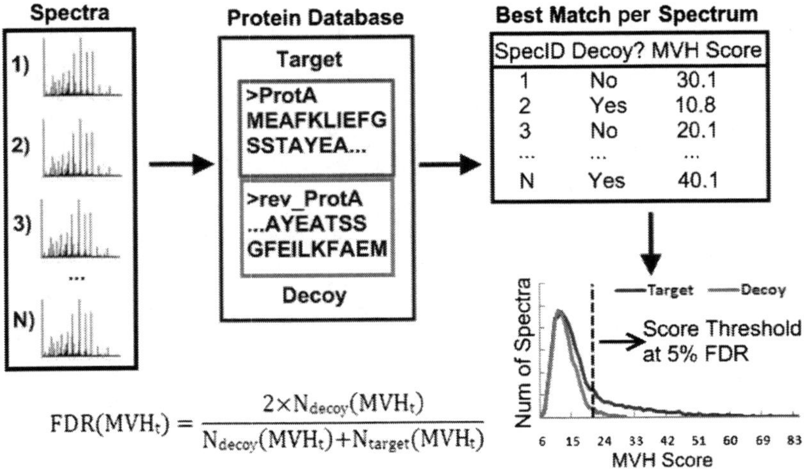

FIGURE 9.8 IDPicker results filtering. Automated validation of peptide identifications in large datasets using false discovery rate (FDR) filtering.

Peptide Identification Filtering IDPicker employs target-decoy database searches for estimating false discovery rates (FDR) of raw identifications [12]. In the FDR-based method, all spectra in a dataset are searched against a composite database containing "target" and "decoy" proteomes (Figure 9.8). A "target" proteome is composed of protein sequences appropriate for sample to be analyzed, whereas a "decoy" proteome contains nonsense protein sequences. Typically decoys are created by simply reversing sequences in the target proteome. Decoys are also distinctly marked such that they can be distinguished from target sequences in search results. IDPicker processes raw identifications from an entire LC-MS/MS analysis as a single unit. For each spectrum result IDPicker extracts the top peptide assignment, corresponding MVH score, and decoy status (Figure 9.8). Next IDPicker computes the number of target and decoy peptides that score above a certain MVH threshold MVH_t as $N_{target}(MVH_t)$ and $N_{decoy}(MVH_t)$, respectively. The FDR at MVH_t is computed according to the equation in Figure 9.8. IDPicker reduces the MVH threshold from the maximum value until the allowed FDR is produced (Figure 9.8). For example, when user chooses a 5% FDR, IDPicker automatically determines the MVH threshold that corresponds to 5% FDR and filters peptides in the entire collection by using that threshold. Due to the use of a nonparametric score evaluation approach, IDPicker can process results from arbitrary search engines without requiring retraining. The software can also combine complementary scoring metrics (e.g., XCorr and DeltaCN) reported by a search engine for better discrimination. Where PeptideProphet [13] attempts to estimate each peptide identification's probability of correctness, IDPicker estimates the error rate for the full set of accepted peptide identifications.

Peptide-Protein Assembly A naïve protein identification algorithm simply lists all proteins from which the filtered peptides are derived. This approach overstates the

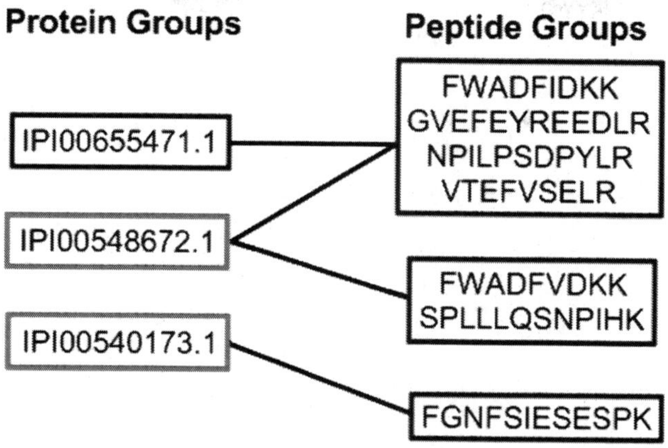

FIGURE 9.9 Parsimony. Only two proteins (in gray boxes) are needed to explain all observed peptides.

number of identified proteins when some of the peptide identifications are shared among multiple proteins. This case often arises when a sample comes from a species that contains homologous proteins or splice variants (e.g., human proteins). For example, Figure 9.9 shows a bipartite graph of different isoforms of glutathione transferase and their corresponding identified peptides. The top peptide group is shared by two isoforms of the protein. A naïve protein identification algorithm would list all three isoforms as present in the sample. However, a parsimonious analysis of the bipartite graph reveals that isoform 2 and 3 are sufficient to explain all observed peptides (Figure 9.9).

IDPicker parsimony analysis proceeds in multiple stages. First, filtered peptide identifications are represented as a bipartite graph consisting of peptide and protein vertices (Figure 9.10*A*). An edge is drawn between a peptide and a protein vertex if the peptide sequence can be considered evidence for the protein sequence. Next, protein vertices that are connected to the same set of peptide vertices are treated as indiscernible and collapsed into a meta-protein group (Figure 9.10*B*). Similarly peptide vertices that connect to the same protein vertices are collapsed into a meta-peptide group (Figure 9.10*B*). The bipartite graph is then decomposed into individual clusters in which every meta-protein is required to share at least one meta-peptide with another meta-protein in the same cluster (Figure 9.10*C*). Finally, IDPicker applies a greedy algorithm to each cluster to derive a minimal list of meta-proteins to explain all observed meta-peptides (Figure 9.10*C*). IDPicker parsimony is transparent to the user. For each cluster IDPicker produces a tabular list of association tables revealing which meta-proteins map to which meta-peptides, and a table (similar to Figure 9.9) illustrating the relationship between proteins and peptides of a cluster.

IDPicker provides its functionality through a flexible graphical user interface (GUI). The GUI guides users through the process of selecting pepXML files from a database search, organizing selected files into a multilayer experimental hierarchy (different samples, 2D separations, technical replicates, etc.; see Section 9.5 for an

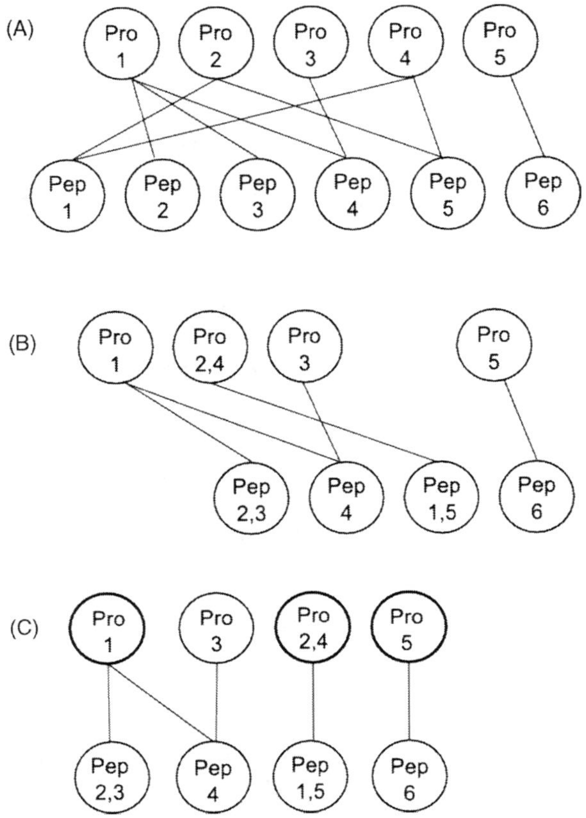

FIGURE 9.10 Bipartite-graph based parsimony. (*A*) Connect peptides to all proteins that contain them. (*B*) Group indistinguishable peptides and proteins to form meta-peptide and meta-protein vertices. (*C*) Decompose the graph into protein clusters that share peptides. Identify proteins (1, 2, 4, and 5) that are sufficient to explain all observed peptides.

example), and configuring thresholds employed for peptide filtering and protein assembly. The software then launches the analysis, informing the user of its progress. The software generates HTML formatted identification reports. The published descriptions of IDPicker [5,6] describe its features more thoroughly.

9.5 RESULTS OF A SHOTGUN PROTEOMICS STUDY

Most shotgun proteomics studies often contain multiple LC-MS/MS experiments (different samples, 2D separations, technical replicates, etc.). Researchers often need to combine search results from multiple experiments into a single identification report for comparison. Figure 9.11 shows an example MyriMatch-IDPicker protein identification report derived from a large proteomics study. The study contains a total of six samples, three from cancer subjects and three from control subjects. Each sample was

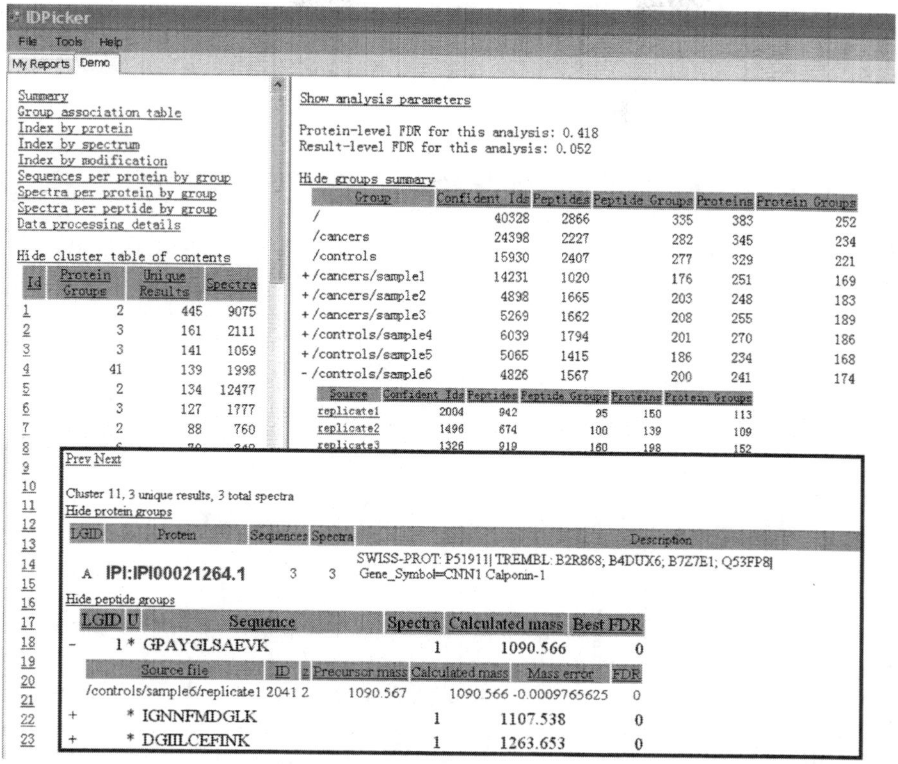

FIGURE 9.11 An example protein identification report. IDPicker protein identification report displaying MyriMatch database search results for three cancer and three control samples.

analyzed in triplicate LC-MS/MS analyses. The identification report in Figure 9.11 captures this experimental design in a tree-based hierarchy. The report summarizes the total numbers of spectral, peptide, and protein identifications at each level of the experimental hierarchy. For instance, the report starts by summarizing identifications at the root level (denoted with "/"). After the root node, the cohorts (cancers vs. controls) are summarized, followed by samples and technical replicates. The identification report also contains a navigation pane on the left side, which lists protein clusters by their index numbers. Users can follow these indecies to examine individual clusters. For instance, the inset of Figure 9.11 shows an example cluster for the Calponin protein. This protein cluster report shows all peptide and spectral identifications of Calponin. For each peptide-spectrum match (PSM), the expected mass, calculated mass, mass error, and FDR value are displayed. Users can also manually evaluate these PSMs using an integrated spectrum viewer.

IDPicker also generates protein, peptide, and modification level summary tables. For instance, the "spectra per protein by group" table summarizes proteins and the distribution of their respective spectral counts across the experimental hierarchy (by condition, samples, replicates, etc.). Such reports are extremely useful to highlight differences between control and disease samples using label-free quantification. The label-free quantification software QuasiTel [14] was designed for the direct use of IDPicker protein and peptide level summary tables.

9.6 IMPROVEMENTS TO MyriMatch DATABASE SEARCH ENGINE

MyriMatch can rapidly assign peptide sequences to product-ion spectra obtained by MS/MS. The performance of MyriMatch has been improved by parallelizing the algorithm to use multi-CPU computers or multi-node clusters. MyriMatch also can handle peptides with specified posttranslational and chemical modifications.

9.6.1 Parallel Processing

Modern mass spectrometers are already very sensitive and becoming more so. MS research cores routinely generate millions of mass spectra per day. Even the latest workstations do not have enough computational power to support the data processing needs of a high-throughput laboratory. A solution to this problem is to harness the power of parallel processing.

MyriMatch software is designed to take advantage of multi-core CPUs and multi-computer cluster configurations. When running in parallel mode, MyriMatch spawns multiple worker processes that are controlled by a single master process. The master initiates a search by reading the protein database and the raw MS files. Each worker is given the entire set of spectra and a unique subset of protein sequences. When a worker is finished searching the set, it asks the master for next set of protein sequences. This process continues until all proteins are searched against the spectra. Finally, the master collects PSMs from all workers and groups them by their spectrum identifier. This type of parallelization is different from that of Parallel Tandem [15] and Sequest-PVM [16], both of which distribute unique subsets of spectra and the entire sequence database between the workers.

The parallelization implemented in MyriMatch adapts well for both multi-CPU computers and multi-node clusters. When using a multi-CPU computer, MyriMatch distributes available CPUs between the master and workers by using threads. When using a multi-node cluster, the master process is retained on the master node of the cluster, and workers are populated on the rest of the nodes by using Message Passing Interface (MPI). MyriMatch allows workers to run asynchronously, which facilitates optimal utilization of mixed nodes available on a cluster. The software provides parallelization in a very transparent manner. When using a multi-CPU computer, the user simply instructs MyriMatch how many CPUs to use for search. When using a cluster, the user instructs MyriMatch how many nodes to use for search. The software distributes its operations accordingly and produces one output file per one input raw file.

Search times scale inversely to number of CPUs utilized in the cluster (Table 9.3). The table shows run times of searching 1000 spectra against two different databases using 1, 2, 20, and 40, CPUs. All CPUs were part of a cluster of Mac Mini computers connected to a master node. Each cluster node is equipped with a 1.8 GHz Intel Core 2

TABLE 9.3 MyriMatch Semitryptic Search Times

	1 CPU	2 CPUs	20 CPUs	40 CPUs
Swiss-Prot Human (20, 348 proteins)	02:24:54	01:30:23	00:07:28	00:03:57
IPI Human (87,925 proteins)	06:51:21	04:31:19	00:22:15	00:11:06

Note: Time measurements were taken on Mac Mini cluster in HH:MM:SS format. 1000 spectra searched against two databases using a parent mass tolerance of 1.25 *m/z*.

Duo processor and 2 GB RAM. The master node is equipped with two quad-core 2.4 GHz Intel Xeon processors and 10 GB RAM, and it communicates with cluster nodes over a dedicated gigabit network.

9.6.2 Protein Modification Analysis

Post-translational modifications (PTMs) to proteins are very common in any biological sample. They are often associated with biological processes like protein folding, signal transduction, protein degradation, and molecular timing. Database search engines are routinely used to check samples for suspected PTMs. (Tsarbopoulos and Bazoti in Chapter 12 of this volume discuss three common PTMs and their analysis.)

MyriMatch supports two types of modifications: static and dynamic. Static modifications are the simplest kind, altering all occurrences of a residue in the sequence database. For example, carbamidomethylation causes cysteines to gain 57 Da, increasing their mass from 103 to 160 Da. Thus, when cysteine carbamidomethylation is specified as a static modification in a search, all cysteine residues in any database sequence are treated as having the higher mass of 160 Da. Static modifications are useful when a particular residue in all proteins is expected to have a different mass than usual. For example, enrichment of cysteine containing peptides via isotopic coded affinity tag (ICAT) labeling permanently alters the mass of all cysteine residues in the captured peptides.

Dynamic modifications are more powerful because they are treated as potential modifications to database residues rather than omnipresent modifications. When potential modification residues are present in a database peptide, MyriMatch searches all possible combinations of modified and unmodified forms of the peptide. Identification of deamidated peptides is a textbook application of a dynamic modification search. Deamidation has to be specified as a potential modification in these searches because not all asparagine and glutamine residues can be deamidated. Other applications of dynamic modification search include phosphorylation, methylation, oxidation, and so forth. Thus a dynamic modification search is necessary when both modified and unmodified peptide forms are present in a sample. Unlike static modifications, dynamic modifications increase search times. This is because MyriMatch tests several different forms of a database peptide when it contains a potentially modified residue. MyriMatch was improved to consider a large set of modifications per search. MyriMatch does not impose limits on either number of modifications allowed per residue or number of potential modifications tested per peptide, which is in contrast to the workings of other search engines like Sequest and X! Tandem.

9.7 APPLICATIONS OF MyriMatch-IDPicker PIPELINE

Tandem mass spectrometers can accept peptides from a variety of experimental sources. 1D and 2D gel electrophoresis experiments enable the selection of a small number of proteins from a complex background. Strong cation exchange (SCX) and isoelectric focusing (IEF) experiments provide sensitive peptide separations required for an in depth analysis of samples. Immuno-precipitation (IP) "pull downs" are useful for isolating particular protein(s) of interest from a whole sample. MyriMatch can identify peptides produced from any of these strategies [17,18].

9.7.1 Characterizing Protein–Protein Interactions

IP pull downs coupled with MS/MS analysis are routinely used to characterize protein–protein interactions. An example of the use of this capability is the identification of yeast Mot1P interacting partners [19]. Mot1P is transcription regulator that can both repress and activate mRNA gene transcription. Mot1P complex can be isolated from yeast whole-cell lystates by using IP pull downs. Precipitated protein complexes are digested, and resulting peptides are fractionated using SCX chromatography. These peptide fractions are subjected to LC-MS/MS analysis. A MyriMatch search identified a range of transcriptional cofactors and chromatin-remodeling proteins that were not previously known to interact with Mot1P. The IDPicker reports produced from these identifications enabled statistical evaluation of protein associations from the IP data. Independent experimentation confirmed the potential novel interactions identified by the database search.

9.7.2 Characterizing Yeast Proteome on Diverse Instrument Platforms

Modern mass spectrometers can be purchased or assembled in a variety of configurations. MyriMatch can process spectral data from diverse instruments without requiring any retraining. An example of using this capability is the development of a yeast performance standard to benchmark the performance of various LC-MS platforms [20]. A yeast cell lysate standard was generated by National Institute of Standards and Technology (NIST). This yeast standard was sent to several different laboratories for characterization. Participating sites also received a standard operating procedure detailing how the sample was to be analyzed. Spectral data from all participating sites was gathered and analyzed by using MyriMatch database search engine. Several performance metrics were extracted from the corresponding IDPicker reports, which were used to characterize the yeast standard and evaluate identification variability between diverse LC-MS platforms.

9.7.3 Characterizing DNA-Protein Crosslinks

MyriMatch can be configured to consider a large set of modifications per search. An example of this capability is the determination whether or not bis-electrophile diepoxybutane adducts crosslink DNA to human histones [21]. Diepoxybutane is an oxidative by-product of two mutagens, 1,2-dibromomethane and 1,3-butadiene, that have been widely used in petrochemical and pesticide industries. Today these mutagens are commonly found in automobile exhaust and cigarette smoke. Upon treatment of isolated histones with diepoxybutane in the presence of short DNA strands a wide palette of adducts on the histones were formed. Proteins were subsequently digested and subjected to tandem mass spectrometry. The resulting spectra were searched with MyriMatch, configured to look for peptide adducts. Histone H2b isoform showed an increased affinity for DNA when treated with 1,2-dibromomethane, which was later verified with independent experimentation.

9.8 CONCLUSIONS

The MyriMatch-IDPicker pipeline has a lot to offer for mass spectrometrists, biologists, and clinicians. For mass spectrometrists, this pipeline can process MS/MS data from diverse instrument platforms and experimental conditions. For biologists, this pipeline abstracts away the complexity of peptide identification via shotgun proteomics and provides final identification reports that can be directly used for quantification. For clinicians, the final protein identification reports enable summarization of large cohorts.

Protein and peptide identification has come a long way since its origins in the 1980s. Although the technologies of shotgun proteomics have taken backstage to the science they have enabled, automating the identification process makes these techniques more accessible to biologists and clinicians. Shotgun proteomics can now be routinely employed for studying complex biological systems and the proteomic underpinnings of pathological conditions.

ACKNOWLEDGMENTS

The authors acknowledge funding from NIH Grants R01 CA126218 and U24 CA126479.

REFERENCES

1. Wysocki, V. H., Tsaprailis, G., Smith, L. L., Breci, L. A. (2000). Mobile and localized protons: A framework for understanding peptide dissociation. *J Mass Spectrom, 35*, 1399–1406.
2. Paizs, B., Suhai, S. (2005). Fragmentation pathways of protonated peptides. *Mass Spectrom Rev 24*, 508–548.
3. Altschul, S. F., Gish, W., Miller, W., Myers, E. W., Lipman, D. J. (1990). Basic local alignment search tool. *J Mol Biol 215*, 403–410.
4. Tabb, D. L., Fernando, C. G., Chambers, M. C. (2007). MyriMatch: Highly accurate tandem mass spectral peptide identification by multivariate hypergeometric analysis. *J Proteome Res 6*, 654–61.
5. Zhang, B., Chambers, M. C., Tabb, D. L. (2007). Proteomic parsimony through bipartite graph analysis improves accuracy and transparency. *J Proteome Res 6*, 3549–3557.
6. Ma, Z., Dasari, S., Chambers, M. C., Litton, M. D., et al. (2009). IDPicker 2.0: Improved protein assembly with high discrimination peptide identification filtering. *J Proteome Res 8*, 3872–3881.

7. Kessner, D., Chambers, M., Burke, R., Agus, D., Mallick, P. (2008). ProteoWizard: Open source software for rapid proteomics tools development. *Bioinformatics 24*, 2534–2536.
8. ProteoWizard Technical Documentation: Installation and Data Formats.
9. Cox, J., Mann, M. (2008). MaxQuant enables high peptide identification rates, individualized p.p.b.-range mass accuracies and proteome-wide protein quantification. *Nat Biotechnol 26*, 1367–1372.
10. http://sashimi.sourceforge.net/schema_revision/pepXML/pepXML_v114.xsd, pepXML data format.
11. Tabb, D. L., Friedman, D. B., Ham, A. L. (2006). Verification of automated peptide identifications from proteomic tandem mass spectra. *Nat Protoc 1*: 2213–2222.
12. Elias, J. E., Gygi, S. P., (2010). Target-decoy search strategy for mass spectrometry-based proteomics. *Meth Mol Biol 604*, 55–71.
13. Keller, A., Nesvizhskii, A. I., Kolker, E., Aebersold, R. (2002). Empirical statistical model to estimate the accuracy of peptide identifications made by MS/MS and database search. *Anal Chem 74*: 5383–5392.
14. Li, M., Gray, W., Zhang, H., Chung, C. H., Billheimer, D., Yarbrough, W. G., Liebler, D. C., Shyr, Y., Slebos, R. J. C. (2010). Comparative shotgun proteomics using spectral count data and quasi-likelihood modeling. *J Proteome Res 9*, 4295–4305.
15. Duncan, D. T., Craig, R., Link, A. J. (2005). Parallel tandem: A program for parallel processing of tandem mass spectra using PVM or MPI and X!Tandem. *J Proteome Res 4*, 1842–1847.
16. Sadygov, R. G., Eng, J., Durr, E., Saraf, A., McDonald, H., MacCoss, M. J., Yates III, J. R. (2002). Code developments to improve the efficiency of automated MS/MS spectra interpretation. *J Proteome Res 1*, 211–215.
17. Slebos, R. J. C., Brock, J. W. C., Winters, N. F., Stuart, S. R., Martinez, M. A., Li, M., Chambers, M. C., Zimmerman, L. J., Ham, A. J., Tabb, D. L., Liebler, D. C. (2008). Evaluation of strong cation exchange versus isoelectric focusing of peptides for multidimensional liquid chromatography-tandem mass spectrometry. *J Proteome Res 7*, 5286–5294.
18. McConnell, R. E., Higginbotham, J. N., Shifrin, D. A., Tabb, D. L., Coffey, R. J., Tyska, M. J. (2009). The enterocyte microvillus is a vesicle-generating organelle. *J Cell Biol 185*, 1285–1298.
19. Arnett, D. R., Jennings, J. L., Tabb, D. L., Link, A. J., Weil, P. A. (2008). A proteomics analysis of yeast Mot1p protein–protein associations: Insights into mechanism. *Mol Cell Proteomics 7*, 2090–2106.
20. Paulovich, A. G., Billheimer, D., Ham, A. L., Vega-Montoto, L. J., Rudnick, P. A., Tabb, D. L., Wang, P., Blackman, R. K., Bunk, D. M., Cardasis, H. L., Clauser, K. R., Kinsinger, C. R., Schilling, B., Tegeler, T. J., Variyath, A. M., Wang, M., Whiteaker, J. R., Zimmerman, L. J., Fenyo, D., Carr, S. A., Fisher, S. J., Gibson, B. W., Mesri, M., Neubert, T. A., Regnier, F. E., Rodriguez, H., Spiegelman, C., Stein, S. E., Tempst, P., Liebler, D. C. (2009). A CPTAC inter-laboratory study characterizing a yeast performance standard for benchmarking LC-MS platform performance. *Mol Cell Proteomics 9*(2): 242–254.
21. Loecken, E. M., Dasari, S., Hill, S., Tabb, D. L., Guengerich, F. P., The bis-electrophile diepoxybutane cross-links DNA to human histones but does not result in enhanced mutagenesis in recombinant systems. (2009). *Chem Res Toxicol 22*, 1069–1076.

PART II

APPLICATIONS

CHAPTER 10

Mass Spectrometry-Based Screening and Characterization of Protein–Ligand Complexes in Drug Discovery

CHRISTINE L. ANDREWS, MICHAEL R. ZIEBELL, ELLIOTT NICKBARG, and XIANSHU YANG

10.1 INTRODUCTION

Despite its long, well-established history, drug discovery today is evolving at a rapid rate as it strives to deal with the growing number of diseases and disease targets. More than 100 years ago, Emil Fisher wrote "the enzyme and glucoside must fit together like a key and a lock in order to initiate a chemical action upon each other [1]." In1913, Paul Ehrlich stated that "*corpora non agunt nisi fixata*" (a drug will not work unless it is bound) [2]. The lock–key and receptor–ligand theories hold true today and provide the prevalent model for understanding the molecular basis of drug action. These theories are the foundation for a broad range of techniques that interrogate the binding of small molecules to target proteins.

A key challenge for modern drug discovery is the need to survey many receptor–ligand interactions for potential disease interventions. This is made all the more challenging by the exponential expansion of possible disease-related protein targets, on the one hand, and the synthesis of millions of new chemical entities, on the other. The growth in number of disease targets has arisen out of a deeper understanding of human physiology, largely driven by public and private research efforts over the past 60 years. Target validation utilizes molecular biology to validate proteins whose function alters or fails as possible causes of a disease and, thus, creates specific targets for drug therapy. These targets are further isolated, expressed in other hosts, and purified as single proteins for *in vitro* studies. The three-dimensional structures of

Protein and Peptide Mass Spectrometry in Drug Discovery, Edited by Michael L. Gross, Guodong Chen, and Birendra N. Pramanik.

proteins are determined using X-ray crystallography, NMR spectroscopy, electron microscopy, and other methodologies.

The number of characterized protein sequences continues to grow exponentially every year. As of June 2010, there were 65,968 characterized protein structures in the protein database [3]. As the number of protein targets has increased, so too has the number of possible lead compounds. In recent years combinatorial and automated parallel syntheses have generated literally millions of new compounds that have the potential to be new drugs. The rapid and reliable identification of potent and high-affinity ligands from large compound collections with the large number of potential targets is a key challenge for drug discovery.

In the past two decades high-throughput screening (HTS) has revolved around the combined world of multiple-well microplates and robotic processing. Many pharmaceutical companies have set up higher density and lower volume formats (e.g., 384- and 1536-well microplates) Currently most *in vitro* functional activity-based HTS methods discover active hits by detecting a change in the activity of a target in the presence of potential inhibitors or activators. These changes in activity are commonly detected using substrates that have been labeled with either radioisotopes or fluorescent probes. Detection methods include fluorescence intensity, fluorescence polarization, homogeneous time-resolved fluorescence (HTRF), and Förster resonance energy transfer (FRET). The need for multiple reagents, however, limits the flexibility of a label-based assay and speed, particularly in HTS formats.

Label-free HTS technologies offer a number of unique advantages over label-dependent assays. In particular, they provide direct monitoring of analyte binding without modifying either the target or drug. Some label-free technologies include impedance, optical biosensor-based technologies (surface plasmon resonance and waveguides), and liquid chromatography-mass spectrometry, (LC-MS)-based techniques [4].

The mass spectrometer detects and quantifies analytes based on mass-to-charge ratio and in so doing attaches a signature to each molecule present in a mixture. Its speed, sensitivity and specificity have made MS a universal technique. Coupled with chromatography, MS can routinely detect samples of less than 1 pmol, automatically analyze hundreds of samples per hour, and provide accurate mass measurements (thus increasing the specificity and certainty in identifications). The variety of MS ionization techniques available offers enough flexibility that most organic molecules are amenable to some form of MS analysis. LC-MS already plays a key role in drug discovery, encompassing monitoring of compound synthesis, purification of lead products, ADME profiling [5], and proteomics studies [6]. LC-MS is now recognized as a versatile, sensitive and automatic label-free technique for binding and functional HTS in drug discovery. This chapter focuses on the unique roles that chromatography and mass spectrometry play in high throughput screening and looks to the future of HTS methods based on chromatography and MS.

10.2 AFFINITY SELECTION MASS SPECTROMETRY (AS-MS)

Functional-based HTS requires a unique set of characterized targets, substrates, and readouts for each screen. In contrast, affinity-based HTS involving MS detection can be set up as a general screening platform because its readout is universal—the molecular weights of small molecule binders or protein–ligand complexes. Ligands can bind at multiple distinct sites on a target and can bind via multiple modes. Therefore affinity-based screening can identify a set of ligands exhibiting multiple activities on the same target, such as agonists and antagonists. Unlike functional-based HTS, affinity-based HTS does not require knowledge of the structure or function of the target or the development of a target-specific assay [7,8]. This simplifies assay development for affinity selection—mass spectrometry (AS-MS) HTS.

The AS-MS screening method begins with small-molecule incubation with a purified disease target under physiological binding conditions to allow the formation of noncovalent protein–ligand complexes. The protein–ligand complex can then be directly detected by infusion-MS. Alternatively, a variety of front-end affinity selections techniques in conjunction with MS can be used to identify the ligand alone (indirect detection methods).

10.2.1 Direct Detection of Noncovalent Protein–Ligand Complexes

AS-MS techniques can be broadly divided into direct and indirect methods. Direct detection of noncovalent complexes requires that protein and ligand be prepared in a volatile, MS-compatible buffer such as 10 mM ammonium acetate. Noncovalent protein–ligand complexes are detected by MS without prior separation of bound and unbound ligands, and the mass of bound, small molecule adds to the mass of the protein, while nonbinders are detected at much lower mass-to-charge ratios. This method can provide information on the stability of protein–ligand complex and its stoichiometry. It can also be used to study DNA–ligand complexes, RNA–ligand complexes, and protein–protein interactions [7,9–11].

Although direct detection techniques can be informative, there exist at least three limitations for their use in screening. First is the need for MS-compatible binding buffer solutions that almost certainly differ from physiologically relevant ones. Proteins often require nonvolatile salts, cofactors, additives, and neutral pH to maintain their native conformation and activity. Native proteins and protein–ligand complexes have large masses and have few charges, so they are only detected at high m/z values. Second is the need for high mass accuracy and mass resolving power for direct detection to be effective. To distinguish between a protein alone and a protein–ligand complex, a mass spectrometer with high mass accuracy and resolving power is required; examples are quadrupole time-of-flight (Q-TOF) mass spectrometers, Fourier transform ion cyclotron resonance (FT-ICR), and orbitrap mass spectrometers [12]. Data analysis is another challenge, and spectrum deconvolution is required to extract ligand information from multiple charge states of protein–ligand complex. This represents the final challenge, especially in a HTS setting. Owing to these concerns, the use of direct detection of noncovalent protein–ligand complexes has not yet proved successful as a large-scale screening platform.

10.2.2 Indirect Detection of Noncovalent Protein–Ligand Complexes

The promise of small-molecule ligand-directed drug discovery has catalyzed significant activity within the biotechnological and pharmaceutical industries and led to the development of high throughput affinity screening technologies and complementary small-molecule libraries. The most powerful indirect detection discovery platforms capitalize on recent advances in small molecule pooling into large collections, analytical instrumentation, and data-analysis technologies. Criteria that will ensure a high probability of finding chemical ligands for any potential binding site on a protein include [13]: (1) an inclusive set of highly diverse compounds that are encoded according to mass into mixtures; (2) an efficient, automated and robust system to handle sample preparation, separation of bound ligands from unbound small molecules, and detection of bound ligands; and (3) sophisticated bioinformatics and cheminformatics, including a fully automatable program to accept MS raw data and to compare the observed *m/z* values with known molecule masses from the screened library.The bioinformatics also serves to filter and triage the primary hit lists.

Unlike direct detection methods, indirect detection measures only previously bound ligands or, alternatively, unbound ligands. This avoids the problem of keeping the protein–ligand complex intact through the MS ionization process. In most cases a purified and soluble protein is incubated with a mixture library of small organic compounds to form protein–ligand complexes prior to analysis. The complexes are then isolated from unbound ligands by using ultrafiltration or size exclusion chromatography, they are then dissociated under various denaturing conditions (e.g., acid with organic solvents and/or raising the HPLC column temperature), and the released ligands are subsequently detected by MS. In a different manifestation, Frontal affinity chromatography/ESI-MS (FAC-MS) [14] immobilizes a target protein on a chromatographic support, and the libraries are continuously infused into the columns (described below).

Several pharmaceutical and biotechnological companies have developed high-throughput affinity screening technologies, including ALIS at Schering-Plough (now Merck), SpeedScreen at Novartis Pharmaceutical Corporation, uFAS-AS at Abbott Labs, GPC-spin-columns coupled with LC-MS at Wyeth Pharmaceuticals (now Pfizer) and Amgen, and FAC-MS at Transition Therapeutics. These systems will be described in the following sections.

10.3 SOLUTION-BASED AS-MS AS SCREENING TECHNOLOGIES

Solution-based AS-MS methods do not require immobilization of the target proteins or ligands on a solid support. These platforms avoid possible alterations of protein properties by tagging or chemical linkage and minimize the risk of inactivation of the protein. They allow investigation of greater compound diversity because immobilizing functional groups is not necessary. AS-MS can be utilized as a generic platform to screen many targets with very large libraries on the basis of affinity. The protein target and library are homogeneous (in solution), and any cofactors, metal ion, buffer reagents, or detergents necessary for proper protein folding and stability can also be included in the solution. Library components can act through all binding sites. Targets can also be screened in a ligand-bound state with an inhibitor cofactor to discover new hits that bind to allosteric sites.

Small-molecule libraries are either mixtures produced by parallel synthetic approaches such as combinatorial chemistry or mixtures produced by pooling individual discrete compounds, or natural product extracts [15]. In the first step of AS-MS HTS, the library and target protein are incubated with the appropriate binding buffer at room temperature for 30 to 60 min to form any protein–ligand complexes. Following incubation, a variety of techniques can be applied to separate protein–ligand complexes from unbound small molecules.

10.3.1 Automated Ligand Identification System (ALIS)

In the Automated Ligand Identification System (ALIS) [16,17], size exclusion chromatography (SEC) is integrated online with an LC-MS system to separate complexes from unbound constituents and detect binding compounds. At Schering-Plough (now Merck), ALIS is routinely applied to four phases of lead discovery: (1) biochemical validation of the target before HTS screening; (2) HTS screening of diverse small molecule libraries; (3) characterization of newly discovered hits, their binding affinity and modes; and (4) affinity ranking and dissociation rate determinations to support lead optimization [17]. The ALIS process (Figure 10.1) includes several stages: (1) affinity

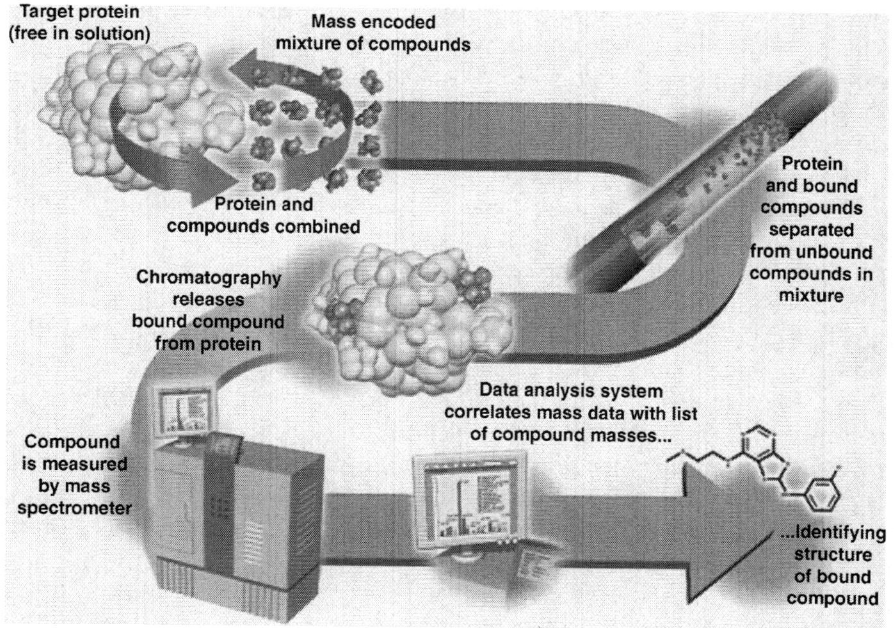

FIGURE 10.1 Schematic of ALIS, the automated ligand identification system that utilizes size-exclusion chromatography coupled online to LC-MS for the study of protein–ligand interactions. Reprinted with permission from [27]. (*See the color version of this figure in Color Plates section.*)

selection, (2) separation, (3) reversed-phase chromatography, and (4) ligand detection. In general, during the affinity selection stage, 1 μL of 1 μM to 10 μM target protein is mixed with 1 μL of a mass-encoded library (approximately containing 1500–2500 compounds, each at 1 μM) in a suitable, physiologically relevant buffer. These samples are placed in 96-well plates and incubated for 30 min at room temperature so to allow protein–ligand complexes to form. The plates are then chilled to 4°C and loaded into an ALIS system.

During the second stage, a re-usable size-exclusion column is used to separate rapidly the protein–ligand complexes from the unbound compounds; this requires less than 20 sec. Protein–ligand complexes in the SEC eluent are monitored directly by inline UV absorbance at 230 nm, and an inline valve system is used to direct automatically the protein–ligand complex to a C-18 reversed-phase column. Here the third stage occurs and the protein–ligand complexes are dissociated by raising the column temperature to 60°C and making the solution acidic with the mobile phase used for RP-HPLC. The fourth and final stage of ALIS occurs when the dissociated ligands are eluted into an ESI-MS system (e.g., Waters LCT time-of-flight mass spectrometer) and the *m/z* values of their molecular ions detected. The raw data are analyzed, and the binding ligands are identified according to their molecule weights by a customized software program [18].

The ALIS system is fully integrated through specialized valving schemes. This arrangement reduces the loss of protein and compounds owing to offline liquid transfer steps. Online UV detectors, which monitor the separation of protein–ligand complex with the rest of unbound compounds, isolate the protein–ligand complexes and eliminate false positive hits from unbound small molecules. Another advantage of ALIS is that the SEC and RPC columns are fully regenerated and ready for the next sample during one cycle time.

One of the important considerations for the success of ALIS is the mixture-based combinatorial chemistry approach termed "NeoMorph." This "core plus building block" approach enables a template structure to be adorned with a variety of chemical moieties that ensure good coverage of structure diversity [19]. For example, one core with three connection sites is coupled to a set of 15 diverse amine building blocks to furnish a mixture combinatorial library of 3375 compounds. During library design, software algorithms minimize the amount of mass redundancy present at both the library stage and library pooling stage. Each library member is self-encoded by its molecular weight [20,21]. With this method one can generate and screen the multimillion compound libraries by using the ALIS affinity screen against several targets spanning multiple therapeutic areas. Adam and coworkers [15] used a similar system to screen polyadenosine polymerase with natural product fungal extracts containing known parnafungins.

A single ALIS experiment that deals with over 2500 compounds is completed in less than 10 min, and one ALIS system can screen up to 375,000 compounds per day. This occurs with minimal protein and compound consumption. The approach is productive, having been used to screen a variety of target proteins [16,17,22–26], including integral membrane proteins (e.g., M2R, the M2 acetylcholine receptor) [23].

Ligand Binding Affinity Measurement by ALIS Another elegant feature of the ALIS platform is that it not only can be used for HTS, but also can be used to characterize protein–ligand interactions. Once ALIS hits are confirmed, their binding affinity can be determined by using the ALIS titration method. In general, a fixed concentration (1–10 μM) of a target protein and serially increasing concentrations of a test ligand are incubated and then analyzed by ALIS. The MS data for the bound ligand are then plotted versus the total ligand concentration to yield a saturation binding curve and generate a K_d value by nonlinear regression analysis [27].

ALIS Competition-Based Binding Mode Determination One advantage of AS-MS is the ability to discover several possible classes of ligands (e.g., orthosteric agonists, reverse agonists, antagonists, and allosteric modulators) in single experiment. In ALIS ligand–ligand competition experiments [24], the target protein is first equilibrated with a constant concentration of test ligand and then titrated with serially increasing concentrations of a known competitor ligand (titrant). The mass response signals of a test ligand and competitor are analyzed, and the ability of the titrant to displace the target-bound ligands reveals the binding site classification (competitive vs. noncompetitive). The ratios of the titrant-to-ligand mass response will generate a straight line plot with increasing titrant concentration if these two compounds bind the same site, whereas allosteric (noncompetitive) ligands will yield a hyperbolically curved ratio plot because saturating concentration of the titrant will not completely displace the allosteric ligand.

ALIS Affinity Ranking in Mixtures ALIS direct competition can be expanded from one test ligand to a mixture of compounds, enabling affinity ranking of a compound mixture. In this approach [25] the target protein is first equilibrated with a mixture of compounds and then titrated with serially increasing concentration of a known competitor ligand. The mass response of each ligand is normalized for the entire titration curve by dividing each response of each ligand by its highest value. The titration data are then fit to a variable slope sigmoidal dose–response curve using GraphPad Prism. The titrant concentration that displaces 50% of another ligand is defined as the affinity competition experiment 50% (ACE_{50}) value of the ligand. A higher ACE_{50} value means that a higher concentration of titrant is required to compete off the specific ligand (i.e., the specific ligand has a higher affinity than the other members of the mixture, as shown in Figure 10.2). In an ACE_{50} study with CDK2, the ALIS affinity ranking values correlate well with both ALIS K_d values and corresponding biochemical data [25]. Other examples illustrating this approach involve Akt-1 [24], lipid phosphatase SHIP2 [28], and M2 acetylcholine receptor [23]. ALIS affinity ranking is an excellent tool for mixture-based hit optimization and identifies quickly compounds with improved affinity relative to their progenitor.

A similar approach was taken in the affinity ranking of dipeptidyl peptidase IV DPP-4 [29]. The ligand mixtures were pre-incubated with 10 μM DPP-4 at concentrations ranging from 0.5 to 32 μM per library member and analyzed with the AS-MS platform. Mass spectrometric responses were normalized with the highest MS response for each individual mixture component. The normalized MS response versus

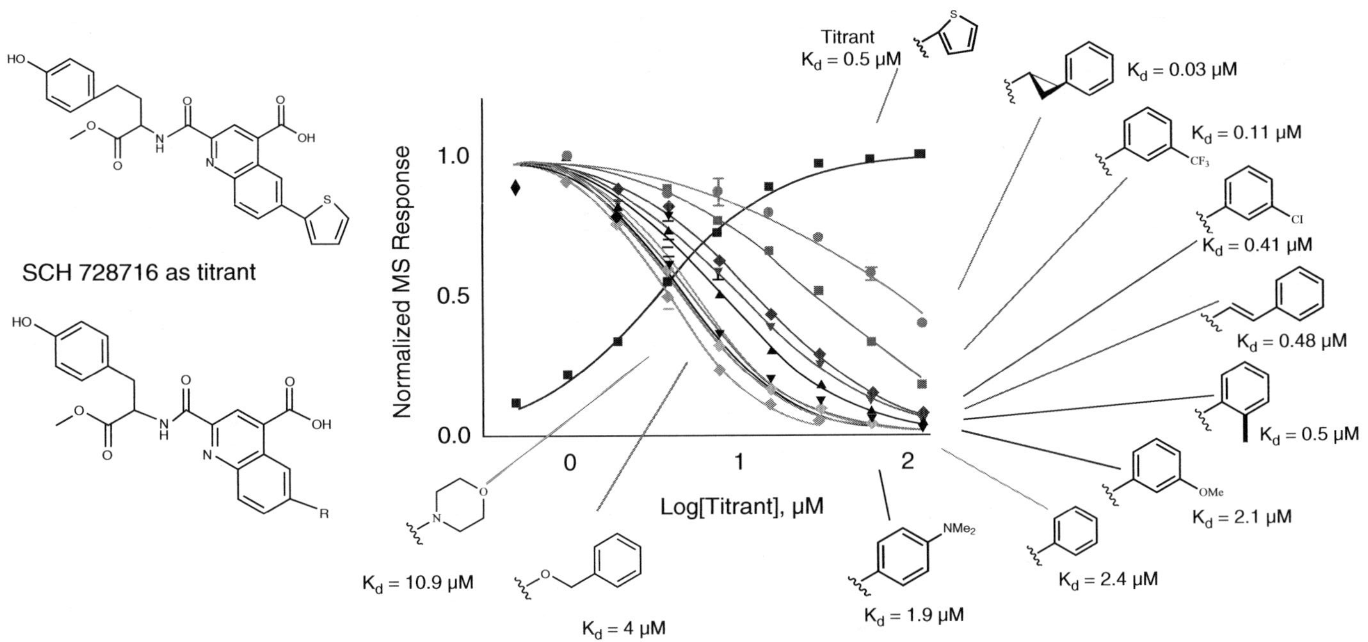

FIGURE 10.2 Identification of optimized CDK2 ligands using the calibrated ACE50 method. Reprinted with permission from [25].

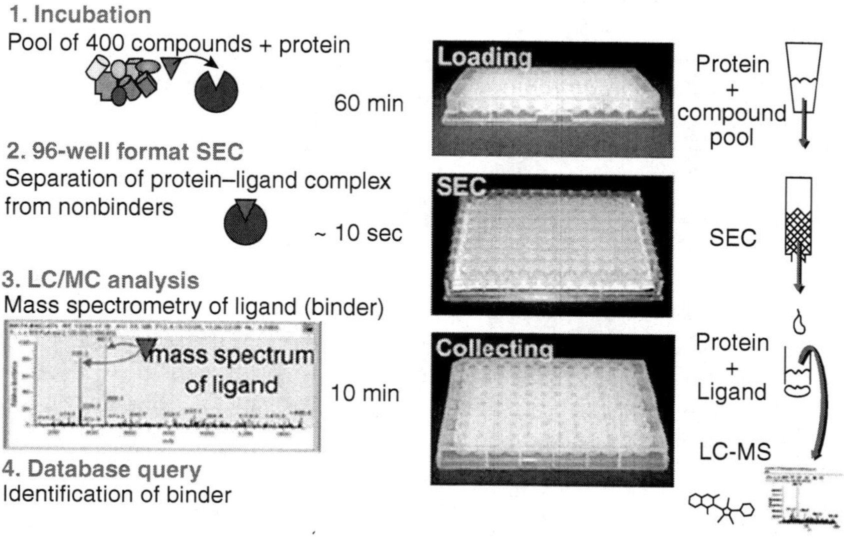

FIGURE 10.3 The basic principles of SpeedScreen technology. The left panel describes the four process steps of incubation, 96w-SEC, LC/MS-analysis, and database query. The right panel depicts the material used for these process steps. Reprinted with permission from [31]. (*See the color version of this figure in Color Plates section.*)

the compound concentration was plotted, and the ligands were ranked from highest to lowest affinity, which correlated well with their IC_{50} values.

10.3.2 SpeedScreen

Another successful HTS approach, termed SpeedScreen [30,31], was developed at Novartis Pharmaceuticals and is based on the separation of protein–ligand complexes from unbound small molecules in 96-well plates filled with pre-swollen Sephadex G-25 beads. This method provides fast, parallel separations of protein–ligand complexes from unbound compounds and can accommodate discrete libraries containing more than 600,000 compounds for the primary SpeedScreen. As shown in Figure 10.3, each mixture library, containing 400 compounds, is aliquoted into a 0.5-μL volume and incubated with 24.5 μL of 10 μM of target protein in the 96-well screening plate. The final concentration of individual compound is 7 μM. The separation of protein–ligand complexes from unbound compounds is executed by fast SEC via centrifugation of the SpeedScreen "sandwich," which consists of the incubation, separation, and collection plates. The collected protein–ligand complexes are dissociated during the HPLC separation with a solvent gradient from 5% to 95% acetonitrile in a 10-min experiment and are eluted to ESI-ion trap or other suitable mass spectrometer. The bound ligands are identified by comparison of their *m/z* values with those of the specific compound list of particular library. For a primary screening campaign of a 500,000-compound library with SpeedScreen, approximately 9 days and 12.5 mg of a 25 kDa protein are needed. According to recent literature, SpeedScreen has screened more than 26 targets. From the analysis of the molecular properties of hits obtained from the screened 26 targets [32], SpeedScreen is a demonstrated robust screening technology that does not accumulate frequent hitters, promiscuous binders, or potential covalent binders.

Paul Vouros at Northeastern University in the US is one of the first researchers to characterize combinatorial libraries with LC-MS [33], and also to screen combinatorial libraries against biological targets. The work utilizes G25 SEC spin columns to separate protein–ligand complexes from unbound compounds and to analyze the protein–ligand complexes by either LC-MS or capillary electrophoresis (CE)-MS [34]. In 2005 Flarakos et al. designed a custom-made SEC cartridge packed with Sephadex G-25 media that could be integrated with a UV detector and LC-MS to form an online SEC-UV-HPLC-MS system [35]. In this system the separation of protein–ligand complex can be monitored by the UV detector, and one custom-made cartridge can be used for more than 25 injections.

10.3.3 Ultracentrification Coupled to Mass Spectrometry

Pulsed Ultrafiltration-MS (PUF-MS) was first described by Van Breeman and associates [36] and differs from the methods previously described in that it relies on ultrafiltration, rather than size exclusion chromatography, to separate a bound complex from bulk unbound constituents. Although a chromatography step is often used subsequent to the ultrafiltration isolation of the complex, infusion of the intact complex and direct detection of the ligand noncovalently bound to the protein is possible. This method has been utilized to investigate proteins such as RXRα [37], as well as several earlier examples reviewed elsewhere [36]. Abbott Laboratories successfully expanded on ultrafiltration as a screening methodology in drug discovery with their approach called affinity selection/mass spectrometry, or ASMS [38]. ASMS involves several selection steps that both wash away unbound ligands and concentrate the bound ligand–protein complex. It has been successfully applied to several biological systems, including MurF [38], Chk1 [39], and Bcl-xL [40].

10.3.4 Gel Filtration–MS Platform

Wyeth was a pioneering force in testing the use of affinity selection as a HTS approach in the late 1990s. In 1997 size-exclusion separation by GPC spin columns and ultrafiltration microconcentrators was coupled with a mass spectrometer to study human cytomegalovirus protease inhibitors [41]. A few years later it was modified with a gel filtration-MS platform, including a custom-built robot for sample automation [42]. The robot automated the entire screening process, including the preparation of compound and protein solutions, the combination and incubation of the protein-compound mixture, the preparation and loading of 96-well plate GPC spin columns, the centrifugation of the 96-well plates, and the collection of the eluents. The eluents then were analyzed by HPLC-MS.

Approximately 32,000 compounds were selected as a subset of the Wyeth chemical collection and pooled into sample libraries, with each library containing

10 compounds. Approximately 0.1 mM of target protein solution was incubated with each compound mixture; the resulting compound concentration of 1 mM. The protein–ligand complexes were separated from unbound compounds with the robotic system, and approximately 25 μL of eluent was analyzed by an HPLC-MS system. The MS raw data were interpreted with custom-built software [43]. The literature describes two targets, MMP-1 and RGS4 that were screened with this platform.

10.3.5 Frontal Affinity Chromatography–Mass Spectrometry (FAC-MS)

Frontal chromatography was developed in 1975 as an extension of frontal analysis, which characterizes the interaction of an analyte with a column sorbent [44]. Frontal affinity chromatography–mass spectrometry (FAC-MS) was established in 1998 [14]. Unlike the methods described thus far, FAC-MS immobilizes the protein in a column and utilizes the elution volume of a compound as a measure of its binding affinity. This technique can be broadly applied to several challenges in the drug discovery process and can be utilized for a wide range of biological targets [45]. Through the use of a void volume marker (i.e., a nonbinding compound) and MS detection, several compounds can be evaluated simultaneously, allowing for mixture-based analysis and affinity ranking. Figure 10.4 shows the basic premise of the FAC-MS process and its use in mixture based analyses [46].

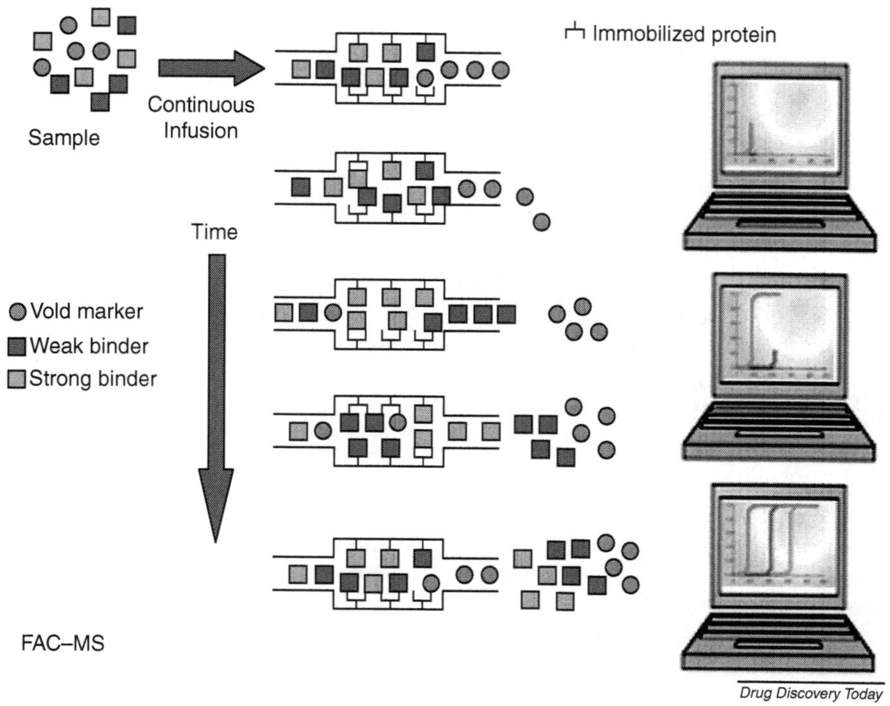

FIGURE 10.4 Frontal affinity chromatography–mass spectrometry (FAC-MS). Reprinted with permission from [46].

FAC-MS was successfully applied to several biological systems, including the epidermal growth factor receptor and the EphB2 tyrosine kinase receptor [47]; further, it was used for global kinase screening [45]. Several analytical advances (i.e., microaffinity columns [48] and protein-doped monolithic columns [49]) can be incorporated in FAC-MS to expand its capability.

10.3.6 Indirect Detection AS-MS

In contrast to the previously described methods that detect binding compounds, some methods measure unbound compounds after separation from the protein–ligand complex [50]. The unbound compounds are eluted with organic solvent and analyzed by LC-MS. If an eluted compound is detected, it is designated as a weak or nonbinder, whereas the absence of compound signal indicates a strong binder; it cannot elute from the SEC column because it bound to the protein. This method requires the efficient recovery of unbound compounds from SEC spin columns and may be challenging to accomplish in a high-throughput manner. Researchers at DuPont Pharmaceuticals implemented an online SEC-LC-ESI-MS system [51], where a Phenomenex BioSep SEC column is used to separate protein–ligand complexes from unbound compounds. This platform was utilized to screen Matrix Metalloprotease (MMP3) protein with a small combinatorial mixture (36 members) and also was able to provide K_D values and off-rate measurements.

10.3.7 Emerging Technology

Affinity-selection mass spectrometry provides many advantages over traditional HTS and is easily developed to ultra HTS. Even so, researchers are still looking for higher resolution and faster separation of protein–ligand complexes from unbound constituents. Two new technologies are discussed below.

Two-Dimensional Turbulent Flow Chromatography—LC-MS Platform HPLC systems can be operated at low flow rates with small-size media to enhance separation. If the flow rate is increased to a very high level, the flow will start to exhibit turbulent-flow properties. The higher the flow rate, the more turbulent the flow will become. Turbulent-flow chromatography refers to separation systems of microbore columns containing large packing materials (30–60 μm in diameter) and operating at higher flow rates to exhibit turbulent flow. The high-molecular-weight analytes and salts are rapidly washed off, while the low-molecular-weight analytes are retained on column. This approach can also be used as a high-throughput pharmacokinetic screening method [52,53]. One example of this approach for an affinity-selection determination involves rapid separation of protein–ligand complexes from unbound small molecules [54]. The system can screen AChE and BChE with 27-member steroidal alkaloid libraries (20 μM each). All active compounds were confirmed in the assay and the 2D-TFC-LC-MS provided a fast separation of under 15 s.

Proton Affinity Column (PAC)—SPE-LC-MS Platform Recently the Bio-Molecular Analysis Group of the Free University of Amsterdam (VU) developed a new platform that couples a protein affinity column (PAC) with SPE-LC-MS to form a PAC-SPE-LC-MS system [55]. The PAC contained iminodiacetate silica particles (50–80 μM) loaded with Ni^{2+} (Ni-IDA) ions that retain His-tagged proteins by an electrostatic bonding. The coordinate bonding allowed the PAC to separate His-tagged protein–ligand complexes in a process that involved several steps. In step 1, sample preparation, the His-tagged proteins are incubated with a mixture of compounds for 30 min at room temperature and then transferred to the autosampler and stored at 4°C prior to (step 2) the separation stage, where His-tagged proteins and His-tagged protein–ligand complexes are trapped by bonding with the Ni^{2+} ions, allowing unbound compounds to be washed off the column. Step (3) involves the dissociation of protein–ligand complexes, wherein the protein–ligand complexes are eluted, dissociated with a 10-mM glycine-HCl solution at pH 2.0, and transferred online to a C-18 solid-phase extraction cartridge. Glycine-HCl and salts are washed from the C-18 cartridge, and (step 4) the separation and detection of previously bound ligands via an online LC-MS system.

A modification of this approach was implemented whereby cobalt-loaded Dynabeads (TALON), instead of a protein affinity column, were added to pre-incubated protein–ligand mixtures [56]. A 1.4 T permanent neodymium magnet is used to trap the beads to which the His-tagged protein–ligand complexes are attached. The separation of His-tagged protein–ligand complexes from unbound small molecules and the transfer of the complexes are achieved by changing the magnetic field. Two sets of SPE-LC systems were configured to increase throughput. Both systems may be able to detect very weak binders (100 μM to 1 mM affinity) and were used to screen a variety of His-tagged target proteins.

10.4 GAS-PHASE INTERACTIONS

In addition to high-throughput screening methods based on solution affinities, several methods exist to screen intact complexes by MS in the gas phase. Not long after the development of ESI-MS, its utility for direct analysis of relatively fragile biomolecular complexes in the gas phase was demonstrated. The original work described complexes of enzymes with small-molecule substrates, products, and inhibitors and determined that intact biologically relevant noncovalent complexes can be detected in the gas phase [57,58]. Subsequently detection of noncovalent gas phase biomolecular complexes by using ESI-MS was further expanded into a variety of categories including large protein complexes [59], oligonucleotides with proteins, metals or small molecules [60,61], and the utility of mass spectrometry for structural analysis of protein complexes was described in numerous reviews [59,62,63]. On the basis of these experiences, it is now clear that even in the gas phase, the protein conformation can often be preserved sufficiently to generate results that are relevant to solution-based efforts.

Direct ESI-MS analysis of the protein is relatively sensitive compared to detection by NMR and microcalorimetry. Furthermore the use of MS does not require specific modification or labeling of the target protein or of the ligands, as do many fluorescent-based techniques. The mass spectrometer offers the additional dimension of molecular characterization—it directly records the molecular mass or masses of the interacting species with much higher accuracy and mass resolving power than traditional approaches for complex analysis (e.g., size-exclusion chromatography (SEC), and polyacrylamide gel electrophoresis [64]) and analyzes samples with higher speed and greater sensitivity than NMR spectroscopy or analytical ultracentrifugation.

As discussed previously, limitations with detection of protein–ligand complexes in the gas phase are the need to maintain proper protein folding and activity under MS conditions, which are not ideal for the protein. Proteins typically require nonvolatile salts, additives, detergents, or cofactors, and usually need neutral pH to maintain their native state. These needs limit the list of targets that can be analyzed to those that can maintain solubility and proper folding under conditions that promote effective ionization (e.g., relatively low ionic strength, buffers that decompose into volatile components). MS-based assays that depend on these nonsolution conditions for protein activity must be validated to show that the protein can maintain proper folding and that artifacts caused by the ionization process can be excluded. The ESI process itself can induce conformational changes that lead to protein unfolding and structural rearrangements [65], and that may alter the stability of protein–ligand complexes [66]. The transfer of a protein from solution to the gas phase involves desolvation that weakens hydrophobic bonding and strengthens electrostatic effects. The different types of ESI-based techniques (e.g., standard ESI, chip-based NanoESI, and electrosonic spray ionization) and their effects on noncovalent complex formation were compared by Jecklin and coworkers [67], who found that the electrosonic technique is relatively insensitive to experimental parameters and provided affinity binding results that were in good agreement with solution-based values (these various ionization methods are described in Chapter 1 by Cotte-Rodriguez, Zhang, Miao and Chen in this volume) In conclusion, to paraphrase McLafferty and Brueker [68], the real question is not whether the tertiary structure is preserved but for how much, for how long, and under what conditions?

With these considerations in mind, we consider some of the successes of screening for complexes by using gas-phase protein–ligand complexes generated by electrospray and other sprays; some of these successes were previously reviewed [9]. ESI combined with FT-ICR mass spectrometry, for example, can be used to screen noncovalent complexes of the HCK Src Homology 2 (SH2) domains against combinatorial libraries consisting of small 5-mer peptides [69], and also can be applied to screening natural product extracts [70,71]. Zenobi and coworkers [72] used nanospray-MS methods to compare the affinities of some known kinase ligands to their protein targets, and the results are in good agreement with literature values. Efforts to stabilize noncovalent complexes during ionization can be effective when using alternative spray approaches. An example is an automated microchip device that was used to demonstrate effective binding of thyroxine and other small ligands to transthyretin. The reproducibility of this device suggests that it can be part of an effective approach to high-throughput screening [73]. A chip-based nanospray system can also used to screen the human estrogen receptor protein to identify compounds that may interfere with activity. This highlights the potential utility of chip-based system especially considering the low sample consumption, high speed, and automation of the method [74]. Electrosonic spray coupled with precursor ion scanning tandem MS may also be a useful strategy to improve specific detection of protein–ligand complexes and to differentiate those from nonspecific complexes formed by the electrospray process itself [75].

The second dominant "soft" ionization technique for biomolecules is MALDI (matrix-assisted laser desorption ionization). Although it offers clear advantages in terms of throughput (<10 s/sample of analysis time), MALDI has not been as widely utilized as ESI for the analysis of intact noncovalent complexes. Its rather sparse use may be due to the requirement that much of UV-MALDI is done with acidic matrices, which are protein denaturing. This limitation has been overcome by special sample preparation techniques, such as neutral pH matrices [76], or by relying on the first laser pulses collected on fresh samples [77]. MALDI mass spectrometry can be used, in special cases, to detect noncovalent complexes in the gas phase; examples are protein–antibody complexes [78], and peptide–peptide complexes [79].

Intensity fading (IF) MALDI is a means of examining mixtures of a target protein plus potential ligands by comparing the molecular ion signals of ligands that bind to a target protein to those from a mixture without target; it can be applied to soluble as well as immobilized protein targets. The abundances of binders decrease relative to those of nonbinders (hence the term "intensity fading") [80,81]. Infrared laser (IR) MALDI techniques, using Nd-YAG lasers and relying on water as a matrix, can also be used to analyze noncovalent complexes; one example is peptide–sugar complexes [82]. These techniques have the potential of providing a new avenue for high-throughput applications [83]. Another new approach, laser-induced liquid bead ion desorption (LILBID), combines IR laser pulses to desolvate spray droplets, allowing samples under relatively physiological salt conditions to be analyzed in a time-of-flight MS system. Sequence-specific detection of oligonucleotides and a noncovalent protein-oligonucleotide complex (the P50 subunit of NFκB with double stranded-DNA) was demonstrated [84]. The LILBID technique offers the possibility of sensitive analyses under more salt-tolerant conditions than traditional laser-based MS approaches.

10.4.1 Ion-Mobility Mass Spectrometry (IMS)

Ion-mobility mass spectrometry (IMS) is emerging to have promising implications for protein structural analysis and screening applications. IMS utilizes an inert gas cell in an electric field to resolve ions based on their average cross section. A mass spectrometer equipped with an IMS cell can analyze complex mixtures based on molecular shape as well as mass and charge [85]. IMS may have applications for protein–protein complexes; examples are the hemoglobin tetramer [86] and chaperonin-substrate complexes [87]. IMS can be employed to determine stability differences in folding of tetrameric thioredoxin and to isolate the effects of its natural small-molecule ligand, thyroxine, on the unfolding and dissociation of the protein complex [62]. Although IMS has not yet been used in protein–ligand screening applications, its ability to add an additional dimension of molecular analysis may become important.

10.4.2 Hydrogen–Deuterium Exchange (H/DX) (Including SUPREX and PLIMSTEX)

The dynamic processes involving protein structure and protein–ligand complexes in solution can also be probed using reaction rate dependent covalent labeling strategies. The most common is hydrogen/deuterium exchange (H/DX), which is applicable to micromolar and lower concentrations of proteins. The method offers advantages over NMR and X-ray crystallography for determination of ligand binding kinetics and thermodynamics, and for mapping binding sites [88,89]. Furthermore H/DX is more sensitive and has higher throughput. Although some application of H/DX can give ligand affinities, screening applications are limited because labeling often requires complex sample preparation and data analysis, which limit throughput compared to the other MS-based methods [90].

A fully automated H/DX system has been applied to screening of the peroxisome proliferator-activated receptor (PPARγ) against a limited set of 10 ligands, requiring 24 h of analysis [91]. Another approach, PLIMSTEX, compares the amounts of deuterium uptake of the protein and its peptide fragments as a function of the ratios of ligand to protein to afford protein/ligand binding constants [92]. A variant of the HDX technique, SUPREX, measures the amount of solvent exchange into an intact protein using increasing concentrations of chemical denaturants (e.g., urea or guanidine-HCl) and is adapted to higher throughput MS analysis using MALDI [93]. Although SUPREX does not have the capability of furnishing detailed structural information about ligand binding, it does have the capability of assaying ligand binding in protein mixtures under near physiological conditions at better than 3 min per ligand using conventional MALDI, and at less than 20 s per ligand using instruments equipped with high repetition rate lasers and automatic positioning systems [94].

10.4.3 Crosslinking (Including Inhibition of Complex Formation)

Chemical crosslinking, a validated methodology for defining protein–protein interaction partners, can be applied to a wide variety of protein interaction systems [95]. It can be combined with MS to facilitate detection of low-affinity binding partners that would otherwise be lost during ionization and MS detection [96]. The most notable technique, termed "tethering," can be utilized to identify low-affinity fragments. Tethering was used to screen a large fragment library containing compounds with reactive thiol groups against a target protein that contains a reactive thiol on an amino acid near the binding site of the protein. In cases where the natural protein lacks such a group, the reactive site must be introduced by site-directed mutagenesis. The tethering approach can be applied to a number of targets, including cytokines and a variety of enzymes [96–98]. One advantage of crosslinking is that it can be used to stabilize noncovalent complexes and thus allow the use of more energetic detection methods of analysis. This approach can be used to develop a MALDI-based screening system from the disruption of protein–protein interactions. By this method the effects of small-molecule ligands on the protein–protein complex can be assessed by equilibrating the protein complex with compounds of interest, stabilizing the protein–protein partners using crosslinking, then measuring the amount of remaining protein–protein complex using the MALDI instrument. Although limited to biological systems that involve protein–protein interaction partners, this approach can be done under near physiological conditions, compared to "gentle" ESI or Nano-ESI techniques, and has the potential for extremely high throughput [99,100].

There are several other contributions of MS to early compound discovery and drug development. One is the use of MS detection in enzyme assays for screening or confirming drug candidates. These contributions will be described in the following section.

10.5 ENZYME ACTIVITY ASSAYS USING MS FOR SCREENING OR CONFIRMING DRUG CANDIDATES

MS-based biochemical assays have the capability to follow the time-dependent increase in product/reporter signal and/or a concomitant decrease in substrate signal without labeling of either component (Figure 10.5) [101]. Such an approach bypasses the need for reporter agents, including radioactivity, allowing MS-based functional assays to serve as a screening platform, as hit-evaluation technique, or both. Although the first reports of MS to evaluate substrate turnover date back to 1972 (using FAB-MS and GC-MS), the watershed for these methods occurred with the advent of the soft ionization techniques of electrospray and MALDI. Since these breakthroughs, MS-based assays have been used in steady-state kinetics to

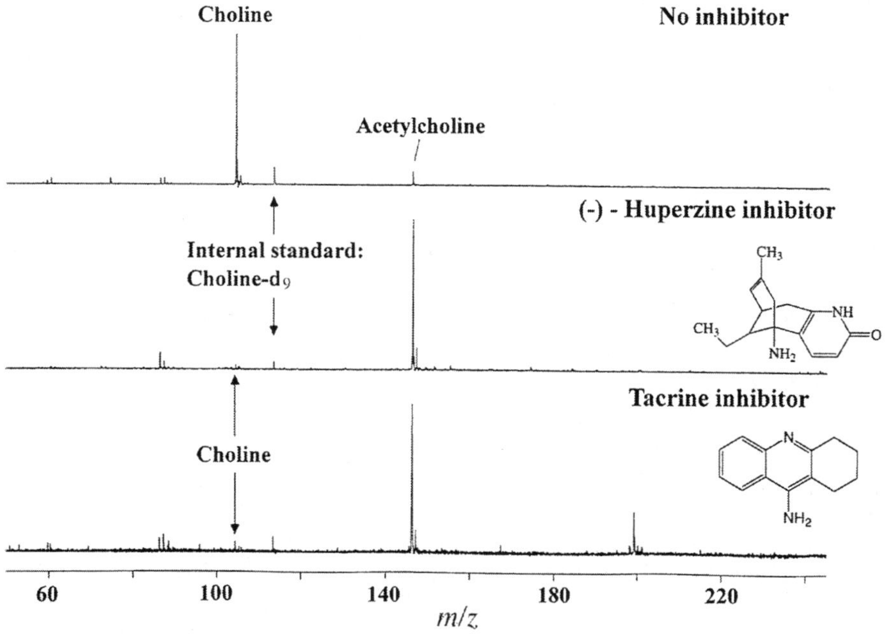

FIGURE 10.5 Enzyme turnover. Reprinted with permission from [101].

determine Michaelis–Menten constants [102–107], and in pre–steady-state kinetics where information about enzyme intermediates can be derived [108,109].

10.5.1 MS to Measure Substrate Turnover

A good example of a functional assay using MS evaluates pharmaceutical hits identified from primary HTS against target enzyme *Escherichia coli* UDP-N-acetyl-muramyl-*L*-alanine ligase (MurC) [110]. This new LC-MS-based assay was compared against the conventional spectrophotometric Malachite Green (MG) assay, which detected the phosphate produced in the reaction. The results demonstrate that the LC-MS assay, which determined the specific ligase activity of MurC, offered several advantages including a lower background (0.2% vs. 26%), higher sensitivity $\geq$10 fold), and capacity for determining IC_{50} values and percent inhibition.

Another example of substrate turnover measured by MS is the activity screening of non-ribosomal peptide synthetases whereby a signature mass shift resulting from the acylation of the carrier domain can be identified [111]. These biosynthetic pathways are of interest to academic and industrial efforts because the compounds produced by the non-ribosomal peptide synthetase (NRPS) and the polyketide synthase (PKS) have potent bioactivity. This MS application could replace the traditional assay, which utilizes radioactivity and thus requires allocated resources and specially certified individuals.

10.5.2 Multiple Component Measurements

One benefit of MS as a detection method is its ability to monitor multiple components simultaneously. This can be exploited by a development of a multiplex MS assay to study enzyme/substrate specificity, from which multiple substrates can be evaluated simultaneously [112]. Similarly an immunopreciptation-LC-MS assay can be used to support in vitro drug efficacy studies. An example from research on Alzheimer's disease shows different isoforms of the amyloid beta peptide can be detected in patients' sera that would not be observed by using other methods. Selected reaction monitoring (SRM) leads to increased assay sensitivity via its selective detection for activity screening [113]. Here SRM is applied to the detection of enzymatic products and the selective monitoring of specific mass-to-charge ratios is used as the readout. SRM can also monitor several products simultaneously (i.e., the use of multiple drug targets and/or substrate). This is particularly useful when drug targets are present in complex environments; an example is the screening cytochrome P450 in liver microsomal preparations [114].

10.5.3 Continuous Flow Screening

In contrast to one well–one reaction platforms, continuous flow screening (CFS) is performed in an online, open-tubular system comprising sample injection, HPLC separation, online biochemical assay, and MS detection [115,116]. Assay reagents are continuously added and mixed with the HPLC eluent, resulting in continuous MS data output that reflects the enzymatic activity (Figure 10.6). Libraries of potential inhibitors separated by HPLC can bind to the drug target and decrease the product

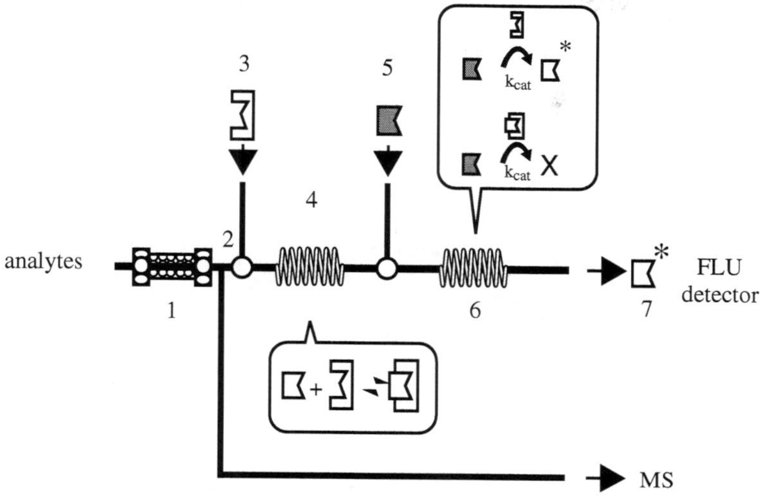

FIGURE 10.6 Continuous flow system for detection of inhibitors of enzyme activity. (1) Separation of analytes by HPLC; (2) HPLC effluent split toward biochemical assay and mass spectrometer; (3) addition of enzyme; (4) continuous incubation of the analytes and enzyme in an open-tubular reactor; (5) addition of substrate; (6) continuous conversion of substrate by the enzyme in an open-tubular reactor; (7) detection of reaction product by fluorescence detection. Reprinted with permission from [116].

formation, resulting in a decrease in the product signal, while MS data from the inhibitor are obtained simultaneously. CFS can be used to screen inhibitors of adenosine deaminase and AChE [117]. Screens such as this take advantage of several benefits of CFS, including: (1) detection of all products, providing a reliable assay readout, and (2) comparison of the retention times of the chromatographic peaks of the active compounds. A liability is the short reaction time, which causes compounds with slow kinetics of binding to exhibit potentially lower inhibition in this format.

10.5.4 Immobilized Enzyme Reactor (IMER)

A technique related to CFS, called an immobilized enzyme reactor (IMER), encapsulates enzymes in a gel through which substrate +/− inhibitors are flowed. A subsequent fractionation step coupled to MS detects turnover and potential inhibitors [118,119]. Among the benefits are reduced reagent consumption, the ability to evaluate binding affinities to the protein of interest, and a means to study inhibition of activity. A variation of IMER makes use of immobilized enzymes on a monolithic column to screen inhibitors [117]. The system includes a two-channel nanoLC to infuse substrate or substrate/inhibitor mixtures through the enzyme reactor. The ratio of the product and substrate was measured as they elute from the enzyme reactor into a mass spectrometer. The relative decrease of the product/substrate ratio is proportional to the inhibitory potency of the compounds present.

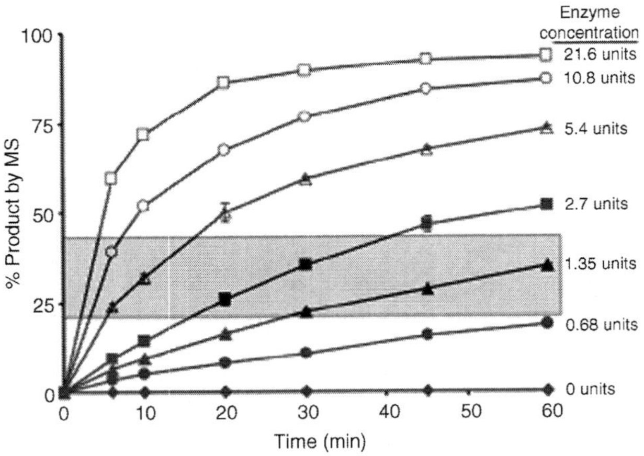

FIGURE 10.7 PKA enzyme titration and time-course at fixed substrate and ATP. To select an enzyme concentration and standard reaction time within an approximate linear range of the reaction (initial rate), a time course at various PKA concentrations was evaluated. MALDI-TOF MS data were collected at the indicated time points. The data points represent the average and standard deviation of triplicate samples. The shaded area on the graph represents the target range (20–40%) for measuring the percent of product formed by MALDI-TOF MS. For this kinase, 2.7 units of enzyme for 30 min were chosen as the standard reaction conditions for further inhibitor screening. Reprinted with permission from [123].

10.5.5 Application of MALDI to High–Throughput Enzyme Assays

Although ESI is commonly used in the various assays, it usually requires a pre-chromatographic separation to accommodate ESI's lower tolerance of signal-suppressing buffer components. The use of MALDI, which is more tolerant of salts, offers to overcome the bottlenecks of serial LC-based assays; multiplexing of sample handling and deposition on MALDI targets and data acquisitions can be done in times as short as 10 s/sample. A primary concern for MALDI, however, is quantification given its inconsistent signal intensity and the shot-to-shot variability [120]. An example of the utility of MALDI-TOF MS is the quantitative analysis of enzyme activity for the lipase catalyzed transesterification reaction and for pyruvate decarboxylase, where the development of product relative to a deuterated internal standard was measured [121]. A MALDI-TOF-based assay can also be used to evaluate the kinetic parameters of glucosamine-6P synthase [122].

Peptide substrates are particularly amenable to MALDI analysis because interference from the matrix signal is minimal in the detection range of the analytes. As a demonstration, kemptide substrate was incubated in the presence or absence of protein kinase A (PKA) and spotted onto a MALDI target plate [123]. A mass increase of 80 Da was detected, correlated to the phosphorylation reaction of expected enzyme activity. Similar results were obtained when a hydroxylase enzyme was incubated with its peptide substrate, and a mass increase of 16 Da, corresponding to the hydroxylation reaction, was detected [114].

10.5.6 Ratiometric Assays Using MALDI

Eliminating the need for internal standards for quantification by MALDI can be accomplished by creating ratiometric assays. In this experiment, signals from both substrate(s) and product(s) are monitored simultaneously, and the ratio of these signals provides an estimate of substrate turnover under different conditions. This approach bypasses the need to quantify a single component in an enzymatic assay. MALDI-TOF MS measurements of substrate-to-product ratios over a relevant concentration range can deliver a linear relationship [123].

10.5.7 Self-assembled Monolayers for MALDI-MS (SAMDI)

An alternative form of MALDI, called self-assembled monolayers for MALDI-MS (SAMDI), can also be applied to enzyme studies. SAMDI utilizes an immobilized substrate on a self-assembled monolayer and couples it to MALDI-MS detection [113,124]. The enzyme substrate is immobilized within SAMs on the surface of a target plate, and the surface exposure orients the substrate so that it is readily accessible to the enzyme [125]. Enzyme activity and subsequent inhibition of the activity can be measured in terms of the consumption of the substrate and/or the increase of product as judged by the MALDI-TOF readout. This approach was first validated by measuring the activity of the enzyme protein arginine methyltransferase [126], which adds methyl groups to the side chain of arginine, causing mass changes of 14 Da with each methyl addition. An arginine-containing peptide was immobilized within the SAM and incubated with the enzyme to permit detection of these characteristic mass shifts.

10.5.8 Desorption/Ionization Process Off of Porous Silicon (DIOS) and Carbon Nanotubes

Improvements for quantification and monitoring enzyme kinetics using MALDI may come from matrix-free applications that reduce chemical noise and aid in shot-to-shot reproducibility. Several technologies address the challenge of omitting matrix; two are discussed here. The first is called desorption/ionization process off of porous silicon (DIOS) in which UV laser energy is captured by the silicon target and transferred directly to an analyte [8,127,128]. Despite the absence of matrix, DIOS-MS produces little or no fragmentation, tolerates moderate amounts of contaminants commonly found in biological samples, and can accommodate the screening of individual compounds against acetylcholinesterase, for example [101]. The presence of an inhibitor can be detected by a decrease of product signal and by a change in the substrate-to-product ratio. This approach can be used to screen rapidly for inhibition in various enzyme assays [129]. Unfortunately, DIOS targets are not commercially available, limiting its widespread use for screening. Another matrix free approach is the use of carbon nanotubes coating the MALDI target, and this approach is promising in an AChE MALDI assay [130].

10.5.9 Overcoming Low Serial Throughput by Rapid Chromatography

For ESI-MS systems, throughput can be greatly increased by performing chromatography in parallel, either by staggering or multiplexing analyses or by decoupling the chromatography step [131]. A successful implementation of parallel analyses was reported by Biotrove (now Biocius) (http://www.biocius.com) [4]. Here a rapid separation step using solid phase extraction (SPE) is directly coupled to MS. Using miniaturization, automation and an ultra-fast sample switching inlet, their RapidFire platform feeds samples directly to the mass spectrometer at 5 to 8 s/sample. Applications include screening acetylcholinesterase [132], Akt [4], and stearoyl-CoA desaturase [133]. The potential for even higher throughput can be achieved by using multiple LC systems interfaced to a single mass spectrometer, and by multiplexing two or more similar enzymes in a single reaction mixture.

By decoupling the purification/fractionation step from detection in MS-based assays, the time-consuming sample handling steps (reagent combination, incubation time and fractionation/purification) are performed in parallel and, therefore, are not time bound by sample size [4]. The purified samples can be analyzed using ESI-MS via direct injection, and the purification can be accomplished by SPE, reverse-phase or size exclusion chromatography. Carbon nanotubes can also be considered as extraction platforms (separate from matrix replacements in MALDI) [134]. Parallel SPE may be viewed as a complicated multi-step process requiring plate conditioning, loading of samples, washing, and sample elution. Depending on the application, it may also be necessary to concentrate the SPE eluent to achieve the required sensitivity in the MS analysis.

10.5.10 MALDI–Triple Quadrupole Mass Spectrometry (MALDI-3Q)

Another approach that overcomes some of the limitations of MALDI and allow for quantification of small molecules while still retaining the beneficial speed of analysis combines MALDI with triple quadrupole MS (MALDI-3Q) [115,135]. The FlashQuant™ system from Applied Biosystems (www.appliedbiosystems.com) and MDS Sciex (www.mdssciex.com) uses MALDI-3Q to analyze samples fractionated by parallel SPE. Throughputs of 5 min per 96-well plate are possible for FlashQuant. Quantitative data comparable to an ESI method can be achieved for a wide range of pharmaceutical products, suggesting that MALDI-3Q may offer a good alternative or enhancement to ESI methods.

10.6 CONCLUSIONS AND FUTURE DIRECTIONS

As MS instrumentation and applications continue to improve and expand, the role MS plays in drug discovery continues to adapt and increase. For drug discovery applications, it is becoming possible to screen more samples faster owing to improvements in MS sensitivity and concurrent miniaturization of liquid-handling hardware. Strategies to increase throughput will enhance the ability of functional

MS screens to aid the discovery and early development of new inhibitors. New ionization methods that expand the scope of compatible buffer systems may allow for increased use in functional assays. Ion mobility, H/DX, and other forms of chemical footprinting when coupled with MS will expand its role for characterizing protein conformations and protein–ligand complexes. These approaches will provide useful data on the effects of small molecules on proteins, complementing other discovery approaches. From the human genome project, the functions of 40% of the human genome comprising open reading frames (ORFs) are known, which should provide potential new targets for drug discovery [136]. Due to its universal nature, AS-MS can provide the solution to study the function of these orphan targets.

Further benefits will be gained by strengthening quantitative aspects of MS-based activity assays, which is particularly true for MALDI-MS. Several techniques highlighted in this chapter are currently on the cutting edge of technology and should provide some of these advances in the short term. After over a hundred years of drug discovery and instrumentation advances, MS is firmly established as an important screening and characterization tool, but with aspects that must be considered as exciting and evolving, MS is still truly an emerging technology.

REFERENCES

1. Fischer, E. (1894). Transl. of "Einfluss der Configuration auf die wirkung der Enzyme" from *Berichte der deutschen Chemischen Gesellschaft 27*, 2985. In *Readings in Pharmacology*, B. H. a. G., Liljestrand, ed. 1963, p. 251.
2. Kaufmann, S. H. E. (2008). Editorial: Paul Ehrlich, founder of chemotherapy. *Nat. Revi. Drug Discov*, *7*, 373.
3. PDB Protein Data Bank: An Information Portal to Biological Macromolecular Structures. http://www.pdb.org/pdb/home/home.do. (accessed 6/16/2010).
4. Shiau, A. K., Massari, M. E., Ozbal, C. C. (2008). Back to basics: Label-free technologies for small molecule screening. *Comb Chem High Throughput Screen 11*, 231–237.
5. Wan, H., Holmen, A. G. (2009). High throughput screening of physicochemical properties and in vitro ADME profiling in drug discovery. *Comb Chem High Throughput Screen 12*, 315–329.
6. Chen, G., Pramanik, B. N. (2009). Application of LC/MS to proteomics studies: current status and future prospects. *Drug Discov Today 14*, 465–471.
7. Siegel, M. M. (2002). Early discovery drug screening using mass spectrometry. *Curr Top Med Chem 2*, 13–33.
8. Deng, G., Sanyal, G. (2006). Applications of mass spectrometry in early stages of target based drug discovery. *J Pharm Biomed Anal 40*, 528–538.
9. Hofstadler, S. A., Sannes-Lowery, K. A. (2006). Applications of ESI-MS in drug discovery: Interrogation of noncovalent complexes. *Nat Rev Drug Discov 5*, 585–595.
10. Zehender, H., Mayr, L. M. (2007). Application of mass spectrometry technologies for the discovery of low-molecular weight modulators of enzymes and protein–protein interactions. *Curr Opin Chem Biol 11*, 511–517.

11. Daniel, J. M., Friess, S. D., Rajagopalan, S., Wendt, S. Zenobi, R. (2002). Quantitative determination of noncovalent binding interactions using soft ionization mass spectrometry. *Int J Mass Spectrom 216*, 1–27.

12. Hu, Q., Noll, R. J., Li, H., Makarov, A., Hardman, M., Graham Cooks, R. (2005). The orbitrap: A new mass spectrometer. *J Mass Spectrom 40*, 430–443.

13. Falb, D., Jindal, S. (2002). Chemical genomics: Bridging the gap between the proteome and therapeutics. *Curr Opin Drug Discov Devel 5*, 532–539.

14. Hindsgaul, O., Schriemer, D. C. (1998) *Apparatus for Screening Compound Libraries Using Frontal Chromatography in Combination with Mass Spectrometry*. Synsorb Biotech Inc., Can. Application: EP 26 pp.

15. Adam, G. C., Parish, C. A., Wisniewski, D., Meng, J., Liu, M., Calati, K., Stein, B. D., Athanasopoulos, J., Liberator, P., Roemer, T., Harris, G., and Chapman, K. T. (2008). Application of affinity selection/mass spectrometry to determine the structural isomer of parnafungins responsible for binding polyadenosine polymerase. *J Am Chem Soc 130*, 16704–16710.

16. Annis, D. A., Athanasopoulos, J., Curran, P. J., Felsch, J. S., Kalghatgi, K., Lee, W. H., Nash, H. M., Orminati, J.-P. A., Rosner, K. E., Shipps, Jr., G. W., Thaddupathy, G.R.A., Tyler, A. N., Vilenchik, L., Wagner, C. R., and Wintner, E. A. (2004). An affinity selection-mass spectrometry method for the identification of small molecule ligands from self-encoded combinatorial libraries: Discovery of a novel antagonist of E. coli dihydrofolate reductase. *Int J Mass Spectrom 238*, 77–83.

17. Annis, D. A., Nickbarg, E., Yang, X., Ziebell, M. R., Whitehurst, C. E. (2007). Affinity selection-mass spectrometry screening techniques for small molecule drug discovery. *Curr Opin Chem Biol 11*, 518–526.

18. Annis, D. A., Birnbaum, M., Birnbaum, S. N., Tyler, A. N. (2003) *Method for Identifying Compounds in a Chemical Mixture*. Neogenesis Pharmaceuticals, Inc., USA. Application: US 26 pp, Cont -in-part of U S Ser No *233*, 794.

19. Lenz, G. R., Nash, H. M., Jindal, S. (2000). Chemical ligands, genomics and drug discovery. *Drug Discov Today 5*, 145–156.

20. Makara, G. M., Nash, H., Zheng, Z., Orminati, J. P., Wintner, E. A. (2003). A reagent-based strategy for the design of large combinatorial libraries: a preliminary experimental validation. *Mol Divers 7*, 3–14.

21. Nash, H. M., Birnbaum, S., Wintner, E. A., Kalghatgi, K., Shipps, G., Jindal, S. (1999) *Method for Producing Mass-Coded Combinatorial Libraries*. Neogenesis, Inc., USA. Application: WO 129 pp.

22. Whitehurst, C. E., Annis, D. A. (2008). Affinity selection-mass spectrometry and its emerging application to the high throughput screening of G protein-coupled receptors. *Comb Chem High Throughput Screen 11*, 427–438.

23. Whitehurst, C. E., Nazef, N., Annis, D. A., Hou, Y., Murphy, D. M., Spacciapoli, P., Yao, Z., Ziebell, M. R., Cheng, C. C., Shipps, G. W., Jr., Felsch, J. S., Lau, D., and Nash, H. M. (2006). Discovery and characterization of orthosteric and allosteric muscarinic M2 acetylcholine receptor ligands by affinity selection-mass spectrometry. *J Biomol Screen 11*, 194–207.

24. Annis, D. A., Nazef, N., Chuang, C. C., Scott, M. P., Nash, H. M. (2004). A general technique to rank protein–ligand binding affinities and determine allosteric versus direct binding site competition in compound mixtures. *J Am Chem Soc 126*, 15495–15503.

25. Annis, D. A., Shipps, G. W., Jr., Deng, Y., Popovici-Muller, J., Siddiqui, M. A., Curran, P. J., Gowen, M., Windsor, W. T. (2007). Method for quantitative protein–ligand affinity measurements in compound mixtures. *Anal Chem 79*, 4538–4542.

26. Coburn, C. A., Stachel, S. J., Li, Y. M., Rush, D. M., Steele, T. G., Chen-Dodson, E., Holloway, M. K., Xu, M., Huang, Q., Lai, M. T., DiMuzio, J., Crouthamel, M. C., Shi, X. P., Sardana, V., Chen, Z., Munshi, S., Kuo, L., Makara, G. M., Annis, D. A., Tadikonda, P. K., Nash, H. M., Vacca, J. P., and Wang, T. (2004). Identification of a small molecule nonpeptide active site beta-secretase inhibitor that displays a nontraditional binding mode for aspartyl proteases. *J Med Chem 47*, 6117–6119.

27. Annis, D. A., Chuang, C. C., Nazef, N. (2007). *ALIS: An Affinity Selection–Mass Spectrometry System for the Discovery and Characterization of Protein–Ligand Interactions in Mass Spectrometry in Medicinal Chemistry: Applications in Drug Discovery*, K. Wanner and G. Höfner, ed. Wiley-VCH, Weinheim, Germany, pp 121–184.

28. Annis, D. A., Cheng, C. C., Chuang, C. C., McCarter, J. D., Nash, H. M., Nazef, N., Rowe, T., Kurzeja, R. J., Shipps, G. W., Jr. (2009). Inhibitors of the lipid phosphatase SHIP2 discovered by high-throughput affinity selection-mass spectrometry screening of combinatorial libraries. *Comb Chem High Throughput Screen 12*, 760–771.

29. Adam, G. C., Meng, J., Athanasopoulos, J., Zhang, X., Chapman, K. T. (2007). Affinity-based ranking of ligands for DPP-4 from mixtures. *Bioorg Med Chem Lett 17*, 2404–2407.

30. Muckenschnabel, I., Falchetto, R., Mayr, L. M. Filipuzzi, I. (2004). SpeedScreen: Label-free liquid chromatography-mass spectrometry-based high-throughput screening for the discovery of orphan protein ligands. *Anal Biochem 324*, 241–249.

31. Zehender, H., Le Goff, F., Lehmann, N., Filipuzzi, I., Mayr, L. M. (2004). SpeedScreen: The "missing link" between genomics and lead discovery. *J Biomol Screen 9*, 498–505.

32. Brown, N., Zehender, H., Azzaoui, K., Schuffenhauer, A., Mayr, L. M. Jacoby, E. (2006). A chemoinformatics analysis of hit lists obtained from high-throughput affinity-selection screening. *J Biomol Screen 11*, 123–130.

33. Carell, T., Wintner, E. A., Sutherland, A. J., Rebek, J., Jr., Dunayevskiy, Y. M., Vouros, P. (1995). New promise in combinatorial chemistry: synthesis, characterization, and screening of small-molecule libraries in solution. *Chem Biol 2*, 171–183.

34. Dunayevskiy, Y. M., Lai, J.-J., Quinn, C., Talley, F., Vouros, P. (1997). Mass spectrometric identification of ligands selected from combinatorial libraries using gel filtration. *Rapid Commun Mass Spectrom 11*, 1178–1184.

35. Flarakos, J., Morand, K. L., Vouros, P. (2005). High-throughput solution-based medicinal library screening against human serum albumin. *Anal Chem 77*, 1345–1353.

36. Johnson, B. M., Nikolic, D., van Breemen, R. B. (2002). Applications of pulsed ultrafiltration-mass spectrometry. *Mass Spectrom Rev 21*, 76–86.

37. Liu, D., Guo, J., Luo, Y., Broderick, D. J., Schimerlik, M. I., Pezzuto, J. M., van Breemen, R. B. (2007). Screening for ligands of human retinoid X receptor-alpha using ultrafiltration mass spectrometry. *Anal Chem 79*, 9398–9402.

38. Comess, K. M., Schurdak, M. E., Voorbach, M. J., Coen, M., Trumbull, J. D., Yang, H., Gao, L., Tang, H., Cheng, X., Lerner, C. G., McCall, J. O., Burns, D. J., and Beutel, B. A. (2006). An ultraefficient affinity-based high-throughout screening process: Application to bacterial cell wall biosynthesis enzyme MurF. *J Biomol Screen 11*, 743–754.

39. Comess, K. M., Trumbull, J. D., Park, C, Chen, Z., Judge, R. A., Voorbach, M. J., Coen, M., Gao, L., Tang, H., Kovar, P., Cheng, X., Schurdak, M. E., Zhang, H., Sowin, T., and

Burns, D. J. (2006). Kinase drug discovery by affinity selection/mass spectrometry (ASMS): application to DNA damage checkpoint kinase Chk1. *J Biomol Screen 11*, 755–764.

40. Qian, J., Voorbach, M. J., Huth, J. R., Coen, M. L., Zhang, H., Ng, S. C., Comess, K. M., Petros, A. M., Rosenberg, S. H., Warrior, U., and Burns, D. J. (2004). Discovery of novel inhibitors of Bcl-xL using multiple high-throughput screening platforms. *Anal Biochem 328*, 131–138.
41. Siegel, M. M., Tabei, K., Bebernitz, G. A., Baum, E. Z. (1998). Rapid methods for screening low molecular mass compounds non-covalently bound to proteins using size exclusion and mass spectrometry applied to inhibitors of human cytomegalovirus protease. *J Mass Spectrom 33*, 264–273.
42. Moy, F. J., Haraki, K., Mobilio, D., Walker, G., Powers, R., Tabei, K., Tong, H., Siegel, M. M. (2001). MS/NMR: A structure-based approach for discovering protein ligands and for drug design by coupling size exclusion chromatography, mass spectrometry, and nuclear magnetic resonance spectroscopy. *Anal Chem 73*, 571–581.
43. Tong, H., Bell, D., Tabei, K., Siegel, M. M. (1999). Automated data massaging, interpretation, and e-mailing modules for high throughput open access mass spectrometry. *J Am Soc Mass Spectrom 10*, 1174–1187.
44. Calleri, E., Temporini, C., Caccialanza, G., Massolini, G. (2009). Target-based drug discovery: The emerging success of frontal affinity chromatography coupled to mass spectrometry. *Chem Med Chem 4*, 905–916.
45. Slon-Usakiewicz, J. J., Dai, J. R., Ng, W., Foster, J. E., Deretey, E., Toledo-Sherman, L., Redden, P. R., Pasternak, A., Reid, N. (2005). Global kinase screening: Applications of frontal affinity chromatography coupled to mass spectrometry in drug discovery. *Anal Chem, 77*, 1268–1274.
46. Slon-Usakiewicz, J. J., Ng, W., Dai, J. R., Pasternak, A., Redden, P. R. (2005). Frontal affinity chromatography with MS detection (FAC-MS) in drug discovery. *Drug Discov Today 10*, 409–416.
47. Slon-Usakiewicz, J. J., Ng, W., Foster, J. E., Dai, J. R., Deretey, E., Toledo-Sherman, L., Redden, P. R., Pasternak, A., Reid, N. (2004). Frontal affinity chromatography with MS detection of EphB2 tyrosine kinase receptor. 1. Comparison with conventional ELISA. *J Med Chem 47*, 5094–5100.
48. Chan, N. W., Lewis, D. F., Rosner, P. J., Kelly, M. A., Schriemer, D. C. (2003). Frontal affinity chromatography-mass spectrometry assay technology for multiple stages of drug discovery: Applications of a chromatographic biosensor. *Anal Biochem 319*, 1–12.
49. Kovarik, P., Hodgson, R. J., Covey, T., Brook, M. A., Brennan, J. D. (2005). Capillary-scale frontal affinity chromatography/MALDI tandem mass spectrometry using protein-doped monolithic silica columns. *Anal Chem 77*, 3340–3350.
50. Wabnitz, P. A., Loo, J. A. (2002). Drug screening of pharmaceutical discovery compounds by micro-size exclusion chromatography/mass spectrometry. *Rapid Commun Mass Spectrom 16*, 85–91.
51. Blom, K. F., Larsen, B. S., McEwen, C. N. (1999). Determining affinity-selected ligands and estimating binding affinities by online size exclusion chromatography/liquid chromatography-mass spectrometry. *J Comb Chem 1*, 82–90.
52. Xu, R. N., Fan, L., Rieser, M. J., El-Shourbagy, T. A. (2007). Recent advances in high-throughput quantitative bioanalysis by LC-MS/MS. *J Pharm Biomed Anal 44*, 342–355.

53. Mullett, W. M. (2007). Determination of drugs in biological fluids by direct injection of samples for liquid-chromatographic analysis. *J Biochem Biophys Meth. 70*, 263–273.
54. Zhou, J.-L., An, J.-J., Li, P., Li, H.-J., Jiang, Y., Cheng, J.-F. (2009). Two-dimensional turbulent flow chromatography coupled on-line to liquid chromatography-mass spectrometry for solution-based ligand screening against multiple proteins. *J Chromatogr A 1216*, 2394–2403.
55. Jonker, N., Kool, J., Krabbe, J. G., Retra, K., Lingeman, H., Irth, H. (2008). Screening of protein–ligand interactions using dynamic protein-affinity chromatography solid-phase extraction-liquid chromatography-mass spectrometry. J Chromatogr *A1205*, 71–77.
56. Jonker, N., Kretschmer, A., Kool, J., Fernandez, A., Kloos, D., Krabbe, J. G., Lingeman, H., Irth, H. (2009). Online magnetic bead dynamic protein-affinity selection coupled to LC-MS for the screening of pharmacologically active compounds. *Anal Chem 81*, 4263–4270.
57. Ganem, B., Li, Y. T., Henion, J. D. (1991). Detection of noncovalent receptor-ligand complexes by mass spectrometry. *J Am Chem Soc 113*, 6294–6296.
58. Ganem, B., Li, Y. T., Henion, J. D. (1991). Observation of noncovalent enzyme-substrate and enzyme-product complexes by ion-spray mass spectrometry. *J Am Chem Soc 113*, 7818–7819.
59. Heuvel, R. H. v. d., Heck, A. J. (2004). Native protein mass spectrometry: from intact oligomers to functional machineries. *Curr Opin Chem Biol 8*, 519–526.
60. Hofstadler, S. A., Griffey, R. H. (2001). Analysis of noncovalent complexes of DNA and RNA by mass spectrometry. *Chem Rev 101*, 377–390.
61. Beck, J. L., Colgrave, M. L., Ralph, S. F., Sheil, M. M. (2001). Electrospray ionization mass spectrometry of oligonucleotide complexes with drugs, metals, and proteins. *Mass Spectrom Rev 20*, 61–87.
62. Hyung, S. J., Robinson, C. V., Ruotolo, B. T. (2009). Gas-phase unfolding and disassembly reveals stability differences in ligand-bound multiprotein complexes. *Chem Biol 16*, 382–390.
63. Sharon, M., Robinson, C. V. (2007). The role of mass spectrometry in structure elucidation of dynamic protein complexes. *Annu Rev Biochem, 76*, 167–193.
64. Pramanik, B. N., Bartner, P. L., Mirza, U. A., Liu, Y. H., Ganguly, A. K. (1998). Electrospray ionization mass spectrometry for the study of non-covalent complexes: an emerging technology. *J Mass Spectrom 33*, 911–920.
65. Nemes, P., Goyal, S. Vertes, A. (2008). Conformational and noncovalent complexation changes in proteins during electrospray ionization. *Anal Chem 80*, 387–395.
66. Yin, S., Loo, J. A. (2009). Mass spectrometry detection and characterization of noncovalent protein complexes. *Meth Mol Biol 492*, 273–282.
67. Jecklin, M. C., Touboul, D., Bovet, C., Wortmann, A., Zenobi, R. (2008). Which electrospray-based ionization method best reflects protein–ligand interactions found in solution? A comparison of ESI, nanoESI, and ESSI for the determination of dissociation constants with mass spectrometry. *J Am Soc Mass Spectrom 19*, 332–343.
68. Breuker, K., McLafferty, F. W. (2008). Stepwise evolution of protein native structure with electrospray into the gas phase, 10(-12) to 10(2) s. *Proc Nat Acad Sci USA 105*, 18145–18152.
69. Wigger, M., Eyler, J. R., Benner, S. A., Li, W., Marshall, A. G. (2002). Fourier transform–ion cyclotron resonance mass spectrometric resolution, identification, and screening

of non-covalent complexes of Hck Src homology 2 domain receptor and ligands from a 324-member peptide combinatorial library. *J Am Soc Mass Spectrom 13*, 1162–1169.

70. Poulsen, S.-A., Davis, R. A., Keys, T. G. (2006). Screening a natural product-based combinatorial library using FTICR mass spectrometry. *Bioorg Med Chem 14*, 510–515.
71. Vu, H., Pham, N. B., Quinn, R. J. (2008). Direct screening of natural product extracts using mass spectrometry. *J Biomol Screen 13*, 265–275.
72. Jecklin, M. C., Touboul, D., Jain, R., Toole, E. N., Tallarico, J., Drueckes, P., Ramage, P., Zenobi, R. (2009). Affinity classification of kinase inhibitors by mass spectrometric methods and validation using standard IC(50) measurements. *Anal Chem 81*, 408–419.
73. Keetch, C. A., Hernanndez, H., Sterling, A., Baumert, M., Allen, M. H., Robinson, C. V. (2003). Use of a microchip device coupled with mass spectrometry for ligand screening of a multi-protein target. *Anal Chem 75*, 4937–4941.
74. Bovet, C., Wortmann, A., Eiler, S., Granger, F., Ruff, M., Gerrits, B., Moras, D., Zenobi, R. (2007). Estrogen receptor-ligand complexes measured by chip-based nanoelectrospray mass spectrometry: An approach for the screening of endocrine disruptors. *Protein Sci 16*, 938–946.
75. Czuczy, N., Katona, M., Takats, Z. (2009). Selective detection of specific protein–ligand complexes by electrosonic spray-precursor ion scan tandem mass spectrometry. *J Am Soc Mass Spectrom 20*, 227–237.
76. Salih, B., Masselon, C., Zenobi, R. (1998). Matrix-assisted laser desorption/ionization mass spectrometry of noncovalent protein-transition metal ion complexes. *J Mass Spectrom 33*, 994–1002.
77. Wortmann, A., Pimenova, T., Alves, S., Zenobi, R. (2007). Investigation of the first shot phenomenon in MALDI mass spectrometry of protein complexes. *Analyst 132*, 199–207.
78. Schlosser, G., Pocsfalvi, G., Malorni, A., Puerta, A., de Frutos, M., Vekey, K. (2003). Detection of immune complexes by matrix-assisted laser desorption/ionization mass spectrometry. *Rapid Commun Mass Spectrom 17*, 2741–2747.
79. Woods, V., Hamuro, Y. (2001). High resolution, high-throughput amide deuterium exchange mass spectrometry (DXMS) determination of protein binding site structure and dynamics: utility in pharmaceutical design. *J Cell Biochem Suppl 37*, 89–98.
80. Yanes, O., Nazabal, A., Wenzel, R., Zenobi, R., Aviles, F. X. (2006). Detection of noncovalent complexes in biological samples by intensity fading and high-mass detection MALDI-TOF mass spectrometry. *J Proteome Res 5*, 2711–2719.
81. Yanes, Ó., Villanueva, J., Querol, E., Aviles, F. X. (2004). Intensity-fading MALDI-TOF-MS: novel screening for ligand binding and drug discovery. *Drug Discov Today TARGETS 3*, 23–30.
82. Von Seggern, C. E., Cotter, R. J. (2004). Study of peptide-sugar non-covalent complexes by infrared atmospheric pressure matrix-assisted laser desorption/ionization. *J Mass Spectrom 39*, 736–742.
83. Li, Y., Shrestha, B., Vertes, A. (2007). Atmospheric pressure molecular imaging by infrared MALDI mass spectrometry. *Anal Chem 79*, 523–532.
84. Hoffmann, J., Schmidt, T. L., Heckel, A., Brutschy, B. (2009). Probing the limits of liquid droplet laser desorption mass spectrometry in the analysis of oligonucleotides and nucleic acids. *Rapid Commun Mass Spectrom 23*, 2176–2180.
85. Ruotolo, B. T., Benesch, J. L., Sandercock, A. M., Hyung, S. J., Robinson, C. V. (2008). Ion mobility-mass spectrometry analysis of large protein complexes. *Nat Protoc 3*, 1139–1152.

86. Scarff, C.A., Patel, V. J., Thalassinos, K., Scrivens, J. H. (2009). Probing hemoglobin structure by means of traveling-wave ion mobility mass spectrometry. *J Am Soc Mass Spectrom 20*, 625–631.

87. van Duijn, E., Barendregt, A., Synowsky, S., Versluis, C., Heck, A. J. (2009). Chaperonin complexes monitored by ion mobility mass spectrometry. *J Am Chem Soc 131*, 1452–1459.

88. Sinz, A. (2007). Investigation of protein–ligand interactions by mass spectrometry. *Chem Med Chem*, *2*, 425–431.

89. Yan, X. Maier, C. S. (2009). Hydrogen/deuterium exchange mass spectrometry. *Methods Mol Biol 492*, 255–271.

90. Fitzgerald, M. C., West, G. M. (2009). Painting proteins with covalent labels: What's in the picture? *J Am Soc Mass Spectrom 20*, 1193–1206.

91. Chalmers, M. J., Busby, S. A., Pascal, B. D., Southern, M. R., Griffin, P. R. (2007). A two-stage differential hydrogen deuterium exchange method for the rapid characterization of protein/ligand interactions. *J Biomol Tech 18*, 194–204.

92. Zhu, M. M., Chitta, R., Gross, M. L. (2005). PLIMSTEX: a novel mass spectrometric method for the quantification of protein–ligand interactions in solution. *Int J Mass Spectrom 240*, 213–220.

93. Powell, K. D., Fitzgerald, M. C. (2004). High-throughput screening assay for the tunable selection of protein ligands. *J Comb Chem 6*, 262–269.

94. Hopper, E. D., Roulhac, P. L., Campa, M. J., Patz, E. F., Jr., Fitzgerald, M. C. (2008). Throughput and efficiency of a mass spectrometry-based screening assay for protein–ligand binding detection. *J Am Soc Mass Spectrom 19*, 1303–1311.

95. Tang, X., Bruce, J. E. (2009). Chemical cross-linking for protein-protein interaction studies. *Methods Mol Biol 492*, 283–293.

96. Erlanson, D. A., Hansen, S. K. (2004). Making drugs on proteins: site-directed ligand discovery for fragment-based lead assembly. *Curr Opin Chem Biol 8*, 399–406.

97. Cancilla, M. T., He, M. M., Viswanathan, N., Simmons, R. L., Taylor, M., Fung, A. D., Cao, K., Erlanson, D. A. (2008). Discovery of an Aurora kinase inhibitor through site-specific dynamic combinatorial chemistry. *Bioorg Med Chem Lett 18*, 3978–3981.

98. Yang, W., Fucini, R. V., Fahr, B. T., Randal, M., Lind, K. E., Lam, M. B., Lu, W., Lu, Y., Cary, D. R., Romanowski, M. J., Colussi, D., Pietrak, B., Allison, T. J., Munshi, S. K., Penny, D. M., Pham, P., Sun, J., Thomas, A. E., Wilkinson, J. M., Jacobs, J. W., McDowell, R. S., and Ballinger, M. D. (2009). Fragment-based discovery of nonpeptidic BACE-1 inhibitors using tethering. *Biochemistry 48*, 4488–4496.

99. Bovet, C., Plet, B., Ruff, M., Eiler, S., Granger, F., Panagiotidis, A., Wenzel, R., Nazabal, A., Moras, D., and Zenobi, R. (2009). Towards high-throughput identification of endocrine disrupting compounds with mass spectrometry. *Toxicol In Vitro 23*, 704–709.

100. Bovet, C., Ruff, M., Eiler, S., Granger, F., Wenzel, R., Nazabal, A., Moras, D., Zenobi, R. (2008). Monitoring ligand modulation of protein-protein interactions by mass spectrometry: Estrogen receptor alpha-SRC1. *Anal Chem 80*, 7833–7839.

101. Thomas, J. J., Shen, Z., Crowell, J. E., Finn, M. G., Siuzdak, G. (2001). Desorption/ionization on silicon (DIOS): A diverse mass spectrometry platform for protein characterization. *Proc Natl Acad Sci USA 98*, 4932–4937.

102. Bothner, B., Chavez, R., Wei, J., Strupp, C., Phung, Q., Schneemann, A., Siuzdak, G. (2000). Monitoring enzyme catalysis with mass spectrometry. *J Biol Chem 275*, 13455–13459.

103. Ge, X., Sirich, T. L., Beyer, M. K., Desaire, H., Leary, J. A. (2001). A strategy for the determination of enzyme kinetics using electrospray ionization with an ion trap mass spectrometer. *Anal Chem 73*, 5078–5082.
104. Hsieh, F. Y., Tong, X., Wachs, T., Ganem, B., Henion, J. (1995). Kinetic monitoring of enzymatic reactions in real time by quantitative high-performance liquid chromatography-mass spectrometry. *Anal Biochem 229*, 20–25.
105. Newton, R. P., Evans, A. M., Langridge, J. I., Walton, T. J., Harris, F. M., Brenton, A. G. (1995). Assay of adenosine 3′, 5′-cyclic monophosphate-dependent protein kinase activity by quantitative fast atom bombardment mass spectrometry. *Anal Biochem 224*, 32–38.
106. Norris, A. J., Whitelegge, J. P., Faull, K. F., Toyokuni, T. (2001). Kinetic characterization of enzyme inhibitors using electrospray-ionization mass spectrometry coupled with multiple reaction monitoring. *Anal Chem 73*, 6024–6029.
107. Pi, N., Armstrong, J. I., Bertozzi, C. R., Leary, J. A. (2002). Kinetic analysis of NodST sulfotransferase using an electrospray ionization mass spectrometry assay. *Biochemistry 41*, 13283–13288.
108. Houston, C. T., Taylor, W. P., Widlanski, T. S., Reilly, J. P. (2000). Investigation of enzyme kinetics using quench-flow techniques with MALDI-TOF mass spectrometry. *Anal Chem 72*, 3311–3319.
109. Zechel, D. L., Konermann, L., Withers, S. G., Douglas, D. J. (1998). Pre-steady state kinetic analysis of an enzymatic reaction monitored by time-resolved electrospray ionization mass spectrometry. *Biochemistry 37*, 7664–7669.
110. Deng, G., Gu, R. F., Marmor, S., Fisher, S. L., Jahic, H., Sanyal, G. (2004). Development of an LC-MS based enzyme activity assay for MurC: Application to evaluation of inhibitors and kinetic analysis. *J Pharm Biomed Anal 35*, 817–828.
111. Dorrestein, P. C., Blackhall, J., Straight, P. D., Fischbach, M. A., Garneau-Tsodikova, S., Edwards, D. J., McLaughlin, S., Lin, M., Gerwick, W. H., Kolter, R., Walsh, C. T., and Kelleher, N. L. (2006). Activity screening of carrier domains within nonribosomal peptide synthetases using complex substrate mixtures and large molecule mass spectrometry. *Biochemistry 45*, 1537–1546.
112. Leary, J. A., Pi, N., Yu, Y. (2004). Mass spectrometry: A chemical-biology tool for probing enzymatic systems. *Adv Mass Spectrom 16*, 323–334.
113. de Boer, A. R., Lingeman, H., Niessen, W. M. A., Irth, H. (2007). Mass spectrometry-based biochemical assays for enzyme-inhibitor screening. *TrAC Trends in Analytical Chemistry 26*, 867–883.
114. Ansede, J. H., Thakker, D. R. (2004). High-throughput screening for stability and inhibitory activity of compounds toward cytochrome P450-mediated metabolism. *J Pharm Sci 93*, 239–255.
115. Greis, K. D. (2007). Mass spectrometry for enzyme assays and inhibitor screening: an emerging application in pharmaceutical research. *Mass Spectrom Rev 26*, 324–339.
116. van Elswijk, D. A., Diefenbach, O., van der Berg, S., Irth, H., Tjaden, U. R., van der Greef, J. (2003). Rapid detection and identification of angiotensin-converting enzyme inhibitors by on-line liquid chromatography-biochemical detection, coupled to electrospray mass spectrometry. *J Chromatogr A1020*, 45–58.
117. Sharma, J., Besanger, T. R., Brennan, J. D. (2008). Assaying small-molecule-receptor interactions by continuous flow competitive displacement chromatography/mass spectrometry. *Anal Chem 80*, 3213–3220.

118. Besanger, T. R., Hodgson, R. J., Green, J. R. A., Brennan, J. D. (2006). Immobilized enzyme reactor chromatography: Optimization of protein retention and enzyme activity in monolithic silica stationary phases. *Anal Chim Acta Anal Meth Proteomics 564*, 106–115.

119. Lebert, J. M., Forsberg, E. M., Brennan, J. D. (2008). Solid-phase assays for small molecule screening using sol-gel entrapped proteins. *Biochem Cell Biol 86*, 100–110.

120. Xu, Z., Yao, S., Wei, Y., Zhou, J., Zhang, L., Wang, C., Guo, Y. (2008). Monitoring enzyme reaction and screening of inhibitors of acetylcholinesterase by quantitative matrix-assisted laser desorption/ionization fourier transform mass spectrometry. *J Am Soc Mass Spectrom 19*, 1849–1855.

121. Bungert, D., Heinzle, E., Tholey, A. (2004). Quantitative matrix-assisted laser desorption/ionization mass spectrometry for the determination of enzyme activities. *Anal Biochem 326*, 167–175.

122. Maillard, L. T., Guérineau, V., Badet-Denisot, M. A., Badet, B., Laprévote, O., Durand, P. (2006). Monitoring enzyme-catalyzed production of glucosamine-6P by matrix-assisted laser desorption/ionization time-of-flight mass spectrometry: A new enzymatic assay for glucosamine-6P synthase. *Rapid Commun. Mass Spectrom FIELD 20*, 666–672.

123. Greis, K. D., Zhou, S., Burt, T. M., Carr, A. N., Dolan, E., Easwaran, V., Evdokimov, A., Kawamoto, R., Roesgen, J., and Davis, G. F. (2006). MALDI-TOF MS as a label-free approach to rapid inhibitor screening. *J Am Soc Mass Spectrom 17*, 815–822.

124. Min, D. H., Tang, W. J., Mrksich, M. (2004). Chemical screening by mass spectrometry to identify inhibitors of anthrax lethal factor. *Nat Biotechnol 22*, 717–723.

125. Mrksich, M. (2008). Mass spectrometry of self-assembled monolayers: A new tool for molecular surface science. *ACS Nano 2*, 7–18.

126. Min, D. H., Yeo, W. S., Mrksich, M. (2004). A method for connecting solution-phase enzyme activity assays with immobilized format analysis by mass spectrometry. *Anal Chem 76*, 3923–3929.

127. Northen, T. R., Woo, H.-K., Northen, M. T., Nordström, A., Uritboonthail, W., Turner, K. L., Siuzdak, G. (2007). High surface area of porous silicon drives desorption of intact molecules. *J Am Soc Mass Spectrom 18*, 1945–1949.

128. Lewis, W. G., Shen, Z., Finn, M. G., Siuzdak, G. (2003). Desorption/ionization on silicon (DIOS) mass spectrometry: Background and applications. *Int J Mass Spectrom 226*, 107–116.

129. Shen, Z., Go, E. P., Gamez, A., Apon, J. V., Fokin, V., Greig, M., Ventura, M., Crowell, J. E., Blixt, O., Paulson, J. C., Stevens, R. C., Finn, M. G., and Siuzdak, G. (2004). A mass spectrometry plate reader: Monitoring enzyme activity and inhibition with a desorption/ionization on silicon (DIOS) platform. *Chembiochem 5*, 921–927.

130. Hu, L., Jiang, G., Xu, S., Pan, C., Zou, H. (2006). Monitoring enzyme reaction and screening enzyme inhibitor based on MALDI-TOF-MS platform with a matrix of oxidized carbon nanotubes. *J Am Soc Mass Spectrom 17*, 1616–1619.

131. Kahmann, J., Winkler, D., Hesterkamp, T. (2011). The impact of physical based methods screening and their delivery of better quality hits. *Drug Discov Today Technol 5*, e15–22.

132. Ozbal, C. C., LaMarr, W. A. Linton, J. R. Green, D. F. Katz, A. Morrison, T. B. Brenan, C. J. (2004). High throughput screening via mass spectrometry: A case study using acetylcholinesterase. *Assay Drug Dev Technol 2*, 373–381.

133. Soulard, P., et al. (2008). Development of a high-throughput screening assay for stearoyl-CoA desaturase using rat liver microsomes, deuterium labeled stearoyl-CoA and mass spectrometry. *Anal Chim Acta 627*, 105–111.

134. Pan, C., Xu, S., Hu, L., Su, X., Ou, J., Zou, H., Guo, Z., Zhang, Y., Guo, B. (2005). Using oxidized carbon nanotubes as matrix for analysis of small molecules by MALDI-TOF MS. *J Am Soc Mass Spectrom 16*, 883–892.

135. Rathore, R., Corr, J. J., Lebre, D. T., Seibel, W. L., Greis, K. D. (2009). Extending matrix-assisted laser desorption/ionization triple quadrupole mass spectrometry enzyme screening assays to targets with small molecule substrates. *Rapid Commun Mass Spectrom 23*, 3293–3300.

136. Venter, J. C., Adams, M. D., Myers, E. W., Li, P. W., Mural, R. J., Sutton, G. G., Smith, H. O., Yandell, M., Evans, C. A., Holt, R. A., Gocayne, J. D., Amanatides, P., Ballew, R. M., Huson, D. H., Wortman, J. R., Zhang, Q., Kodira, C. D., Zheng, X. H., Chen, L., Skupski, M., Subramanian, G., Thomas, P. D., Zhang, J., Gabor Miklos, G. L., Nelson, C., Broder, S., Clark, A. G., Nadeau, J., McKusick, V. A., Zinder, N., Levine, A. J., Roberts, R. J., Simon, M., Slayman, C., Hunkapiller, M., Bolanos, R. Delcher, A., Dew, L, Fasulo, D., Flanigan, M., Florea, L., Halpern, A., Hannenhalli, S., Kravitz, S., Levy, S., Mobarry, C., Reinert, K., Remington, K., Abu-Threideh, J., Beasley, E., Biddick, K., Bonazzi, V., Brandon, R., Cargill, M., Chandramouliswaran, I., Charlab, R., Chaturvedi, K., Deng, Z., Di Francesco, V., Dunn, P., Eilbeck, K., Evangelista, C., Gabrielian, A. E., Gan, W., Ge, W., Gong, F., Gu, Z., Guan, P., Heiman, T. J., Higgins, M. E., Ji, R. R., Ke, Z., Ketchum, K. A., Lai, Z., Lei, Y., Li, Z., Li, J., Liang, Y., Lin, X., Lu, F., Merkulov, G. V., Milshina, N., Moore, H. M., Naik, A. K., Narayan, V. A., Neelam, B., Nusskern, D., Rusch, D. B., Salzberg, S., Shao, W., Shue, B., Sun, J., Wang, Z., Wang, A., Wang, X., Wang, J., Wei, M., Wides, R., Xiao, C., Yan, C., Yao, A., Ye, J., Zhan, M., Zhang, W., Zhang, H., Zhao, Q., Zheng, L., Zhong, F., Zhong, W., Zhu, S., Zhao, S., Gilbert, D., Baumhueter, S., Spier, G., Carter, C., Cravchik, A., Woodage, T., Ali, F., An, H., Awe, A., Baldwin, D., Baden, H., Barnstead, M., Barrow, I., Beeson, K., Busam, D., Carver, A., Center, A., Cheng, M. L., Curry, L., Danaher, S., Davenport, L., Desilets, R., Dietz, S., Dodson, K., Doup, L., Ferriera, S., Garg, N., Gluecksmann, A., Hart, B., Haynes, J., Haynes, C., Heiner, C., Hladun, S., Hostin, D., Houck, J., Howland, T., Ibegwam, C., Johnson, J., Kalush, F., Kline, L., Koduru, S., Love, A., Mann, F., May, D., McCawley, S., McIntosh, T., McMullen, I., Moy, M., Moy, L., Murphy, B., Nelson, K., Pfannkoch, C., Pratts, E., Puri, V., Qureshi, H., Reardon, M., Rodriguez, R., Rogers, Y. H., Romblad, D., Ruhfel, B., Scott, R., Sitter, C., Smallwood, M., Stewart, E., Strong, R., Suh, E., Thomas, R., Tint, N. N., Tse, S., Vech, C., Wang, G., Wetter, J., Williams, S., Williams, M., Windsor, S., Winn-Deen, E., Wolfe, K., Zaveri, J., Zaveri, K., Abril, J. F., Guigo, R., Campbell, M. J., Sjolander, K. V., Karlak, B., Kejariwal, A., Mi, H., Lazareva, B., Hatton, T., Narechania, A., Diemer, K., Muruganujan, A., Guo, N., Sato, S., Bafna, V., Istrail, S., Lippert, R., Schwartz, R., Walenz, B., Yooseph, S., Allen, D., Basu, A., Baxendale, J., Blick, L., Caminha, M., Carnes-Stine, J., Caulk, P., Chiang, Y. H., Coyne, M., Dahlke, C., Mays, A., Dombroski, M., Donnelly, M., Ely, D., Esparham, S., Fosler, C., Gire, H., Glanowski. S., Glasser, K., Glodek, A., Gorokhov, M., Graham, K., Gropman, B., Harris, M., Heil, J., Henderson, S., Hoover, J., Jennings, D., Jordan, C., Jordan, J., Kasha, J., Kagan, L., Kraft, C., Levitsky, A., Lewis, M., Liu, X., Lopez, J., Ma, D., Majoros, W., McDaniel, J., Murphy, S., Newman, M., Nguyen, T., Nguyen, N., Nodell, M., Pan, S., Peck, J., Peterson, M., Rowe, W., Sanders, R., Scott, J., Simpson, M., Smith, T., Sprague, A., Stockwell, T., Turner, R., Venter, E., Wang, M., Wen, M., Wu, D., Wu, M., Xia, A., Zandieh, A., and Zhu, X. (2001). The sequence of the human genome. *Science 291*, 1304–1351.

CHAPTER 11

Utilization of Mass Spectrometry for the Structural Characterization of Biopharmaceutical Protein Products

AMARETH LIM and CATHERINE A. SREBALUS BARNES

11.1 INTRODUCTION

Biopharmaceutical or biotechnology-derived products are recombinant proteins produced by recombinant DNA technologies for human therapeutic use. Since the successful regulatory approval of recombinant human insulin (Humulin®) for treating diabetes in 1982, more and more recombinant proteins are being used to treat various diseases. More than 165 biopharmaceutical products have gained approval for human use. These products include recombinant hormones and growth factors, monoclonal antibody-based products and therapeutic enzymes, recombinant blood factors, recombinant vaccines, and nucleic acid–based products [1]. Market analysis projects the sale of recombinant therapeutic proteins to top $50 billion by 2010 [2].

Trends for new therapeutic protein products approved between 2003 and 2006 indicate that monoclonal antibodies, growth factors, and enzymes were approved in proportionately larger numbers than other biopharmaceuticals [1]. Since the introduction of the first marketed monoclonal antibody (muromonab, OKT3®) for treating acute allograft rejection in 1986, both physicians and patients have begun to accept monoclonal antibodies as novel therapeutics [3]. Monoclonal antibodies are currently the fastest growing [4], and they had an annual global sale of $20.6 billion in 2006 [5].

Monoclonal antibodies and other recombinant therapeutic proteins are produced by using complex, multi-step processes that have many variables impacting the product and its impurity profile. Historically process control and process validation were emphasized to ensure that a given process provides a protein comparable to that used in clinical trials; the underlying idea is that it is not possible to discern all of the molecule's quality attributes using analytical testing ("the process is the product").

Protein and Peptide Mass Spectrometry in Drug Discovery, Edited by Michael L. Gross, Guodong Chen, and Birendra N. Pramanik.

As analytical technologies improve, more comprehensive physicochemical, conformational, and biological characterizations of recombinant proteins become possible. Analytical approaches are routinely used to demonstrate product comparability following significant process changes (e.g., scale or facility changes). Kozlowski and Swann describe the desired product quality profile for therapeutic monoclonal antibodies as an iceberg where routine product testing reveals its tip, and extensive characterization tests, which are not used to evaluate every lot, reveal the middle space. The base of the iceberg corresponds to process control and process validation, which can be used to ensure that the process to make the therapeutic protein yields product quality attributes (even those that may not be apparent during analytical testing) that are consistent with the lots evaluated in clinical trials [6].

Recombinant therapeutic proteins must be characterized thoroughly during clinical development and manufacture of marketed products to satisfy the rigorous requirements set by global regulatory agencies [7–9]. This extensive characterization establishes product identity, purity, potency, strength, safety, and stability [10,11]. Advances in analytical techniques make it possible to characterize fully recombinant proteins [12,13]. Mass spectrometry (MS) is now a powerful analytical tool for the structural characterization of recombinant proteins [14–16].

In a previous review we used examples from the literature to detail the applications of MS for the structural characterization of recombinant therapeutic proteins [17]. For this chapter we selected examples from our own laboratories to illustrate the utilization of MS to ensure the quality of monoclonal antibodies and antibody-based Fc fusion protein products during cell culture development, purification development, formulation development, analytical method development, and clinical trial development. Specific examples include glycosylation profiling during clone selection, column profiling during purification development, identification of primary degradation pathways using forced degradation during formulation development, unknown eluent identification to aid analytical method development, and confirmation of structure/product comparability assessment during clinical trial development.

11.2 MS-BASED APPROACH FOR THE CHARACTERIZATION OF RECOMBINANT THERAPEUTIC PROTEINS

A general MS-based strategy for characterization of recombinant proteins (shown schematically in Figure 11.1) is similar to that used by Gibson and Biemann to verify the cDNA-deduced amino-acid sequences [18]. Typically the molecular mass of an intact protein is measured by using MS (intact analysis). If the protein is glycosylated, it may be necessary to remove the oligosaccharide moieties to help simplify data analysis and eliminate mass spectral overlap between the glycosylated structures and other potential modifications of interest. *N*-Linked oligosaccharides usually can be removed via treatment with peptide-*N*-glycosidase (PNGase F) whereas *O*-linked oligosaccharides can be released using β-elimination, hydrazinolysis, or *O*-glycosidase. For a monoclonal antibody or an antibody-based Fc fusion protein, it is necessary to obtain the mass of the light chain and heavy chain, or in the case of an Fc fusion protein, the single chain,

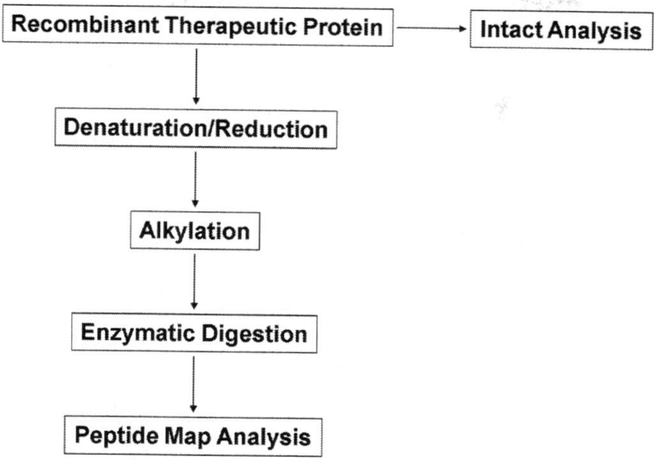

FIGURE 11.1 A schematic diagram of the general MS-based work flow for protein characterization. Detailed description is provided in the text.

using partial reduction. Partial reduction is a technique used to reduce the inter-chain disulfide bonds by treating the protein with dithiothreitol (DTT) in the absence of denaturing agent [19]. If the observed mass of the reduced protein chain is different than the theoretical mass from the gene-deduced amino-acid sequence, an amino-acid substitution, an amino-acid modification (e.g., post-translational modifications or amino acid degradation products), a translational error (e.g., alternative splicing, insertion/deletion), or a chemical/proteolytic cleavage is indicated.

To determine the region that contains the modification, the protein is subjected to peptide map analysis. The protein is usually prepared for digestion via denaturation using guanidine hydrochloride (GdnHCl) or an acid-labile anionic surfactant [20], reduced using DTT or tris(2-carboxyethyl)phosphine (TCEP), and alkylated using iodoacetamide, iodoacetic acid, *N*-ethylmaleimide (NEM), or 4-vinylpyridine. Prior to enzymatic digestion, these reagents are removed using buffer exchange (e.g., dialysis or gel filtration) or their concentrations are reduced by dilution. For enzymatic digestion, trypsin is most commonly used, but Lys-C, Asp-N, and Glu-C can also be utilized, depending on the protein sequence. The resulting peptide mixture is then separated using reversed-phase (RP) HPLC followed by both UV (LC-UV) and mass spectrometric (LC-MS) detection. If necessary, the position and identity of the modification can be determined using tandem mass spectrometry (MS/MS) [21], via offline nanoelectrospray MS/MS or online LC-MS/MS analysis.

For MS protein characterization, we use matrix-assisted laser desorption/ionization (MALDI) time of flight (TOF), electrospray ionization (ESI) triple quadrupole, ESI quadrupole ion trap, ESI orthogonal acceleration TOF (oaTOF), ESI/MALDI quadrupole orthogonal acceleration TOF (hybrid qTOF), and ESI Orbitrap mass spectrometers. Samples are introduced into an ESI mass spectrometer by using nano, capillary, or standard online LC, depending on the amount of sample available. For offline analysis, samples are introduced using nanoelectrospray ionization [22].

TABLE 11.1 Performance Characteristics of a Hybrid qTOF Mass Spectrometer

Characteristics[a]	Results
Better duty cycle	All ions entering the TOF chamber after pulsing into the pusher region will reach the detector resulting in better sensitivity.
Better resolution (>8000 full width half maximum, FWHM)	The charge states of multiply charged peptide ions (+2 to +8) can be determined from the isotopic spacing resulting in better mass accuracy.
High data acquisition rate (>10 spectra/s)	It is suitable to obtain multiple spectra across a chromatographic peak during online LC-MS analysis.
Good mass accuracy and calibration stability	Mass accuracies of at least 5 parts per million (ppm) can be routinely achieved for peptide analysis or at least 20–50 ppm for protein analysis with external calibration.
Extended *m/z* mass range (5–40,000 *m/z*)	It is suitable to analyze very large proteins whose charge states fall outside the 2000–4000 *m/z* range.
	The low end of this extended mass range makes the hybrid qTOF suitable for MS/MS analysis. For example, MS/MS analysis using a quadrupole ion trap can be limited by the "1/3" rule, meaning that fragment ions that have *m/z* values less than one third of the precursor *m/z* value are not detected.

[a]In comparison to other mass spectrometers, such as a quadrupole ion trap or a triple quadrupole, the hybrid qTOF exhibits these performance characteristics that make it the instrument of choice for protein characterization.

Since its introduction in 1996 [23], the ESI hybrid qTOF mass spectrometer has served well especially in the biotechnology industry. The performance characteristics of an ESI hybrid qTOF mass spectrometer (Table 11.1) have made it the instrument of choice for MS protein characterization, especially for recombinant therapeutic proteins [24–27]. The hybrid qTOF is the workhorse in our laboratories.

We are also incorporating an Orbitrap mass spectrometer with electron transfer dissociation (ETD) into our MS arsenal for protein characterization. Introduced in 2003 as a new ESI mass spectrometer with excellent sensitivity, mass resolving power (capable of achieving 150,000 FWHM), and mass accuracy (1–2 ppm) [28], the Orbitrap now plays an important role in MS protein characterization [29–32]. When combined with ETD [33], the Orbitrap has the capability to sequence larger peptides and detect labile post-translational modifications [34–36].

11.3 CELL CULTURE DEVELOPMENT

Early development of therapeutic proteins often includes the evaluation of multiple cell lines (e.g., different host cells) and numerous clones to find the recombinant production system that provides optimal productivity, genetic stability, and appropriate product quality. Once a specific cell line and single-cell clone are selected, it is essential to evaluate cell culture conditions (e.g., seeding density, pH, temperature, and concentrations of nutrients and metabolites) to optimize productivity and cell growth. All of these factors (e.g., cell line type, clone, and cell culture process) can affect the glycosylation profile of the recombinant protein [37,38]. Given that the glycosylation profile can have an effect on the protein's effector function, efficacy, clearance, and immunogenicity [39,40], it is critical to assess the impact of cell culture conditions on the glycoform distribution during cell line, clone selection, and process optimization studies to ensure that the protein is produced with appropriate and consistent glycosylation.

One of our recombinant therapeutic proteins in development is a recombinant IgG4 two-chain Fc fusion protein. This protein is made using a murine myeloma NS0 cell line, which is known to produce nonhuman glycoforms including *N*-linked oligosaccharides containing the Galα(1 $\rightarrow$ 3)Gal linkage (αGal) and terminal *N*-glycolylneuraminic acid (NeuGc). These nonhuman glycoforms are linked to immunogenic responses in humans [41–43]. During early development of this Fc fusion protein, different master wells (fractionated pools of cells from bulk transfections), from which single-cell clones were derived, were evaluated to identify the master well(s) and clones that produced the Fc fusion protein with the lowest levels of αGal and NeuGc glycoforms. Four master wells were identified as lead candidates for single-cell cloning. From these four wells, 21 clones were carried forward into shake-flask studies to evaluate productivity, cleavage products, and glycosylation profile. The clones that produced the lowest levels of αGal and NeuGc, cleavage products, and acceptable productivity were further developed.

To determine the levels of αGal and NeuGc, samples of cell-free media from each shake-flask culture were subjected to a small-scale protein A affinity capture followed by partial reduction and LC-MS analysis. The deconvoluted ESI mass spectrum (Figure 11.2*A*) obtained in an LC-MS experiment of the Fc fusion protein produced by clone A1 (master well A, clone 1) displays proteins of molecular mass 25,804 Da, 29,838 Da, and 31,283–32,077 Da. The 29,838 Da species is the intact, nonglycosylated single-chain protein whereas those at 31,283–32,077 Da are various glycoforms of the intact, glycosylated single-chain protein, including the glycoforms containing the αGal and NeuGc. For instance, the 31,444-Da protein is likely the intact, glycosylated, single-chain protein containing a fucosylated, asialo biantennary glycan with one galactose residue (G1F). The protein at 25,804 Da is a cleavage product (cleavage product A) with the G1F carbohydrate moiety.

The ESI mass spectrum of the Fc fusion protein produced by clone D5 (Figure 11.2*B*) displays, in addition to the proteins observed in Figure 11.2*A*, a set of proteins of mass around 28,658 Da that are likely cleavage products (cleavage product B) possessing the G1F carbohydrate structure. To elucidate the origin of these cleavage products, we varied both the cone voltage and the collision energy in an ESI experiment. The ESI mass spectra (Figure 11.2 and Figure 11.3*A*), obtained using a cone voltage setting of 50 V and a collision energy setting of 10 eV, show that the 25,804-Da component disappeared when both the cone voltage and the collision energy were lowered to 30 V and 5 eV, respectively (Figure 11.3*B*). The relative

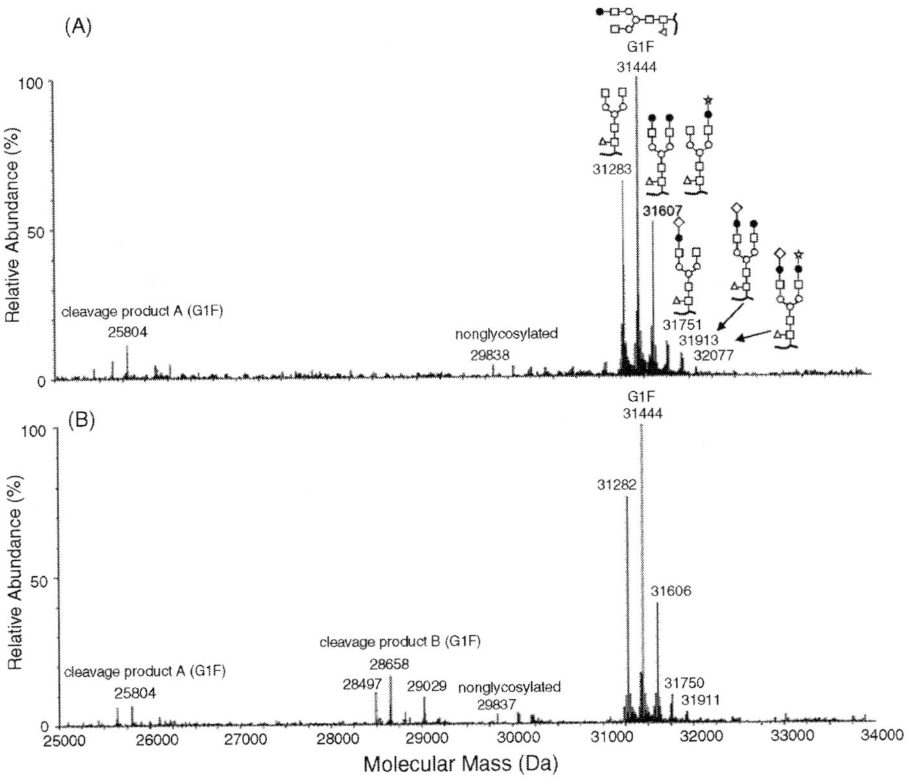

FIGURE 11.2 Deconvoluted ESI mass spectra obtained from the TIC of the LC-MS analysis of the recombinant IgG4 Fc fusion protein produced by (*A*) clone A1 and (*B*) clone D5. The Fc fusion protein was purified from cell culture using rProtein A sepharose (Amersham Pharmacia), released from the protein A using 0.15% formic acid, and partially reduced with DTT in 3 M Tris HCl, pH 8. The sample was analyzed using an Agilent SB-CN Zorbax column (2.1 × 50 mm, 5 μ, 80 Å) with a Waters Alliance Model 2695 HPLC system coupled to a Waters/Micromass Q-Tof micro hybrid qTOF mass spectrometer. HPLC solvents A and B were 0.15% formic acid in water and 0.12% formic acid in acetonitrile, respectively. The LC flow rate was 200 μL/min, and the column temperature was maintained at 60°C. Representation of each oligosaccharide residue: square = *N*-acetylglucosamine; triangle = fucose; open circle = mannose; close circle = galactose; star = Galα(1 → 3)Gal linkage (αGal); and diamond = *N*-glycolylneuraminic acid (NeuGc).

abundance of this material, however, increased dramatically when both the cone voltage and collision energy were increased to 75 V and 15 eV, respectively (Figure 11.3*C*). These results suggest that the cleavage product A is likely an artifact of fragmentation within the mass spectrometer. The cleavage product A originated from the cleavage of the Fc fusion protein on the *N*-terminal side of a Pro residue. This cleavage is characteristic of peptides [44,45] and proteins [46,47]. The relative abundance of the 28,658-Da cleavage product B was not affected by the cone voltage and collision energy (data not shown). This cleavage product was present in samples

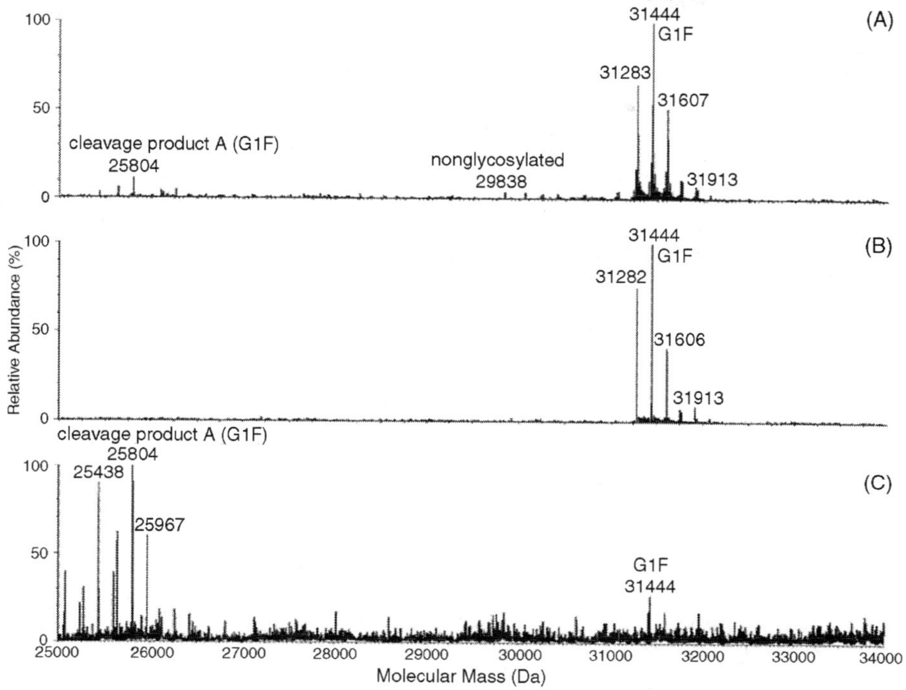

FIGURE 11.3 Deconvoluted ESI mass spectra extracted from the LC-MS analysis of the recombinant IgG4 Fc fusion protein produced by clone A1 using (*A*) cone voltage 50 V, collision energy 10 eV, (*B*) cone voltage 30 V, collision energy 5 eV, and (*C*) cone voltage 75 V, collision energy 15 eV. The samples were analyzed as described in Figure 11.2.

produced only by certain clones. Product *B* resulted from the cleavage of the Fc fusion protein on the *C*-terminal side of a Trp residue, most likely from proteolytic clipping.

One can calculate from the areas of the peaks in the deconvoluted ESI mass spectrum the relative abundance of each glycoform of the Fc fusion protein, affording a graphical representation of the glycoform distribution of the Fc fusion protein from the LC-MS glycosylation profile analysis (Figure 11.4). The LC-MS data demonstrate that glycosylation is highly variable among the clones; that is, the glycoform distributions are dependent on each specific clone. The LC-MS results suggest that clones B1, B2, B3, B4, B5, B6, B7, B8, and B9 produce the Fc fusion protein with the lowest levels of αGal and NeuGc. Clones C1, C3, and C4 produce the Fc fusion protein with the highest levels of αGal and NeuGc. Furthermore, although clone D5 produces the Fc fusion protein with relatively low levels of αGal and NeuGc, it also gives a significant amount of cleavage products (Figure 11.2*B*). Based on the LC-MS results (glycosylation profiling and cleavage products) and titer, we selected clones B1, B3, B6, B7, and B9 for further process development.

We then implemented a method involving small-scale protein A affinity capture followed by partial reduction and LC-MS analysis to obtain glycosylation profiling and sequence integrity of a recombinant Fc fusion protein during clone selection.

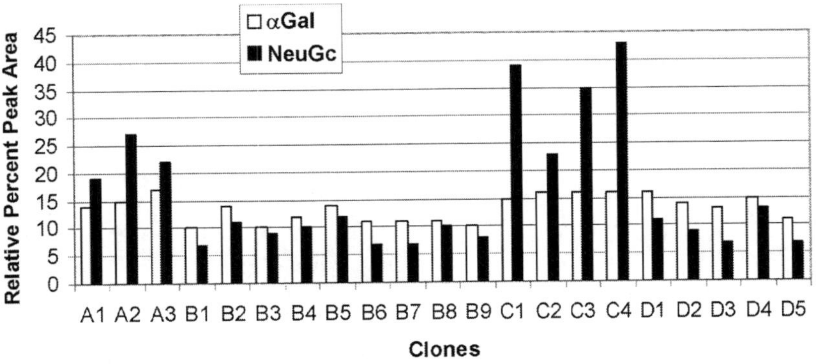

FIGURE 11.4 A graphical representation, obtained from the LC-MS glycosylation profiling of the 21 clone samples, showing the glycoform distribution produced by each clone. The letters and numbers represent the four master wells and various clones, respectively.

Alternative non-MS approaches for monitoring carbohydrate profiles would not have identified the clone-dependent proteolysis described here. Recently, we introduced an LC-MS-based method to obtain site-specific glycosylation profile of a highly sialylated monoclonal antibody with two *N*-linked glycosylation sites on the heavy chain [48]; not only does this methodology provide information on site-specific glycosylation, but also it informs on glycation in the light chain and *C*-terminal Lys processing of the heavy chain. Our LC-MS methods provide only relative estimation (not absolute) of the glycosylation profile. They do give, however, sufficient information to assess the glycosylation profile of monoclonal antibodies and Fc fusion proteins during cell line, clone selection, and process optimization studies to select the top clones for further development.

11.4 PURIFICATION DEVELOPMENT

Post-translational modifications have a wide range of effects on the properties of a recombinant protein, from its biological activity to its interaction with other proteins [49]. (Readers interested in this subject should refer to Chapter 12 by Tsarbopoulous and Bazoti in this volume, which provides a review of glycosylation, phosphorylation, and disulfide-bond formation.) For recombinant therapeutic proteins, post-translational modifications can be detrimental if they alter the biological activity or affect safety and immunogenicity of the product. Regulatory authorities have set rigorous requirements for the characterization of recombinant therapeutic proteins to establish the product identity, purity, potency, strength, safety, and stability [10,11]. Process and product-related substances and impurities of the recombinant proteins can be removed during purification and isolated for further characterization by LC-MS. We selected two examples to illustrate the utilization of MS during the development of the purification process for a recombinant IgG4 Fc fusion protein and an IgG1 monoclonal antibody with two *N*-linked glycosylation sites.

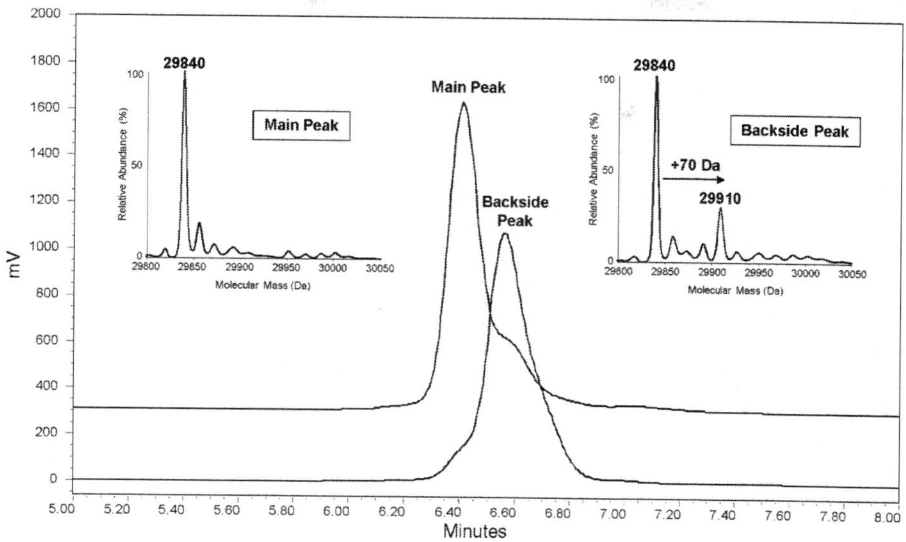

FIGURE 11.5 Analytical RP-HPLC chromatograms of the recombinant IgG4 Fc fusion protein, showing the UV traces of the main peak (starting material) and the backside peak (purified material). The insets show the deconvoluted ESI mass spectra of the protein from each peak after *N*-linked deglycosylation and partial reduction. The samples were analyzed using a Vydac 214MS5205 (2.1 × 50 mm, 5 μ, 300 Å) HPLC column with an Agilent 1100 HPLC system coupled to an Applied Biosystems QSTAR XL hybrid qTOF mass spectrometer. HPLC solvents A and B were 0.05% TFA in water and 0.04% TFA in acetonitrile, respectively. The LC flow rate was 200 μL/min, and the column temperature was maintained at 45°C. The QSTAR XL was equipped with an IonSpray source and operated in the positive-ion mode. The LC flow rate was split 1:20 using an LC Packings Acurate flow splitter in order for the flow rate to be compatible with the flow rate normally introduced into the mass spectrometer with an IonSpray source.

11.4.1 Identification of a Pyruvic Acid Modification Covalently Linked at the *N*-Terminus of a Recombinant IgG4 Fc Fusion Protein

During process development of a recombinant IgG4 Fc fusion protein, analytical RP-HPLC analysis of preparative anion exchange (AEX) column material showed a backside peak accompanying the main peak. The component corresponding to the backside peak was further enriched by using preparative AEX followed by preparative RP-HPLC; Figure 11.5 shows the analytical RP-HPLC chromatograms of the starting material (labeled as main peak) and purified material (labeled as backside peak). Both components were treated with PNGase F to remove the *N*-linked glycosylation, partially reduced with DTT to cleave the inter-chain disulfide bonds, and the products analyzed by LC-MS. The deconvoluted ESI mass spectrum of the Fc fusion protein (main peak) shows that the deglycosylated, single-chain protein has a mass of 29,840 Da (inset of Figure 11.5), consistent with the mass calculated from the amino acid sequence. The ESI mass spectrum of the Fc fusion protein corresponding to the backside peak shows not only the presence of the Fc fusion protein with mass

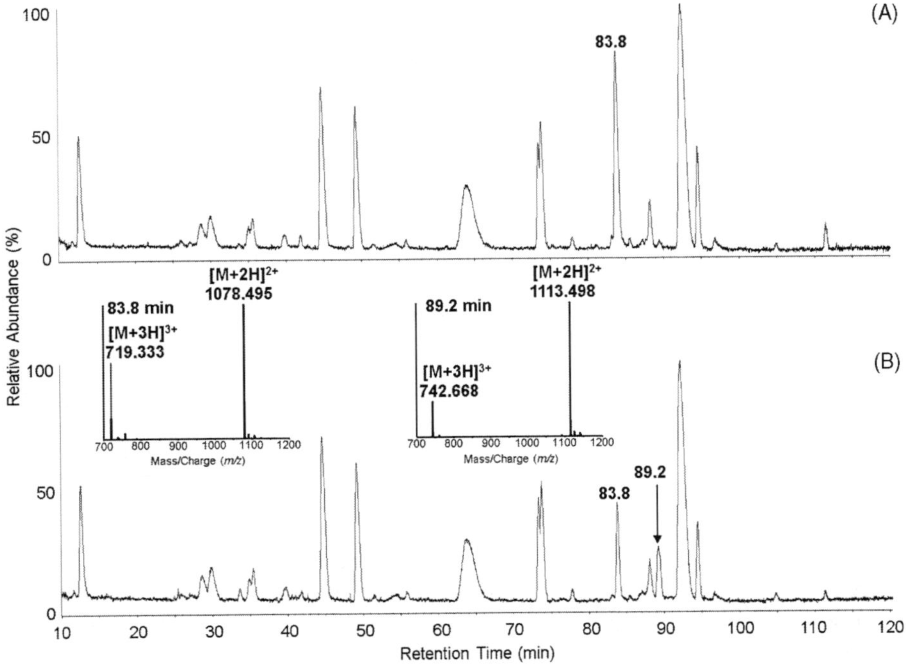

FIGURE 11.6 LC-MS TICs of the tryptic digest of the recombinant IgG4 Fc fusion protein from the (*A*) main peak and (*B*) backside peak. The insets show the raw mass spectra obtained from the TIC peaks at 83.8 min and 89.2 min. The Fc fusion protein was denatured using 0.1% RapiGest SF (Waters), reduced with 10-fold molar excess of DTT over disulfides, *S*-carboxyamidomethylated with 5-fold molar excess of iodoacetamide over total thiols, and digested with trypsin (1:40, enzyme:substrate). The resulting tryptic digest was analyzed using a Vydac 218MS52 (2.1 × 250 mm, 5 μ, 300 Å) with the Agilent 1100–Applied Biosystems QSTAR XL LC-MS system. HPLC solvents A and B were 0.05% TFA in water and 0.04% TFA in acetonitrile, respectively. The LC flow rate was 200 μL/min, and the column temperature was maintained at 45°C. The QSTAR XL was equipped with an IonSpray source and operated in the positive ion mode. The LC flow rate was split 1:20 using an LC Packings Acurate flow splitter in order for the flow rate to be compatible with the flow rate normally introduced into the mass spectrometer with an IonSpray source.

29,840 Da but also a new constituent at 29,910 Da, which is 70 Da higher in mass than that expected for the Fc fusion protein (inset of Figure 11.5).

To identify this +70 Da modification, the Fc fusion protein in each fraction was submitted to LC-MS peptide mapping; Figure 11.6 shows the TIC of the tryptic digest of the Fc fusion protein from the main peak and backside peak. The mass spectrum of the constituent eluting at 83.8 min (inset of Figure 11.6) displays $[M + 2H]^{2+}$ of *m/z* 1078.495 and $[M + 3H]^{3+}$ of *m/z* 719.333, affording a molecular mass of 2154.977 Da, which is consistent with the molecular mass of the *N*-terminal peptide composed of residues 1–20 (theoretical mass 2154.971 Da). In contrast, the mass spectrum obtained for the constituent eluting at 89.2 min (inset of Figure 11.6) shows two ions with $[M + 2H]^{2+}$ of *m/z* 1113.498 and $[M + 3H]^{3+}$ of *m/z* 742.668, giving a molecular mass of 2224.981 Da, which is 70 Da higher in mass than the mass of the 1–20 peptide. The molecular mass is suggestive that the + 70 Da modification is located in the *N*-terminal portion of the Fc fusion protein. To identify the position of the +70-Da modification, an aliquot of the tryptic digest was submitted to offline nanoelectrospray MS/MS analysis. MS/MS analysis of the $[M + 3H]^{3+}$ of *m/z* 719.3 for the *N*-terminal peptide and $[M + 3H]^{3+}$ of *m/z* 742.7 for the *N*-terminal peptide +70 Da identified the position of the modification at the first (*N*-terminal) amino-acid residue (data not shown). A $+70 \pm 1$ Da modification can result from an amino acid substitution (e.g., Ser → Arg, Gly → Gln/Lys, Asp → Trp) or a post-translational modification (e.g., dehydroalanine, sarcosyl, pyruvate). The nature of the first amino-acid residue of the Fc fusion protein is not compatible with the +70-Da modification as an amino-acid substitution or a dehydroalanine or sarcosyl modification. That +70-Da modification, however, can result from a pyruvic acid modification at the *N*-terminus. The observed mass of the peptide (1–20) +70 Da is within 2 ppm of the mass 2224.976 Da, calculated for the peptide (1–20) with a pyruvic acid modification at the *N*-terminus.

To confirm this identification, the Fc fusion protein was incubated with 100 mM pyruvic acid in 100 mM Tris HCl, pH 9 at 37°C for 49 h. An aliquot was treated with PNGase F and submitted to LC-MS analysis. The data show the Fc fusion protein with a mass of 59,678 Da (two-chain) and its modified forms with observed masses of 59,748 Da (+ 70 Da, pyruvic acid on one chain) and 59,817 Da (+ 139 Da, pyruvic acid on both chains) (data not shown). These results confirm that the +70-Da modification is a covalent attachment of pyruvic acid to the *N*-terminus of the Fc fusion protein. A proposed mechanism of the reaction of pyruvic acid with the *N*-terminus of the protein is shown in Figure 11.7. Analysis of purified fractions of

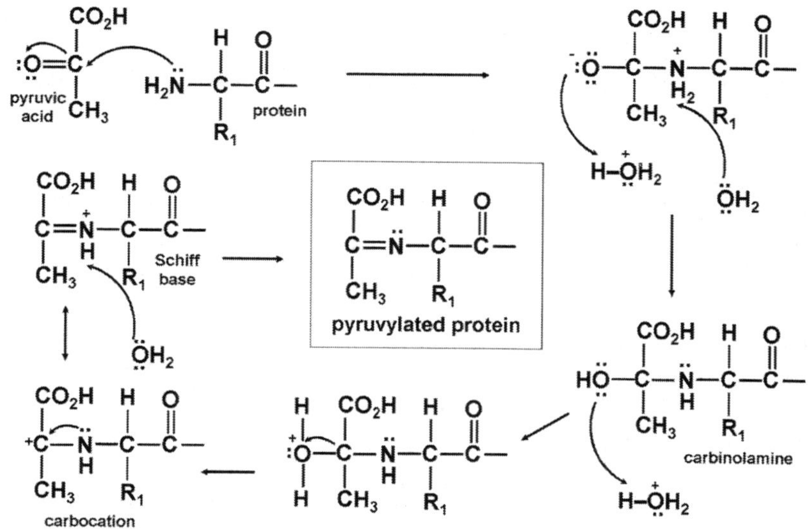

FIGURE 11.7 A proposed mechanism of the reaction of pyruvic acid with the N-terminus of the recombinant IgG4 Fc fusion protein.

the modified protein by using a cell-based bioassay showed that this modification decreases the bioactivity of the Fc fusion protein. With this modification on one of the two Fc fusion protein "arms," the relative potency is only 41%, suggesting complete inactivation of the modified "arm." This modification was also observed in human adult hemoglobin [50] and in the DNA-binding protein Ner [51].

11.4.2 Identification of Hinge Region Cleavage in an IgG1 Monoclonal Antibody with Two *N*-Linked Glycosylation Sites

During purification development of an IgG1 monoclonal antibody with two *N*-linked glycosylation sites in the heavy chain, analytical size exclusion chromatography (SEC) analysis of hydrophobic interaction chromatography (HIC) fractions showed that earlier eluting fractions from a preparative HIC column contained additional constituents seen on the backside of the monomer peak (data not shown). In an attempt to identify these components, the IgG1 molecule was introduced to a HIC column and eluted using a sodium sulfate gradient. Fractions were collected throughout the elution to obtain an enriched fraction corresponding to the late-eluting components. Analytical SEC analysis of one of the HIC fractions (fraction 12, data not shown) showed two SEC backside constituents present at 3% and 12%, respectively. Fraction 12 was submitted for further characterization by LC-MS.

This IgG1 monoclonal antibody possesses two *N*-linked glycosylation sites in the heavy chain. One site is located in the Fab region, and the other is in the Fc region (Figure 11.8*A*). The glycoforms in the Fab region are highly sialylated. The combination of the high sialylation and the two *N*-linked glycosylation sites results in extensive heterogeneity and makes analytical characterization challenging. To identify the late-eluting SEC forms of the antibody, the HIC fraction 12 sample was treated with PNGase F to remove *N*-linked glycosylation and submitted to LC-MS analysis. The results showed that this molecule was still quite heterogeneous, making the identification of the components corresponding to the backside peaks impractical.

Given that this standard LC-MS approach for intact analysis did not provide sufficient information to identify the components corresponding to the SEC backside peaks, we sought a different strategy (Figure 11.8*B*). Both the HIC fraction 12 and a control sample were subjected to denaturation and reduction in the presence of 6 M GdnHCl and DTT at 50°C for 20 min. Each sample was then *S*-carboxyamidomethylated with iodoacetamide in the dark at room temperature for 30 min.

LC-MS analysis of the control sample after denaturation, reduction, and *S*-carboxyamidomethylation is shown in Figure 11.9*A*. The deconvoluted ESI mass spectrum of the component eluting at 16.9 min showed that it contains the light chain (observed molecular mass of 24,324 Da, which is consistent with the theoretical mass of 24,323 Da with all five Cys residues *S*-carboxyamidomethylated). The ESI mass spectrum obtained for the component eluting at 17.6 min showed that it contains the glycosylated heavy chain. For example, the observed mass of 51,540 Da is that of the heavy chain with G0F carbohydrate structure at each glycosylation site, and the observed mass of 51,994 Da is that of the heavy chain with G0F carbohydrate structure at the Fc glycosylation site and G1F + NeuAc at the Fab glycosylation

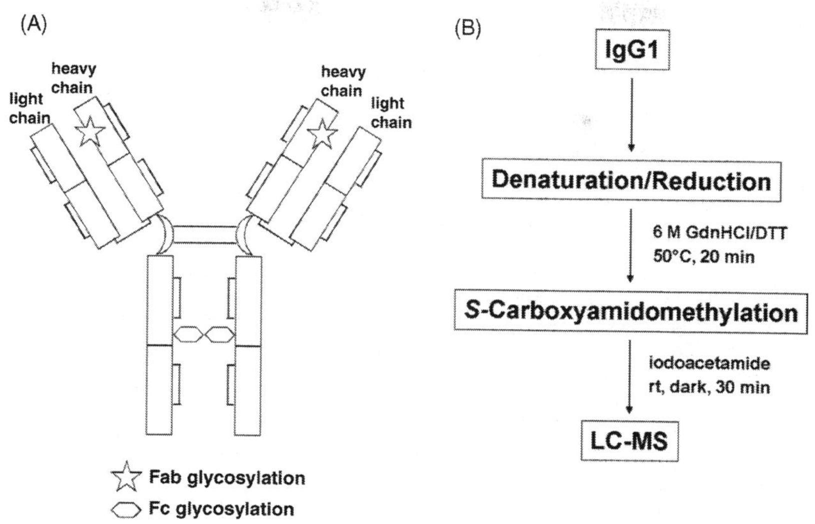

FIGURE 11.8 (*A*) Schematic representation of an IgG1 monoclonal antibody with two *N*-linked glycosylation sites in the heavy chain, one site in the Fab region and the other site in the Fc region. (*B*) An LC-MS-based approach for the identification of the components in the analytical SEC backside peaks of this highly sialylated monoclonal antibody with two *N*-linked glycosylation sites. Briefly, 100 μg of each sample was dried using a SpeedVac concentrator (Thermo Scientific Savant) and then subjected to denaturation and reduction in the presence of 6 M GdnHCl and 50 μg DTT at 50°C for 20 min. Each sample was then *S*-carboxyamidomethylated with 1.3 μL of 500 mM iodoacetamide in the dark at room temperature (rt) for 30 min. These samples were submitted for LC-MS analysis as described in the figure caption for Figure 11.9.

site. Each observed mass of the glycosylated heavy chain includes *S*-carboxyamidomethylation of the eleven Cys residues in the heavy chain.

Similarly LC-MS analysis was done on the HIC fraction 12 sample after denaturation, reduction, and *S*-carboxyamidomethylation (Figure 11.9*B*). Like the TIC of the control sample (Figure 11.9*A*), the components eluting at 16.9 and 17.6 min contain the light chain and glycosylated heavy chain, respectively (data not shown). Unlike the TIC of the control sample, however, the TIC of HIC fraction 12 also showed a broad peak with retention time of 15.8 min. The ESI mass spectrum of the components eluting at this time (inset of Figure 11.9*B*) indicated that their masses are 26,864 Da and 27,026 Da, which are the molecular masses of the G0F and G1F glycoforms of a heavy chain fragment containing residues 219–441 (theoretical masses 26,863 Da and 27,025 Da), respectively. Each mass includes *S*-carboxyamidomethylation of the six Cys residues in each heavy chain fragment.

LC-MS analysis of the HIC fraction 12 sample, using the new approach shown in Figure 11.8*B*, indicated the presence of the G0F and G1F glycoforms of the heavy chain fragment containing residues 219–441. The presence of this fragment suggests that the IgG1 antibody molecule undergoes a cleavage between residues Thr218 and His219 to generate the Fab + Fc and Fab fragments, as illustrated in Figure 11.10. These residues

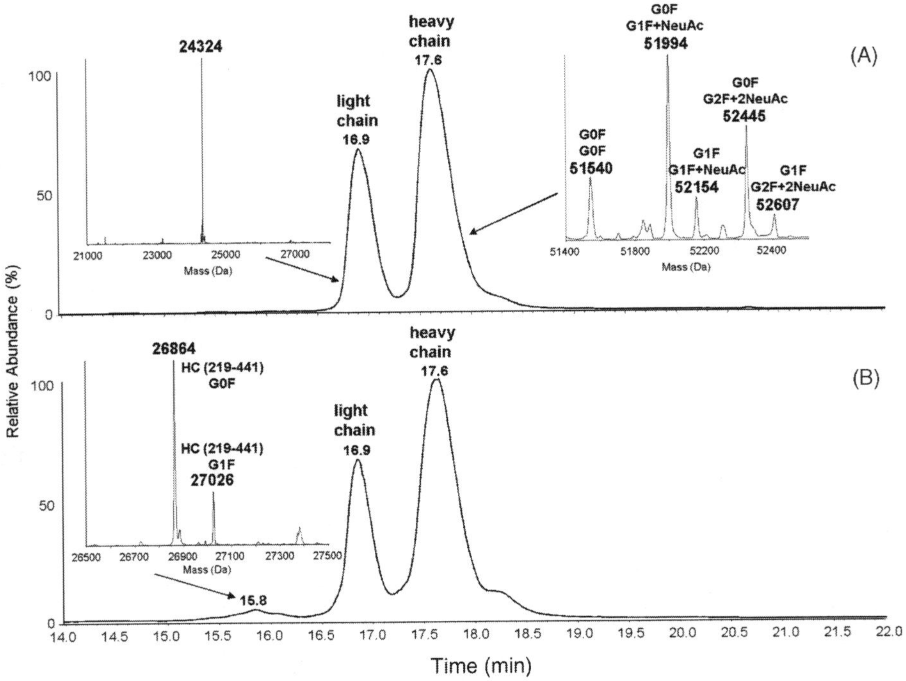

FIGURE 11.9 TICs of the (*A*) control sample and (*B*) HIC fraction 12 after denaturation, reduction, and *S*-carboxyamidomethylation. For the heavy chain, the carbohydrate nomenclature shown on top represents the carbohydrate structure located in the constant region of the heavy chain while the carbohydrate nomenclature shown at the bottom represents the carbohydrate structure located in the variable region. G0F and G1F represent fucosylated, complex biantennary glycans with 0 and 1 galactose residue, and NeuAc denotes *N*-acetylneuraminic acid. For each sample, 3 μg was loaded onto a Polymer Laboratories PLRP-S HPLC column (2.1 × 50 mm, 5 μ, 1000 Å) and analyzed using the Agilent 1100–Applied Biosystems QSTAR XL LC-MS system. HPLC solvents A and B were 0.05% TFA/0.1% formic acid in water and 0.05% TFA/0.1% formic acid in acetonitrile, respectively. The LC flow rate was 200 μL/min, and the column temperature was maintained at 60°C. The QSTAR XL was equipped with a TurboIonSpray source and operated in the positive-ion mode.

are in the hinge region of the IgG1 antibody molecule. Given that SEC separates on the basis of size, the first backside peak in the analytical SEC chromatogram most likely corresponds to the Fab + Fc fragment whereas the second backside peak corresponds to the Fab fragment, both resulting from a cleavage in the hinge region. Hinge-region cleavage of IgG1 monoclonal antibodies is known, is nonenzymatic, and is sensitive to the presence of Cu^{2+} ions, temperature, and hydroxyl radicals [52–55].

11.5 FORMULATION DEVELOPMENT

Modifications that affect the potency, purity, safety, and stability of recombinant therapeutic proteins include deamidation, aggregation, isomerization, cleavage, and

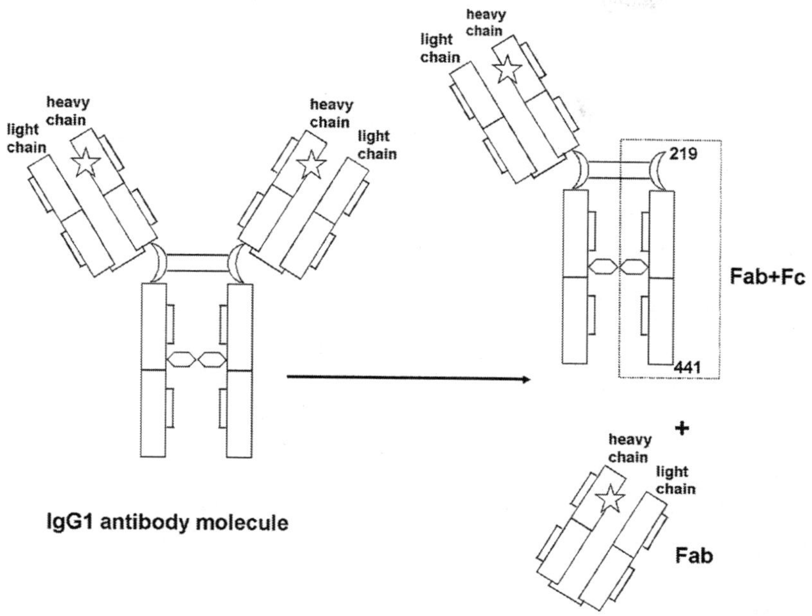

FIGURE 11.10 A schematic diagram illustrating the generation of the Fab + Fc and Fab fragments from hinge region cleavage of the IgG1 monoclonal antibody with two *N*-linked glycosylation sites in the heavy chain.

oxidation. According to ICH Q1A(R2), stress studies should be performed to understand the potential degradation pathways of therapeutic proteins [56]. During formulation development, it is essential to perform forced degradation studies to pinpoint those regions that are susceptible to degradation. The following examples illustrate how LC-MS plays an important role in the identification of potential degradation pathways during formulation development of one of our recombinant IgG4 monoclonal antibodies under development.

To understand the susceptibility to oxidation of an IgG4 recombinant monoclonal antibody in development, the protein was subjected to forced degradation in the presence of UV light, trace metal ions (Fe^{3+}, Ni^{2+}, and Cu^{2+}), and hydrogen peroxide. Before analyzing the stressed samples by LC-MS, we treated each sample with PNGase F to remove *N*-linked glycosylation from the IgG4 antibody molecule and then partially reduced with DTT. In the TIC, the raw mass spectra for the light chain and the heavy chain were summed together and deconvoluted to obtain molecular mass information (data not shown). The relative amounts of each oxidized form were assigned by dividing the peak area of the oxidized form by the sum of the peak areas of the oxidized and the unmodified forms taken from the deconvoluted mass spectra. The results indicate that this IgG4 monoclonal antibody is not susceptible to oxidation in the presence of the trace metal ions (Fe^{3+}, Ni^{2+}, and Cu^{2+}). It is, however, susceptible to oxidation in the presence of UV light and H_2O_2. The samples stressed by UV light and H_2O_2 were peptide mapped by LC-MS to obtain information on Met oxidation. This IgG4 monoclonal antibody has a Met residue in one of the complementarity determining regions (CDR) of the heavy chain. Oxidation of this Met residue could affect the binding affinity of the molecule and, therefore, compromise the efficacy of the therapeutic antibody.

Protein oxidation often occurs at Cys, Trp, Tyr, Phe, His, Ile, Leu, Pro, and Met residues; of these residues, Met is most susceptible. This IgG4 monoclonal antibody has a total of 14 Met residues. Each light chain has a Met residue at position 4, and each heavy chain has 6 Met residues at positions 48, 81, 99, 252, 358, and 428. The LC-MS peptide map data provides information on which of these Met residues are susceptible to oxidation. The LC-UV chromatogram (Figure 11.11) from the LC-MS peptide map analysis of the sample that had been treated with 5 ppm H_2O_2 and incubated at 25°C for 2 weeks is similar to the TIC (data not shown). The major difference is the peaks shown in the TIC have elution times about 1 min later, which is the time difference between arrival at the UV detector and the mass spectrometer. Mass spectra obtained from the peptide map TIC peaks at retention times 45.0 min (UV peak 44.1 min) and 55.8 min (UV peak at 54.8 min) are shown as insets in Figure 11.11. The component eluting at 55.8 min gives an $[M + H]^+$ of *m/z* 835.443, in close agreement with a peptide mass of 834.436 Da which corresponds to the tryptic peptide of this IgG4 monoclonal antibody with residues 249–255. The ion of *m/z* 857.431 is the sodiated form of this peptide. Given that this peptide contains a Met at position 252, one expects its oxidized form to be detected as an $[M + H]^+$ of *m/z* 851.426. This *m/z* was seen, but the oxidized peptide co-eluted with the unmodified peptide, indicating that oxidation resulted in the ESI process [57]. The mass spectrum of the component eluting at 45.0 min showed an $[M + H]^+$ of *m/z* 851.440; this ion corresponds to the oxidized form of the peptide. This oxidation was most likely caused by H_2O_2. Oxidized peptides are less hydrophobic than their unmodified counterparts and tend to elute from a reversed-phase column at earlier retention time [58].

An estimation of each oxidized peptide, expressed as the percentage of each oxidized peptide divided by the sum of the percentages of the oxidized and unmodified peptide (Figure 11.12), illustrates those Met residues that are susceptible to oxidation by UV light and H_2O_2. As indicated in Figure 11.12, only the heavy chain Met252- and Met358-containing tryptic peptides were slightly oxidized in the control sample; the light chain Met4- and the heavy chain Met48-, Met81-, Met99-, and Met428-containing tryptic peptides were not oxidized. When the IgG4 monoclonal antibody was exposed to UV light for 20 h, however, the light chain Met4- and the heavy chain Met252-, Met358-, and Met428-containing tryptic peptides were oxidized. A similar result was observed when the IgG4 monoclonal antibody was exposed to 5 ppm H_2O_2 and incubated at 25°C for 2 weeks, except that the amount of oxidation was much higher.

The oxidation of the light chain Met4- and the heavy chain Met252-, Met358-, and Met428-containing tryptic peptides increased dramatically when the IgG4

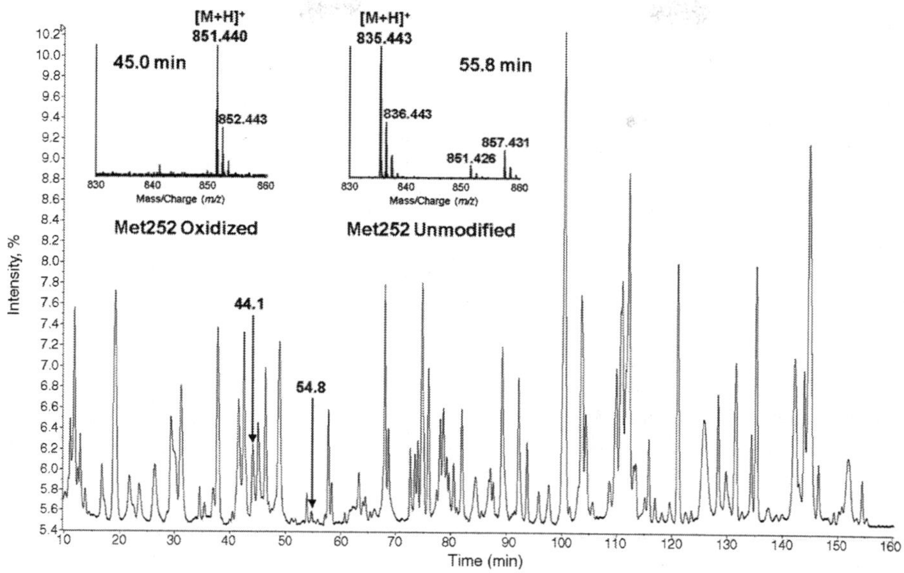

FIGURE 11.11 LC-UV chromatogram of the LC-MS peptide map analysis of the IgG4 monoclonal antibody stressed sample (incubating with 5 ppm H_2O_2 at 25°C for 2 weeks). The stressed sample was denatured using 0.2% RapiGest SF (Waters) in 50 mM NH_4HCO_3, pH 8, reduced with 10-fold molar excess of DTT over expected disulfides at 65°C for 45 min, *S*-carboxyamidomethylated with 5-fold molar excess of iodoacetamide over total thiols for 30 min at room temperature in the dark, and digested with trypsin at 37°C overnight using a 1:20 enzyme substrate ratio. The resulting tryptic digest was submitted for LC-MS analysis using a Vydac 218MS52 (2.1 × 250 mm, 5 μ, 300 Å) with the Agilent 1100–Applied Biosystems QSTAR XL LC-MS system. HPLC solvents A and B were 0.05% TFA in water and 0.04% TFA in acetonitrile, respectively. The LC flow rate was 200 μL/min, and the column temperature was maintained at 45°C. The QSTAR XL was equipped with an IonSpray source and operated in the positive ion mode. The LC flow rate was split 1:20 using an LC Packings Acurate flow splitter in order for the flow rate to be compatible with the flow rate normally introduced into the mass spectrometer with an IonSpray source.

monoclonal antibody was exposed to UV light for 20 h or incubated with 5 ppm H_2O_2 at 25°C for 2 weeks. These results suggest that the light chain Met4 and the heavy chain Met252, Met358, and Met428 residues are accessible to solvent. In contrast, the oxidation of the heavy chain Met48, Met81, and Met99 residues was not significantly affected by either the UV light or H_2O_2, suggesting that these Met residues are not accessible to solvent. In general, Met residues that are more exposed to solvent are more susceptible to oxidation whereas less-exposed Met residues are less susceptible [59–62]. Oxidation of the conserved Fc Met residues does affect the structure and stability of the CH2 and CH3 domains of human IgG1 [63] and the ability of the antibody to bind to FcRn receptor [64]. Thus it is important to develop conditions to minimize Met oxidation during formulation development.

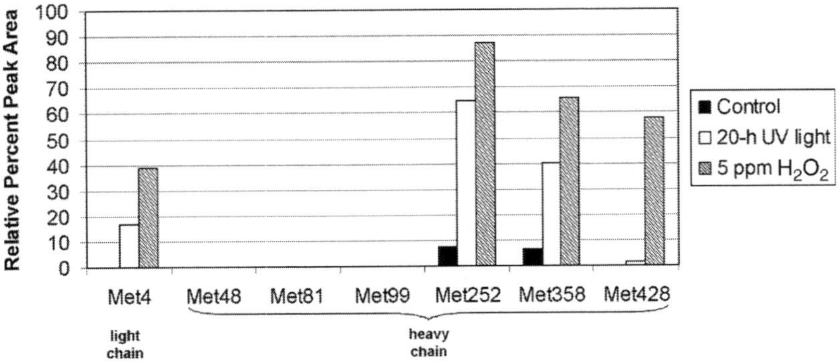

FIGURE 11.12 A graphical representation, taken from the LC-MS peptide map analysis of the IgG4 monoclonal antibody stressed samples, showing which Met residues are susceptible to oxidation. The relative level of each Met oxidation was determined by dividing the percent peak area of each oxidized Met-containing tryptic peptide by the sum of the percent peak areas of the oxidized and unmodified Met-containing tryptic peptides.

11.6 ANALYTICAL METHOD DEVELOPMENT

Analytical control strategies for well-characterized recombinant therapeutic protein products generally require the use of multiple, orthogonal methods for monitoring lot-to-lot differences in product/process impurities and differences in degradation profiles throughout the lifetime of the product. The utilization of MS has led to a much broader understanding of the modifications to which recombinant proteins are susceptible during cell culture production, purification, formulation, and storage. We selected two specific examples to illustrate how MS can play an important role during the development of recombinant therapeutic proteins, giving identification of unknowns seen as chromatographic peaks and elucidation of degradation mechanisms.

11.6.1 Utilization of Partial Reduction and LC-MS to Distinguish an IgG4 Monoclonal Antibody Charge Variants That Co-elute in Cation Exchange HPLC

Charge-based separation techniques including isoelectric focusing polyacrylamide gel analysis (IEF), cation exchange HPLC (CEX-HPLC), capillary isoelectric focusing (cIEF), and capillary zone electrophoresis (CZE) can monitor monoclonal antibody charge heterogeneity. The various charged forms of monoclonal antibodies produced by mammalian cell culture typically include the C-terminal lysine variants [65,66], antibody fragments [53,67], *N*-terminal glutamine/pyroglutamate variants [27,68], Fc Met oxidized forms [61], Trp oxidized forms [69,70], *C*-terminal amidated forms [71], deamidation products [72,73], succinimide forms [72,73], glycated forms [74,75], and glycosylated forms with varying degrees of sialyation [48,76].

CEX-HPLC and other charge-based separations for monitoring charge heterogeneity are an integral part of the analytical control strategy for most therapeutic monoclonal antibodies. These approaches are used at the time of product release to monitor lot-to-lot consistency and to evaluate rates of antibody degradation during drug stability studies. As part of our CEX-HPLC method development, we typically use LC-MS to analyze isolated CEX chromatographic fractions of stressed and nonstressed antibody samples to identify charge variant peaks in the CEX chromatogram. The approach involves using partial reduction to reduce the monoclonal antibody into the light and heavy chains and, in some cases, *N*-linked deglycosylation using PNGase F to remove the *N*-linked oligosaccharides prior to LC-MS analysis. Additional analysis may include obtaining an LC-MS peptide map of the isolated CEX fractions to confirm the nature and location of the modifications within the protein.

CEX-HPLC analysis of one of our IgG4 monoclonal antibodies in development, expressed in a CHO fed-batch culture, showed an approximately twofold increase in the relative abundance of a specific basic variant (BV3) relative to a previous process that used an NS0 cell line. LC-MS analysis following partial reduction of the BV3 CEX-HPLC fractions from the NS0- and CHO-derived lots indicated that the twofold increase was due to an incomplete pyroglutamate (Pyr) formation at the *N*-terminal glutamine (Gln) residue of the heavy chain in the CHO-derived material. LC-MS analysis after partial reduction of the drug products from both lots indicated that the products produced by the NS0 cell line contained about 2% incomplete Pyr (antibody with Gln at the *N*-terminus) whereas that derived from the CHO cell line had about 10% incomplete Pyr (data not shown).

In addition stability studies on the NS0-derived material showed increasing levels of BV3 after storing at accelerated temperatures. In contrast, comparative stability studies of the CHO-derived material showed decreasing levels of BV3 under identical storage conditions. To understand the changes in the CEX profiles for the materials produced using the two different cell lines, we performed LC-MS analysis after partial reduction analysis of the isolated CEX-HPLC fractions from the stability samples of the IgG4 monoclonal antibody drug product derived from both the NS0 and CHO processes.

We obtained the TIC for the CHO-derived IgG4 monoclonal antibody drug product after storing at 25°C for 9 months (Figure 11.13*A*) and compared it to the corresponding LC-UV chromatogram (Figure 11.13*B*). The TIC (Figure 11.13*A*) shows four distinct peaks at 14.7, 16.3, 17.4, and 18.0 min; their deconvoluted ESI mass spectra are shown as insets. The mass spectrum of the low abundance component eluting at 18.0 min is not shown; this material contains the heavy chain with an additional reduction of one of the intra-chain disulfide bonds, resulting from an artifact of the sample preparation. Partial reduction should only reduce the inter-chain disulfide bonds between the light chain and heavy chain and between the two heavy chains, leaving all the intra-chain disulfide bonds intact. It is possible, however, for the DTT to reduce some of the intramolecular disulfide bonds during sample preparation.

The ESI mass spectrum obtained for the proteins eluting at 16.3 min (inset of Figure 11.13) shows their masses to be 23,607 and 23,770 Da, corresponding to those of the unmodified light chain and the glycated light chain (theoretical masses 23,607 and 23,770 Da), respectively. The ESI mass spectrum obtained for the eluent at 14.7 min shows a single component of mass 23,589 Da, which is 18 Da lower in mass

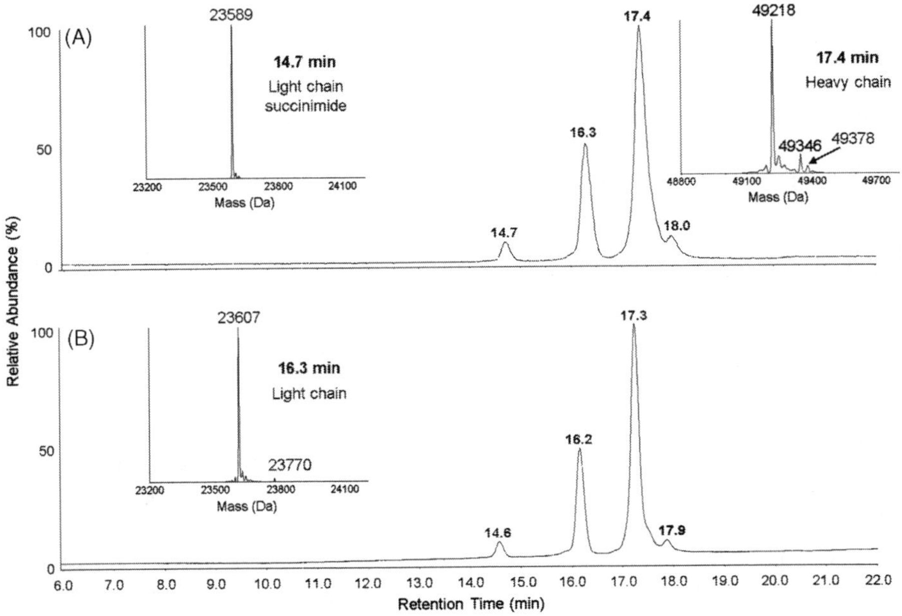

FIGURE 11.13 LC-MS partial reduction analysis of the IgG4 monoclonal antibody stressed sample (incubating at 25°C for 9 months), showing the (*A*) LC-MS TIC and (*B*) LC-UV chromatograms. Deconvoluted ESI mass spectra taken from the TIC at various retention times are shown as insets. The IgG4 monoclonal antibody was subjected to *N*-linked deglycosylation and partial reduction. The sample was analyzed using a Polymer Laboratories PLRP-S HPLC column (2.1 × 50 mm, 5 μ, 1000 Å) with the Agilent 1100–Applied Biosystems QSTAR XL LC-MS system. HPLC solvents A and B were 0.05% TFA/0.1% formic acid in water and 0.05% TFA/0.1% formic acid in acetonitrile, respectively. The LC flow rate was 200 μL/min, and the column temperature was maintained at 60°C. The QSTAR XL was equipped with a TurboIon-Spray source and operated in the positive-ion mode.

than that of the light chain. There are multiple reactions in proteins that lead to a loss of a water molecule. One is an Asp isomerization in the light chain to form isoAsp and Asp through a cyclic imide (succinimide) intermediate (theoretical mass 23,589 Da). The light chain of this monoclonal antibody has an Asp-Gly consensus sequence for an Asp isomerization. LC-MS peptide mapping confirmed this assignment. Similarly LC-MS analysis after partial reduction of the isolated CEX-HPLC BV3 fractions from the NS0- and CHO-derived drug substance lots at the initial time point showed comparable levels of the succinimide form of the light chain (data not shown). The LC-MS data also indicated that the succinimide form of the light chain in both lots consistently increased during accelerated storage conditions.

The ESI mass spectrum for the eluent at 17.4 min (inset of Figure 11.13) shows three components of masses 49,218, 49,346, and 49,378 Da, which are the masses of the heavy chain, the heavy chain with a *C*-terminal Lys, and the heavy chain with glycation (theoretical masses 49,216, 49,344, and 49,379 Da), respectively. These results confirm that the most abundant form of the heavy chain (49,218 Da) in the accelerated stability sample is one in which the heavy chain with its *N*-terminal Gln residue is converted to Pyr. There is no significant peak in the deconvoluted mass spectrum for this sample at 49,233 Da (+17 Da), corresponding to the expected mass of the heavy chain with Gln at the *N*-terminus.

Cyclization of Gln to Pyr at the *N*-terminus is a spontaneous reaction that is highly dependent on temperature and buffer conditions. Mammalian cell culture process conditions for most monoclonal antibodies accelerate the non-enzymatic conversion of the *N*-terminal Gln to Pyr in the heavy chain; most processes yielding almost complete conversion (>95%) to the Pyr form [77]. *N*-Terminal Gln residues that are not converted to Pyr during bioprocessing contribute to charge heterogeneity observed in charge-based separations. Unlike the *N*-terminal Pyr containing a neutral ring amide, the *N*-terminal Gln has a free amino terminus, making the antibody more basic than the dominant Pyr form. Similar to the facile conversion of the expected antibody *N*-terminal Gln forms to Pyr forms during mammalian cell culture production, the conversion of any *N*-terminal Gln forms remaining in the drug substance following cell culture production and purification may occur during storage of the drug, especially under accelerated conditions.

The MS data obtained for the component eluting at 17.4 min (inset of Figure 11.13) indicate that the *N*-terminal Pyr form is the most abundant form of the antibody heavy chain in the CHO-derived accelerated stability sample. Similar LC-MS data obtained for an unstressed sample of the same CHO-derived lot showed significantly higher levels (approximately 10%) of the *N*-terminal Gln form of the heavy chain. These results indicate that the *N*-terminal Gln residue is converted to Pyr, as expected, in the CHO-derived material upon storage of the drug product under accelerated conditions.

The conversion of the *N*-terminal Gln to Pyr in the heavy chain for the CHO-derived drug product occurs gradually with storing at various accelerated conditions (Figure 11.14*A*, from LC-MS peptide map analysis). The succinimide on the light chain also forms slowly for the same material stored under identical conditions (from LC-MS of partially reduced material, Figure 11.14*B*). The data indicate that the Asp residue in the Asp-Gly sequence in the light chain readily isomerizes to the succinimide at accelerated conditions (e.g., 40°C) whereas the *N*-terminal Gln residue in the heavy chain slowly converts to Pyr under less aggressive conditions (e.g., 5°C or short-term 25°C storage). Given that CEX-HPLC BV3 is comprised of both the light chain succinimide and the *N*-terminal Gln antibody variants, it is not surprising that we observed net decreases in BV3 for CHO-derived drug product under milder storage conditions and net increases in BV3 at more accelerated conditions. These results also provide an explanation for the observed differences in BV3 rate of increase and decrease for the NS0- and CHO-derived antibody products, respectively. Because the NS0-derived material do not contain appreciable levels of the heavy chain *N*-terminal Gln variant, the changes in CEX-HPLC BV3 result only from increases in the light chain succinimide antibody variant. Conversely, changes in BV3 for the CHO-derived materials result from both formation of the light chain succinimide variant and conversion of the heavy chain N-terminal Gln variant to the more abundant Pyr form. The net increase/decrease in the modified forms of the CHO-derived antibody product may differ depending on the storage conditions and duration, as shown in Figure 11.14.

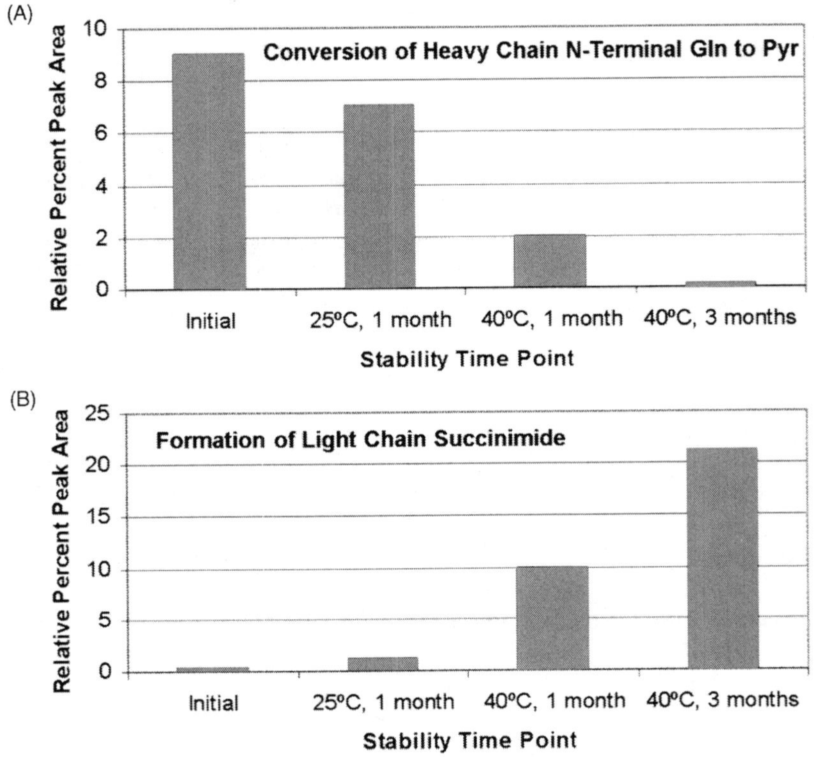

FIGURE 11.14 A graphical representation taken from the (*A*) LC-MS peptide map analysis of the IgG4 monoclonal antibody stability samples (showing the conversion of the *N*-terminal Gln to Pyr in the heavy chain) and (*B*) LC-MS partial reduction analysis of the IgG4 monoclonal antibody stability samples (showing succinimide formation of the light chain as a result of Asp isomerization). The LC-MS partial reduction conditions are described in the figure caption of Figure 11.13, and the LC-MS peptide mapping conditions were similar to those described in the caption of Figure 11.11. The data for the formation of succinimide in the light chain were taken from the LC-MS partial reduction rather than the LC-MS peptide map experiment because the succinimide intermediate is more stable in the light chain protein than in the light chain tryptic peptide. For either experiment, the monoclonal antibody was treated with an enzyme (either trypsin or PNGase F) at pH 8 for 15 h. Under these conditions the succinimide intermediate was more stable in the light chain protein (partial reduction analysis) than in the light chain tryptic peptide (peptide map analysis) because, conformationally, the succinimide intermediate was not as solvent accessible when it was present in the light chain protein but more solvent accessible when it was present in the light chain tryptic peptide. In contrast, the data for the conversion of Gln to Pyr in the heavy chain were taken from the LC-MS peptide map experiment because the peptide map offered better chromatographic separation of the *N*-terminal Pyr form from the *N*-terminal Gln form.

11.6.2 Development of an RP-HPLC Method for Monitoring an IgG4 Fc Fusion Protein Post-Translational Modifications

We implemented online RP-HPLC MS to optimize an RP-HPLC-UV release and stability method for resolution of post-translationally modified forms of an IgG4 Fc fusion protein produced by CHO fed-batch cell culture. The RP-HPLC-UV method included reduction of the homodimeric Fc fusion protein by using DTT under denaturing conditions prior to RP-HPLC analysis. Denaturation and reduction of the two-chain Fc fusion protein into its individual chains resulted in a significant improvement in the RP-HPLC resolution of the post-translationally modified forms of the protein. MS was employed during method development for the identification of unknown components eluting from the RP-HPLC. The outcomes of the analysis are a TIC obtained by LC-MS analysis of the reduced Fc fusion protein following storage at 5°C for 12 weeks (Figure 11.15*A*) and an LC-UV chromatogram (Figure 11.15*B*). Knowing the time difference between the UV and MS detectors (approximately 0.1 min), we could readily compare a series of peaks in both chromatograms. The MS data, shown as insets, are effective for identifying the various eluents. For example, the ESI mass spectrum obtained for components eluting at 13.5 min shows that the dominant material has a molecular mass of 31,287 Da. This mass is consistent with that of the reduced Fc fusion protein with a G0F carbohydrate structure (theoretical mass 31,287 Da) at the conserved Asn-Xxx-Thr site in the Fc region, where Xxx can be any amino acid residue except Pro. The single-chain Fc fusion protein containing the G1F carbohydrate moiety was also detectable; its observed mass is 31,448 Da (theoretical 31,449 Da).

Similarly analysis of the MS data (insets of Figure 11.15) indicated that the additional components eluting at 12.9, 14.2, and 14.8 min (Figure 11.15) are the oxidized/hydroxylated, phosphorylated, and nonglycosylated forms of the Fc fusion protein, respectively. Separate experiments utilizing LC-MS peptide mapping confirmed that the oxidation sites are a Trp residue in the binding region of the Fc fusion protein and Met residues in the Fc region. Note that we have not been able to differentiate Trp oxidation from Trp hydroxylation in this sample. Like oxidation of Trp, hydroxylation of Trp also gives rise to a + 16 Da mass shift. Hydroxylation of Trp involves the incorporation of a single oxygen atom into the Trp side chain and can occur enzymatically during cell culture production [78]. The low-level Trp/Met oxidations of the Fc-fusion protein may result from the fed-batch CHO cell culture expression of the recombinant Fc fusion protein. These modifications likely arise from the reaction of the Trp and Met residues with reactive oxygen species generated by oxidative stress during cell culture production [79,80]. The ESI mass spectrum obtained for the components at 14.2 min (inset of Figure 11.15) shows that the dominant species has a mass at 31,367 Da, which is 80 Da higher than that of the unmodified reduced form of the Fc fusion protein. This 80 Da modification may result from phosphorylation or sulfation. The species of mass 31,269 Da, 98 Da lower in mass than the mass at 31,367 Da, may arise by a facile loss of H_3PO_4 from a phosphorylated Ser or Thr [81,82]. LC-MS analysis of an aliquot of the sample after treatment with alkaline phosphatase showed the disappearance of the +80 Da

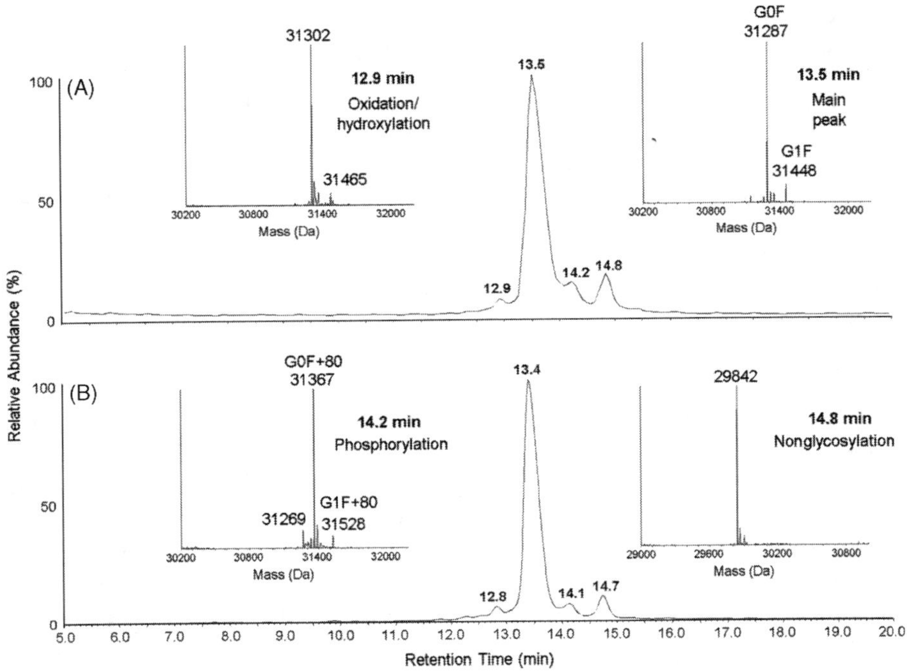

FIGURE 11.15 LC-MS analysis of the recombinant IgG4 Fc fusion protein, showing the (*A*) TIC and (*B*) UV chromatograms. Deconvoluted ESI mass spectra taken from the TIC at various retention times are shown as insets. The IgG4 Fc fusion protein was treated with DTT in 6 M GdnHCl/0.1 M ammonium bicarbonate, pH 8 and incubated at 37°C for 30 min. The sample was analyzed using an Advanced Chromatography Technologies ACE 3 C4-300 (4.6 × 150 mm, 3 μ, 300 Å) with a Waters Alliance Model 2795 HPLC system coupled to a Waters LCT Premier oaTOF mass spectrometer. HPLC solvents A and B were 0.05% TFA in 20% acetonitrile, and 0.05% TFA in 90% acetonitrile, respectively. The LC flow rate was 1000 μL/min, and the column temperature was maintained at 60°C. For the LC flow rate to be compatible with the flow rate normally introduced into the mass spectrometer, the LC flow rate was reduced to approximately 200 μL/min using a Valco stainless steel flow splitting tee.

modification, providing supporting evidence that this modification is due to phosphorylation and not sulfation (data not shown). Using LC-MS peptide mapping, we located the phosphorylation site at the linker region that connects the binding region of the therapeutic protein to the IgG4 Fc sequence. Offline nanoelectrospray MS/MS analysis identified the site of phosphorylation to be Ser46 in the linker region (data not shown). A final modified form of the Fc fusion protein, which elutes at 14.8 min, is non-glycosylated that does not contain the expected *N*-linked oligosaccharides at the Asn-Xxx-Thr consensus site. Nonglycosylated forms of monoclonal antibodies in which the Fc heavy chain *N*-linked glycans are not incorporated during cell culture expression were documented by others [83].

Tentative RP-HPLC identifications, assignment based on the ESI mass spectra for the reduced Fc fusion protein, were confirmed in subsequent LC-MS peptide mapping. These experiments involved trypsin digestion of enriched fractions from the RP-HPLC separation followed by LC-MS analysis of the digest. When necessary, online LC-MS/MS or offline nanoelectrospray MS/MS analysis was employed to identify the position of the modification. Online LC-MS for the initial development of an initial RP-HPLC method for an Fc fusion protein is essential for understanding the various modified forms of the protein. The RP-HPLC method was used subsequently for process optimization studies to evaluate the impact of cell culture and purification process conditions on the levels of the various modified forms in the final drug substance.

Incorporation of orthogonal MS-based approaches into the development of standard methods for monitoring therapeutic recombinant protein product quality is essential for identifying the substances and impurities that are responsible for lot-to-lot differences and profile changes on stability. The two examples highlighted above demonstrate the effectiveness of integrating MS during the initial development of analytical methods for recombinant proteins to identify unknown components in the complex chromatographic profiles of these products.

11.7 CONFIRMATION OF STRUCTURE/PRODUCT COMPARABILITY ASSESSMENT

According to ICH Q6B, extensive characterization of recombinant therapeutic proteins must be performed throughout clinical/commercial development and after significant process changes [84]. Throughout the useful lifetime of a recombinant therapeutic protein, it is often necessary to make changes to improve product quality and yield. These changes may involve using a different cell line, modifying cell culture conditions, adding additional purification step(s), or using a different formulation. These changes can affect product quality and result in lots with different levels of protein microheterogeneity. As a result major process changes require a comparability assessment using an assortment of orthogonal analytical methods to demonstrate product comparability between the lots produced before and after the changes [13,85,86]. In addition, according to ICH Q5E, it is essential to perform lot-to-lot characterization to help ensure product comparability/consistency and to demonstrate that those changes have not affected the product identity, purity, potency, strength, safety, and stability [87].

This extensive characterization includes verifying the primary structure of the protein and assessing any microheterogeneity. For recombinant monoclonal antibodies and antibody-based Fc fusion proteins, microheterogeneity may arise from *N*-linked glycosylation, *N*-terminal pyroglutamation, *C*-terminal Lys processing, *C*-terminal amidation, Met oxidation, Asn deamidation, Asp isomerization, unpaired Cys residues, cleavage products, and glycation [88–90].

MS plays an important role during the comparability assessment. It affords the molecular mass of the intact protein for the lots produced before and after the changes to assess the overall protein sequence and microheterogeneity. To compare the old and new lots, the proteins from each lot are enzymatically digested. The resulting digests, along with a 1:1 (volume/volume, v/v) co-mixture, are analyzed using RP-HPLC with both UV and MS detection. A 1:1 (v/v) co-mixture is included in the analysis to determine whether the presence of a new peak in the peptide map of one lot is due to system variability or whether it is evidence of product incomparability [91]. Characterization of each component in the peptide map is performed for one of the lots using the MS data. The LC-UV chromatographic profiles are then visually examined for consistency in the overall elution pattern with respect to the number of peaks and the relative peak retention time, peak shape, and peak intensity. This visual comparison helps establish whether the new lot is comparable to the old lot. If the retention time, peak shape, and peak intensity are slightly different, then analysis of the MS data of the peptide map of the suspected lot is performed.

Visual examination of the LC-UV chromatographic profiles is preferred over the TIC profiles because the UV data are for peptide bond absorption at 214 nm. This absorption leads to larger relative peak intensities for larger peptides than for smaller peptides, making the UV data more reliable for indicating the relative weights of each peptide. In addition LC-UV is more routine, once all the peaks have been identified by MS. The TIC chromatographic profiles are usually not used because these ESI MS data suffer from ionization efficiency differences; that is, some peptides ionize more favorably giving artificially higher abundances. For example, smaller peptides usually ionize better than larger peptides. Furthermore, unlike UV, which measures mainly peptide bonds at 214 nm, MS detects any ionizable substance, affording additional peaks not related to the protein. The following example illustrates how LC-MS plays an important role during the product comparability assessment of a recombinant IgG4 Fc fusion protein after implementation of a new cell line.

This recombinant IgG4 Fc fusion protein was initially produced using an NS0 cell line. The Fc protein produced from the NS0 cell line had been fully characterized using MS, showing that the experimental molecular mass of the protein from the LC-MS analysis of the intact protein was consistent with the theoretical mass. LC-MS peptide mapping of the NS0-derived material provided 100% sequence coverage. During the development of this Fc fusion protein, it was necessary to switch the cell line from NS0 to CHO. As a result of the switch, it was essential to perform a product comparability assessment for the lots produced by the two different cell lines.

LC-MS analysis was performed on the intact protein from each lot after treatment with PNGase F to remove the *N*-linked glycosylation. The ESI mass spectrum of the protein produced from each lot showed a peak at 59,675 Da, which is the mass of the two-chain Fc fusion protein without the *N*-linked glycosylation (theoretical mass 59,673 Da). LC-MS analysis was also performed on each lot after treatment with DTT to obtain information on the single-chain protein with *N*-linked glycosylation. LC-MS after partial reduction of the protein produced using the NS0 cell line yielded masses corresponding to the various glycoforms of the single-chain protein. The predominant masses were 31,283, 31,445, and 31,607 Da, which are consistent with the single-chain protein containing the G0F, G1F, and G2F carbohydrate structures (theoretical masses 31,283, 31,445, and 31,607 Da), respectively. In addition masses of low abundance products were observed at 31,770 and 31,915 Da, which are those of the nonhuman glycoforms G2F + αGal and G2F + NeuGc (theoretical masses 31,769 and 31,914 Da), respectively. Similarly LC-MS following partial reduction of the protein from the CHO cell line yielded masses at 31,283, 31,445, and 31,607 Da, which agree well with the masses of the single-chain protein containing the G0F, G1F, and G2F carbohydrate structures, respectively. Components with masses of the nonhuman glycoforms were not observed in the CHO-derived lot.

To assess the structural comparability of the CHO-derived protein versus the NS0-derived protein, both lots were subjected to trypsin digestion followed by RP-HPLC with both UV and MS detection (Figure 11.16). Visual examination of the LC-UV chromatographic profiles shows that the overall elution pattern is similar with respect to the number of peaks, relative peak retention time, relative peak shape, and relative peak intensity, except for the set of peaks around 50 min (marked with an arrow). This set of peaks corresponds to glycosylated peptides. Analysis of the corresponding mass spectra indicated that the observed differences can be attributed to the known differences in the carbohydrate structures between the NS0- and CHO-derived lots. Complete data analysis of the peptide map for the CHO-derived lot yielded information accounting for 100% sequence coverage of the 275 expected amino-acid residues of the primary sequence and confirmed the *N*-linked glycosylation at Asn126. In addition the LC-MS peptide map detected very low levels of post-translational and chemical modifications of the protein, including partial hydroxylation of Lys28; partial hydroxylation and glycosylation in the form of α-1,2-glucosylgalactosyl-*O*-hydroxylysine of Lys28; partial phosphorylation of Ser46; partial oxidation of Met81 and Met 257; partial nonglycosylation of Asn126; partial deamidation of Asn144, Asn213, Asn218, and Asn250; partial isomerization of Asp230; and desGly275 with subsequent *C*-terminal amidation of Leu274. The Ser46 phosphorylation and Leu274 *C*-terminal amidation were not detected in the NS0-derived lot. The partial oxidation, deamidation, and isomerization observed in the peptide map could result during sample preparation because the digestion was carried out overnight (~15 h) at 37°C. Hydroxylation of a Lys residue followed by glycosylation is known to occur in recombinant adiponectin [92].

11.8 CONCLUSIONS

Recombinant therapeutic proteins for human use must be characterized thoroughly to establish quality, safety, and efficacy and to satisfy the rigorous requirements set by global regulatory agencies. It is generally necessary to use multiple/orthogonal techniques for the comprehensive characterization of complex and heterogeneous recombinant proteins. Advances in various analytical techniques have made it possible to better characterize recombinant proteins. MS continues to be one of the key analytical techniques for protein characterization.

In this review we selected examples from our own laboratories to illustrate how MS can be utilized to help ensure the quality of biopharmaceutical protein products

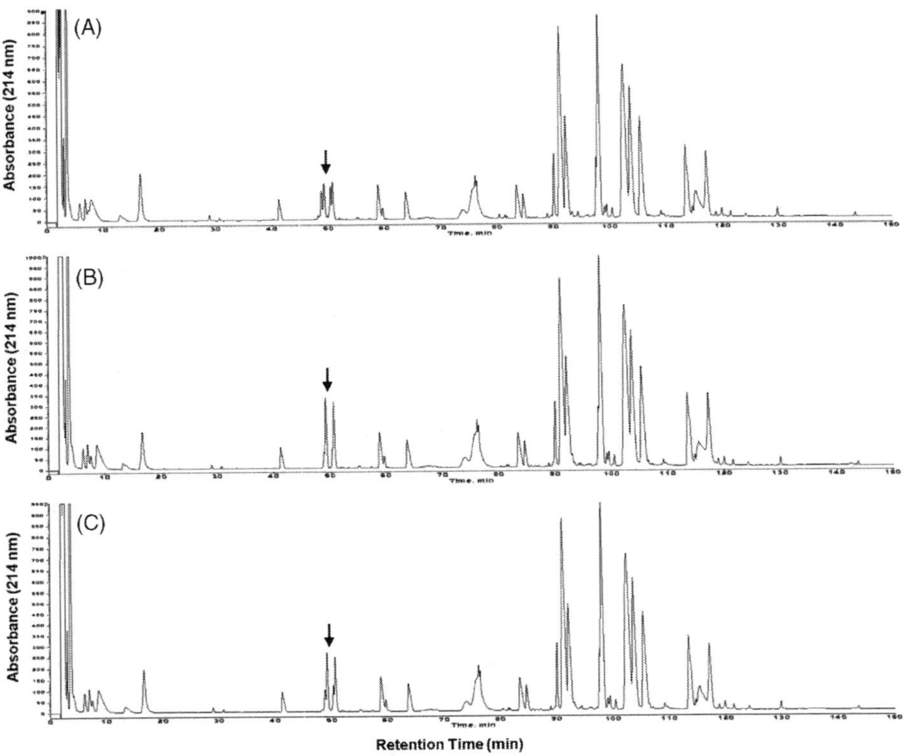

FIGURE 11.16 LC-UV chromatographic profiles taken from the LC-MS peptide map analysis of the recombinant IgG4 Fc fusion protein from the (*A*) NS0-derived lot, (*B*) CHO-derived lot, and (*C*) 1:1 co-mixture of the NS0- and CHO-derived lot. The Fc fusion protein was denatured using 6 M GdnHCl/0.25 M TrisHCl, pH 8, reduced with 10-fold molar excess of DTT over disulfides, *S*-carboxyamidomethylated with 2-fold molar excess of iodoacetamide over total thiols, and digested with trypsin (1:25, enzyme:substrate). The resulting tryptic digest was analyzed using a Waters XBridge C18 HPLC column (1.0 × 150 mm, 3.5 μ, ~137 Å) with an Agilent 1100 capillary–Applied Biosystems QSTAR XL LC-MS system. HPLC solvents A and B were 0.05% TFA in water and 0.04% TFA in acetonitrile, respectively. The LC flow rate was 50 μL/min, and the column temperature was maintained at 50°C. The QSTAR XL was equipped with a TurboIonSpray source and operated in the positive-ion mode.

during the developments stages of cell culture, purification, formulation, analytical methods, and clinical trial. The integration of MS as part of the arsenal of analytical tools for protein characterization is essential to select the top clones that provide optimal product quality and productivity, to optimize purification processes to remove product-related substances and impurities, to understand the degradation mechanisms, to identify unknown chromatographic peaks, to understand analytical method specificity, and to assess product comparability. This information is critical for ensuring the quality, safety, and efficacy of therapeutic proteins.

ACKNOWLEDGMENTS

We thank Monica Myers, Laura Ashey, Dr. Peter Lambooy, Steve Plichta, Barbara Williams, Rick Meyer, Sarah Demmon, John Richardson, Evgenia Pindel, Arati Pradhan, Dr. Kingman Ng, Andrew Werner, and Patrick Donovan for providing the various samples for LC-MS analysis. We also are grateful to Drs. Michael DeFelippis, Bryan Harmon, and Peter Lambooy for reviewing an early version of this chapter.

REFERENCES

1. Walsh, G. (2006). Biopharmaceutical benchmarks—2006. *Nat Biotechnol 24*, 769–776.
2. Pavlou, A. K., Reichert, J. M. (2004). Recombinant protein therapeutics—Success rates, market trends and values to 2010. *Nat Biotechnol 22*, 1513–1519.
3. Reichert, J. M. (2008). Monoclonal antibodies as innovative therapeutics. *Curr Pharmaceut Biotechnol, 9*, 423–430.
4. Aires da Silva, F., Corte-Real, S., Goncalves, J. (2008). Recombinant antibodies as therapeutic agents: pathways for modeling new biodrugs. *BioDrugs 22*, 301–314.
5. Maggon, K. (2007). Monoclonal antibody "gold rush." *Curr Med Chem 14*, 1978–1987.
6. Kozlowski, S., Swann, P. (2006). Current and future issues in the manufacturing and development of monoclonal antibodies. *Adv Drug Deliv Rev 58*, 707–722.
7. Hayakawa, T. (1997). Global perspective on specifications for biotechnology products—Perspective from Japan. *Developm Biol Standardiz 91*, 15–23.
8. Murano, G. (1997). FDA perspective on specifications for biotechnology products—From IND to PLA. *Developm Biol Standardiz 91*, 3–13.
9. Trouvin, J. H. (1997). Development of specifications for biotechnology products—Perspective from Europe. *Developm Biol Standardiz 91*, 25–30.
10. Federici, M. M. (1994). The quality control of biotechnology products. *Biologicals 22*, 151–159.
11. Zoon, K., Garnick, R. (1998). Definition of a well-characterized biotechnology product. *Developm Biol Standardiz 96*, 191–197.
12. Garnick, R. L., Solli, N. J., Papa, P. A. (1988). The role of quality control in biotechnology: An analytical perspective. *Anal Chem 60*, 2546–2557.
13. Chirino, A. J., Mire-Sluis, A. (2004). Characterizing biological products and assessing comparability following manufacturing changes. *Nat Biotechnol 22*, 1383–1391.
14. Carr, S. A., Hemling, M. E., Bean, M. F., Roberts, G. D. (1991). Integration of mass spectrometry in analytical biotechnology. *Anal Chem 63*, 2802–2824.
15. Nguyen, D. N., Becker, G. W., Riggin, R. M. (1995). Protein mass spectrometry: Applications to analytical biotechnology. *J Chromatogr A705*, 21–45.
16. Andersen, J. S., Svensson, B., Roepstorff, P. (1996). Electrospray ionization and matrix assisted laser desorption/ionization mass spectrometry: Powerful analytical tools in recombinant protein chemistry. *Nat Biotechnol 14*, 449–457.
17. Srebalus Barnes, C. A., Lim, A. (2007). Applications of mass spectrometry for the structural characterization of recombinant protein pharmaceuticals. *Mass Spectrom Revi 26*, 370–388.

18. Gibson, B. W., Biemann, K. (1984). Strategy for the mass spectrometric verification and correction of the primary structures of proteins deduced from their DNA sequences. *Proc Nat Acad Sci USA 81*, 1956–1960.
19. Smales, C. M., James, D. C., eds. (2005). *Therapeutic Proteins: Methods and Protocols.* Humana Press, Clifton, pp. 411–419.
20. Yu, Y. Q., Gilar, M., Lee, P. J., Bouvier, E. S., Gebler, J. C. (2003). Enzyme-friendly, mass spectrometry-compatible surfactant for in-solution enzymatic digestion of proteins. *Anal Chem 75*, 6023–6028.
21. Biemann, K., Scoble, H. A. (1987). Characterization by tandem mass spectrometry of structural modifications in proteins. *Science 237*, 992–998.
22. Wilm, M., Mann, M. (1996). Analytical properties of the nanoelectrospray ion source. *Anal Chem 68*, 1–8.
23. Morris, H. R., Paxton, T., Dell, A., Langhorne, J., Berg, M., Bordoli, R. S., Hoyes, J., Bateman, R. H. (1996). High sensitivity collisionally-activated decomposition tandem mass spectrometry on a novel quadrupole/orthogonal-acceleration time-of-flight mass spectrometer. *Rapid Commun Mass Spectrom 10*, 889–896.
24. Nemeth-Cawley, J. F., Rouse, J. C. (2002). Identification and sequencing analysis of intact proteins via collision-induced dissociation and quadrupole time-of-flight mass spectrometry. *J Mass Spectrom 37*, 270–282.
25. Rouse, J. C., McClellan, J. E., Patel, H. K., Jankowski, M. A., Porter, T. J. (2005). Top-down characterization of protein pharmaceuticals by liquid chromatography/mass spectrometry: Application to recombinant factor IX comparability- a case study. *Meth Mol Biol 308*, 435–460.
26. Gadgil, H. S., Bondarenko, P. V., Pipes, G. D., Dillon, T. M., Banks, D., Abel, J., Kleemann, G. R., Treuheit, M. J. (2006). Identification of cysteinylation of a free cysteine in the Fab region of a recombinant monoclonal IgG1 antibody using Lys-C limited proteolysis coupled with LC/MS analysis. *Anal Biochem 355*, 165–174.
27. Lyubarskaya, Y., Houde, D., Woodard, J., Murphy, D., Mhatre, R. (2006). Analysis of recombinant monoclonal antibody isoforms by electrospray ionization mass spectrometry as a strategy for streamlining characterization of recombinant monoclonal antibody charge heterogeneity. *Anal Biochem 348*, 24–39.
28. Hardman, M., Makarov, A. A. (2003). Interfacing the orbitrap mass analyzer to an electrospray ion source. *Anal Chem 75*, 1699–1705.
29. Macek, B., Waanders, L. F., Olsen, J. V., Mann, M. (2006). Top-down protein sequencing and MS3 on a hybrid linear quadrupole ion trap–orbitrap mass spectrometer. *Mol Cell Proteomics 5*, 949–958.
30. Scigelova, M., Makarov, A. (2006). Orbitrap mass analyzer—Overview and applications in proteomics. *Proteomics 6* (Suppl 2), 16–21.
31. Zhang, Z., Shah, B. (2007). Characterization of variable regions of monoclonal antibodies by top-down mass spectrometry. *Anal Chem 79*, 5723–5729.
32. Bondarenko, P. V., Second, T. P., Zabrouskov, V., Makarov, A. A., Zhang, Z. (2009). Mass measurement and top-down HPLC/MS analysis of intact monoclonal antibodies on a hybrid linear quadrupole ion trap–orbitrap mass spectrometer. *J Am Soc Mass Spectrom 20*, 1415–1424.
33. Syka, J. E., Coon, J. J., Schroeder, M. J., Shabanowitz, J., Hunt, D. F. (2004). Peptide and protein sequence analysis by electron transfer dissociation mass spectrometry. *Proc Nati Acad Sci USA 101*, 9528–9533.

34. McAlister, G. C., Phanstiel, D., Good, D. M., Berggren, W. T., Coon, J. J. (2007). Implementation of electron-transfer dissociation on a hybrid linear ion trap-orbitrap mass spectrometer. *Anal Chem 79*, 3525–3534.

35. McAlister, G. C., Berggren, W. T., Griep-Raming, J., Horning, S., Makarov, A., Phanstiel, D., Stafford, G., Swaney, D. L., Syka, J. E., Zabrouskov, V., Coon, J. J. (2008). A proteomics grade electron transfer dissociation-enabled hybrid linear ion trap–orbitrap mass spectrometer. *J Proteome Res 7*, 3127–3136.

36. Wang, H., Straubinger, R. M., Aletta, J. M., Cao, J., Duan, X., Yu, H., Qu, J. (2009). Accurate localization and relative quantification of arginine methylation using nanoflow liquid chromatography coupled to electron transfer dissociation and orbitrap mass spectrometry. *J Am Soc Mass Spectrom 20*, 507–519.

37. Goochee, C. F., Gramer, M. J., Andersen, D. C., Bahr, J. B., Rasmussen, J. R. (1991). The oligosaccharides of glycoproteins: Bioprocess factors affecting oligosaccharide structure and their effect on glycoprotein properties. *Biotechnology 9*, 1347–1355.

38. Jenkins, N., Parekh, R. B., James, D. C. (1996). Getting the glycosylation right: Implications for the biotechnology industry. *Nat Biotechnol 14*, 975–981.

39. Jefferis, R. (2005). Glycosylation of recombinant antibody therapeutics. *Biotechnol Progr 21*, 11–16.

40. Jefferis, R. (2009). Glycosylation as a strategy to improve antibody-based therapeutics. *Nat Rev Drug Discov 8*, 226–234.

41. Galili, U. (1993). Interaction of the natural anti-Gal antibody with alpha-galactosyl epitopes: A major obstacle for xenotransplantation in humans. *Immunol Today 14*, 480–482.

42. Noguchi, A., Mukuria, C. J., Suzuki, E., Naiki, M. (1995). Immunogenicity of *N*-glycolylneuraminic acid-containing carbohydrate chains of recombinant human erythropoietin expressed in Chinese hamster ovary cells. *J Biochem 117*, 59–62.

43. Sheeley, D. M., Merrill, B. M., Taylor, L. C. (1997). Characterization of monoclonal antibody glycosylation: Comparison of expression systems and identification of terminal alpha-linked galactose. *Anal Biochem 247*, 102–110.

44. Hunt, D. F., Yates, J. R. 3rd, Shabanowitz, J., Winston, S., Hauer, C. R. (1986). Protein sequencing by tandem mass spectrometry. *Proc Nat Acad Sci USA 83*, 6233–6237.

45. Tang, X. J., Thibault, P., Boyd, R. K. (1993). Fragmentation reactions of multiply-protonated peptides and implications for sequencing by tandem mass spectrometry with low-energy collision-induced dissociation. *Anal Chem 65*, 2824–2834.

46. Loo, J. A., Edmonds, C. G., Smith, R. D. (1993). Tandem mass spectrometry of very large molecules. 2. Dissociation of multiply charged proline-containing proteins from electrospray ionization. *Anal Chem 65*, 425–438.

47. Schaaff, T. G., Cargile, B. J., Stephenson, J. L., Jr, McLuckey, S. A. (2000). Ion trap collisional activation of the $(M + 2H)^{2+}$ - $(M + 17H)^{17+}$ ions of human hemoglobin beta-chain. *Anal Chem 72*, 899–907.

48. Lim, A., Reed-Bogan, A., Harmon, B. J. (2008). Glycosylation profiling of a therapeutic recombinant monoclonal antibody with two *N*-linked glycosylation sites using liquid chromatography coupled to a hybrid quadrupole time-of-flight mass spectrometer. *Anal Biochem 375*, 163–172.

49. Walsh, G., Jefferis, R. (2006). Post-translational modifications in the context of therapeutic proteins. *Nat Biotechnol 24*, 1241–1252.

50. Prome, D., Blouquit, Y., Ponthus, C., Prome, J. C., Rosa, J. (1991). Structure of the human adult hemoglobin minor fraction A1b by electrospray and secondary ion mass

spectrometry. Pyruvic acid as amino-terminal blocking group. *J Biol Chem 266*, 13050–13054.

51. Rose, K., Simona, M. G., Savoy, L. A., Regamey, P. O., Green, B. N., Clore, G. M., Gronenborn, A. M., Wingfield, P. T. (1992). Pyruvic acid is attached through its central carbon atom to the amino terminus of the recombinant DNA-derived DNA-binding protein Ner of bacteriophage Mu. *J Biol Chem 267*, 19101–19106.
52. Smith, M. A., Easton, M., Everett, P., Lewis, G., Payne, M., Riveros-Moreno, V., Allen, G. (1996). Specific cleavage of immunoglobulin G by copper ions. *Int J Peptide Protein Res 48*, 48–55.
53. Cordoba, A. J., Shyong, B. J., Breen, D., Harris, R. J. (2005). Non-enzymatic hinge region fragmentation of antibodies in solution. *J Chromatogr B Anal Technol Biomed Life Sci 818*, 115–121.
54. Cohen, S. L., Price, C., Vlasak, J. (2007). Beta-elimination and peptide bond hydrolysis: Two distinct mechanisms of human IgG1 hinge fragmentation upon storage. *J Am Chem Soc 129*, 6976–6977.
55. Yan, B., Yates, Z., Balland, A., Kleemann, G. R. (2009). Human IgG1 hinge fragmentation as the result of H_2O_2-mediated radical cleavage. *J Biol Chem 284*, 35390–35402.
56. ICH, Q1A(R2). (2003). Guidance on Q1A(R2) stability testing of new drug substances and products. *International Conference on Harmonisation.*
57. Morand, K., Talbo, G., Mann, M. (1993). Oxidation of peptides during electrospray ionization. *Rapid Commun Mass Spectrom 7*, 738–743.
58. Griffiths, S. W., Cooney, C. L. (2002). Development of a peptide mapping procedure to identify and quantify methionine oxidation in recombinant human alpha1-antitrypsin. *J Chromatogr A942*, 133–143.
59. Kroon, D. J., Baldwin-Ferro, A., Lalan, P. (1992). Identification of sites of degradation in a therapeutic monoclonal antibody by peptide mapping. *Pharmaceut Res 9*, 1386–1393.
60. Marshak, D. R., ed. (1996). *Techniques in Protein Chemistry VII.* Academic Press, New York, pp. 275–284.
61. Chumsae, C., Gaza-Bulseco, G., Sun, J., Liu, H. (2007). Comparison of methionine oxidation in thermal stability and chemically stressed samples of a fully human monoclonal antibody. *J Chromatogr B Anal Technol Biomed Life Sci 850*, 285–294.
62. Liu, H., Gaza-Bulseco, G., Zhou, L. (2009). Mass spectrometry analysis of photo-induced methionine oxidation of a recombinant human monoclonal antibody. *J Am Soc Mass Spectrom 20*, 525–528.
63. Liu, D., Ren, D., Huang, H., Dankberg, J., Rosenfeld, R., Cocco, M. J., Li, L., Brems, D. N., Remmele, R. L., Jr. (2008). Structure and stability changes of human IgG1 Fc as a consequence of methionine oxidation. *Biochemistry 47*, 5088–5100.
64. Bertolotti-Ciarlet, A., Wang, W., Lownes, R., Pristatsky, P., Fang, Y., McKelvey, T., Li, Y., Li, Y., Drummond, J., Prueksaritanont, T., Vlasak, J. (2009). Impact of methionine oxidation on the binding of human IgG1 to Fc Rn and Fc gamma receptors. *Mol Immunol 46*, 1878–1882.
65. Harris, R. J. (1995). Processing of *C*-terminal lysine and arginine residues of proteins isolated from mammalian cell culture. *J Chromatogr A705*, 129–134.
66. Dick, L. W., Jr, Qiu, D., Mahon, D., Adamo, M., Cheng, K. C. (2008). *C*-Terminal lysine variants in fully human monoclonal antibodies: Investigation of test methods and possible causes. *Biotechnol Bioeng 100*, 1132–1143.

67. Valliere-Douglass, J., Wallace, A., Balland, A. (2008). Separation of populations of antibody variants by fine tuning of hydrophobic-interaction chromatography operating conditions. *J Chromatogr A1214*, 81–89.

68. Moorhouse, K. G., Nashabeh, W., Deveney, J., Bjork, N. S., Mulkerrin, M. G., Ryskamp, T. (1997). Validation of an HPLC method for the analysis of the charge heterogeneity of the recombinant monoclonal antibody IDEC-C2B8 after papain digestion. *J Pharmaceut Biomed Anal 16*, 593–603.

69. Wei, Z., Feng, J., Lin, H. Y., Mullapudi, S., Bishop, E., Tous, G. I., Casas-Finet, J., Hakki, F., Strouse, R., Schenerman, M. A. (2007). Identification of a single tryptophan residue as critical for binding activity in a humanized monoclonal antibody against respiratory syncytial virus. *Anal Chem 79*, 2797–2805.

70. Yang, J., Wang, S., Liu, J., Raghani, A. (2007). Determination of tryptophan oxidation of monoclonal antibody by reversed phase high performance liquid chromatography. *J Chromatogr A1156*, 174–182.

71. Johnson, K. A., Paisley-Flango, K., Tangarone, B. S., Porter, T. J., Rouse, J. C. (2007). Cation exchange-HPLC and mass spectrometry reveal C-terminal amidation of an IgG1 heavy chain. *Anal Biochem 360*, 75–83.

72. Harris, R. J., Kabakoff, B., Macchi, F. D., Shen, F. J., Kwong, M., Andya, J. D., Shire, S. J., Bjork, N., Totpal, K., Chen, A. B. (2001). Identification of multiple sources of charge heterogeneity in a recombinant antibody. *J Chromatogr B Biomed Sci Appl 752*, 233–245.

73. Terashima, I., Koga, A., Nagai, H. (2007). Identification of deamidation and isomerization sites on pharmaceutical recombinant antibody using H(2)(18)O. *Anal Biochem 368*, 49–60.

74. Brady, L. J., Martinez, T., Balland, A. (2007). Characterization of nonenzymatic glycation on a monoclonal antibody. *Anal Chem 79*, 9403–9413.

75. Quan, C., Alcala, E., Petkovska, I., Matthews, D., Canova-Davis, E., Taticek, R., Ma, S. (2008). A study in glycation of a therapeutic recombinant humanized monoclonal antibody: Where it is, how it got there, and how it affects charge-based behavior. *Anal Biochem 373*, 179–191.

76. Santora, L. C., Krull, I. S., Grant, K. (1999). Characterization of recombinant human monoclonal tissue necrosis factor-alpha antibody using cation-exchange HPLC and capillary isoelectric focusing. *Anal Biochem 275*, 98–108.

77. Dick, L. W., Jr, Kim, C., Qiu, D., Cheng, K. C. (2007). Determination of the origin of the *N*-terminal pyro-glutamate variation in monoclonal antibodies using model peptides. *Biotechnol Bioeng 97*, 544–553.

78. Kuhn, D. M., Ruskin, B., Lovenberg, W. (1980). Tryptophan hydroxylase: The role of oxygen, iron, and sulfhydryl groups as determinants of stability and catalytic activity. *J Biol Chem 255*, 4137–4143.

79. Halliwell, B. (2003). Oxidative stress in cell culture: An under-appreciated problem? *FEBS Lett 540*, 3–6.

80. Halliwell, B. (2007). Biochemistry of oxidative stress. *Biochem Soc Trans 35*, 1147–1150.

81. Huddleston, M. J., Annan, R. S., Bean, M. F., Carr, S. A. (1993). Selective detection of phosphopeptides in complex mixtures by electrospray liquid chromatography/mass spectrometry. *J Am Soc Mass Spectrom 4*, 710–717.

82. Annan, R. S., Carr, S. A. (1996). Phosphopeptide analysis by matrix-assisted laser desorption time-of-flight mass spectrometry. *Anal Chem 68*, 3413–3421.

83. Rustandi, R. R., Washabaugh, M. W., Wang, Y. (2008). Applications of CE SDS gel in development of biopharmaceutical antibody-based products. *Electrophoresis 29*, 3612–3620.

84. ICH, Q6B. (1999). Guidance on specifications: Test procedures and acceptance criteria for biotechnological/biological products. *International Conference on Harmonisation.*

85. Maiorella, B. L., Ferris, R., Thomson, J., White, C., Brannon, M., Hora, M., Henriksson, T., Triglia, R., Kunitani, M., Kresin, L., Dollinger, G., Jones, R., Senyk, G., Young, J., Moyer, B., McEntire, J., Dougherty, J., Monica, T., Goochee, C. (1993). Evaluation of product equivalence during process optimization for manufacture of a human IgM monoclonal antibody. *Biologicals 21*, 197–205.

86. Brass, J. M., Krummen, K., Moll-Kaufmann, C. (1996). Quality assurance after process changes of the production of a therapeutic antibody. *Pharmaceut Acta Helv 71*, 395–403.

87. ICH, Q5E. (2005). Guidance on Q5E comparability of biotechnological/biological products subject to changes in their manufacturing process. *International Conference on Harmonisation.*

88. Harris, R. J. (2005). Heterogeneity of recombinant antibodies: Linking structure to function. *Developm Biol (Basel) 122*, 117–127.

89. Liu, H., Gaza-Bulseco, G., Faldu, D., Chumsae, C., Sun, J. (2008). Heterogeneity of monoclonal antibodies. *J Pharmaceut Sci 97*, 2426–2447.

90. Vlasak, J., Ionescu, R. (2008). Heterogeneity of monoclonal antibodies revealed by charge-sensitive methods. *Curr Pharmaceut Biotechnol 9*, 468–481.

91. Allen, D., Baffi, R., Bausch, J., Bongers, J., Costello, M., Dougherty, J., Jr, Federici, M., Garnick, R., Peterson, S., Riggins, R., Sewerin, K., Tuls, J. (1996). Validation of peptide mapping for protein identity and genetic stability: Biologics and biotechnology section, pharmaceutical research and manufacturers of America. *Biologicals 24*, 255–274.

92. Wang, Y., Xu, A., Knight, C., Xu, L. Y., Cooper, G. J. (2002). Hydroxylation and glycosylation of the four conserved lysine residues in the collagenous domain of adiponectin: Potential role in the modulation of its insulin-sensitizing activity. *J Biol Chem 277*, 19521–19529.

CHAPTER 12

Post-translationally Modified Proteins: Glycosylation, Phosphorylation, and Disulfide Bond Formation

ANTHONY TSARBOPOULOS and FOTINI N. BAZOTI

12.1 INTRODUCTION

The decoding of the human genome has enabled the search and production of therapeutic proteins through the large-scale identification of biologically relevant human proteins. These macromolecules are produced in large quantities by recombinant DNA technology, and their structure can be elucidated by a combination of separation techniques and mass spectrometry (MS). MS is invaluable towards the evaluation of these compounds as drugs directed at clinically relevant disease targets or pathways leading to a disease. In the 2000s, several protein drugs have become a growing segment of the vaccine and diagnostic agent market and an important resource in the physician's arsenal.

Development of protein drugs, however, can be complicated for many reasons. One reason addressed in the chapter is the propensity of protein drugs to undergo post-translational modification. A majority of all proteins undergo co- and/or post-translational modifications (PTMs), resulting in heterogeneity of the product; the extent of PTMs depends on the nature of the host cell and the conditions of the fermentation and recovery processes. These modifications include alteration of the protein's polypeptide chain by proteolytic cleavages at the *C*- and *N*-termini, deamidation, aggregation, formation of disulfide bonds, cysteine (C) and methionine (M) oxidation, and covalent attachment of phosphate, sulfate, alkyl groups, carbohydrates, lipids, and other groups. Nearly 200 structurally distinct covalent modifications have been identified so far, ranging in size and complexity [1]. The presence of PTMs

Protein and Peptide Mass Spectrometry in Drug Discovery, Edited by Michael L. Gross, Guodong Chen, and Birendra N. Pramanik.

is often required for normal biological function or tissue disposition of the protein, although in many cases the role of the modification is as of yet unknown. Therefore knowledge of these modifications is extremely important, because they may alter physical and chemical properties, folding, conformation distribution, stability, activity, which in turn may affect cellular processes, in which the protein is involved [2–4]. Examples of the latter can be signal transduction in case of phosphorylation, proteolysis in case of ubiquitination [5], targeting and cell-matrix interaction in case of glycosylation, and protein–ligand binding in case of carboxylation. Therefore, thorough understanding of a specific protein structure–function relationship requires not only detailed information on amino-acid sequence, which can be inferred by the corresponding DNA sequence, but also the structure elucidation of the protein modifications. Moreover PTM analysis of proteins is essential for determining their exact structure, which is required for the production, registration to regulatory agencies, and product quality of therapeutic pharmaceutical proteins.

The presence of PTMs often complicates or even prevents the use of classical tools for protein sequence analysis (e.g., automated *Edman* degradation). In addition the presence of lipid or carbohydrate on proteins can dramatically decrease the accuracy of the molecular weight (M_r) measurement when using sedimentation velocity, gel permeation, or SDS PAGE analysis.

In the last two decades MS has emerged as an integral part of biological research and has been established as the central technology toward protein mapping and PTM localization, as well as for PTM structure identification and quantification. Electrospray ionization (ESI) MS coupled with online liquid chromatographic separation (LC-MS) [6,7] and matrix-assisted laser desorption ionization (MALDI) [8,9] are the methods of choice for mapping PTMs [10,11]. MS-based approaches are often employed to check the purity of a recombinant therapeutic protein, to maximize its production yield and/or to minimize the undesirable degradation products or PTMs. In addition recent advances in MS instrumentation have made possible data-dependent experiments, such as automatic ion selection for fragmentation followed by monitoring of product ions specific for certain PTMs through tailored tandem MS (MS/MS) scanning methods [12–14]. Similarly MS-based approaches can be employed in the development of therapeutic protein formulations for clinical and eventually commercial use, ensuring product integrity and protection from chemical degradation and/or aggregation.

In this chapter we discuss MS-based methodologies that are employed to detect, identify, and map common PTMs, with an exclusive focus on glycosylation, phosphorylation, and disulfide bonding. Furthermore we highlight the importance of accurate knowledge and monitoring of PTMs in recombinant proteins intended for therapeutic use in humans. Although we cannot cover all the PTMs in a single chapter, the approaches used for the PTMs discussed here are representative of those used for other PTMs.

12.2 GLYCOSYLATION

Glycosylation, like phosphorylation, is a highly biologically relevant and a ubiquitous PTM of proteins. It is estimated that more than half of the proteins in a eukaryotic cell are glycosylated [15], and glycosylation represents the most common PTM for recombinant protein products expressed in mammalian and insect cell lines. The carbohydrate moieties of glycoproteins participate in many biological processes, such as circulation, molecular and immune recognition, which in turn affect intracellular signaling, fertilization, embryonic development, immune defense, inflammation, cell adhesion, division processes, and viral replication. In addition glycan-chain modification can significantly impact protein solubility, stability, resistance to proteolysis, and immunogenicity. Carbohydrate modifications can also considerably alter protein conformation, which may consequently modulate the functional activity of the protein, especially in its interactions with other proteins or ligands. Carbohydrate modification can be used toward the production of "custom-made" glycoproteins tailored for specific therapeutic use. At this point it should be emphasized that the structure of the recombinant glycoprotein is dependent not only on the polypeptide backbone but also on a number of other variables, especially the presence of processing enzymes (*glycosidases* and *glycosyltransferases*) [16,17], which may change during cell growth, differentiation, and development [18–20]. Possible variation of a protein's glycosylation patterns often leads to changes in their function and can be associated with a disease. Therefore complete structural analysis of a glycoprotein end product will require not only the determination of the primary peptide sequence but also detailed information on the glycosylation sites, the glycosylation patterns, and the structure elucidation of the attached carbohydrates (*glycoproteome*). A thorough understanding of a drug protein glycosylation will be invaluable for gaining insight into their involvement in disease mechanisms and the potential for novel therapeutic interventions [21]. Characterizing the glycoproteome, however, is a challenging and daunting task because the structural heterogeneity of these glycans is vast, necessitating the development of highly sensitive and efficient analytical methods for detection, separation, and structural investigation of glycoproteins.

12.2.1 MS Detection of Glycoproteins

The first step in the characterization of glycoproteins is often the separation by two-dimensional (2D) gel electrophoresis, where characteristic spots reflecting their different isoelectric points and molecular weights (M_r) of different glycoforms can be seen. Subsequent detection of the glycosylation pattern of the electroblotted glycoproteins may be performed by lectin affinity chromatography (LAC) [22–25], where carbohydrate-specific lectins can be used to probe defined oligosaccharide structures (*motifs*). In addition this affinity purification can be employed as an enrichment method for the glycosylated peptides and proteins. A serious drawback of the 2D gel electrophoresis approach is the low solubility of the membrane glycoproteins, which results in their poor detection. An alternative method of higher resolving potential is capillary electrophoresis (CE), where the various glycoforms are detected even though no information on the nature of the attached glycans is revealed [26]. ESI and MALDI MS are the common tools for these large biomolecules, providing a basis for analysis of intact glycoproteins. In the case of glycoprotein analysis by MALDI MS, signals corresponding to protonated molecules ($M + H^+$)

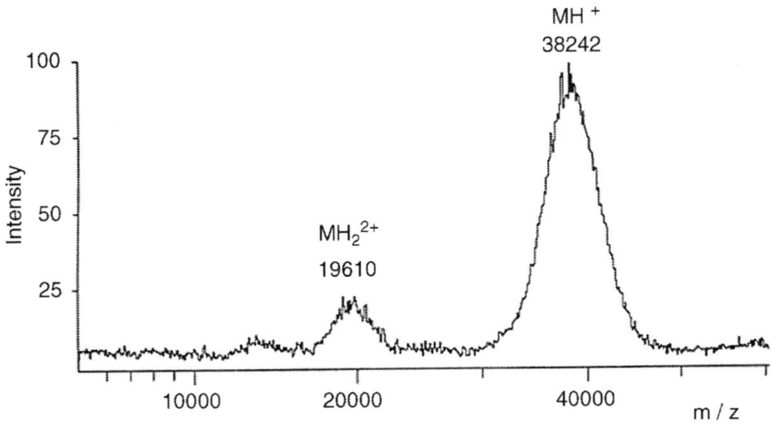

FIGURE 12.1 Positive-ion MALDI TOF mass spectrum of CHO-derived interleukin-4 receptor (IL-4R) using a 3-hydroxypicolinic acid matrix. Reprinted with permission from [28].

of the individual glycoforms are produced, even though the limited mass resolution of the mass analyzer prevents the determination of the heterogeneity for glycoproteins with M_r over 30 kDa and a relatively high percentage of carbohydrate content [27]. This is clearly shown in the MALDI mass spectra of CHO interleukin-4 receptor (IL-4R) [28] (Figure 12.1) and Sf9-derived interleukin-5 receptor α-subunit (IL-5Rα) [29], where the 35% and 17% carbohydrate content, respectively, along with the possible presence of salt adducts, prevents the recording of the individual glycoform signals.

The linear time-of-flight (TOF), and hybrid quadrupole TOF (qTOF) mass spectrometers provide better mass resolving power, sensitivity, and extended mass range for this type of analysis, whereas the highest mass resolving power and mass accuracy is available with the Fourier-transform ion cyclotron resonance (FT ICR) analyzer, but this instrumentation has not been extensively evaluated in that mass range. It should be noted that the choice of an appropriate MALDI matrix is a critical step toward achieving the optimum mass resolving power and separation of the individual glycoform signals [27,30]. The recent introduction of the ion mobility (IM) analyzer [31] seems promising toward improving the specificity of the MALDI analysis without compromising the mass-range capabilities.

On the contrary, ESI MS analysis of intact glycoproteins has better success for detecting individual glycoforms, although the mass range of the ESI method is smaller and more congested for the inherently heterogeneous glycoproteins. Other limitations are the poorly efficient desolvation of glycoproteins during ionization and the formation of multiple adducts. These limitations are especially serious in the analysis of high M_r glycoproteins, where formation of higher charge state ions and of salt adducts often results in unresolved weak signals or none at all. The latter is less of a problem with nanoelectrospray ionization (nESI) [32], where the improved desolvation efficiency of glycoproteins, owing to the generation of smaller size spray droplets, improves the sensitivity of the ESI MS analysis [33]. This is shown in the nESI mass spectrum of

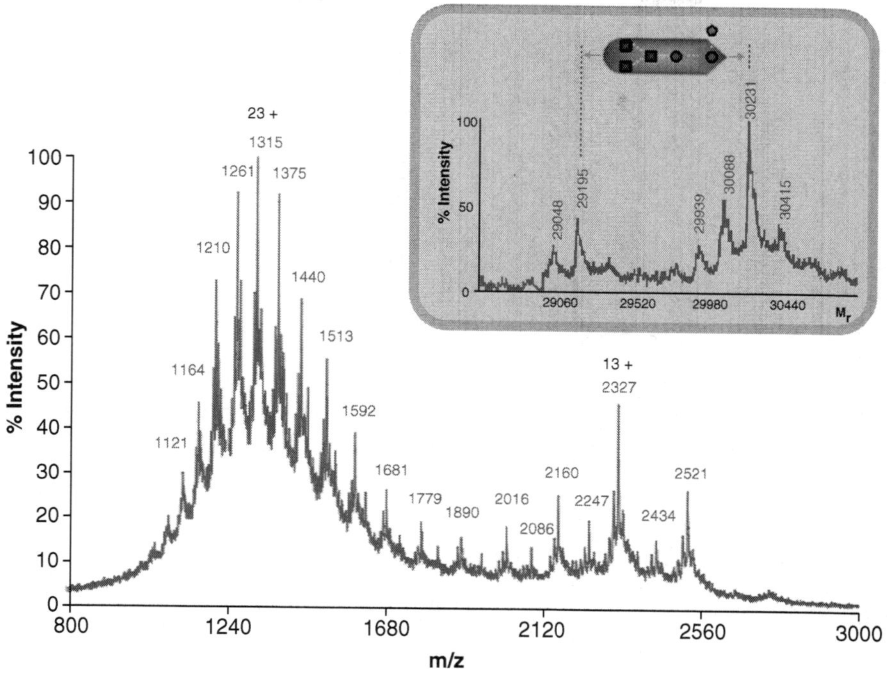

FIGURE 12.2 Positive-ion nESI mass spectrum of Sf9-derived interleukin 4 receptor (IL-4R) obtained on a qTOF instrument. The deconvoluted mass spectrum (shown in the inset) indicates the presence of two high-mannose glycoforms with mass separation corresponding to a fucosylated $Man_3(GlcNAc)_2$  moiety. Reprinted with permission from [37].

Sf9-derived IL-4R obtained on an orthogonal TOF mass spectrometer, where two sets of glycoform signals, with reasonable separation and mass measurement accuracy, can be observed (Figure 12.2). The deconvoluted mass spectrum (Figure 12.2 inset) indicates the presence of two high-mannose glycoforms; the mass separation between peaks corresponds to a fucosylated $Man_3(GlcNAc)_2$ structure.

The high mass resolving power of the analyzer allowed the nESI-generated signals corresponding to their sodium adducts to be sufficiently separated. This advantage, along with the improved mass measurement accuracy and extended mass range, has made the orthogonal qTOF and recently the IM TOF as the analyzers of choice for nESI MS analysis of glycoproteins. The limited mass range of quadrupole, quadrupole ion trap and even the orbitrap analyzers [34], where the mass range is limited to around *m*/*z* 3000, is a significant drawback when larger glycoproteins or noncovalent complexes thereof must be detected. In such cases an upper mass limit greater than even *m*/*z* 10,000 may be required [35,36]. This is shown in the nESI mass spectrum of the intact IgG1 monoclonal antibody, obtained on an IM TOF mass spectrometer (Figure 12.3A).

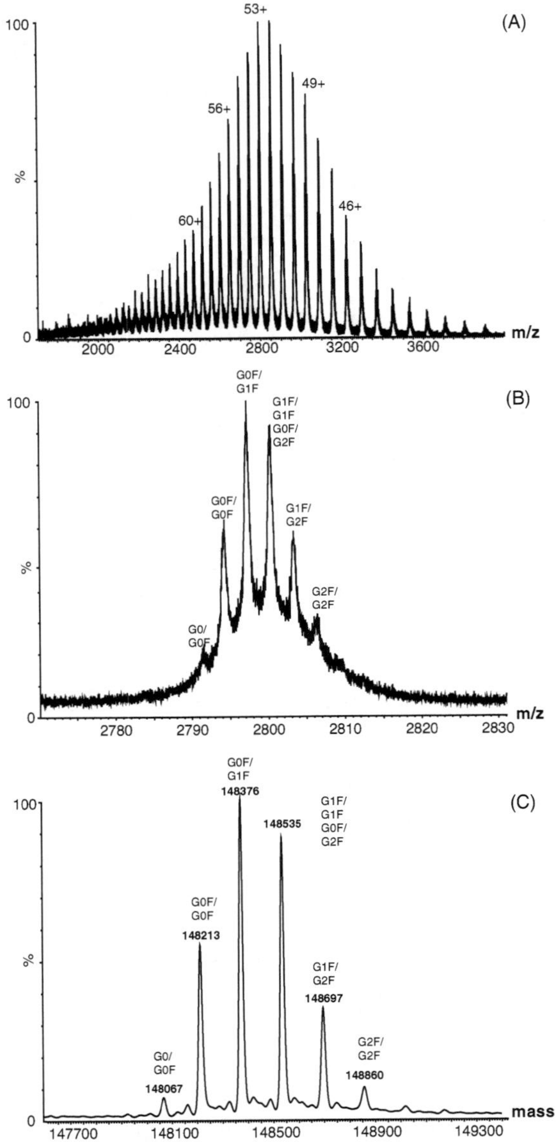

FIGURE 12.3 Positive-ion nESI ion-mobility (IM) TOF mass spectrum of the intact IgG1 *(A)*; the 53+ charged ion with the signals corresponding to various glycoforms as annotated *(B)*; the deconvoluted mass spectrum showing the glycosylation heterogeneity of the monoclonal antibody arising from variations in the hexose and fucose content *(C)*. Reprinted with permission from [35].

A series of ions, up to *m/z* 3800, can be observed in the nESI mass spectrum, and they are successive charge states of the intact IgG1, with sufficient separation to reveal the presence of six glycoform variants, as shown by the close-up of the 53 + charge state (Figure 12.3B) and the respective deconvoluted mass spectrum (Figure 12.3C). The separation of the glycoform signals is also significantly affected by the presence of satellite signals corresponding to various adducts with sulfuric/phosphoric acid. This adduct formation can be minimized, or even eliminated, by a simple desalting sample pretreatment on a ZipTip$_{C18}$ pipette tip or exchange with ammonium salts. Another way to achieve reduction of salt-adduct formation is by the employment of an elevated entrance potential (i.e., nozzle-skimmer or orifice) in the nESI MS analysis and/or by the use of low pH solvents (10% formic acid vs. 0.1% formic acid) [37].

Yet the biggest challenge for the analysis of glycoproteins is their low abundance compared to that of unmodified proteins, and the resulting low signal intensities of the mass spectral peaks. The latter are mainly due to both the low ionization efficiency of glycoproteins and the distribution of their signal among the various glycoforms sharing a common peptide sequence, rendering their detection a daunting task. This can be improved by coupling ESI-MS directly to a separation device, either nLC [38] or capillary zone electrophoresis (CZE) [39,40]. Another promising solution to this problem is to perform an *enrichment step* for the glycoproteins (or glycosylated peptides). This eliminates the most abundant unmodified proteins from competing for charge during the ionization process and results in higher ionization efficiencies and increased probability for detecting glycoproteins. Several approaches are available for glycoprotein enrichment, including LAC for glycoprotein isolation [41] and various chemical derivatization methods combined with affinity purification [42,43]. The latter approach makes use of reactions that are specific for the glycan of interest, such as the tagging-via-substrate strategy for identifying *O-GlcNAc* glycosylated proteins [44–46]. Overall, the rapid assessment of glycosylation at the molecular level is invaluable in glycoprotein screening for certain diseases, such as the differentiation of transferrin in normal and carbohydrate-deficient syndrome patients, who usually lack both *N*-linked disialylated carbohydrate chains [47].

12.2.2 Glycan Identification, Classification, and Heterogeneity

The goals of any PTM analysis are to obtain as complete a map as possible of the modification sites in a protein and to define the structure of the modification at each specific site. Therefore both the structure determination of the carbohydrate moieties and the identification of their attachment sites on the protein backbone are imperative for the analysis of a glycoprotein. There are *O-glycosylation* and *N-glycosylation* modifications. *O*-glycosylation occurs at a later stage during protein processing confined in the *Golgi* apparatus, with the initial addition of *N*-acetylgalactosamine (*GalNAc*) residue to the hydroxyl group of serine (S), threonine (T), and hydroxyproline (Hyp) residues, followed by other carbohydrates.

These *O*-linked glycans contain a number of different core regions, with the mucin-type *O*-glycans being the four most frequently occurring [48–50]. Regarding *N*-glycosylation, the glycan is β-glycosidically attached via *N*-acetylglucosamine (*GlcNAc*) to the amide group of an asparagine (N) residue occurring in the tripeptide sequence Asn-Xaa-Ser/Thr (N–X–S/T sequon), where Xaa is any amino acid except proline (P) or aspartic acid (D) [16,51]. Because there is a common biosynthetic pathway in the endoplasmic reticulum compartment of the cell, all *N*-glycans share a common pentasaccharide core (i.e., the *trimannosyl* core with two *N*-acetylglucosamine residues ($Man_3GlcNAc_2$)). These *N*-linked oligosaccharides can be further divided into three classes: *high mannose*-type, *complex*-type, and *hybrid*-type as shown in Figure 12.4.

The *high mannose*-type glycoproteins (e.g., ovalbumin) contain two to eight mannose residues added to the pentasaccharide core. Glycoproteins containing *complex*-type *N*-glycans (e.g., fetuin) exhibit a higher degree of heterogeneity by having a number of *GlcNAc*, *Gal*, *Fuc* and *NeuAc* (sialic acid) residues attached to the pentasaccharide core, showing the highest structural variation due to the possible extension and/or branching of the outer chains. Finally, the *hybrid*-type glycoproteins combine features from both high mannose and complex-type glycans [16].

We emphasize that the populations of *O*- or *N*-linked sugars attached to an individual protein will depend on the cell type, in which the glycoprotein is expressed, and its development stage. The end result is a diverse degree of occupancy at different *O*- or *N*-linked glycosylation sites leading to a wide variety of structurally different

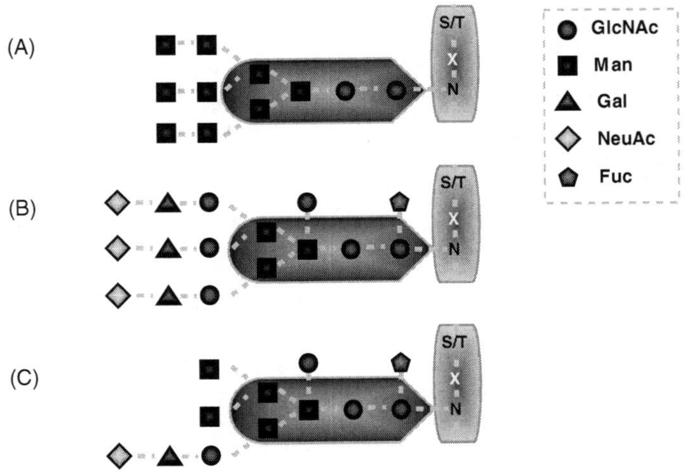

FIGURE 12.4 Classes of *N*-linked carbohydrate structures sharing a common pentasaccharide core structure—the trimannosyl core with two *N*-acetylglucosamine residues ($Man_3GlcNAc_2$) (Man: ■, GlcNAc: ●, Gal: ▲, NeuAc: ◆, Fuc: ⬟). *(A)* High-mannose type; *(B)* complex type (triantennary); *(C)* hybrid type.

oligosaccharides that generate a complex mixture of glycosylated variants (*glycoforms*). Therefore, complete structural characterization of a glycoprotein requires: (1) the determination of the peptide primary sequence and the glycosylation attachment sites, and (2) the definition of the attached oligosaccharides in terms of their linear sequencing, branching, and linkage. Complete characterization is important for understanding the role of glycans in physiopathological processes and designing "custom-made" glycoproteins for drug therapy.

The M_r determination of the intact glycoprotein by either MALDI or ESI MS analysis usually reveals the heterogeneity for glycoproteins with M_r up to 20 to 30 kDa with a relatively low percentage of carbohydrate content (as shown above). ESI MS analysis particularly allows the accurate M_r measurement of each glycoform, rendering this approach useful for monitoring and comparing the glycoform pattern of recombinant glycoproteins. This is nicely shown in the nESI mass spectrum of the CHO-derived interleukin-4 (CHO IL-4), obtained on an orthogonal TOF mass spectrometer; one observes several envelopes of multiply charged signals comprised of well separated peaks corresponding to four distinct glycoforms (Figure 12.5).

Still, the nature of glycan chains and glycosylation sites cannot be derived from this type of analysis. A reasonable strategy for a more complete analysis is: (1) to obtain coverage of the known consensus sites for a given modification, (2) to perform site-specific analysis of the glycan chains (which usually requires enzymatic cleavage of the proteins into peptides of a suitable size for sequence analysis), and (3) to invoke chromatographic or electrophoretic separation [52,53] of the mixture followed by MS analysis. First, coverage of the protein sequence by the aforementioned MS mapping of enzyme-generated peptide mixtures provides not only confirmation of the expected sequence but also identification of any existing modifications, including the glycosylation attachment sites. A more detailed profile of protein glycosylation can be built if new, unexpected signals are seen. That profile includes the determination of the site-specific glycosylation pattern inferred from the signal envelopes corresponding to the specific glycan structures shown in Figure 12.4. Therefore, this MS mapping approach can provide information on the classification and heterogeneity of the attached glycan chains.

12.2.3 Glycoprotein Mapping by LC-ESI and MALDI Tandem MS

Several experimental approaches and protocols can be used for glycoprotein mapping (i.e., identification of the glycosylation sites and structure characterization of the attached glycan chains). One approach for characterization of glycoprotein glycans involves their removal by base-catalyzed β-elimination (for *O*-linked carbohydrates) or *N*-Glycanase/PNGase F (for *N*-linked carbohydrates). Removal of *O*-linked carbohydrates by alkaline β-elimination followed by reduction leads to the simultaneous conversion of S and T residues to A and α-aminobutyric acid sequences, respectively ($\Delta m = -16$ Da). Thus the former *O*-glycosylation sites can be identified by the 16-Da mass shifts. Enzymatic release of *O*-linked sugar chains is often

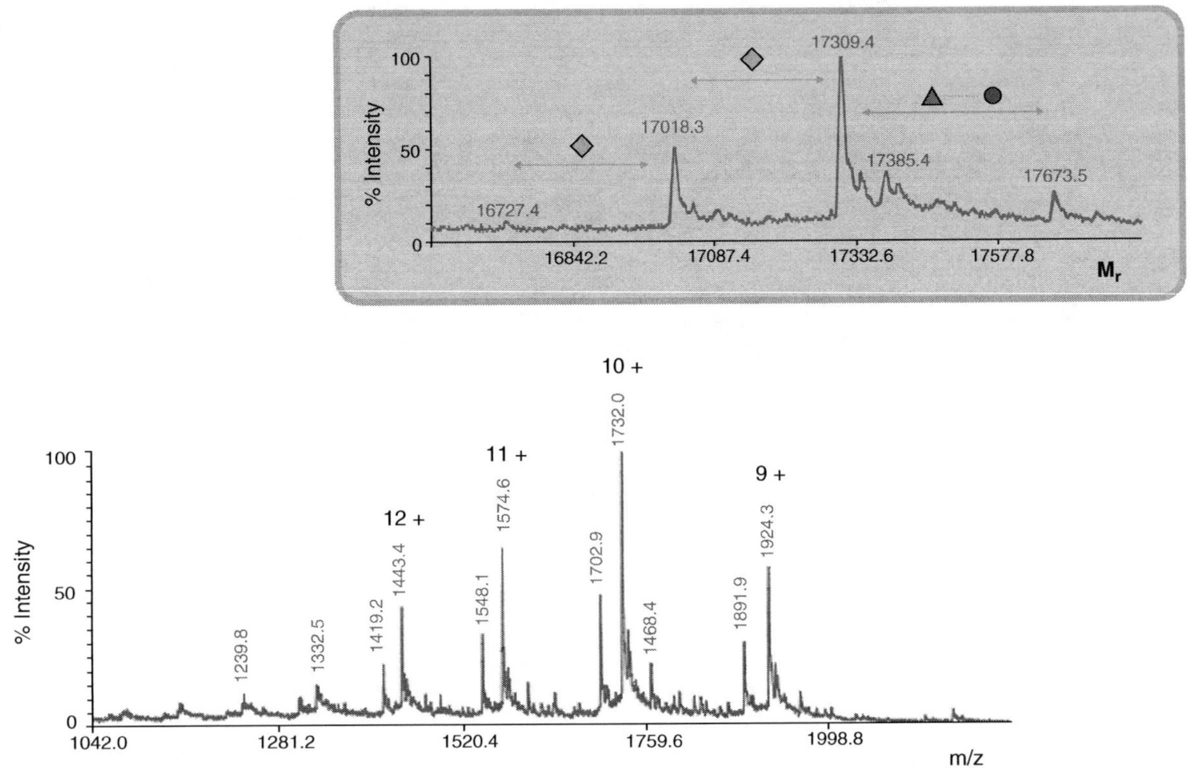

FIGURE 12.5 Positive-ion nESI mass spectrum of CHO-derived interleukin-4 (IL-4) obtained on a qTOF instrument. The inset depicts the deconvoluted mass spectrum where the heterogeneity due to the addition of sialic acid (◆: in-chain mass of 291 Da) and lactosamine (▲······●: in-chain mass of 365 Da) units is clearly illustrated.

incomplete owing to the narrow specificity of the relevant enzymes (e.g., *O*-Glycanase) and the structure variability of the *O*-linked sugars. However, *N*-linked deglycosylation with *N*-Glycanase converts the glycosylated N residues to D sequences, and this change can be detected by the 1 Da mass increase [54]. This mass difference can be magnified by carrying out the *N*-Glycanase reaction in fully or partially (50%) ^{18}O-labeled glycosylated N residues; the result is characteristic doublets separated by 2 Da [55]. These doublets can be used to locate the modification site and to determine the degree of occupancy at each *N*-linked glycosylation site. The resulting mixture of structurally heterogeneous glycans can be characterized by HPLC, LC-ESI MS [26,56,57], LAC [58], CE [59], as well as by MALDI [60,61] and ESI MS mapping [62,63]. The advantage of MALDI-TOF MS is the simplicity of the spectra, which contain usually intense signals corresponding to protonated or sodiated individual glycan molecules. The multiply charged ions generated in ESI-MS result in higher signal dispersion, leading to lower sensitivity and higher complexity of the mass spectra. Either mapping method can be used to obtain compositional information of monosaccharides, even though no distinction between different stereoisomers can be deduced or information on the site and the extent of glycosylation can be made at this point.

An alternative approach to obtain a more detailed structural profile of highly complex protein glycosylation involves enzymatic cleavage of the glycoprotein (usually with trypsin or another endoprotease) followed by LC- or CE-ESI MS/MS [64–66] and MALDI-TOF MS [67] analysis of the resulting glycopeptides as shown in Figure 12.6.

Comparative MS mapping of the glycoprotein before and after enzymatic hydrolysis of the attached *N*- or *O*-linked oligosaccharides may reveal the glycosylation site. Identification of the exact location of the attachment site often requires LC separation of the enzyme-derived protein fragments coupled with online MS or tandem MS analysis (e.g., LC-ESI MS/MS); without separation, the signals for the glycopeptides would be highly suppressed by the presence of other peptides. In the case of CHO IL-4 containing two potential glycosylation sites fulfilling the N-X-S/T sequence motif, comparative mapping of the V8 protease digest prior and after the *N*-Glycanase hydrolysis provides the identification of the glycosylation site (i.e., the N^{38} residue rather than N^{105}, the other potential site). Glycopeptide signals corresponding to the $V_{4,5}$ peptide fragment containing a sialylated biantennary *N*-linked oligosaccharide can be observed (M_r 3929 and 4220), confirming the *N*-glycosylation of the N^{38} residue. These signals were absent in the *N*-Glycanase-treated digest, where multiply charged ions corresponding to the $V_{4,5}$ peptide were detected [68].

In addition to locating the modification site, elucidating the attached glycan structures may be assisted by the fragmentation information derived from the mass spectra and product-ion spectra from MS/MS experiments. This can be facilitated by an enrichment step prior to the MS analysis (Figure 12.6), which usually alleviates the difficulty arising from the presence of multiple glycans present at each glycosylation site. This step could be LAC, size-exclusion chromatography, immunoaffinity chromatography; their use for the characterization of glycans and glycopeptides is described in a number of reviews [57,69,70]. Several MS/MS scanning methods can

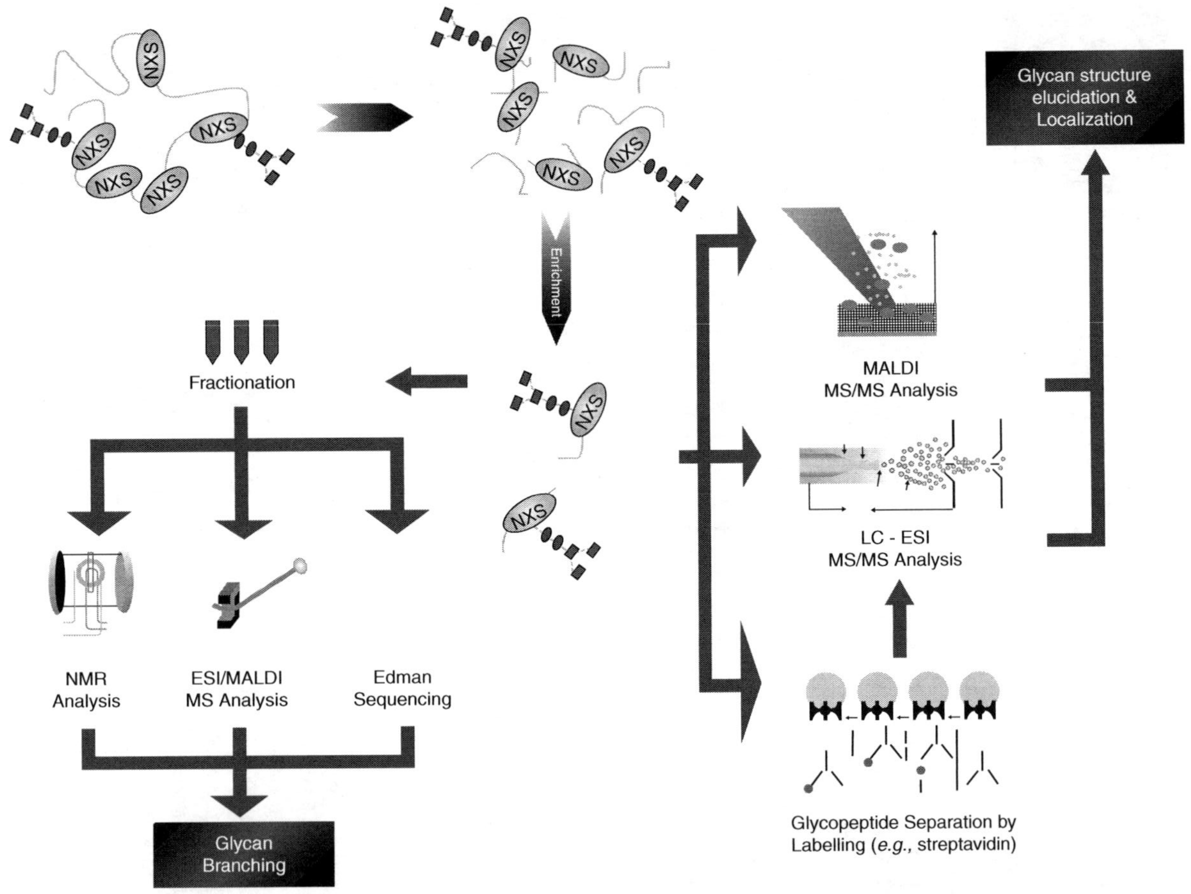

FIGURE 12.6 Scheme of the different analytical approaches employed for the separation and analysis of glycoproteins by LC-ESI and MALDI tandem MS. (*See the color version of this figure in Color Plates section.*)

also be employed for the identification of glycopeptides by using (1) detection of the low mass fragment ions, generated from collision-induced dissociation (CID), which are diagnostic of the sugar modification, (2) "precursor ion" scanning to detect selectively the ions that give rise to these low-mass marker ions, and (3) constant neutral loss scanning of terminal monosaccharide residues to pinpoint the glycopeptides. Alternatively, electron-capture dissociation (ECD) [71–73] and electron-transfer dissociation (ETD) [74] approaches to MS/MS can provide information on the location of the glycan modification. Unlike traditional MS/MS approaches where glycosidic bond cleavage is the dominant fragmentation pathway, both ECD and ETD induce fragmentation of the peptide backbone with minimal loss of the glycan moiety. Therefore both techniques are excellent tools for screening and locating the glycan modification [75] and are complementary to CID methods.

In the traditional MS/MS methods, where glycan modifications are more susceptible to cleavage by CID than the peptide backbone, the product-ion spectra of glycopeptides are mainly dominated by fragment ions produced by cleavages of glycosidic linkages, rendering the peptide sequence determination in *N*-linked glycopeptides a difficult task. The low mass sugar-specific oxonium ions in the product-ion spectra of ESI-produced ions include those at *m*/*z* 162 for Hex^+, *m*/*z* 204 for $HexNAc^+$, *m*/*z* 274 and 292 for $NeuAc^+$, *m*/*z* 366 for Hex-$HexNAc^+$, and *m*/*z* 657 for NeuAc-Hex-$HexNAc^+$. In cases where MS/MS is not available, these low-mass marker ions can be generated by either "in-source" fragmentation of ESI-produced ions [68,76] or post-source decay (PSD) of MALDI-produced ions [77,78]. In the former the fragmentation can be induced by increasing the source entrance potential in the mass spectrometer to induce fragmentation in the declustering region. This is nicely illustrated in the LC-ESI MS analysis of the trypsin-treated CHO IL-4 [68], where the glycopeptide-containing fractions are easily identified by the presence of the sugar-diagnostic ions without having to search each individual mass spectrum for glycopeptide-characteristic patterns. Especially the observation of the marker ions at *m*/*z* 274, 366 and 657 in the ESI mass spectrum can reveal the presence of sialylated *N*-linked oligosaccharide of the complex-type in the specific chromatographic fraction. Moreover the mass separation of the signals within the triply and quadruply multiply charged ion envelopes show the presence of mono- and di-sialylated glycoforms (291 Da apart) along with higher M_r components containing additional lactosamine units (Hex-HexNAc, having mass differences of 365 Da) owing to the presence of extended arms or branching (Figure 12.7).

We note that the presence of carbohydrate often provides shielding of a neighboring proteolytic site, leading to the (occasional) incorporation of the adjacent peptide fragment; an example is the incorporation of the T_5 glycopeptide into the adjacent disulfide-linked peptide T_4-T_{10} [68].

Similarly this rapid glycopeptide screening approach can be applied to other mammalian-cell derived proteins, as for Sf9-derived IL-5Rα, which contained 17% carbohydrate [29]. Choosing LC-ESI MS analysis of the IL-5Rα tryptic digest can allow the identification of all glycopeptide-containing fractions (Figure 12.8), where four glycosylation sites out of the six potential sites fulfilling the glycosylation sequence motif were revealed in this example [29].

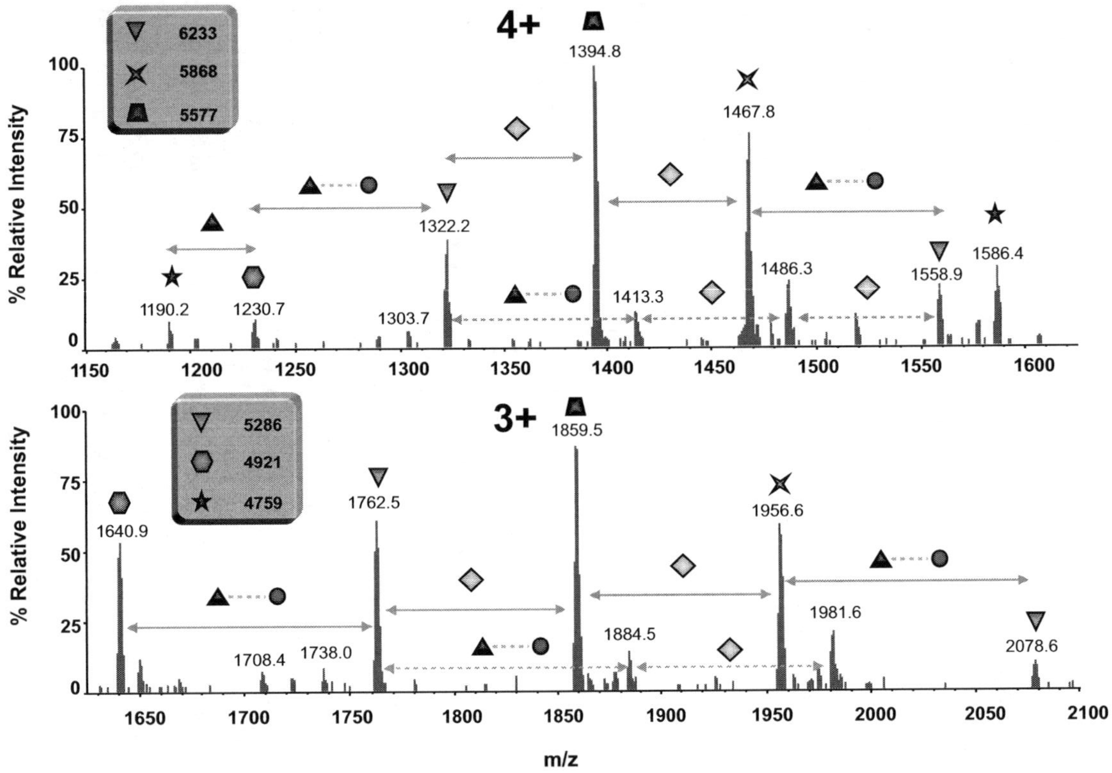

FIGURE 12.7 Positive-ion ESI mass spectrum of CHO-derived interleukin-4 (IL-4) tryptic glycopeptide fraction using "in-source" fragmentation, showing the glycoform signals in the triply (*lower panel*) and quadruply (*upper panel*) charged ion envelopes arising from carbohydrate heterogeneity. The corresponding M_r values are shown in the insets. Reprinted with permission from [68].

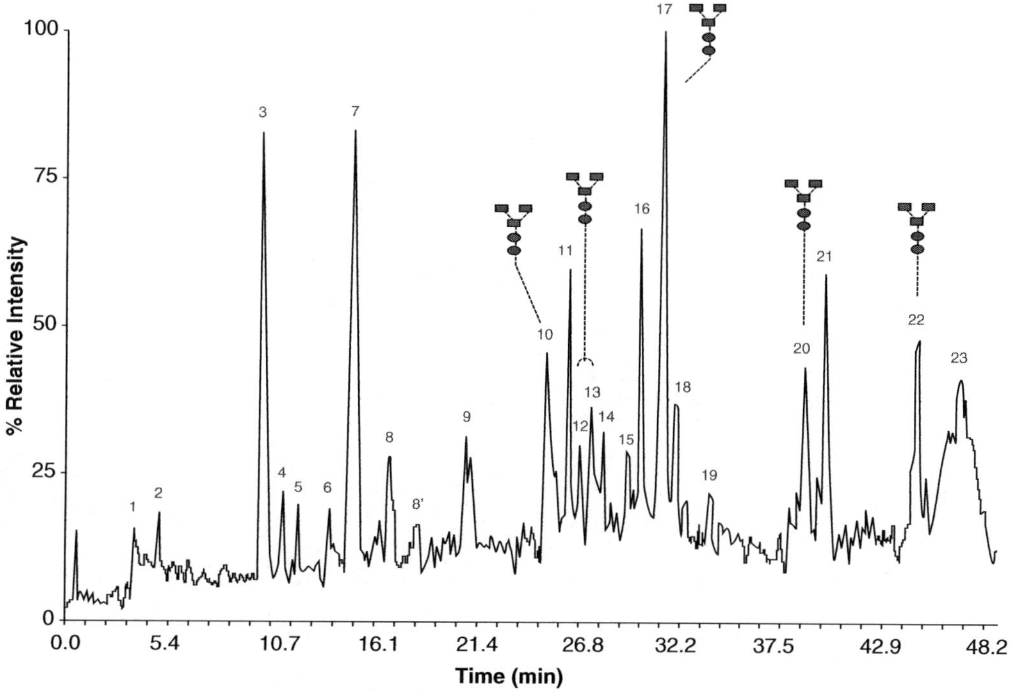

FIGURE 12.8 Total ion current (TIC) chromatogram obtained from the LC-ESI MS analysis of the Sf9 IL-5Rα tryptic digest using "in-source" fragmentation. The glycopeptide-containing tryptic fractions are labeled, as inferred by the detection of the low-mass diagnostic ions at m/z 162, 204, 274, 366, and 657.

The ESI mass spectrum of one glycopeptide-containing fraction (Figure 12.8, peak 20) shows signals corresponding to triply and quadruply charged molecules that are two glycopeptides containing a $Man_3(GlcNAc)_2$ high-mannose carbohydrate with an intersecting fucose residue (Figure 12.9).

This LC-ESI MS approach proved to be very useful for detecting several glycoforms and providing insights into the glycosylation map of IL-5Rα. In the case of insect cell-derived glycoproteins (e.g., Sf9, Sf21), the expected mass of the defined high-mannose glycan structures can readily suggest the structures of the attached oligosaccharides. In addition the excellent mass measurement accuracy provided by ESI MS analysis (usually within 1 Da from the expected mass value) affords a high degree of confidence in the assignment of the glycan structures. Nevertheless, further information on the linkage, branching points, and configuration of the constituent monosaccharides (*microheterogeneity*) can be provided by 2D nuclear magnetic resonance (NMR) analysis [79].

12.2.4 Glycosylation Site Quantitation

Although MS-based methods are now invaluable means toward glycoprotein characterization, the quantitation of the attached glycans remains a challenging task, mainly owing to their vast structural heterogeneity and the low ionization efficiency of the individual glycoforms.

There are two approaches for quantifying glycosylation on proteins. One may either quantify the glycans after enzymatic or chemical hydrolysis, or quantify the glycopeptides formed by digestion. In the former approach, which does not provide site-specific information about the glycosylation profiles, MS techniques can enable the detection of glycans released from glycoproteins without derivatization [80–82]. Derivatization of oligosaccharides by permethylation, however, is often performed before MS analysis to (1) ensure a more uniform ionization of oligosaccharides especially of sialic acid residues, (2) facilitate the separation of methylated glycans from salts and other impurities that may affect their mass spectral signals [83,84], and (3) yield more predictable MS/MS fragmentation of the methylated glycans than that of their native counterparts, leading to more certain structural assignments [85–88]. Another way to improve quantitation is by adding an isotopically labeled (^{13}C, D, ^{15}N, etc.) internal standard, and then measuring the analyte's response relative to that of the standard. In addition *isotopic labeling,* where one of the samples is modified with a "light" tag and another with a "heavy" tag, can be used for comparative glycomics, making it possible to determine relative changes in the abundances of specific oligosaccharide structures in complex glycoprotein mixtures obtained from biological samples [89,90]. One approach entails permethylation using *heavy/light* methyl iodide [$^{13}CH_3I$ vs. $^{12}CH_3I$] prior to MS analysis [91,92], but this approach does not provide a measure of the relative abundance of the individual glycans. In a modified approach, termed quantitation by isobaric labeling (QUIBL) [93], labeling with $^{13}CH_3I$ or $^{12}CH_2DI$ generates isobaric pairs of per-*O*-methylated glycans of enhanced abundance because they appear at the same nominal *m/z* value. The small mass difference between isobars, however, requires separation of the $^{13}CH_3$ and $^{12}CH_2D$

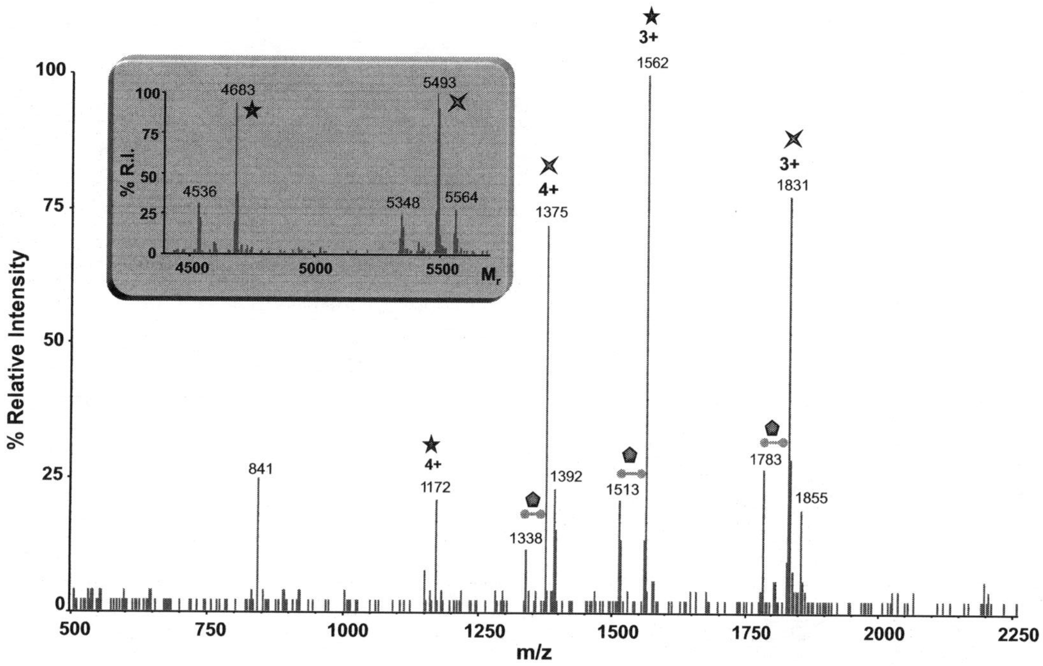

FIGURE 12.9 Positive-ion ESI mass spectrum of Sf9 IL-5Rα tryptic glycopeptide component (Figure 12.8, TIC peak 20). This glycopeptide contains a high-mannose carbohydrate component with M_r values of 4683 and 5493 (inset). Glycoform heterogeneity due to variations in the fucose (⬟) content is also indicated in both triply and quadruply charged ESI ion envelopes.

labeled glycans by a high mass-resolving power analyzer (e.g., FT ICR), and further MS^n analysis allows the relative quantitation of isomeric glycans.

The use of *lectins* to bind glycoproteins in complex samples and the subsequent detection of different glycosylation owing to differential binding of the lectins is another traditional quantitation strategy that uses labeling and optical detection. For example, it is possible to distinguish different types of *N*-linked glycosylation (e.g., *high mannose* and *complex-type*) in glycoproteins/glycopeptides by using different lectins [94] or by tagging either the lectins or the analyte with a fluorophore, followed by monitoring of the fluorescence change upon binding [95,96]. Even though this lectin microarrays method has low detection limits, it is not capable of detecting subtle glycosylation changes, such as a change in the number of mannoses present on a high-mannose glycan.

Quantitation of glycopeptides by label-free approaches is another strategy, where normalization of data alleviates the variation due to changes in MS response among samples [97–100]. A relatively recent normalization method takes the ion abundance from each glycopeptide and divides by the total intensity of all glycopeptide peaks present in a given spectrum, accounting for the weak glycopeptide ionization and any variability arising from nonglycosylated interferences [101].

12.3 PHOSPHORYLATION

12.3.1 MS Detection of Phosphorylation

Approximately one-third of eukaryotic proteins are phosphorylated, by way of a reversible process catalyzed by protein kinases and phosphatases [102]. Protein phosphorylation plays a central role in intracellular communication [103,104] and regulates signal transduction and a wide variety of cellular events such as growth, metabolism, proliferation, and differentiation [105]. Irregular phosphorylation events result in deregulated cell growth and apoptosis, and play a crucial role in many human diseases including various cancers. Therefore the determination of phosphorylation sites in proteins is required for understanding specific cellular regulation pathways, and defining new drug targets in drug discovery.

Protein phosphorylation occurs mostly on S (pS, ~90%), T (pT, ~10%), and Y (pY, ~0.05%) [106], but it can also happen to a lesser extent on the H, R, K, C, E, and D residues. Complete characterization of a phosphorylated protein should include the identification of the phosphorylation sites and the quantitation of the extent of phosphorylation. A preliminary view and a rapid assessment of the number of the phosphate groups present in a protein can be provided by M_r measurement of the intact phosphoprotein, as in the case of mitogen-activated kinase (MEK1) where MALDI MS analysis indicated the presence of one phosphate group [107]. There are a number of methods developed recently for further characterization and identification of the phosphorylation sites. A well-established method involving analysis of ^{32}P-labeled phosphoproteins [108] by *Edman* degradation [109,110] followed by 2D phosphopeptide mapping is usually constrained by the poor solubility of the phosphoamino-acid products and the radioactivity-related inconvenience. In addition these methods have the disadvantage of long analysis time due to the necessity to obtain highly purified phosphopeptides. However, MS has emerged as a sensitive and dependable method for locating protein phosphorylation sites [111,112]; MS has high throughput, can accommodate automation, and has no need for radioactivity. Protein phosphorylation events are detected by an 80-Da increase in the residual amino-acid mass, corresponding to the addition of HPO_3.

The MS-based phosphorylation analysis usually begins with proteolytic digestion of the phosphoprotein of interest to generate peptide mixture. Trypsin is the most commonly used enzyme, although endoproteinase Glu-C can be used prior to immobilized metal affinity chromatography (IMAC) enrichment to reduce significantly the nonspecific binding of acidic nonphosphorylated peptides [113]. The second step is purification, isolation, and enrichment of phosphopeptides from the proteolytic peptide mixture; these steps enhance the phosphopeptide signals. The third step is desalting and concentrating the eluted peptides by using reversed phase (RP) chromatography, and this step is followed by MS and tandem MS analysis of the phosphopeptides to determine the phosphorylation site(s).

Both MALDI and ESI ionization sources are used as ionization methods in phosphopeptide analysis, whereas the use of qTOF, orbitrap, and IM TOF analyzers has prevailed over the traditional triple quadrupole mass analyzer for mass analysis. The use of nESI [32] has notably increased the ionization efficiency and sensitivity of analysis, as is the case for analysis of glycoproteins [33,114]. Chromatographic separation prior to ionization can also reduce ion suppression, remove salt interferences, and concentrate dilute samples.

Upon proteolytic digestion, one generates peptides, many of which contain one or more potential phosphorylation sites even though the phosphopeptide measured M_r value only reveals one net phosphorylation. Therefore one carries out tandem MS analysis of phosphopeptides to assign the protein sequence and to establish those amino-acid residues that are modified. The signals for the modified and unmodified forms of each phosphopeptide can be used to estimate the stoichiometry of the modification at each site [1]. The presence of phosphopeptides in a mixture can also be confirmed by reacting the peptides with phosphatase and observing a downshift of 80 Da (or multiples of 80) in the respective signals; the downshifts are due to the loss of HPO_3 group(s) [115,116]. This determination has remained a challenging task, mainly because the stoichiometry of phosphorylation is low as is the ionization efficiency of phosphopeptides relative to their nonmodified counterparts. One way to enhance the ionization efficiency of pS/pT containing peptides is to replace the phosphate groups on S and T residues via β-elimination and *Michael* addition reaction by a group that gives rise to higher MALDI and ESI signal intensities [117,118]. This strategy combined with the enhanced production of informative product ions in tandem MS analysis of the chemically modified peptides facilitates identification and quantitation of S/T phosphorylation in proteins [118]. However, a plethora of different enrichment methods can be used to compensate for the low abundance and stoichiometry of phosphoproteins or phosphopeptides; these approaches are discussed next.

12.3.2 Enrichment of Phosphorylated Peptides and Proteins

The enrichment of phosphopeptides from the total digest is essential especially for large proteins, where the enzyme-derived peptide mixture is extremely complex. The most successful enrichment methods prior to MS analysis are the *antibody*- and *affinity*-based methods. Another approach utilizes *chemical* methods through covalent coupling although such methods suffer from nonquantitative yields and side reactions. The most commonly used methods are discussed below.

Immunopurification Immunopurification (IP), originally described by Vandekerckhove and coworkers in 1999 [119], is a well-established enrichment strategy for peptides phosphorylated on tyrosine (pY); it uses immobilized antiphosphotyrosine antibodies. The method has been significantly improved since its introduction, and a variety of high affinity and specificity antibodies for pY residues can be used to map thousands of Y phosphorylation sites in different cancer cell lines [120–122]. This antibody-based pY enrichment strategy can be also applied to isolate pY-containing proteins; an example is the IP/SDS-PAGE study of proteins involved in epidermal growth factor signaling [123]. Similarly IP-capable phosphor-specific antibodies can be used for the identification of S and T phosphorylated proteins [124]. Although protein phosphorylation on S or T residues comprises the bulk of cellular phosphorylation sites, there are no antibodies with good specificity, motivating the use of alternative methods for their enrichment.

Immobilized Metal Affinity Chromatography The immobilized metal affinity chromatography (IMAC) [125–128] enrichment method incorporates the high-affinity binding of the negatively charged phosphate of the peptide with certain trivalent metal ions (i.e., Fe^{3+}, Ga^{3+} and Al^{3+}) [129–131]. The drawback is the nonspecific binding of peptides containing abundant acidic residues (D or E), especially when loading samples at pHs outside the 2 to 3.5 range [126]. This problem seems to have been diminished or even eliminated by esterifying the carboxylic acidic groups prior to IMAC enrichment [132]. The incomplete derivatization of the carboxylic groups along with the partial deamidation and subsequent methylation of N and glutamine (Q) residues, however, compromise the sensitivity of this enrichment technique and further complicate the data interpretation [133]. Nevertheless, the IMAC enrichment approach is well matched for phosphopeptide detection using online LC-MS and MS/MS analysis [132,134].

Metal Oxide Affinity Chromatography The aforementioned drawbacks of the IMAC method, and the difficulty in adaptation of the IMAC protocols among different laboratories, point to the need for other enrichment methods with high specificity. In that respect metal oxide affinity chromatography (MOAC) [135] is a powerful method for phosphopeptide enrichment. Titanium dioxide (TiO_2) packed columns can be used to enrich phosphopeptides from tryptic digest mixtures both offline [136] and on line [135,137] for analysis by LC-ESI MS and MALDI MS [138]. Nevertheless, the high chemical stability and affinity of the TiO_2 packing material is compromised by the simultaneous adsorption of other acidic amino-acid residues (D or E). This problem can be avoided by using 2,5-dihydroxy-benzoic acid (DHB) as a competitive binder for other carboxylic acids, enhancing the selectivity of TiO_2 for phosphopeptides [135]. Furthermore zirconium dioxide (ZrO_2) microtips can be employed successfully [139] for phosphopeptide isolation prior to MS analysis. The ZrO_2 can be more selective for the enrichment of singly phosphorylated peptides, whereas TiO_2 preferentially enriches multiple phosphorylated peptides. For small hydrophilic phosphopeptides, one should consider the use of *graphite* powder microcolumns for improved detection [140,141].

Strong Cation Exchange Chromatography The use of strong cation exchange (SCX) resins is another valuable approach for enrichment of phosphopeptides. This approach is based on charge separation between the phosphorylated and nonphosphorylated peptides [142]. The phosphorylated tryptic peptides can be easily separated by SCX chromatography because they carry a reduced positive charge state compared to nonphosphorylated ones (i.e., reduction of one charge per phosphate group). The presence of multiple phosphate groups, however, yields a zero or negative charge resulting in no retention on the SCX column. In fact this unbound phosphopeptide-rich SCX fraction can be further enriched by IMAC or TiO_2 enrichment [143,144].

Hydrophilic Interaction Chromatography Hydrophilic interaction chromatography (HILIC) should also be considered for phosphopeptide enrichment [145]. In this case retention is based on hydrophilicity as opposed to hydrophobicity, which is the basis for RP chromatographic separation of peptides prior to MS analysis. The decreased solubility of longer peptides ($M_r > 2000$ Da) in the organic solvent and the strong interactions of multiply phosphorylated peptides with the stationary phase constitute the limitations of this method.

Chemical Tagging Methods Another highly successful approach for affinity purification and quantitation of phosphorylated peptides is β-elimination of pS and pT residues followed by *Michael* addition of various nucleophilic tags, such as a *biotin affinity tag* combined with an avidin column [146,147]. A known caveat of this method is the β-elimination of other PTMs, especially *O*-linked oligosaccharides, which may lead to data misinterpretation and false assignment of glycosylation sites to be phosphorylation sites [148]. In addition protection of C residues should be carried out to prevent adverse side reactions [149]. The low yield of this reaction usually requires an increased amount of protein.

12.3.3 Phosphorylation Site Identification

The large complexity of phosphopeptide mixtures generated from the proteolytic digestion of a phosphorylated protein requires the use of a chromatographic separation (usually on a C_{18} RP column) prior to ESI MS analysis. Several MS-based approaches can be employed for the characterization of phosphorylated proteins and the identification of the phosphorylation sites, and their respective merits and contributions were reviewed by Loyet et al. [112]. These approaches are based on the lability of the phospho moiety of the modified peptides upon CID, where characteristic fragmentation pathways produce signature ions, allowing their distinction from nonphosphorylated peptides. In the positive-ion mode, pS- and pT-containing phosphopeptides undergo a neutral loss of H_3PO_4 (98 Da) via β-elimination, whereas loss of HPO_3 (80 Da) occurs usually for pS, pT and pY-containing phosphopeptides [150,151]. In addition the presence of pY residues can be revealed by the characteristic immonium ion at *m/z* 216.043 [152]. In this case the use of high mass-resolving power and mass accuracy instruments is necessary (e.g., qTOF, FT ICR and orbitrap), to differentiate it from other peptide fragment ions with the same nominal mass, and thus provide unequivocal assignment of the pY residues [153].

The fragment ions in the negative-ion mode, however, are most commonly employed for a more sensitive detection of phosphopeptides than for positive ions. Negative-ion, product-ion spectra from MS/MS of phosphorylated S, T, and Y residues show fragments at *m/z* 79 (PO_3^-) and 63 (PO_2^-) [154]. Therefore *precursor ion scanning* of *m/z* 79 in the negative-ion mode can be employed to detect selectively phosphopeptides that fragment to produce this marker ion [155,156]; the approach is applicable for identifying phosphorylation occurring on all S, T, and Y residues. The detection of the *m/z* 79 marker ion employing "in-source" fragmentation in an ESI (or nESI) source can also be used for screening phosphopeptide-containing ions in LC-MS [157], an approach similar to that employed for the detection of glycopeptides (Section 12.3). Likewise *neutral-loss scanning* of H_3PO_4 (e.g., 97.97 from singly charged or 48.99 from doubly charged precursors) can identify peptides containing pS and pT residues.

Precursor ion and neutral-loss scans can be incorporated in various LC-MS/MS experimental strategies, including preliminary precursor ion scanning (*m/z* 79) in the negative-ion mode followed by polarity switching and *product-ion scanning* of the detected phosphopeptide candidates [158,159], or by high-energy C-trap dissociation (HCD) on an orbitrap mass spectrometer [160]. Another approach of obtaining sequence-specific information is the selection of the neutral-loss precursor ion, which undergoes further (MS^3) fragmentation on a linear trap quadrupole (LTQ), LTQ-FT or LTQ orbitrap mass spectrometer [14,161,162]. All the aforementioned LC-MS/MS scanning approaches can be carried out in an automatic data-dependent acquisition (DDA) mode, on any number of ions in the mass spectrum based on a pre-set intensity threshold, and/or *m/z* value or charge state.

Alternatives to the low-energy CID fragmentation of phosphorylated peptides are ECD [71,73] and ETD [74], where fragmentation occurs through peptide interactions with low-energy electrons (ECD) or radical anions (ETD) that provide electron transfer. Both ECD and ETD have advantages over CID for phosphopeptide sequencing and phosphorylation site localization because they induce fragmentation of the peptide backbone without any significant loss of the phosphate moiety from the fragment ions [74,163,164]. As opposed to b- and y-type ions that are usually generated by CID of phosphopeptides (Figure 12.10), a series of c- and z-type sequence ions are produced and seen in the respective ECD and ETD spectra.

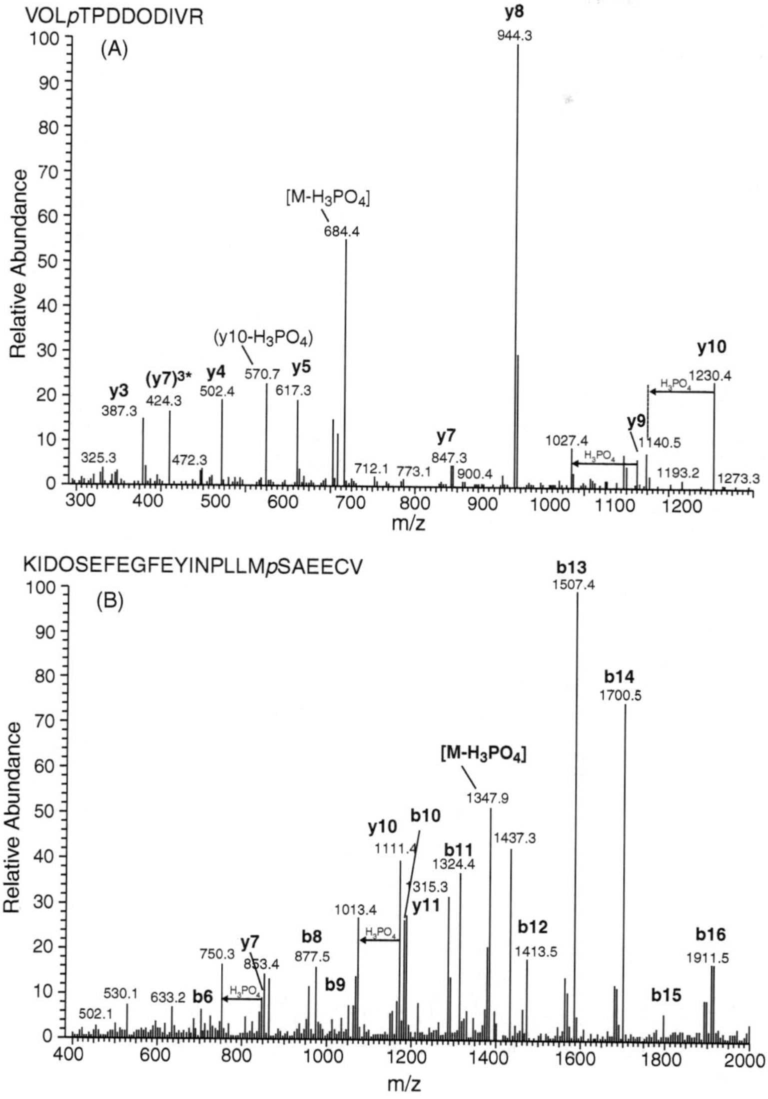

FIGURE 12.10 Positive-ion nESI mass spectra of protein kinase Cι (PKCι) tryptic phosphopeptides: *(A)* VQLpTPDDDDIVR, where b and y fragment ions point to modification at the T[555] site, and *(B)* KIDQSEFEGFEYINPLLMpSAEECV where the y_7 and y_{10} fragment ions, with concomitant neutral losses of phosphoric acid point to S[582] as the modification site. Neutral-loss-dependent MS^3 scans were triggered for all precursor ions, showing a loss corresponding to the mass of phosphoric acid during CID. The precursor ions were measured with a mass deviation of 1.5 ppm and 2.4 ppm, respectively. Reprinted with permission from [162].

This is nicely shown in the ECD FTICR mass spectrum of the triply charged phosphorylated tryptic peptide of α_{S2} case in precursor, where the series of c- and z-type ions arise from extensive peptide backbone cleavages (Figure 12.11). The complete amino-acid sequence along with the phosphorylation site at S^{46} of the 40-NM$_{ox}$AINPpSKENLCSTFCK-56 peptide is determined by the mass difference between the c_6 and c_7 ions.

The use of IM MS [165] can also facilitate phosphorylation studies by reducing the complexity of the mass spectra and allowing for lower abundant species to be detected [166]. IM MS enables the separation of phosphorylated peptides from nonphosphorylated peptides of the same *m/z* because the phosphorylated peptides seem to have smaller cross sections. More confidence for this conclusion will come when more phosphorylated peptides have been reported by using ESI in the positive-ion mode [167–169].

MALDI MS can also be used to study phosphorylation of proteins, and it has high sensitivity and tolerance to sample composition. Phosphorylated peptides can be specifically recognized by the characteristic mass shift of 80 (or multiples of 80) Da in the MALDI TOF mass spectra upon alkaline phosphatase treatment arising from the loss of HPO_3 group(s) [170,171]. In addition, phosphopeptides are identified in the positive-ion MALDI TOF reflectron mass spectrum by the presence of $(M + H - H_3PO_4)^+$ and $(M + H - HPO_3)^+$ fragment ions, which are indicative of phosphorylation on S/T and Y residues, respectively [172]. Similarly MALDI combined with PSD can distinguish between S/T and Y phosphorylation and in many cases provide sequence information on the site of phosphorylation [173,174]. Sequencing of phosphopeptides can also be performed by coupling of MALDI to hybrid analyzers, such as qTOF, TOF-TOF, ion trap or LTQ orbitrap, which allow rapid identification of phosphopeptides together with locating the phosphorylation sites [175–180].

The aforementioned MS and MS/MS-based approaches are now established as effective tools for phosphopeptide fingerprinting (i.e., identifying and locating the phosphorylation sites within a peptide sequence). The analysis can be assisted by using database search tools, such as *Turbo SEQUEST* [181,182], *Mascot* [183], and *Protein Prospector*, [184]. Comprehensive identification of phosphorylation sites or other PTMs, especially in a high-throughput manner, remains challenging, however, because product-ion spectra interpretation can be difficult, the peptide fragmentation can be poor, and unexpected modifications can be encountered [185]. Analysis is especially difficult for phosphorylated peptides with multiple and neighboring sites (i.e., pinpointing the exact position of the modification site. Unrestrictive PTM search algorithms (e.g., *MS-Alignment* [186] and *MOD^i* [187]) can facilitate the product-ion spectral interpretation of peptides with multiple and unrestricted PTMs. This problem can also be addressed by searching through phosphopeptide databases listing MS-deduced phosphorylation sites associated with particular proteins (e.g., *Phosphosite* (http://www.phosphosite.org), *SwissProt* (http://ca.expasy.org/sprot), *Human Protein Reference Database* (http://www.hprd.org), and *PhosIDA* (http://www.phosida.com), which also predicts phosphosites [188]). In addition all the MS-based phosphorylation information has promoted the development of phospho-prediction tools (e.g., *Scansite* (http://scansite.mit.edu) that search for motifs within

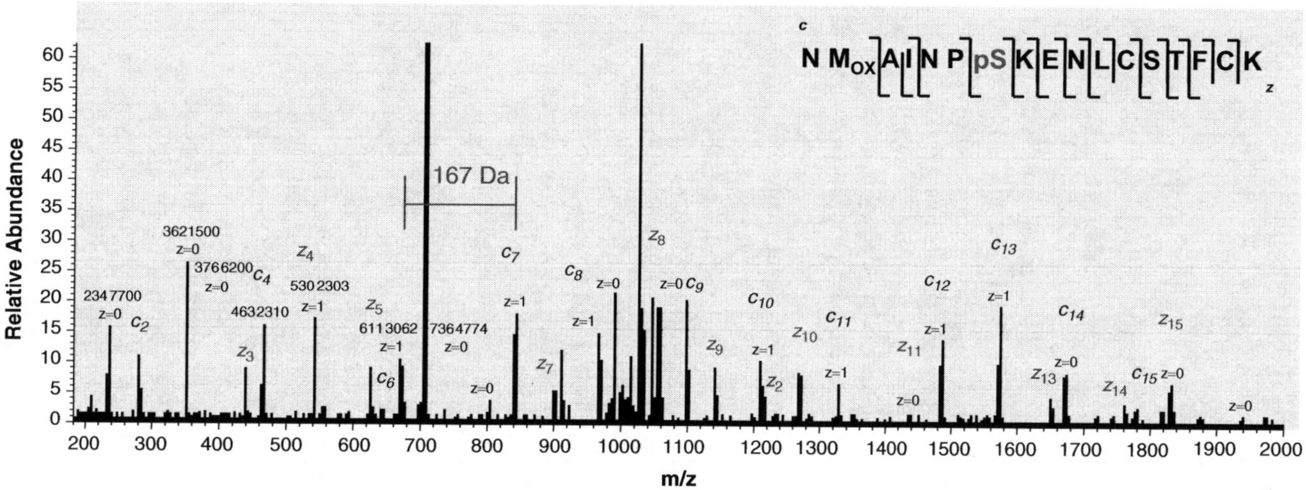

FIGURE 12.11 Electron capture dissociation (ECD) FT ICR mass spectrum of the triply protonated monophosphorylated peptide 40-NM_{ox}AINPpS-KENLCSTFCK-56 (M_r 2110.841) generated from tryptic digestion of a_{S2} casein. An extensive series of c- and z-type fragment ions were produced, which determined the complete amino acid sequence and the phosphorylation site at S^{46} as determined by the mass difference of 167 Da between the c_6- and c_7-ions. Spectrum kindly provided by Professor R. Zubarev and Dr. M. Nielsen.

proteins that are likely to be phosphorylated by specific protein kinases or bind to certain domains such as SH2 domains. However, determining the location of phosphorylation sites and other PTMs can sometimes be deduced from product-ion mass spectra of intact proteins ("top-down" approach) by using the *ProSight* algorithm [189].

12.3.4 Phosphopeptide Quantitation

Although MS has demonstrated strength in phosphopeptide fingerprinting (i.e., identifying and locating phosphorylation sites within peptide sequences), the low occupancy of the phosphorylation sites is an obstacle to determining site stoichiometry. This is particularly important because phosphorylation regulates signal transduction pathways in a reversible way (i.e., through successive phosphorylation and dephosphorylation events at multiple sites). Moreover a certain protein might be involved in more than one signaling pathway with different stimuli inducing diverse and overlapping patterns of phosphorylation. Therefore quantitation of phosphorylation and determination of the ratio of a protein phosphorylation on multiple residues is difficult but crucial.

The approaches that were traditionally used for quantitation of phosphorylation are phosphoamino-acid analysis and *Edman* degradation, as well as *ELISA* and other phosphorylation-site specific antibody-based methods [190,191], with the latter being superior in terms of sensitivity. Several MS-based approaches are rapidly evolving as alternatives for phosphopeptide quantitation. These methods generally involve *stable isotope labeling,* where samples from two cell states are labeled with isotope-derivatized moieties, mixed, and then analyzed simultaneously by LC-MS/MS. Each phosphopeptide then appears as doublet in the mass spectrum, and the relative abundance of the two peaks reflects the relative amount of the peptide in each sample. The signals suffer little, if any, suppression in the ionization. The label can be introduced during cell culture, such as in stable isotope labeling by amino acids in cell culture (SILAC) [192], an approach involving metabolic labeling of proteins using amino acids labeled with ^{13}C for ^{12}C or ^{15}N for ^{14}N (usually R and K in case of proteolysis by trypsin). SILAC results in easily interpretable isotope patterns, making it ideal to study PTMs such as phosphorylation changes [193] both in vitro and in vivo. Other metabolic labeling techniques include phosphopetide quantitation in yeast using ^{15}N-incorporation [194] and enzymatic ^{18}O-labeling [195].

Another quantitation strategy involves *chemical labeling of reactive peptide groups*, such as amine, sulfhydryl, and carboxylic groups [196]. This labeling can occur either on the protein level with *isotope-coded affinity tags* (ICAT) [197] or on the enzyme-generated peptides by using *tandem mass tags* (iTRAQ) [198]. The ICAT reagent consists of a biotin group followed by a linker bound to C residues. Proteins isolated from healthy and diseased cells, for example, can be treated with the light- and heavy-form (e.g., deuterium) ICAT reagent, respectively, mixed and digested with trypsin. The ICAT-labeled peptides are separated by affinity chromatography using an avidin column, and analyzed by LC-MS, allowing both identification and quantitation of proteins contained in the original samples. Still this approach may yield erroneous results due to poor binding or recovery of phosphopeptides. However, the iTRAQ approach utilizes isobaric *tandem mass tags* containing amine-specific groups linked to a fragmentable tag (reporter and balance group). Quantitation is performed by using the product-ion (MS/MS) spectra rather than the mass spectra [198]. When the isobaric tags react with the enzyme-generated peptides to form an amide linkage to either the *N*-terminal or the ε-amino group of K, the resulting derivatized peptides are indistinguishable either by their mass spectra or by the backbone fragmentation upon CID. Their product-ion spectra exhibit abundant low-mass *tag* ions with different *m/z* values corresponding to the different reporter groups. Measurement of the relative abundances of these diagnostic reporter ion signals allows quantitation of protein, as well as determination of the relative phosphopeptide abundances. Although SILAC and iTRAQ are the most frequently used techniques in quantitative MS-based phosphoproteomics, only the latter can be directly applied to tissue samples.

12.4 DISULFIDE BOND DETECTION AND MAPPING

12.4.1 MS Detection

Despite the undisputed "popularity" of glycosylation and phosphorylation in the PTM literature, *disulfide bonding* in proteins is one of the most frequently encountered PTMs of proteins. These crosslinkages result from oxidation of the sulfhydryl groups of two C residues, and play important roles in establishing and maintaining the 3D structure of extracellular proteins [199–202]. *Intramolecular* disulfide-bonds stabilize the tertiary structures of proteins, while *intermolecular* disulfide bonds are involved in stabilizing quaternary structure [203]. Therefore it is important to have reliable methods for locating and determining disulfide bond arrangement of proteins. The extent and location of disulfide bond formation provide insights into protein activity relationships and guide further structural determination by NMR and X-ray crystallography. The number of disulfides in a given protein can sometimes be deduced by a simple MS analysis before and after reduction. This is nicely illustrated in the ESI MS analysis of rhGM-CSF, where the 4-Da shift in the measured M_r upon reduction with β-mercaptoethanol indicated the presence of two disulfide bonds [204]. Following the determination of the number of disulfide linkages, mapping of the protein's primary sequence can be carried out to locate the existing disulfide arrangement.

12.4.2 Disulfide Mapping

The methodology for disulfide mapping [205] involves either enzymatic or chemical cleavage of the protein backbone between half-cystine residues to produce peptides containing one disulfide bond, followed by analysis of the disulfide-linked peptides before and after reduction. MS has played a central role in the analysis of the resulting disulfide-linked peptides, and the determination of the disulfide linkages in proteins;

initially by fast atom bombardment (FAB)/liquid secondary ion (LSI) [206,207] and plasma desorption (PD) [208,209] and later by the more sensitive methods of MALDI [8] and ESI [6] combined with qTOF [210] or LIFT-TOF/TOF MS [211] analysis. The disulfide-linked peptides yield unique mass spectral signals, which are assembled to specific segments of the protein. The conditions for the cleavage between half-cystine residues must be carefully controlled because disulfide interchange (sometimes called disulfide scrambling) can occur at neutral and alkaline pH. Consequently *pepsin* has favorable properties for generating disulfide-linked peptides [212] because its acidic pH optimum allows proteins to unfold to conformations that are more accessible to cleavage and also precludes disulfide bond interchange [213,214]. Nevertheless, its broad specificity [215] leads to complex digests that contain overlapping peptides owing to "ragged" cleavage. One of the first reports on disulfide mapping [212] demonstrated the use of FAB MS in combination with *Edman* analysis for disulfide bond analysis. Identification of intramolecular disulfides was achieved by the 2-Da increase upon reduction of their constituent half-cystines with β-mercaptoethanol or dithiothreitol. In the case of intermolecular disulfide bonds, a mass spectrometric analysis (here by FAB/LSI MS) takes advantage of protonated molecules $(M + H^+)$ for both the constituent half-cystine-containing peptide fragments [204,216], as this is nicely shown in the LSI mass spectrum of the V8 protease disulfide-linked peptide of rhGM-CSF (Figure 12.12).

Similar intermolecular disulfide bond reduction can occur in disulfide mapping by MALDI MS [217,218], which is further promoted by in situ reduction in MALDI matrices using tris(2-carboxyethyl)phosphine (TCEP) [219]. On the contrary, ESI MS analysis of disulfide-linked peptides does not yield half-cystinyl peptide signals arising from partial disulfide bond reduction, as shown in the ESI mass spectrum of the disulfide-linked tryptic peptide T_{20}-T_{25} of IL-5Rα (Figure 12.13).

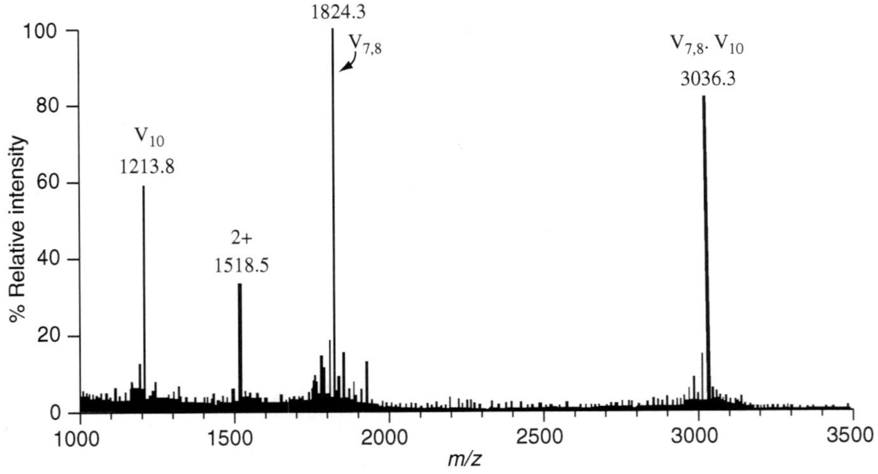

FIGURE 12.12 Positive-ion Cs^+ LSI mass spectrum of the HPLC-isolated fraction from the V8 protease digest of rhGM-CSF containing the disulfide-linked peptide $V_{7,8}$-V_{10}. Reprinted with permission from [204].

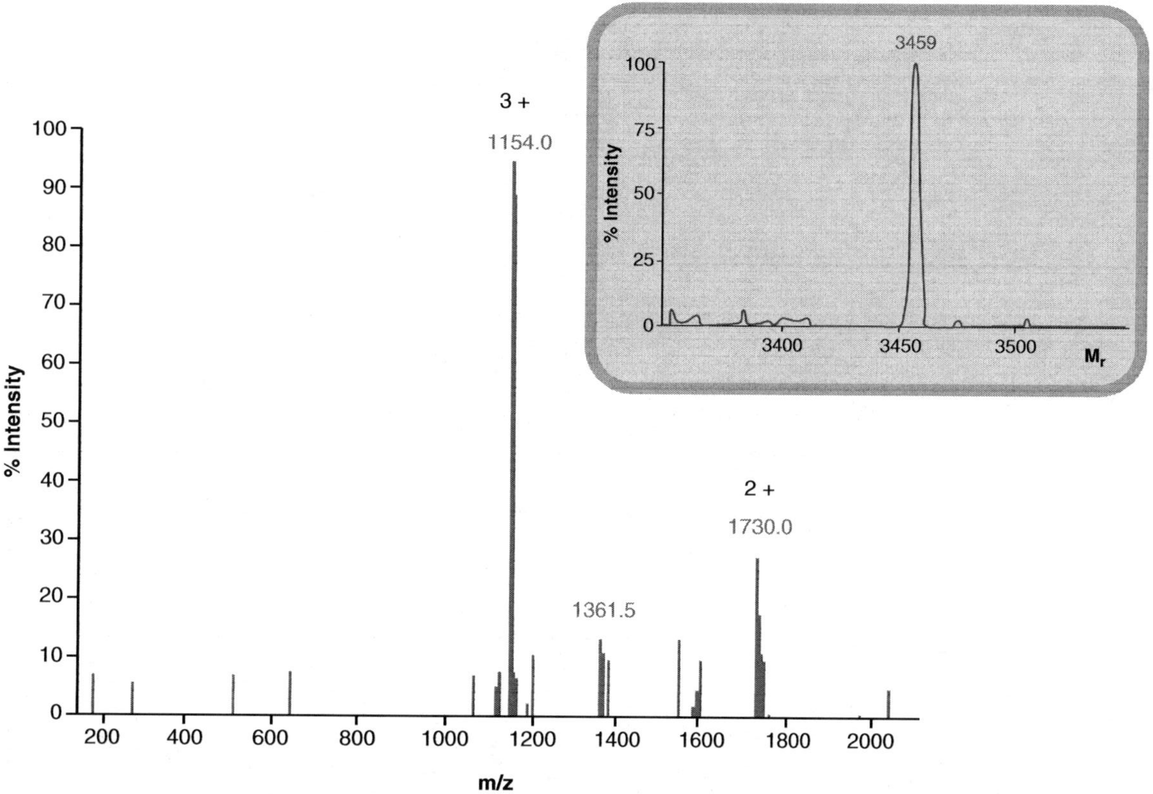

FIGURE 12.13 Positive-ion ESI mass spectrum of Sf9 IL-5Rα tryptic fraction (Figure 12.8, TIC peak 14) containing the disulfide-bonded peptides T_{20} and T_{25}.

When protein chains are linked and no further proteolysis is possible, identification of the exact location of the disulfide linkage often requires (1) successive proteolytic digestions, as was demonstrated for interleukin-13 (IL-13) where chymotrypsin followed by *S. aureus* V8 protease enzymatic treatment was employed [220], or (2) chromatographic separation of the enzyme-derived protein fragments coupled with online MS/MS analysis (e.g., LC-ESI MS/MS), and/or offline MS/MS analysis and *Edman* sequencing [221,222]. This is essential for proteins where three proteolytic fragments linked by intermolecular disulfides are not susceptible to further proteolysis between half-cystines, or where two peptide chains are linked and one of the chains contains an intramolecular disulfide and no further proteolysis is possible. The existence of disulfide bonds is usually confirmed by fragmentation of putatively disulfide-linked peptides by MS/MS analysis following ionization by FAB [223], ESI [224], or MALDI PSD TOF [225]. The latter has the advantage of high mass range, acceptable mass measurement, and the formation of a characteristic triplet of ions separated by 33–Da arising from cleavage at the cystine C–S bond with a concomitant proton transfer [226]. This ion-triplet formation can be used as a diagnostic tool for the location and identification of disulfide-paired peptides, even from complex digest mixtures of proteins.

An approach minimizing sample handling is to cleave disulfide bonds by ECD [227,228] in a "top-down" proteomics experiment. Another approach for disulfide-linked and protease-resistant proteins, especially in chains with adjacent half-cystines, is to reduce with TCEP and alkylate under acidic conditions in the presence of denaturing agents, where disulfide bonds are partially reduced [229]. The alkylating agent is either iodoacetamide or 1-cyano-4-(dimethylamino)pyridinium tetrafluoroborate (CDAP), which cyanylate half-cystine residues at acidic pH, efficiently minimizing any potential interchange during the alkylation [229]. The protein derivatives generated by partial reduction and alkylation are amenable to proteolysis and LC-MS analysis [230,231].

Finally, *stable isotope-labeling of peptides with*^{18}O greatly facilitate the identification and characterization of disulfide-linked peptides [232]. Isotope profiles of peptides produced in 1:1 mixture of $H_2{}^{16}O/H_2{}^{18}O$ (arising from a combination of isotope ratios and average mass increases) enable disulfide-linked peptides to be identified in complex peptic digests or chromatographic fractions thereof. This procedure allows the production and isolation of disulfide-linked peptides to be performed at an acidic pH in order to preclude disulfide rearrangement, and it may also be used to aid the interpretation of product-ion spectra of peptides with multiple disulfide bonds.

12.5 FUTURE PERSPECTIVES

To date, over one hundred recombinant proteins are indicated for several medical conditions (e.g., anticancer, antidiabetic, antithrombotic, antiviral, hemophilia, hematopoiesis, monoclonal antibodies, antithrombotic, vaccines, and interferons) and have been approved for human therapeutic use [233,234]. Even though recombinant protein therapies offer a high degree of specificity for a number of clinically relevant disease targets, their efficacy and immunogenicity can be highly dependent on the protein sequence and the presence or absence of specific PTMs. Consequently each and every PTM of a protein is of a great concern for regulatory agencies (e.g., the directed modification of protein glycosylation [*glycoengineering*], or the artificial attachment of polymers to therapeutic proteins demand analytical tools for their characterization). Therefore advanced and sensitive analytical methodologies are needed to confirm as unambiguously as possible PTMs in a protein sequence, structure, and activity at each stage of a therapeutic pharmaceutical protein manufacturing process. This, along with the constantly growing knowledge of the biological roles of protein modification, will be invaluable toward the design of novel protein therapeutic agents.

MALDI and ESI MS have emerged as routine and essential analytical tools for the quality control and structure characterization of therapeutic proteins, assuring their batch-to-batch consistency, safety, potency, and stability. This is a requirement imposed by the International Conference on Harmonization (ICH) of technical requirements for registration of pharmaceuticals for human use (ICH Q6B guidance); the document states that "these forms may be active and their presence may have no deleterious effect on the safety and efficacy of the product. The manufacturer should define the pattern of heterogeneity of the desired product and demonstrate consistency with those of lots used in preclinical and clinical studies. If a consistent pattern of product heterogeneity is demonstrated, an evaluation of the activity, efficacy, and safety (including immunogenicity) of individual forms may not be necessary." Nevertheless, it is a major and challenging task to confirm PTM site(s) and to understand their structure–function relationship (i.e., ranking the importance of several PTMs occurring in a protein toward regulating certain cellular events). For example, the glycan profile of a glycoprotein may be responsible for the biological activity or absence thereof [235], as in the case of erythropoietin (glycoprotein hormone), where removal of the terminal sialic acid residues from the carbohydrate chains results in loss of activity in vivo [236]. Moreover changes in levels and types of glycosylation can be associated with certain diseases, making glycoprotein screening invaluable, not only for diagnostic purposes but also for the design of novel therapeutic drugs. This is illustrated by the use of glycosylation mapping as a diagnostic tool for aggressive breast cancer [237], while glycan profiling of normal and diseased forms of a glycoprotein has provided new insights for future research in rheumatoid arthritis, prostate cancer, prion disease, and congenital disorders of glycosylation [238–241].

Moreover identification and quantitation of phosphorylation sites in proteins is vital for gaining an understanding of key transduction pathways and the mechanisms of protein kinases, which are essential elements mediating cell growth and programmed cell death. Deregulation of either protein kinases or protein phosphatases results in deregulated cell growth and abnormal cellular signaling [242], which are associated with human cancers. Therefore a complete list and characterization of phosphorylated proteins (*phosphoproteomics*) will promote the possibility of developing agonists and antagonists of protein kinases, a prominent drug target [243,244]. Several inhibitors targeting protein tyrosine kinases have already demonstrated their value in cancer and hepatitis C virus treatment, and there is a growing interest in the development of drugs targeting protein tyrosine phosphatases as promising novel cancer therapies [245].

Similarly establishment of the correct disulfide bond pattern in a recombinant protein product is essential to ensure its biological activity, as for insulin-like growth factor [246]. In addition the MS-deduced disulfide arrangement in structural domains of serine proteases, such as tissue plasminogen activator [247], can be the basis both for producing new classes of protein drugs and for learning more in structure–function relationship studies.

The ever-increasing analytical capabilities of MS, including improvements in sensitivity and mass measurement accuracy, combined with improved enrichment strategies and bioinformatics, have made it possible to analyze glycoproteomes and phosphoproteomes. Many challenges still remain, mainly in assessing the extent and stoichiometry of phosphorylation of low-abundance proteins, providing real-time system analysis to correlate better protein phosphorylation with biological processes, and analyzing multiple phosphorylation sites (a task for top-down). The main analytical challenge is the limited capacity to perform peptide identification by LC-MS/MS analysis, motivating development of better MS software to ensure optimal peptide selection for MS/MS, higher sensitivity and dynamic range mass spectrometers, and higher scan rates especially with the incorporation of the higher capacity and improved resolution technique of ultra-pressure liquid chromatography [248,249]. Furthermore there are still difficulties in the analysis and the structural characterization of large therapeutic proteins, mainly arising from their heterogeneity, especially in glycoproteins that have a relatively high content of carbohydrate. Improved sensitivity and mass measurement accuracy have rendered qTOF, IM TOF, and orbitrap as the analyzers of choice for the detection of PTMs in proteins, although the orbitrap has a lower mass range than the others. The advantages of the qTOF and IM analyzers are especially relevant in the area of monitoring noncovalent interactions of PTM-modified proteins. In this case detection and stoichiometry of protein–protein or protein–ligand interactions are required as well as the realization of subunit contacts in heterogeneous macromolecular assemblies [250]. Successful detection and examination of these complexes will have a significant impact on drug-discovery research, by revealing protein–receptor and/or receptor–drug interactions, and on functional proteomics. Finally, there is a growing demand to develop methods for simultaneous monitoring of PTM variations within single polypeptides of a protein sequence, revealing PTM hierarchies and establishing the heterogeneity thereof. The emergence of "top-down" MS methodology [189] should help advance the ultimate goal of functional proteomics (i.e., deciphering the PTM function in signaling networks and the inter-relations among PTMs).

ACKNOWLEDGMENTS

We would like to thank Professor Roman Zubarev and Dr. Michael Nielsen for kindly providing Figure 12.11. We also acknowledge the kind permission of the Schering-

Plough Research Institute to reproduce previously reported, but unpublished, data regarding the IL-5Rα.

ABBREVIATIONS

2D	Two-Dimentional
CDAP	1-Cyano-4-(DimethylAmino)Pyridinium tetrafluoroborate
CE	Capillary Electrophoresis
CHO IL-4	CHO-derived Interleukin-4
CID	Collision-Induced Dissociation
CZE	Capillary Zone Electrophoresis
DDA	Data Dependent Acquisition
DHB	2,5-DiHydroxy-Benzoic acid
ECD	Electron Capture Dissociation
ESI	Electrospray Ionization
ETD	Electron Transfer Dissociation
ERPA	Extended Range Proteomic Analysis
FAB	Fast Atom Bombardment
FT ICR	Fourier-Transform Ion Cyclotron Resonance
HCD	High-energy C-trap Dissociation
HILIC	Hydrophilic Interaction Chromatography
ICAT	Isotope-Coded Affinity Tags
IL	Interleukin
IL-4R	Interleukin 4 receptor
IL-5Rα	Interleukin 5 receptor α subunit
IM	Ion Mobility
IMAC	Immobilized Metal Affinity Chromatography
ICH	International Conference on Harmonization
IP	Immunopurification
iTRAQ	isobaric Tag for Relative and Absolute Quantitation
LAC	Lectin Affinity Chromatography
LC	Liquid Chromatography
LSI	Liquid Secondary Ion
LTQ	Linear Trap Quadrupole
MALDI	Matrix-Assisted Laser Desorption Ionization
MOAC	Metal Oxide Affinity Chromatography
M_r	Molecular Mass
MS	Mass Spectrometry
MS/MS	tandem Mass Spectrometry
nESI	nano Electrospray Ionization
NMR	Nuclear Magnetic Resonance
PD	Plasma Desorption
PSD	Post-Source Decay
PTMs	Post-Translational Modifications

qTOF	quadrupole Time of Flight
QUIBL	Quantitation by Isobaric Labeling
RP	Reversed Phase
SCX	Strong Cation Exchange
SILAC	Stable Isotope Labeling by Amino acids in Cell culture
TCEP	Tris(2-CarboxyEthyl)Phosphine
TOF	Time of Flight

REFERENCES

1. Carr, S. A., Annan, R. S., Huddleston, M. J. (2005). Mapping posttranslational modifications of proteins by MS-based selective detection: Application to phosphoproteomics. *Meth Enzymol 405*, 82–115.
2. Dwek, R. A. (1996). Glycobiology: Toward understanding the function of sugars. *Chem Rev 96*, 683–720.
3. Collins, M. O., Yu, L., Choudhary, J. S. (2007). Analysis of protein phosphorylation on a proteome-scale. *Proteomics 7*, 2751–2768.
4. Wedemeyer, W. J., Welker, E., Narayan, M., Scheraga, H. A. (2000). Disulfide bonds and protein folding. *Biochemistry 39*, 4207–4216.
5. Kirkpatrick, D. S., Denison, C., Gygi, S. P. (2005). Weighing in on ubiquitin: The expanding role of mass-spectrometry-based proteomics. *Nat Cell Biol 7*, 750–757.
6. Fenn, J. B., Mann, M., Meng, C. K., Wong, S. F., Whitehouse, C. M. (1989). Electrospray ionization for mass spectrometry of large biomolecules. *Science 246*, 64–71.
7. Fenn, J. B., Mann, M., Meng, C. K., Wong, F. S., Whitehouse, C. M. (1990). Electrospray ionization-principles and practice. *Mass Spectrom Rev 9*, 37–70.
8. Karas, M., Hillenkamp, F. (1988). Laser desorption ionization of proteins with molecular masses exceeding 10,000 Daltons. *Anal Chem 60*, 2299–2301.
9. Hillenkamp, F., Karas, M., Beavis, R. C., Chait, B. T. (1991). Matrix-assisted laser desorption/ionization mass spectrometry of biopolymers. *Anal Chem 63*, 1193A–1203A.
10. Hancock, W. S., Wu, S. L., Shieh, P. (2002). The challenges of developing a sound proteomics strategy. *Proteomics 2*, 352–359.
11. Larsen, M. R., Trelle, M. B., Thingholm, T. E., Jensen, O. N. (2006). Analysis of posttranslational modifications of proteins by tandem mass spectrometry. *BioTechniques 40*, 790–798.
12. Covey, T., Shushan, B., Bonner, R., Schröder, W., Hucho, F. (1991). Methods in protein sequence analysis. In Jörnvall, H., Höög, J. O., Gustavsson, A. M., eds., *LC/MS and LC/MS/MS Screening of the Sites of Posttranslational Modification in Proteins*. Birkhäuser Press, Basel.
13. Bateman, R. H., Carruthers, R., Hoyes, J. B., Jones, C., Langridge, J. I., Millar, A., Vissers, J. P. (2002). A novel precursor ion discovery method on a hybrid quadrupole orthogonal acceleration time-of-flight (Q-TOF) mass spectrometer for studying protein phosphorylation. *J Am Soc Mass Spectrom 13*, 792–803.
14. Schroeder, M. J., Shabanowitz, J., Schwartz, J. C., Hunt, D. F., Coon, J. J. (2004). A neutral loss activation method for improved phosphopeptide sequence analysis by quadrupole ion trap mass spectrometry. *Anal Chem 76*, 3590–3598.

15. Apweiler, R., Hermjakob, H., Sharon, N. (1999). On the frequency of protein glycosylation, as deduced from analysis of the SWISS-PROT database. *Biochim Biophys Acta 1473*, 4–8.

16. Kornfeld, R., Kornfeld, S. (1985). Assembly of asparagines-linked oligosaccharides. *An Rev Biochem 54*, 631–664.

17. Schachter, H. (1995). Glycoproteins. In Montreuil, J., Vliegenthart, J. F. G., Schachter, H., eds., *New Comprehensive Biochemistry*, Vol. *29a*. Elsevier, Amsterdam, pp. 123–454.

18. Helenius, A., Aebi, M. (2001). Intracellular functions of *N*-linked glycans. *Science 291*, 2364–2369.

19. Roth, J. (2002). Protein *N*-glycosylation along the secretory pathway: Relationship to organelle topography and function, protein quality control, and cell interactions. *Chem Rev 102*, 285–303.

20. Spiro, R. G. (2002). Protein glycosylation: Nature, distribution, enzymatic formation, and disease implications of glycopeptide bonds. *Glycobiology 12*, 43R–56R.

21. Axford, J. (2001). The impact of glycobiology on medicine. *Trends Immunol 22*, 237–239.

22. Lis, H., Sharon, N., Katchalski, E. (1969). Identification of the carbohydrate–protein linking group in soybean hemagglutinin. *Biochim Biophys Acta 192*, 364–366.

23. Mortz, E., Sareneva, T., Haebel, S., Julkunen, I., Roepstorff, P. (1996). Mass spectrometric characterization of glycosylated interferon-gamma variants separated by gel electrophoresis. *Electrophoresis 17*, 925–931.

24. Ji, J., Chakraborty, A., Geng, M., Zhang, X., Amini, A., Bina, M., Regnier, F. (2000). Strategy for qualitative and quantitative analysis in proteomics based on signature peptides. *J Chromatogr B. Biomed Sci Appl 745*, 197–210.

25. Nawarak, J., Phutrakul, S., Chen, S. T. (2004). Analysis of lectin-bound glycoproteins in snake venom from the elapidae and viperidae families. *J Proteome Res 3*, 383–392.

26. Mechref, Y., Novotny, M. V. (2002). Structural investigations of glycoconjugates at high sensitivity. *Chem Rev 102*, 321–369.

27. Karas, M., Bahr, U., Strupat, K., Hillenkamp, F., Tsarbopoulos, A., Pramanik, B.N. (1995). Matrix dependence of metastable fragmentation of glycoproteins in MALDI TOF mass spectrometry. *Anal Chem 67*, 675–679.

28. Rajan, N., Tsarbopoulos, A., Kumarasamy, R., O'Donnell, R., Taremi, S. S., Baldwin, S. W., Seelig, G. F., Fan, X., Pramanik, B., Le, H. V. (1995). Characterization of Recombinant Human Interleukin-4 Receptor from CHO cells: Role of N-linked Oligosaccharides. *Biochem Biophys Res Commun 206*, 694–702.

29. Tsarbopoulos, A., Prongay, A., Baldwin, S., Kumarasamy, R., Schwartz, J., Pramanik, B. N., Le, H. V., Karas, M., Hillenkamp, F. (1996). Mass spectrometric analysis of the Sf9 cell-derived interleukin-5 receptor, In *Proceedings of the 44th ASMS Conference on Mass Spectrometry and Allied Topics*, Portland, OR. American Society for Mass Spectrometry, Santa Fe, NM, pp. 359a–b.

30. Giménez, E., Benavente, F., Barbosa, J., Sanz-Nebot, V. (2007). Towards a reliable molecular mass determination of intact glycoproteins by matrix-assisted laser desorption/ionization time-of-flight mass spectrometry. *Rapid Commun Mass Spectrom 21*, 2555–2563.

31. Clemmer, D. E., Jarrold, M. F. (1997). Ion mobility measurements and their applications to clusters and biomolecules. *J Mass Spectrom 32*, 577–592.

32. Wilm, M., Shevchenko, A., Houthaeve, T., Breit, S., Schweigerer, L., Fotsis, T., Mann. M. (1996). Femtomole sequencing of proteins from polyacrylamide gels by nano-electrospray mass spectrometry. *Nature 379*, 466–469.

33. Wilm, M., Mann. M. (1996). Analytical properties of the nanoelectrospray ion source. *Anal Chem 68*, 1–8.

34. Makarov, A. (2000). Electrostatic axially harmonic orbital trapping: A high-performance technique of mass analysis. *Anal Chem, 72*, 1156–1162.

35. Olivova, P., Chen, W., Chakraborty, A. B., Gebler, J. C. (2008). Determination of *N*-glycosylation sites and site heterogeneity in a monoclonal antibody by electrospray quadrupole ion-mobility time-of-flight mass spectrometry. *Rapid Commun Mass Spectrom 22*, 29–40.

36. Gordiyenko, Y., Robinson, C. V. (2008). The emerging role of MS in structure elucidation of protein-nucleic acid complexes. *Biochem Soc Trans 36*, 723–731.

37. Tsarbopoulos, A., Bahr, U., Karas, M., Pramanik, B. N. (2002). Structural analysis of glycoproteins by electrospray ionization mass spectrometry, In Pramanik, B. N., Ganguly, A. K., Gross, M. L., eds., *Applied Electrospray Mass Spectrometry.* Dekker, New York, pp. 283–306.

38. Liu, C. L., Bowers, L. D. (1997). Mass spectrometric characterization of the β-subunit of human chorionic gonadotropin. *J Mass Spectrom 32*, 33–42.

39. Demelbauer, U. M., Plematl, A., Kremser, L., Allmaier, G., Josic, D., Rizzi, A. (2004). Characterization of glyco isoforms in plasma-derived human antithrombin by on-line capillary zone enectrophoresis-electrospray ionization-quadrupole ion trap-mass spectrometry of the intact glycoproteins. *Electrophoresis 25*, 2026–2032.

40. Neusüb, C., Demelbauer, U., Pelzing, M. (2005). Glycoform characterization of intact erythopoietin by capillary electrophoresis-electrospray–time-of-flight mass spectrometry. *Electrophoresis 26*, 1442–1450.

41. West, C. M., Van Der Wel, H., Sassi, S., Gaucher, E. A. (2004). Cytoplasmic glycosylation of protein–hydroxyproline and its relationship to other glycosylation pathways. *Biochim Biophys Acta 1673*, 29–44.

42. Nagy, K., Vekey, K., Imre, T., Ludanyi, K., Barrow, M. P., Derrick, P. J. (2004). Electrospray ionization fourier transform ion cyclotron resonance mass spectrometry of human alpha-1-acid glycoprotein. *Anal Chem 76*, 4998–5005.

43. Schirm, M., Schoenhofen, I. C., Logan, S. M., Waldron, K. C., Thibault, P. (2005). Identification of unusual bacterial glycosylation by tandem mass spectrometry analyses of intact proteins. *Anal Chem 77*, 7774–7782.

44. Yang, Z., Hancock, W. S. (2004). Approach to the comprehensive analysis of glycoproteins isolated from human serum using a multi-lectin affinity column. *J Chromatogr A1053*, 79–88.

45. Wang, Y., Wu, S. L., Hancock, W. S. (2006). Approaches to the study of *N*-linked glycoproteins in human plasma using lectin affinity chromatography and nano-HPLC coupled to electrospray linear ion trap–Fourier transform mass spectrometry. *Glycobiology 16*, 514–523.

46. Madera, M., Mechref, Y., Klouckova, I., Novotny, M.V. (2006). Semiautomated high-sensitivity profiling of human blood serum glycoproteins through lectin preconcentration

and multidimensional chromatography/tandem mass spectrometry. *J Proteome Res 5*, 2348–2363.

47. Nakanishi, T., Shimizu, A., Okamoto, N., Ingendoh, A., Kanai, M. (1995). Electrospray ionization-tandem mass spectrometry analysis of peptides derived by enzymatic digestion of oxidized globin subunits: An improved method to determine amino acid substitution in the hemoglobin "core." *J Am Soc Mass Spectrom 6*, 854–859.
48. Lamblin, G., Degroote, S., Perini, J. M., Delmotte, P., Scharfman, A., Davril, M., Lo-Guidice, J. M., Houdret, N., Dumur, V., Klein, A., Rousse, P. (2001), Human airway mucin glycosylation: A combinatory of carbohydrate determinants which vary in cystic fibrosis. *Glycoconjugate J 18*, 661–684.
49. Hang, H. C., Bertozzi, C. R. (2005). The chemistry and biology of mucin-type *O*-linked glycosylation. *Bioorg Medicinal Chem 13*, 5021–5034.
50. Wopereis, S., Lefeber, D. J., Morava, E., Wevers, R. A. (2006). Mechanisms in protein O-glycan biosynthesis and clinical and molecular aspects of protein *O*-glycan biosynthesis defects: A review. *Clin Chem 52*, 574–600.
51. Vance, B. A., Wu, W., Ribaudo, R. K., Segal, D. M., Kearse, K. P. (1997). Multiple dimeric forms of human CD69 result from differential addition of *N*-glycans to typical (Asn–X–Ser/Thr) and atypical (Asn–X–Cys) glycosylation motifs. *J Biol Chem 272*, 23117–23122.
52. Watty, A., Neubauer, G., Dreger, M., Zimmer, M., Wilm, M., Burden, S. J. (2000). The in vitro and in vivo phosphotyrosine map of activated MuSK. *Proc Nat Acad Sci USA 97*, 4585–4590.
53. Wu, X., Ranganathan, V., Weisman, D. S., Heine, W. F., Ciccone, D. N., O'Neill, T. B., Crick, K. E., Pierce, K. A., Lane, W. S., Rathbun, G., Livingston, D. M., Weaver, D. T. (2000). ATM phosphorylation of Nijmegen breakage syndrome protein is required in a DNA damage response. *Nature 405*, 477–482.
54. Carr, S. A., Roberts, G. D. (1986). Carbohydrate mapping by mass spectrometry: A novel method for identifying attachment sites of Asn-linked sugars in glycoproteins. *Anal Biochem 157*, 396–406.
55. Küster, B., Mann, M. (1999). ^{18}O-labeling of *N*-glycosylation sites to improve the identification of gel-separated glycoproteins using peptide mass mapping and database searching. *Anal Chem 71*, 1431–1440.
56. Novotny, M. V., Mechref, Y. (2005). New hyphenated methodologies in high sensitivity glycoprotein analysis. *J Separation Sci 28*, 1956–1968.
57. Geyer, H., Geyer, R. (2006). Strategies for analysis of glycoprotein glycosylation. *Biochim Biophys Acta 1764*, 1853–1869.
58. Sharon, N., Lis, H. (1989). Lectins as cell recognition molecules. *Science 246*, 227–234.
59. Jorgenson, J. W. (1986). Electrophoresis. *Anal Chem 58*, 743A–760A.
60. Papac, D. I., Wong, A., Jones, A. J. (1996). Analysis of acidic oligosaccharides and glycopeptides by matrix-assisted laser desorption/ionization time-of-flight mass spectrometry. *Anal Chem 68*, 3215–3223.
61. Powell, A. K., Harvey, D. J. (1996). Stabilization of sialic acids in N-linked oligosaccharides and gangliosides for analysis by positive ion matrix assisted laser desorption/ionization mass spectrometry. *Rapid Commun Mass Spectrom 10*, 1027–1032.

62. Wuhrer, M., Koeleman, C. A., Deelder, A. M., Hokke, C. H. (2004). Normal-phase nanoscale liquid chromatography-mass spectrometry of underivatized oligosaccharides at low-femtomole sensitivity. *Anal Chem 76*, 833–838.

63. Wuhrer, M., Koeleman, C. A. M., Hokke, C. H., Deelder, A. M. (2004). Nano-scale liquid chromatography-mass spectrometry of 2-aminobenzamide-labeled oligosaccharides at low femtomole sensitivity. *Int J Mass Spectrom 232*, 51–57.

64. Ling, V., Guzzetta, A. W., Canova-Davis, E., Stults, J. T., Hancock, W. S., Covey, T. R., Shushan, B. I. (1991). Characterization of the tryptic map of recombinant DNA derived tissue plasminogen activator by high-performance liquid chromatography-electrospray ionization mass spectrometry. *Anal Chem 63*, 2909–2915.

65. Zhang, H., Yi, E. C., Li, X. J., Mallick, P., Kelly-Spratt, K. S., Masselon, C. D., Camp, D. G. 2nd, Smith, R. D., Kemp, C. J., Aebersold, R. (2005). High throughput quantitative analysis of serum proteins using glycopeptide capture and liquid chromatography mass spectrometry. *Mol Cell Proteomics 4*, 144–155.

66. Wuhrer, M., Catalina, M. I., Deelder, A. M., Hokke, C. H. (2007). Glycoproteomics based on tandem mass spectrometry of glycopeptides. *J Chrom B., Anal Technol Biomed Life Sci 849*, 115–128.

67. Alving, K., Körner, R., Paulsen, H., Peter-Katalinic, J. (1998). Nanospray-ESI low-energy CID and MALDI post-source decay for determination of O-glycosylation sites in MUC4 peptides. *J Mass Spectrom 33*, 1124–1133.

68. Tsarbopoulos, A., Pramanik, B. N., Nagabhushan, T. L., Covey, T. R. (1995). Structural analysis of the CHO-derived interleukin-4 by liquid-chromatography/electrospray ionization mass spectrometry. *J Mass Spectrom 30*, 1752–1763.

69. Budnik, B. A., Lee, R. S., Steen, J. A. (2006). Global methods for protein glycosylation analysis by mass spectrometry. *Biochim Biophys Acta 1764*, 1870–1880.

70. Temporini, C., Calleri, E., Massolini, G., Caccialanza, G. (2008). Integrated analytical strategies for the study of phosphorylation and glycosylation in proteins. *Mass Spectrom Rev 27*, 207–236.

71. Zubarev, R. A., Kelleher, N. L., McLafferty, F. W. (1998). Electron capture dissociation of multiply charged protein cations. A nonergodic process. *J Am Chem Soc 120*, 3265–3266.

72. Stensballe, A., Jensen, O. N., Olsen, J. V., Haselmann, K. F., Zubarev, R. A. (2000). Electron capture dissociation of singly and multiply phosphorylated peptides. *Rapid Commun Mass Spectrom 14*, 1793–1800.

73. Zubarev, R. A., Horn, D. M., Fridriksson, E. K., Kelleher, N. L., Kruger, N. A., Lewis, M. A., Carpenter, B. K., McLafferty, F. W. (2000). Electron capture dissociation for structural characterization of multiply charged protein cations. *Anal Chem 72*, 563–573.

74. Syka, J. E., Coon, J. J., Schroeder, M. J., Shabanowitz, J., Hunt, D. F. (2004). Peptide and protein sequence analysis by electron transfer dissociation mass spectrometry. *Proc Nat Acad Sci USA 101*, 9528–9533.

75. Mirgorodskaya, E., Roepstorff, P., Zubarev, R. A. (1999). Localization of *O*-glycosylation sites in peptides by electron capture dissociation in a Fourier transform mass spectrometer. *Anal Chem 71*, 4431–4436.

76. Guzzetta, A. W., Basa, L. J., Hancock, W. S., Keyt, B. A., Bennett, W. F. (1993). Identification of carbohydrate structures in glycoprotein peptide maps by the use of LC/MS with selected ion extraction with special reference to tissue plasminogen activator

and a glycosylation variant produced by site directed mutagenesis. *Anal Chem 65*, 2953–2962.

77. Harvey, D. J., Bateman, R. H., Bordoli, R. S., Tyldesley, R. (2000). Ionization and fragmentation of complex glycans with a quadrupole time-of-flight mass spectrometer fitted with a matrix-assisted laser desorption/ionization ion source. *Rapid Commun Mass Spectrom, 14*, 2135–2142.
78. Wheeler, S. F., Harvey, D. J. (2001). Extension of the in-gel release method for structural analysis of neutral and sialylated *N*-linked glycans to the analysis of sulfated glycans: Application to the glycans from bovine thyroid-stimulating hormone. *Anal Biochem 296*, 92–100.
79. Koerner, T. A., Yu, R. K., Scarsdale, J. N., Demou, P. C., Prestegard, J. H. (1988). Analysis of complex carbohydrate primary and secondary structure via two-dimensional proton nuclear magnetic resonance spectroscopy. *Adv Exper Med Biol, 228*, 759–784.
80. Keck, R. G., Briggs, J. B. Jones, A. J. (2005). Oligosaccharide release and MALDI-TOF MS analysis of *N*-linked carbohydrate structures from glycoproteins. *Meth Mol Biol 308*, 381–396.
81. Huang, Y., Mechref, Y., Novotny, M. V. (2001). Microscale nonreductive release of *O*-linked glycans for subsequent analysis through MALDI mass spectrometry and capillary electrophoresis. *Anal Chem 73*, 6063–6069.
82. Papac, D. I., Briggs, J. B., Chin, E. T., Jones, A. J. (1998). A high-throughput microscale method to release *N*-linked oligosaccharides from glycoproteins for matrix-assisted laser desorption/ionization time-of-flight mass spectrometric analysis. *Glycobiology 8*, 445–454.
83. Dell, A., Reason, A. J., Khoo, K. H., Panico, M., McDowell, R. A., Morris, H. R. (1994). Mass spectrometry of carbohydrate-containing biopolymers. *Meth Enzymol 230*, 108–132.
84. Kang, P., Mechref, Y., Klouckova, I., Novotny, M.V. (2005). Solid-phase permethylation of glycans for mass spectrometric analysis. *Rapid Commun Mass Spectrom 19*, 3421–3428.
85. Viseux, N., de Hoffmann, E., Domon, B. (1997). Structural analysis of permethylated oligosaccharides by electrospray tandem mass spectrometry. *Anal Chem, 69*, 3193–3198.
86. Reinhold, V. N., Sheeley, D. M. (1998). Detailed characterization of carbohydrate linkage and sequence in an ion trap mass spectrometer: glycosphingolipids. *Anal Biochem, 259*, 28–33.
87. Sheeley, D. M., Reinhold, V. N. (1998). Structural characterization of carbohydrate sequence, linkage, and branching in a quadrupole Ion trap mass spectrometer: Neutral oligosaccharides and *N*-linked glycans. *Anal Chem 70*, 3053–3059.
88. Morelle, W., Faid, V., Michalski, J. C. (2004). Structural analysis of permethylated oligosaccharides using electrospray ionization quadrupole time-of-flight tandem mass spectrometry and deuteroreduction. *Rapid Commun Mass Spectrom 18*, 2451–2464.
89. Goshe, M. B. Smith, R. D. (2003). Stable isotope-coded proteomic mass spectrometry. *Curr Opin Biotechnol 14*, 101–109.
90. Tao, W. A., Aebersold, R. (2003). Advances in quantitative proteomics via stable isotope tagging and mass spectrometry. *Curr Opin Biotechnol 14*, 110–118.
91. Alvarez-Manilla, G., Warren, N. L., Abney, T., Atwood, J., III, Azadi, P., York, W. S., Pierce, M., Orlando, R. (2007). Tools for glycomics: Relative quantitation of glycans by isotopic permethylation using $^{13}CH_3I$. *Glycobiol 366*, 1266–1281.

92. Aoki, K., Perlman, M., Lim, J. M., Cantu, R., Wells, L., Tiemeyer, M. (2007). Dynamic developmental elaboration of *N*-linked glycan complexity in the *Drosophila melanogaster* embryo. *J Biol Chem 282*, 9127–9142.
93. Atwood, J. A., III, Cheng, L., Alvarez-Manilla, G., Warren, N. L., York, W. S., Orlando, R. (2008). Quantitation by isobaric labeling: Applications to glycomics. *J Proteome Res 7*, 367–374.
94. Wuhrer, M., Catalina, M. I., Deelder, A. M., Hokke, C. H. (2007). Glycoproteomics based on tandem mass spectrometry of glycopeptides. *J Chrom B849* 115–128.
95. Patwa, T. H., Zhao, J., Anderson, M. A., Simeone, D. M., Lubman, D. M. (2006). Screening of glycosylation patterns in serum using natural glycoprotein microarrays and multi-lectin fluorescence detection. *Anal Chem 78*, 6411–6421.
96. Zhao, J., Patwa, T. H., Qiu, W., Shedden, K., Hinderer, R., Misek, D. E., Anderson, M. A., Simeone, D. M., Lubman, D. M. (2007). Glycoprotein microarrays with multi-lectin detection: Unique lectin binding patterns as a tool for classifying normal, chronic pancreatitis and pancreatic cancer sera. *J Proteome Res 6*, 1864–1874.
97. Wang, W., Zhou, H., Lin, H., Roy, S., Shaler, T. A., Hill, L. R., Norton, S., Kumar, P., Anderle, M., Becker, C. H. (2003). Quantification of proteins and metabolites by mass spectrometry without isotopic labeling or spiked standards. *Anal Chem 75*, 4818–4826.
98. Old, W. M., Meyer-Arendt, K., Aveline-Wolf, L., Pierce, K. G., Mendoza, A., Sevinskey, J. R., Resing, K. A., Ahn, N. G. (2005). Comparison of label-free methods for quantifying human proteins by shotgun proteomics. *Mol Cell Proteomics 4*, 1487–1502.
99. Callister, S. J., Barry, R. C., Adkins, J. N., Johnson, E. T., Qian, W.-J., Webb-Robertson, B.-J., Smith, R. D., Lipton, M. S. (2006). Normalization approaches for removing systematic biases associated with mass spectrometry and label-free proteomics. *J Proteome Res 5*, 277–286.
100. Ono, M., Shitashige, M., Honda, K., Isobe, T., Kuwabara, H., Matsuzuki, H., Hirohashi, S., Yamada, T. (2006). Label-free quantitative proteomics using large peptide data sets generated by nanoflow liquid chromatography and mass spectrometry. *Mol Cell Proteomics 5*, 1338–1347.
101. Rebecchi, K. R., Wenke, J. L., Go, E. P., Desaire, H. (2009). Label-free quantitation: A new glycoproteomics approach. *J Am Soc Mass Spectrom* doi: 10.1016/j.jasms.2009.01.013.
102. Hunter, T. (1995). Protein kinases and phosphatases: The yin and yang of protein phosphorylation and signaling. *Cell 80*, 225–236.
103. Graves, J. D., Krebs, E. G. (1999). Protein phosphorylation and signal transduction. *Pharmacol Therapeut 82*, 111–121.
104. Hunter, T. (2000). Signaling-2000 and beyond. *Cell 100*, 113–127.
105. Cohen, P. (2002). The origins of protein phosphorylation. *Nat Cell Biol 4*, E127–E130.
106. Hunter, T. (1991). Protein kinase classification. *Meth Enzymol 200*, 3–37.
107. Wang, S., Liu, Y. H., Smith, C. K., Pramanik, B. N. Mass spectrometric characterization of MEK1 phosphorylation during expression in insect cells. Presented at the 52nd ASMS Conference on Mass Spectrometry and Allied Topics, May 23–27, 2004, Nashville, Tennessee.
108. Engstrom, L., Ekman, P., Humble, E., Ragnarsson, U., Zetterqvist, O. (1984). In Wold, E., Moldave, K., eds., Detection and identification of substrates for protein kinases: use of proteins and synthetic peptides. *Methods in Enzymology*, Vol. *107*. Academic Press, San Diego, p. 130.

109. Annan, W. D., Manson, W., Nimmo, J. A. (1982). The identification of phosphoseryl residues during the determination amino acid sequence in phosphoproteins. *Anal Biochem 121*, 62–68.

110. Meyer, H. E., Hoffmann-Posorske, E., Korte, H., Heilmeyer, L. M. Jr., (1986). Sequence analysis of phosphoserine-containing peptides: Modification for picomolar sensitivity. *FEBS Lett 204*, 61–66.

111. Arnott, D., Gawinowicz, M. A., Grant, R. A., Neubert, T. A., Packman, L. C., Speicher, K. D., Stone, K., Turck, C. W. (2003). ABRF-PRG03: Phosphorylation site determination. *J Biomol Techniq 14*, 205–215.

112. Loyet, K. M., Stults, J. T., Arnott, D. (2005). Mass spectrometric contributions to the practice of phosphorylation site mapping through 2003: A literature review. *Mol Cell Proteomics 4*, 235–245.

113. Seeley, E. H., Riggs, L. D., Regnier, F. E. (2005). Reduction of non-specific binding in Ga(III) immobilized metal affinity chromatography for phosphopeptides by using endoproteinase glu-C as the digestive enzyme. *J Chromatogr, B Anal Technol Biomed Life Sci 817*, 81–88.

114. Körner, R., Wilm, M., Morand, K., Schubert, M., Mann, M. (1996). Nano electrospray combined with a quadrupole ion trap for the analysis of peptides and protein digests. *J Am Soc Mass Spectrom 7*, 150–156.

115. McLachlin, D. T., Chait, B. T. (2001). Analysis of phosphorylated proteins and peptides by mass spectrometry. *Curr Opin Chem Biol 5*, 591–602.

116. Mann, M., Ong, S. E., Grønborg, M., Steen, H., Jensen, O. N., Pandey, A. (2002). Analysis of protein phosphorylation using mass spectrometry: Deciphering the phosphoproteome. *Trends Biotechnol 6*, 261–268.

117. Klemm, C., Schröder, S., Glückmann, M., Beyermann, M., Krause, E. (2004). Derivatization of phosphorylated peptides with *S*- and *N*-nucleophiles for enhanced ionization efficiency in matrix-assisted laser desorption/ionization mass spectrometry. *Rapid Commun Mass Spectrom 18*, 2697–2705.

118. Chen, M., Su, X., Yang, J., Jenkins, C. M., Cedars, A. M., Gross, R. W. (2010) Facile identification and quantitation of protein phosphorylation via β-elimination and Michael addition with natural abundance and stable isotope labeled thiocholine. *Anal Chem 82*, 163–171.

119. De Corte, V., Demol, H., Goethals, M., Van Damme, J., Gettemans, J., Vandekerckhove, J. (1999). Identification of Tyr438 as the major in vitro c-Src phosphorylation site in human gelsolin: A mass spectrometric approach. *Protein Sci 8*, 234–241.

120. Lehr, S., Kotzka, J., Herkner, A., Sikmann, A., Meyer, H. E., Krone, W., Müller-Wieland, D. (2000). Identification of major tyrosine phosphorylation sites in the human insulin receptor substrate Gab-1 by insulin receptor kinase in vitro. *Biochemistry 39*, 10898–10907.

121. Rush, J., Moritz, A., Lee, K. A., Guo, A., Goss, V. L., Spek, E. J., Zhang, H., Zha, X. M., Polakiewicz, R. D., Comb, M. J. (2005). Immunoaffinity profiling of tyrosine phosphorylation in cancer cells. *Nat Biotechnol 23*, 94–101.

122. Rikova, K., Guo, A., Zeng, Q., Possemato, A., Yu, J., Haack, H., Nardone, J., Lee, K., Reeves, C., Li, Y., Hu, Y., Tan, Z., Stokes, M., Sullivan, L., Mitchell, J., Wetzel, R., Macneill, J., Ren, J. M., Yuan, J., Bakalarski, C. E., Villen, J., Kornhauser, J. M., Smith, B., Li, D., Zhou, X., Gygi, S. P., Gu, T. L., Polakiewicz, R. D., Rush, J., Comb, M. J. (2007).

Global survey of phosphotyrosine signaling identifies oncogenic kinases in lung cancer. *Cell 131*, 1190–1203.

123. Pandey, A., Podtelejnikov, A. V., Blagoev, B., Bustelo, X. R., Mann, M., Lodish, H. F. (2000). Analysis of receptor signaling pathways by mass spectrometry: Identification of vav-2 as a substrate of the epidermal and platelet-derived growth factor receptors. *Proc Nat Acad Sci USA 97*, 179–184.
124. Gronborg, M., Kristiansen, T. Z., Stensballe, A., Andersen, J. S., Ohara, O., Mann, M., Jensen, O. N., Pandey, A. (2002). A mass spectrometry-based proteomic approach for identification of serine/threonine-phosphorylated proteins by enrichment with phosphor-specific antibodies: Identification of a novel protein, Frigg, as a protein kinase A substrate. *Mol Cell Proteomics 1*, 517–527.
125. Andersson, L., Porath, J. (1986). Isolation of phosphoproteins by immobilized metal (Fe^{3+}) affinity chromatography. *Anal Biochem 154*, 250–254.
126. Posewitz, M. C., Tempst, P. (1999). Immobilized gallium (III) affinity chromatography of phosphopeptides. *Anal Chem 71*, 2883–2892.
127. Ballif, B. A., Villén, J., Beausoleil, S. A., Schwartz, D., Gygi, S. P. (2004). Phosphoproteomic analysis of the developing mouse brain. *Mol Cell Proteomics 3*, 1093–1101.
128. Ficarro, S. B., Salomon, A. R., Brill, L. M., Mason, D. E., Stettler-Gill, M., Brock, A., Peters, E. C. (2005). Automated immobilized metal affinity chromatography/nano-liquid chromatography/electrospray ionization mass spectrometry platform for profiling protein phosphorylation sites. *Rapid Commun Mass Spectrom 19*, 57–71.
129. Muszyńska, G., Andersson, L., Porath, J. (1986). Selective adsorption of phosphoproteins on gel-immobilized ferric chelate. *Biochemistry 25*, 6850–6853.
130. Scanff, P., Yvon, M., Pelissier, J. P. (1991). Immobilized Fe^{3+} affinity chromatographic isolation of phosphopeptides. *J Chromatogr 539*, 425–432.
131. Macek, B., Mann, M., Olsen, J. V. (2009). Global and site-specific quantitative phosphoproteomics: Principles and applications. *An Rev Pharmacol Toxicol 49*, 199–221.
132. Ficarro, S. B., McCleland, M. L., Stukenberg, P. T., Burke, D. J., Ross, M. M., Shabanowitz, J., Hunt, D. F., White, F. M. (2002). Phosphoproteome analysis by mass spectrometry and its application to *Saccharomyces cerevisiae. Nat Biotechnol 20*, 301–305.
133. Seward, R. J., Perlman, D. H., Berg, E. A., Hu, J., Costello, C. E. (2004). Methyl esterification of peptides prior to immobilized metal-ion affinity chromatography: determination of optimal time course and side products. Presented at 52nd ASMS Conference on Mass Spectrometry and Allied Topics, May 23–27, 2004, Nashville, Tennessee.
134. Nuwaysir, L. M., Stults, J. T. (1993). Electrospray ionization mass spectrometry of phosphopeptides isolated by on-line immobilized metal-ion affinity chromatography. *J Am Soc Mass Spectrom 4*, 662–669.
135. Larsen, M. R., Thingholm, T. E., Jensen, O. N., Roepstorff, P., Jorgensen, T. J. (2005). Highly selective enrichment of phosphorylated peptides from peptide mixtures using titanium dioxide microcolumns. *Mol Cell Proteomics 4*, 873–886.
136. Sano, A., Nakamura, H. (2007). Evaluation of titanium and titanium oxides as chemo-affinity sorbents for the selective enrichment of organic phosphates. *Anal Sci 23*, 1285–1289.
137. Pinkse, M. W., Uitto, P. M., Hilhorst, M. J., Ooms, B., Heck, A. J. (2004). Selective isolation at the femtomole level of phosphopeptides from proteolytic digests using 2D-NanoLC-ESI-MS/MS and titanium oxide precolumns. *Anal Chem 76*, 3935–3943.

138. Röhrig, H., Colby, T., Schmidt, J., Harzen, A., Facchinelli, F., Bartels, D. (2008). Analysis of desiccation-induced candidate phosphoproteins from *Craterostigma plantagineum* isolated with a modified metal oxide affinity chromatography procedure. *Proteomics 8*, 3548–3560.

139. Kweon, H. K., Håkansson, K. (2006). Selective zirconium dioxide-based enrichment of phosphorylated peptides for mass spectrometric analysis. *Anal Chem 78*, 1743–1749.

140. Chin, E. T., Papac, D. I. (1999). The use of a porous graphitic carbon column for desalting hydrophilic peptides prior to matrix-assisted laser desorption/ionization time-of-flight mass spectrometry. *Anal Biochem 273*, 179–185.

141. Larsen, M. R., Graham, M. E., Robinson, P. J., Roepstorff, P. (2004). Improved detection of hydrophilic phosphopeptides using graphite powder microcolumns and mass spectrometry: Evidence for in vivo doubly phosphorylated dynamin I and dynamin III. *Mol Cell Proteomics 5*, 456–465.

142. Beausoleil, S. A., Jedrychowski, M., Schwartz, D., Elias, J. E., Villen, J., Li, J., Cohn, M. A., Cantley, L. C., Gygi, S. P. (2004). Large-scale characterization of HeLa cell nuclear phosphoproteins. *Proc Nat Acad Sci USA 101*, 12130–12135.

143. Gruhler, A., Olsen, J. V., Mohammed, S., Mortensen, P., Faergeman, N. J., Mann, M., Jensen, O. N. (2005). Quantitative phosphoproteomics applied to the yeast pheromone signaling pathway. *Mol Cell Proteomics 4*, 310–327.

144. Villen, J., Beausoleil, S. A., Gerber, S. A., Gygi, S. P. (2007). Large-scale phosphorylation analysis of mouse liver. *Proc Nat Acad Sci USA 104*, 1488–1493.

145. McNulty, D. E., Annan, R. S. (2008). Hydrophilic interaction chromatography reduces the complexity of the phosphoproteome and improves global phosphopeptide isolation and detection. *Mol Cell Proteomics 7*, 971–980.

146. Oda, Y., Nagasu, T., Chait, B. T. (2001). Enrichment analysis of phosphorylated proteins as a tool for probing the phosphoproteome. *Nat Biotechnol 19*, 379–382.

147. Adamczyk, M., Gebler, J. C., Wu, J. (2001). Selective analysis of phosphopeptides within a protein mixture by chemical modification, reversible biotinylation and mass spectrometry. *Rapid Commun Mass Spectrom 15*, 1481–1488.

148. Wells, L., Vosseller, K., Cole, R. N., Cronshaw, J. M., Matunis, M. J., Hart, G. W. (2002). Mapping sites of *O*-GlcNAc modification using affinity tags for serine and threonine post-translational modifications. *Mol Cell Proteomics 1*, 791–804.

149. McLachlin, D. T., Chait, B. T. (2003). Improved beta-elimination-based affinity purification strategy for enrichment of phosphopeptides. *Anal Chem 75*, 6826–6836.

150. Carr, S. A., Huddleston, M. J., Annan, R. S. (1996). Selective detection and sequencing of phosphopeptides at the femtomole level by mass spectrometry. *Anal Biochem 239*, 180–192.

151. Chang, E. J., Archambault, V., McLachlin, D. T., Krutchinsky, A. N., Chait, B. T. (2004). Analysis of protein phosphorylation by hypothesis-driven multi-stage mass spectrometry. *Anal Chem 76*, 4472–4483.

152. Steen, H., Kuster, B., Mann, M. (2001). Quadrupole time-of-flight versus triple-quadrupole mass spectrometry for the determination of phosphopeptides by precursor ion scanning. *J Mass Spectrom 36*, 782–790.

153. Steen, H., Kuster, B., Fernandez, M., Pandey, A., Mann, M. (2002). Tyrosine phosphorylation mapping of the epidermal growth factor receptor signaling pathway. *J Biol Chem 277*, 1031–1039.

154. Huddleston, M. J., Annan, R. S., Bean, M. F., Carr, S. A. (1993). Selective detection of phosphopeptides in complex mixtures by electrospray liquid chromatography/mass spectrometry. *J Am Soc Mass Spectrom 4*, 710–717.
155. Wilm, M., Neubauer, G., Mann, M. (1996). Parent ion scans of unseparated peptide mixtures. *Anal Chem 68*, 527–533.
156. Neubauer, G., Mann, M. (1999). Mapping of phosphorylation sites of gel-isolated proteins by nanoelectrospray tandem mass spectrometry: Potentials and limitations. *Anal Chem 71*, 235–242.
157. Annan, R. S., Huddleston, M. J., Verma, R., Deshaies, R. J., Carr, S. A. (2001). A multidimensional electrospray MS-based approach to phosphopeptide mapping. *Anal Chem 73*, 393–404.
158. Williamson, B. L., Marchese, J., Morrice, N. A. (2006). Automated identification and quantification of protein phosphorylation sites by LC/MS on a hybrid triple quadrupole linear ion trap mass spectrometer. *Mol Cell Proteomics 5*, 337–346.
159. Chen, G., Liu, Y. H., Pramanik, B. N. (2007). LC/MS analysis of proteins and peptides in drug discovery, In Kazakevich, Y., LoBrutto, R., eds. , *HPLC for Pharmaceutical Scientists*. Wiley, New York.
160. Olsen, J. V., Macek, B., Lange, O., Makarov, A., Horning, S., Mann, M. (2007). Higher-energy C-trap dissociation for peptide modification analysis. *Nat Meth 4*, 709–712.
161. Olsen, J. V., Mann, M. (2004). Improved peptide identification in proteomics by two consecutive stages of mass spectrometric fragmentation. *Proc Nat Acad Sci USA 101*, 13417–13422.
162. Macek, B., Benda, C., Jestel, A., Maskos, K., Mann, M., Messerschmidt, A. (2008). Phosphorylation of the human full-length protein kinase Ciota. *J Proteome Res 7*, 2928–2935.
163. Shi, S. D., Hemling, M. E., Carr, S. A., Horn, D. M., Lindh, I., McLafferty, F. W. (2001). Phosphopeptide/phosphoprotein mapping by electron capture dissociation mass spectrometry. *Anal Chem 73*, 19–22.
164. Chi, A., Huttenhower, C., Geer, L. Y., Coon, J. J., Syka, J. E., Bai, D. L., Shabanowitz, J., Burke, D. J., Troyanskaya, O. G., Hunt, D. F. (2007). Analysis of phosphorylation sites on proteins from *Saccharomyces cerevisiae* by electron transfer dissociation (ETD) mass spectrometry. *Proc Nat Acad Sci USA 104*, 2193–2198.
165. Wyttenbach, T., Bowers, M. (2003). Gas-phase conformations: The ion mobility/ion chromatography method. *Topics Curr Chem 225*, 207–232.
166. Valentine, S. J., Kulchania, M., Barnes, C. A. S., Clemmer, D. E. (2001). Multidimensional separations of complex peptide mixtures: a combined high-performance liquid chromatography/ion mobility/time-of-flight mass spectrometry approach. *Int J Mass Spectrom 212*, 97–109.
167. Ruotolo, B. T., Verbeck, G. F. t., Thomson, L. M., Woods, A. S., Gillig, K. J., Russell, D. H. (2002). Distinguishing between phosphorylated and nonphosphorylated peptides with ion mobility-mass spectrometry. *J Proteome Res 1*, 303–306.
168. Ruotolo, B. T., Gillig, K. J., Woods, A. S., Egan, T. F., Ugarov, M. V., Schultz, J. A., Russell, D. H. (2004). Analysis of phosphorylated peptides by ion mobility-mass spectrometry. *Anal Chem 76*, 6727–6733.

169. Thalassinos, K., Grabenauer, M., Slade, S. E., Hilton, G. R., Bowers, M. T., Scrivens, J. H. (2009). Characterization of phosphorylated peptides using traveling wave-based and drift cell ion mobility mass spectrometry. *Anal Chem 81*, 248–254.

170. Yip, T. T., Hutchens, T. W. (1992). Mapping and sequence-specific identification of phosphopeptides in unfractionated protein digest mixtures by matrix-assisted laser desorption/ionization time-of-flight mass spectrometry. *FEBS Lett 308*, 149–153.

171. Liao, P. C., Leykam, J., Andrews, P. C., Gage, D. A., Allison, J. (1994). An approach to locate phosphorylation sites in a phosphoprotein: Mass mapping by combining specific enzymatic degradation with matrix-assisted laser desorption/ionization mass spectrometry. *Anal Biochem 219*, 9–20.

172. Annan, R. S., Carr, S. A. (1996). Phosphopeptide analysis by matrix-assisted laser desorption time-of-flight mass spectrometry. *Anal Chem 68*, 3413–3421.

173. Kaufmann, R., Spengler, B., Lützenkirchen, F. (1993). Mass spectrometric sequencing of linear peptides by product-ion analysis in a reflectron time-of-flight mass spectrometer using matrix-assisted laser desorption ionization. *Rapid Commun Mass Spectrom 7*, 902–910.

174. Patterson, S. D., Thomas, D., Bradshaw, R. A. (1996). Application of combined mass spectrometry and partial amino acid sequence to the identification of gel separated proteins. *Electrophoresis 17*, 877–891.

175. Qin, J., Chait, B. T. (1997). Identification and characterization of posttranslational modifications of proteins by MALDI ion trap mass spectrometry. *Anal Chem 69*, 4002–4009.

176. Jonscher, K. R., Yates, J. R., III (1997). Matrix-assisted laser desorption ionization/quadrupole ion trap mass spectrometry of peptides: Application to the localization of phosphorylation sites on the P protein from *Sendai* virus. *J Biol Chem 272*, 1735–1741.

177. Laiko, V. V., Moyer, S. C., Cotter, R. J. (2000). Atmospheric pressure MALDI/ion trap mass spectrometry. *Anal Chem 72*, 5239–5243.

178. Krutchinsky, A. N., Zhang, W., Chait, B. T. (2000). Rapidly switchable matrix-assisted laser desorption/ionization and electrospray quadrupole-time-of-flight mass spectrometry for protein identification. *J Am Soc Mass Spectrom 11*, 493–504.

179. Shevchenko, A., Loboda, A., Shevchenko, A., Ens, W., Standing, K. G. (2000). MALDI quadrupole time-of-flight mass spectrometry: A powerful tool for proteomic research. *Anal Chem 72*, 2132–2141.

180. Bennett, K. L., Stensballe, A., Podtelejnikov, A. V., Moniatte, M., Jensen, O. N. (2002). Phosphopeptide detection and sequencing by matrix-assisted laser desorption/ionization quadrupole time-of-flight tandem mass spectrometry. *J Mass Spectrom 37*, 179–190.

181. Yates, J. R. III, Eng, J. K., McCormack, A. L. (1995). Mining genomes: Correlating tandem mass spectra of modified and unmodified peptides to sequences in nucleotide databases. *Anal Chem 67*, 3202–3210.

182. Ducret, A., Van Oostveen, I., Eng, J. K., Yates, J. R. III, Aebersold, R. (1998). High throughput protein characterization by automated reverse-phase chromatography/electrospray tandem mass spectrometry. *Protein Sci 7*, 706–719.

183. Perkins, D. N., Pappin, D. J., Creasy, D. M., Cottrell, J. S. (1999). Probability-based protein identification by searching sequence databases using mass spectrometry data. *Electrophoresis 20*, 3551–3567.

184. Clauser, K. R., Baker, P., Burlingame, A. L. (1999). Role of accurate mass measurement (+/− 10 ppm) in protein identification strategies employing MS or MS/MS and database searching. *Anal Chem 71*, 2871–2882.

185. Seo, J., Jeong, J., Kim, Y. M., Hwang, N., Paek, E., Lee, K. J. (2008). Strategy for comprehensive identification of post-translational modifications in cellular proteins, including low abundant modifications: application to glyceraldehyde-3-phosphate dehydrogenase. *J Proteome Res 7*, 587–602.

186. Tsur, D., Tanner, S., Zandi, E., Bafna, V., Pevzner, P. A. (2005). Identification of post-translational modifications by blind search of mass spectra. *Nat Biotechnol 23*, 1562–1567.

187. Kim, S., Na, S., Sim, J. W., Park, H., Jeong, J., Kim, H., Seo, Y., Seo, J., Lee, K. J., Paek, E. (2006). MODi: A powerful and convenient web server for identifying multiple post-translational peptide modifications from tandem mass spectra. *Nucl Acids Res 34*, W258–W263.

188. Gnad, F., Ren, S., Cox, J., Olsen, J. V., Macek, B., Oroshi, M., Mann, M. (2007). PHOSIDA (phosphorylation site database): Management, structural and evolutionary investigation, and prediction of phosphosites. *Genome Biol 8*, R250.

189. Siuti, N., Kelleher, N. L. (2007). Decoding protein modifications using top-down mass spectrometry. *Nat Meth 4*, 817–821.

190. Wulfkuhle, J. D., Aquino, J. A., Calvert, V. S., Fishman, D. A., Coukos, G., Liotta, L. A., Petricoin, E. F., III., (2003). Signal pathway profiling of ovarian cancer from human tissue specimens using reverse-phase protein microarrays. *Proteomics 3*, 2085–2090.

191. Cha, H., Hancock, C., Dangi, S., Maiguel, D., Carrier, F., Shapiro, P. (2004). Phosphorylation regulates nucleophosmin targeting to the centrosome during mitosis as detected by cross-reactive phosphorylationspecific MKK1/MKK2 antibodies. *Biochem J 378*, 857–865.

192. Ong, S. E., Blagoev, B., Kratchmarova, I., Kristensen, D. B., Steen, H., Pandey, A., Mann, M. (2002). Stable isotope labeling by amino acids in cell culture, SILAC, as a simple and accurate approach to expression proteomics. *Mol Cell Proteomics 1*, 376–386.

193. Hinsby, A. M., Olsen, J. V., Mann, M. (2004). Tyrosine phosphoproteomics of fibroblast growth factor signaling: A role for insulin receptor substrate-4. *J Biol Chem 279*, 46438–46447.

194. Oda, Y., Huang, K., Cross, F. R., Cowburn, D., Chait, B. T. (1999). Accurate quantitation of protein expression and site-specific phosphorylation. *Proc Nat Acad Sci USA 96*, 6591–6596.

195. Yao, X., Freas, A., Ramirez, J., Demirev, P. A., Fenselau, C. (2001). Proteolytic ^{18}O labeling for comparative proteomics: Model studies with two serotypes of adenovirus. *Anal Chem 73*, 2836–2842.

196. Julka, S., Regnier, F. (2004). Quantification in proteomics through stable isotope coding: A review. *J Proteome Res 3*, 350–363.

197. Gygi, S. P., Rist, B., Gerber, S. A., Turecek, F., Gelb, M. H., Aebersold, R. (1999). Quantitative analysis of complex protein mixtures using isotope-coded affinity tags. *Nat Biotechnol 17*, 994–999.

198. Ross, P. L., Huang, Y. N., Marchese, J. N., Williamson, B., Parker, K., Hattan, S., Khainovski, N., Pillai, S., Dey, S., Daniels, S., Purkayastha, S., Juhasz, P., Martin, S., Bartlet-Jones, M., He, F., Jacobson, A., Pappin, D. J. (2004). Multiplexed protein

quantitation in *Saccharomyces cerevisiae* using amine-reactive isobaric tagging reagents. *Mol Cell Proteomics 3*, 1154–1169.

199. Creighton, T. E. (1984). Disulfide bond formation in proteins. In Wold, F., Moldave, K., eds., *Methods in Enzymology*, vol. *107*. Academic Press, San Diego, CA, pp. 305–329.
200. Wedemeyer, W. J., Welker, E., Narayan, M., Scheraga, H. A. (2000). Disulfide bonds and protein folding. *Biochemistry 39*, 4207–4216.
201. Matsumura, M., Signor, G., Matthews, B. W. (1989). Substantial increase of protein stability by multiple disulfide bonds. *Nature 342*, 291–293.
202. Thornton, J. M. (1981). Disulphide bridges in globular proteins. *J Mol Biol 151*, 261–287.
203. Welker, E., Raymond, L. D., Scheraga, H. A., Caughey, B. (2002). Intramolecular versus intermolecular disulfide bonds in prion proteins. *J Biol Chem 277*, 33477–33481.
204. Tsarbopoulos, A., Pramanik, B., Labdon, J., Reichert, P., Gitlin, G., Patel, S., Sardana, V., Nagabhushan T. L., Trotta, P. P. (1993). Isolation and characterization of a resistant core peptide of recombinant human granulocyte-macrophage colony-stimulating factor (GM-CSF): Confirmation of the GM-CSF amino acid sequence by mass spectrometry. *Protein Sci 2*, 1948–1958.
205. Gorman, J. J., Wallis, T. P., Pitt, J. J. (2002). Protein disulfide bond determination by mass spectrometry. *Mass Spectrom Rev 21*, 183–216.
206. Barber, M., Bordoli, R. S., Sedgwick, R. D., Tyler, A. N. (1981). Fast atom bombardment of solids (FAB): A new ion source for mass spectrometry. *J Chem Soc Chem Commun 7*, 325–327.
207. Smith, D. L., Zhou, Z. (1990). Strategies for locating disulfide bonds in proteins. In McCloskey, J. A., ed., *Methods in Enzymology*, vol. *193*. Academic Press, New York, pp. 374–389.
208. Sundqvist, B., Roepstorff, P., Fohlman, J., Hedin, A., Håkansson, P., Kamensky, I., Lindberg, M., Sahlepour, M., Säve, G. (1984). Molecular weight determination of proteins by californium plasma desorption mass spectrometry. *Science 226*, 696–698.
209. Tsarbopoulos, A., Becker, G. W., Occolowitz, J. L., Jardine, I. (1988). Peptide and protein mapping by ^{252}Cf-plasma desorption mass spectrometry. *Anal Biochem 171*, 113–123.
210. Burlingame, A. L., Boyd, R. K., Gaskell, S. J. (1998). Mass spectrometry. *Anal Chemistry, 70*, 647R–716R.
211. Schnaible, V., Wefing, S., Resemann, A., Suckau, D., Bücker, A., Wolf-Kümmeth, S., Hoffmann, D. (2002). Screening for disulfide bonds in proteins by MALDI in-source decay and LIFT-TOF/TOF-MS. *Anal Chem 74*, 4980–4988.
212. Morris, H. R., Pucci, P. (1985). A new method for rapid assignment of S-S bridges in proteins. *Biochem Biophys Res Commun 126*, 1122–1128.
213. Sanger, F. (1953). A disulphide interchange reaction. *Nature 171*, 1025–1026.
214. Ryle, A. P., Sanger, F. (1955). Disulphide interchange reactions. *Biochem J 60*, 535–540.
215. Hirs, C. H. W. (1967). Cystine peptides. *Meth Enzymol 11*, 475–478.
216. Yazdanparast, R., Andrews, P., Smith, D. L., Dixon, J. E. (1986). A new approach for detection and assignment of disulfide bonds in peptides. *Anal Biochem 153*, 348–353.
217. Tsarbopoulos, A., Karas, M., Strupat, K., Pramanik, B., Nagabhushan T. L., Hillenkamp, F. (1994). Comparative mapping of recombinant proteins and glycoproteins by plasma desorption and matrix-assisted laser desorption/ionization mass spectrometry. *Anal Chem 66*, 2062–2070.

218. Patterson, S. D., Katta, V. (1994). Prompt fragmentation of disulfide-linked peptides during matrix-assisted laser desorption ionization mass spectrometry. *Anal Chem 66*, 3727–3732.

219. Fischer, W. H., Rivier, J. E., Craig, A. G. (1993). In situ reduction suitable for matrix-assisted laser desorption/ionization and liquid secondary ionization using tris(2-carboxyethyl)phosphine. *Rapid Commun Mass Spectrom 7*, 225–228.

220. Tsarbopoulos, A., Varnerin, J., Cannon-Carlson, S., Wylie, D., Pramanik, B. N., Tang, J., Nagabhushan, T. L. (2000). Mass spectrometric mapping of disulfide bonds in recombinant human interleukin-13. *J Mass Spectrom 35*, 446–453.

221. Carr, S. A., Hemling, M. E., Bean, M. F., Roberts, G. D. (1991). Integration of mass spectrometry in analytical biotechnology. *Anal Chem 63*, 2802–2824.

222. Sun, Y., Bauer, M. D., Keough, T. W., Lacey, M. P. (1996). Disulfide bond location in proteins. *Meth Mol Biol 61*, 181–210.

223. Bean, M. F., Carr, S. A. (1992). Characterization of disulfide positions in proteins and sequence analysis of cystine-bridged peptides by tandem mass spectrometry. *Anal Biochem 201*, 216–226.

224. Bauer, M., Sun, Y., Degenhardt, C., Kozikowski, B. (1993). Assignment of all four disulfide bridges in echistatin. *J Protein Chem 12*, 759–764.

225. Gorman, J. J., Ferguson, B. L., Speelman, D., Mills, J. (1997). Determination of the disulfide bond arrangement of human respiratory syncytial virus attachment (G) protein by matrix assisted laser desorption/ionization time-of-flight mass spectrometry. *Protein Sci 6*, 1308–1315.

226. Tsarbopoulos, A., Bahr, U., Pramanik, B. N., Karas, M. (1997). Glycoprotein analysis by delayed extraction and post-source decay MALDI TOF MS. *Int J Mass Spectrom Ion Processes 169/170*, 251–261.

227. Zubarev, R. A., Kruger, N. A., Fridriksson, E. K., Lewis, M. A., Horn, D. M., Carpenter, B. K., McLafferty, F. W. (1999). Electron capture dissociation of gaseous multiply-charged proteins is favored at disulfide bonds and other sites of high hydrogen atom affinity. *J Am Chem Soc 121*, 2857–2862.

228. Horn, D. M., Ge, Y., McLafferty, F. W. (2000). Activated ion electron capture dissociation for mass spectral sequencing of larger (42 kDa) proteins. *Anal Chem 72*, 4778–4784.

229. Wu, J., Watson, J. T. (1997). A novel methodology for assignment of disulfide bond pairings in proteins. *Protein Sci 6*, 391–398.

230. Bures, E. J., Hui, J. O., Young, Y., Chow, D. T., Katta, V., Rohde, M. F., Zeni, L., Rosenfeld, R. D., Stark, K. L., Haniu, M. (1998). Determination of disulfide structure in agouti-related protein (AGRP) by stepwise reduction and alkylation. *Biochemistry 37*, 12172–12177.

231. van den Hooven, H. W., van den Burg, H. A., Vossen, P., Boeren, S., de Wit, P. J., Vervoort, J. (2001). Disulfide bond structure of the AVR9 elicitor of the fungal tomato pathogen Cladosporium fulvum: evidence for a cystine knot. *Biochemistry, 40*, 3458–3466.

232. Wallis, T. P., Pitt, J. J., Gorman, J. J. (2001). Identification of disulfide-linked peptides by isotope profiles produced by peptic digestion of proteins in 50% (18)O water. *Protein Sci 10*, 2251–2271.

233. Walsh, G. (2003). Biopharmaceutical benchmarks-2003. *Nat Biotechnol 21*, 865–870.

234. Newman, D. J., Cragg, G. M. (2007). Natural products as sources of new drugs over the last 25 years. *J Nat Prod 70*, 461–477.

235. Cumming, D. A. (1991). Glycosylation of recombinant protein therapeutics: Control and functional implications. *Glycobiology 1*, 115–130.

236. Dube, S., Fisher, J. W., Powell, J. S. (1988). Glycosylation at specific sites of erythropoietin is essential for biosynthesis, secretion and biological function. *J Biol Chem 263*, 17516–17521.

237. Dwek, M. V., Ross, H. A., Leathem, A. J. (2001). Proteome and glycosylation mapping identifies post-translational modifications associated with aggressive breast cancer. *Proteomics 1*, 756–762.

238. Rudd, P. M., Elliott, T., Cresswell, P., Wilson, I. A., Dwek, R. A. (2001). Glycosylation and the immune system. *Science 291*, 2370–2376.

239. Peracaula, R., Tabares, G., Royle, L., Harvey, D. J., Dwek, R. A., Rudd, P. M., de Llorens, R. (2003). Altered glycosylation pattern allows the distinction between prostate-specific antigen (PSA) from normal and tumor origins. *Glycobiology 13*, 457–470.

240. Rudd, P. M., Merry, A. H., Wormald, M. R., Dwek, A.R. (2002). Glycosylation and prion protein. *Curr Opin Struc Biol 12*, 578–586.

241. Butler, M., Quelhas, D., Critchley, A. J., Carchon, H., Hebestreit, H. F., Hibbert, R. G., Vilarinto, L., Teles, E., Matthijs, G., Scholler, E., Argibay, P., Harvey, D. J., Dwek, R. A., Jaeken, J., Rudd, P. M. (2003). Detailed glycan analysis of serum glycoproteins of patients with congenital disorders of glycosylation indicates the specific defective glycan processing step and provides an insight into pathogenesis. *Glycobiology 13*, 601–622.

242. Blume-Jensen, P., Hunter, T. (2001). Oncogenic kinase signaling. *Nature 411*, 355–365.

243. Cohen, P. (2002). Timeline: Protein kinases—The major drug targets of the twenty-first century? *Nat Rev Drug Discov 1*, 309–315.

244. Kumar, C. C., Madison, V. (2001). Drugs targeted against protein kinases. *Expert Opin Emerg Drugs 6*, 303–315.

245. Irandoust, M., van den Berg, T. K., Kaspers, G. J., Cloos, J. (2009). Role of tyrosine phosphatase inhibitors in cancer treatment with emphasis on SH2 domain-containing tyrosine phosphatases (SHPs). *Anticancer Agents Med Chem 9*, 212–220.

246. Raschdorf, F., Dahinden, R., Maerki, W., Richter, W. J., Merryweathe, J. P. (1988). Location of disulphide bonds in human insulin-like growth factors (IGFs) synthesized by recombinant DNA technology. *Biol Mass Spectrom 16*, 3–8.

247. Vlahos, C. J., Wilhelm, O. G., Hassell, T., Jaskunas, S. R., Bang, N. U. (1991). Disulfide pairing of the recombinant kringle-2 domain of tissue plasminogen activator produced in *Escherichia coli. J Biol Chem 266*, 10070–10072.

248. MacNair, J. E., Lewis, K. C., Jorgenson, J. W. (1997). Ultrahigh-pressure reversed-phase liquid chromatography in packed capillary columns. *Anal Chem 69*, 983–989.

249. MacNair, J. E., Opiteck, G. J., Jorgenson, J. W., Moseley, M. A. III., (1997). Rapid separation and characterization of protein and peptide mixtures using 1.5 microns diameter non-porous silica in packed capillary liquid chromatography/mass spectrometry. *Rapid Commun Mass Spectrom 11*, 1279–1285.

250. Ruotolo, B. T., Giles, K., Campuzano, I., Sandercock, A. M., Bateman, R. H., Robinson, C. V. (2005). Evidence for macromolecular protein rings in the absence of bulk water. *Science 310*, 1658–1661.

CHAPTER 13

Mass Spectrometry of Antigenic Peptides

HENRY ROHRS

13.1 INTRODUCTION

This chapter focuses on the mass spectrometry (MS) of naturally processed peptides displayed on the surfaces of antigen-presenting cells (APCs) by a protein assembly known as the major histocompatibility complex (MHC). Mass spectrometry has played an important role in elucidating the properties of these immunopeptides since the pioneering work of Hunt in 1992. This area of research has benefited from the rapid advance of technology in proteomics, but it differs in many ways from conventional bottom-up proteomics, which is usually based on controlled enzymatic digestion (e.g., using trypsin) [1]. This chapter will focus on the study of class I and class II peptides by using representative examples of the work of a few researchers. This is not an exhaustive review, and it begins with a short history of the discovery of key features of naturally processed peptides and a brief description of their biological nature.

These antigenic peptides may have an important role in drug discovery. For example, identifying those peptides that trigger the immune response in type 1 diabetes mellitus will lay for researchers a foundation for studies to prevent the disease.

13.1.1 Brief History of MHC Studies

Immunologists made rapid progress in elucidating the T-cell mediated adaptive immunity during the last two decades of the twentieth century. In 1981, Unanue and coworkers [2] showed that processing of antigens by macrophages was necessary and sufficient for stimulating T cells, and in 1985, this group reported that processed

Protein and Peptide Mass Spectrometry in Drug Discovery, Edited by Michael L. Gross, Guodong Chen, and Birendra N. Pramanik.

peptides were bound directly to MHC molecules [3]. In 1987, Wiley and coworkers determined the X-ray crystal structure of an MHC complex, HLA-A2.1, which showed the binding groove as well as electron density in the groove that could not be resolved [4]. This led to the idea that many different peptides are bound by an MHC, and in 1991, Rammensee and his colleagues used Edman degradation to show that pooled class 1 peptides had MHC-allele specific motifs [5]. Many groups began determining the sequence of class 1 and class II MHC peptides by using high-performance liquid chromatography (HPLC) separation and Edman degradation [6–8].

In 1992, Hunt and coworkers pioneered the use of tandem mass spectrometry (MS) to sequence both class I and class II MHC peptides [9,10]. They demonstrated that the MS approach can yield more peptides, and thus they obtained a broader overview of the function of the MHC besides providing motif information. Their work also showed that peptides from both class I and class II MHCs have a distribution of lengths and that the increased number of sequenced peptides provide a better snapshot of their origin. Mass spectrometry soon became the method of choice for mapping MHC peptidomes.

The key developments in mass spectrometry that have benefited MHC study are the same as those that have benefited bottom-up proteomics [11]. These include the development of electrospray ionization [12] and MALDI [13] for the production of biomolecular ions for MS analysis [14], the commercial availability of mass spectrometers with high resolving power and acquisition speed to match the improvements in chromatography, and tandem mass spectrometry [15]. Continual improvements in the throughput and sensitivity of mass spectrometers and the ability to handle large volumes of data have allowed researchers to gain a better understanding of the origins of peptides for a variety of MHCs under normal and pathophysiological conditions. Immunopeptides discovered by mass spectrometry have also been used to elucidate the pathways that are involved in protein processing and peptide display. Several reviews covering the use of mass spectrometry in immunology were written in the last dozen years [11,16–21].

13.1.2 Brief Introduction to Immunobiology

Proteins inside and outside of the cell are constantly being degraded to peptides. For cytosolic proteins this occurs through a proteasomal pathway that eventually leads to the display on the surface of the cell of some of these peptides bound on class I MHC molecules. Peptides derived from foreign proteins inside the cell are identified by an interrogating T cell, and this leads to an immune response. Under normal conditions self-peptides are examined and ignored by the immune system.

Extracellular and membrane proteins are processed through an endosomal pathway. The derived peptides are displayed on class II MHCs, and they are interrogated by CD4 + T cells. There is evidence for cross-presentation, for example, the display of peptides from extracellular sources on class I MHCs [22,23].

In either case the peptides are noncovalently bound in a cleft in the MHC, as shown by the X-ray structure of the murine class II MHC molecule I-A^{g7}; see

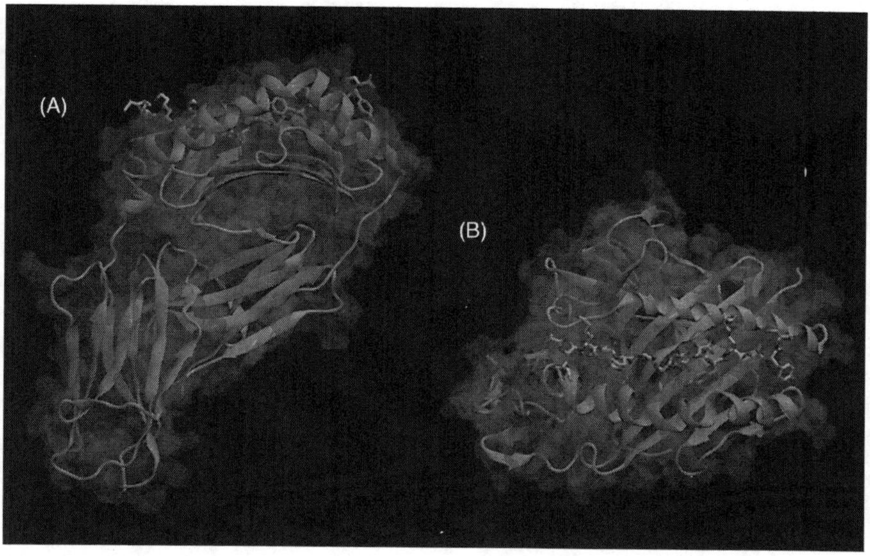

FIGURE 13.1 Murine class II MHC, IAg^7, with an antigenic peptide, HEL11-25, in the binding cleft viewed form the side (*A*) and from above (*B*). The alpha and beta chains of the protein are shown in turquoise and gold, respectively. Note the groove formed by the two alpha helices and the underlying beta sheet. This is structure 1F3J from the Protein Data Bank (www.pdb.org) and was rendered with VMD (www.ks.uiuc.edu/Research/vmd/) by Dr. Manolo Plasencia. (*See the color version of this figure in Color Plates section.*)

Figure 13.1. The class II MHC consists of an alpha and beta chain, both with transmembrane domains. Two membrane-distal alpha helices, one from each chain of the MHC molecule form the sides of the groove that binds the peptide. Residues in the groove allow for interactions (e.g., salt bridges and hydrogen bonding) with the side chains and backbone on the antigenic peptide. These interactions are specific and can lead to binding in the submicromolar range [24]. The class II MHC in the figure, $I\text{-}A^{g7}$, is characterized by a strong preference for an acidic amino acid at the P9 pocket of the groove (where the numbering proceeds from the *N* to *C* terminus of the portion of the peptide that binds in the groove) [25].

Class I MHCs are formed from one alpha chain, which contains the groove and the transmembrane domain, and β2-microglobulin. The polymorphism of MHCs allows them to bind a wide variety of peptides and facilitates rapid immune response to mutating infectious organisms. Each MHC is capable of binding many peptides, and their different structures affect their peptide binding characteristics. For example, class I MHCs are closed at the ends of the cleft and typically bind peptides that are 8 to 10 amino acids long. Class II MHCs are more open at the ends of the cleft, and they bind peptides that are usually between 8 and 20 amino acids long. Even though the peptides vary in length, the core binding region is inside the groove and is usually nine amino acids long.

A wide variety of peptides are bound by an MHC. Estimates of peptide diversity suggest that there are more than 20,000 different class I peptides displayed at greater than one copy per cell. Current separation schemes coupled with an inadequate dynamic range during detection suggest the number could be considerably higher. Thus the class I MHC "peptidome" of an individual animal with six class I MHCs is likely to be greater than 120,000 peptides [16]. Class II MHC polymorphism is more complex, and there are at least as many peptides in the class II peptidome. Those peptides that are discovered in mapping experiments, as well as mutants of these peptides, are often synthesized by researchers for screening the interrogating T cells, determining motifs and their binding, elucidating antigen processing, and measuring the effect of inhibitory or stimulatory molecules on the adaptive immune response. Databases such as SYFPEITHI (www.syfpeithi.de) and the Immune Epitope Database (www.immuneepitope.org) catalog peptides for many MHCs, and this information can be used to learn about the structure and function of the MHCs [26].

For readers interested in more comprehensive treatments of immunology, several references are available [27–30].

13.2 ANALYSIS OF ANTIGENIC PEPTIDES

Antigenic peptides are difficult to study for several reasons. They are usually present at low levels in a complex background containing peptides that are far more abundant. Thus, as in many other proteomics studies, enrichment strategies are usually employed. In addition animal models or cell lines are developed that express only the MHCs of interest. Even so, the number of MHCs per cell is still low.

Each MHC has a preferred motif for binding [5], although many of these motifs remain unknown or only partially determined. A given protein might have only one part of its primary structure that is a candidate for binding. In addition, because the proteins are naturally processed by nonspecific or uncharacterized enzymes, there are no known rules for enzymatic cleavage. In the case of class II peptides, this leads to the presentation of families of peptides. Each family may have several members with the same binding motif. An example is shown in Table 13.1. This family of peptides from integral membrane protein 2B (ITM2B) binds to the murine MHC I-A^{g7} and has a core binding motif of EENIKIFEE. Family members contain flanking residues beyond the core binding motif on either or both the amino and carboxyl termini. Given that this is just one of many binding motifs, there are families of peptides that are found for I-A^{g7}. A selection of peptides and their binding motifs is shown in Table 13.2.

Each peptide comes from a single protein, in contrast to conventional bottom-up proteomics where multiple tryptic peptides are detected for a single protein. Thus the signal for a single protein is diluted among many family members.

In standard bottom-up proteomics, digestion yields peptides with at least two basic sites (Lys or Arg at the *C* terminus and an amino group at the *N* terminus). These sites can accept protons under normal electrospray conditions and become charged for analysis. For an MHC peptide there is no guarantee that the *C*-terminal residue or any

TABLE 13.1 Family of Antigenic Peptides Presented by the Murine Class II MHC I-Ag[7]

									P1	P2	P3	P4	P5	P6	P7	P8	P9				
							T	I	E	E	N	I	K	I	F	E	E				
							T	I	E	E	N	I	K	I	F	E	E	D			
							T	I	E	E	N	I	K	I	F	E	E	D	A		
							T	I	E	E	N	I	K	I	F	E	E	D	A	V	
							T	I	E	E	N	I	K	I	F	E	E	D	A	V	E
						Q	T	I	E	E	N	I	K	I	F	E	E				
						Q	T	I	E	E	N	I	K	I	F	E	E	D			
						Q	T	I	E	E	N	I	K	I	F	E	E	D	A		
						Q	T	I	E	E	N	I	K	I	F	E	E	D	A	V	
						Q	T	I	E	E	N	I	K	I	F	E	E	D	A	V	E
					Y	Q	T	I	E	E	N	I	K	I	F	E	E				
					Y	Q	T	I	E	E	N	I	K	I	F	E	E	D			
					Y	Q	T	I	E	E	N	I	K	I	F	E	E	D	A		
					Y	Q	T	I	E	E	N	I	K	I	F	E	E	D	A	V	E
				R	Y	Q	T	I	E	E	N	I	K	I	F	E	E	D	A		
A	P	A	A	R	Y	Q	T	I	E	E	N	I	K	I	F	E	E	D	A		

Note: These are peptides from the protein ITM2B (UniProt Q89051-1) that are bound to the murine MHC I-Ag[7]. The columns P1 through P9 are the binding pockets on the MHC and the residues labeled in **bold font** (116–124) are the binding motif that fits in the pocket. Sixteen family members are shown in this table. The flanking residues on both the amino (left) and carboxyl (right) termini are a feature characteristic of peptides bound to class II MHC molecules.

TABLE 13.2 Peptides and Binding Motifs from the Murine Class II MHC I-Ag[7]

Protein							P1			P4		P6			P9				
Synaptotagmin				I	Q	E	**A**	H	G	**L**	P	**V**	M	D	**D**	Q			
ODZ				T	P	S	**Q**	Q	A	**A**	K	**S**	F	Y	**D**	R			
Neuromodulin	Q	P	P	T	E	T	**A**	E	S	**S**	Q	**A**	E	E	**E**	K	D	A	
Secretogranin			I	P	E	K	**V**	T	P	**V**	A	**A**	V	Q	**D**	G	F	T	
Chromogranin		R	P	S	S	R	**E**	D	S	**V**	E	**A**	R	S	**D**	F	E	E	
BACE-2				F	A	V	**A**	G	A	**P**	H	**S**	Y	I	**D**	T	Y	F	
Lisch 7		S	G	R	P	R	**A**	R	S	**V**	D	**A**	L	D	**D**	I	N	R	P
Amyloid beta A4						V	**A**	E	E	**I**	Q	**D**	E	V	**D**	E			
NCAM			S	A	P	K	**V**	A	P	**L**	V	**D**	L	S	**D**	T			
ITM2B			Y	Q	T	I	**E**	E	N	**I**	K	**I**	F	E	**E**	D	A		

Note: Each family is represented by one peptide, although in most cases other family members with the same motif were found. The binding region is shown in bold with the contact points for the MHC are **singly and doubly underscored**. The strong preference for an acidic residue in the P9 pocket of the binding groove (shown as **single underscore**) is evident. The other contact points for binding also demonstrate weaker binding preferences.

other residue will be basic, and this impacts the peptides that are detected and their subsequent fragmentation. Many of the peptides are not highly basic and end up singly charged, and as a result they don't fragment extensively and yield poor product-ion spectra for a variety of activation methods.

Another common feature of bottom-up strategies is data-dependent analysis (DDA). In DDA, an initial full mass spectrum from a Q-TOF, ion trap, or Fourier-transform ion cyclotron resonance (FT-ICR) mass spectrometer is used to direct subsequent fragmentation steps. Often 3 to 10 of the most abundant precursor ions in the full mass spectrum are chosen for isolation and activation. Given that the MHC peptides are of low abundance, they are often not selected for fragmentation because they are obscured by more abundant peptides and contaminants.

In directed analyses where quantitation is used to select ions for further study, significant run-to-run variation impacts results. A recent study showed wide variation in detection both within and between laboratories, and new performance benchmarks are being tested [31–33].

Given that peptide sequence is critical for MHC peptide identification, MS^2 data in the form of product-ion spectra are gathered from selected parent ions. The coverage of a particular protein, however, is often sparse and is focused around a primary structure with the necessary binding motif. Strategies that employ scoring based on finding multiple peptides or protein coverage are not generally useful. For example, because class I peptides are only 8 to 10 amino acids long, their detection will often will not lead to a protein score that meets threshold criteria in database search algorithms like those employed by Mascot. Although the sequencing data are useful, extra effort is required to look at individual peptide scores and then at the corresponding spectra to determine those peptides to pursue for further study. Database searching also takes considerably longer because there is no enzymatic constraint, and the possible peptides from a given protein are much larger than that derived from a specific enzymatic digest. Given that the peptides cannot be predicted, the use of directed strategies that generate inclusion lists for targeted proteomics is precluded.

13.2.1 MHC Peptide Analysis in Practice—Sample Preparation

The first and most crucial step in proteomics work directed at finding antigenic peptides is sample preparation. It is essential to isolate MHC peptides from the cells of interest with minimal contamination. Many researchers use variations of the method of Rammensee [34,35]. This approach involves gentle lysis of cells, immunocapture of the MHC-peptide complexes, cleanup, and mild acid elution of the peptides from the bound complex. The generation of a specific monoclonal antibody allows a researcher to focus on a particular MHC. The use of acid-labile or easily removed detergents aids in subsequent mass spectrometric analyses.

If the mixture is to be analyzed by MALDI, either it can be directly spotted with matrix onto a plate or, if a complex mixture, it can be separated by chromatography before spotting. For electrospray ionization, acids and salts must be removed because they suppress ionization. This can be done by using a variety of solid-phase extraction techniques. Examples are reverse-phase binding to immobilized C18 in trap columns or pipet tips (e.g., Waters ZipTip, Glygen NuTip).

13.2.2 MHC Peptide Analysis in Practice—HPLC Separation

Given that MHC peptides are generally between 8 and 25 amino acids long, they can be separated by reverse-phase HPLC, usually using C18 columns with 3- to 5-μm particles with 100- to 300-Å pore sizes. The use of small diameter columns and low flow rates (i.e., nanospray or nanoelectrospray ionization) improves sensitivity. The columns may be purchased, or pulled and packed by the user. Laboratories equipped with a column puller (e.g., from Sutter Instruments of Novato, CA) and a pressure vessel (e.g., from Computech of Kansas City, MO) can produce packed 75- to 100-μm diameter columns for a few dollars each [36]. This allows one to tune the column for each application. Typical column lengths are 10 to 30 cm. Pulled columns or commercial columns can be used for nanoESI sources or for automated spotting of MALDI plates [37].

The wide variety of column materials allows researchers to create multiphasic columns for advanced separations. Examples of this include multidimensional protein identification technology (MUDPIT) [36] and online enrichment strategies such as titanium dioxide bed volumes for trapping phosphopeptides [38].

The peptides are often eluted from the column/spray tip by reversed phase HPLC directly into the mass spectrometer. For nanospray, the low flow rates are achieved by splitting the flow from a capillary HPLC, or by using direct nanoflow HPLCs available from many manufacturers (e.g., Eksigent, Waters, Agilent). These operate at pressures up to about 4000 psi and can achieve flow rates well below 1 μL/min. Chromatographic peaks from these systems are typically 10 to 30 s wide at baseline.

The recent development of UPLCs, which operate at 10,000 to 15,000 psi and produce 1- to 2-s wide peaks, leads to better separation and should improve performance provided the UPLCs are matched with mass spectrometers that scan sufficiently rapidly to take advantage of these narrow peaks. Special fittings must be used at these pressures, and all connections assembled with special care, as any dead volume (e.g., caused by cleaving silica tubing at a slight angle) can broaden the peaks.

13.2.3 MHC Peptide Analysis in Practice—Mass Spectrometers

Most mass spectrometers directed at biomolecule analysis come equipped with an electrospray or a MALDI ion source. For ESI, the source consists of a charged needle and annular space for sheath gas. The charged droplets that exit the needle are directed into the mass spectrometer [39,40] (see Chapter 1 by Cotte-Rodrigues, Zhang, Miao, and Chen of this volume). One can also purchase nanospray sources (e.g., from New Objective in Woburn, MA or Phoenix S&T Technology) that are designed to accommodate capillary columns with small diameters (e.g., ~75 μm) [41]. The nanosprays typically are operated at 2 kV, and they are charged by using an electrode that contacts the spray solution through a microbore tee. Some companies have also developed microfluidic chips that incorporate tips for nanospray in addition to an integrated column (e.g., see Agilent Chip Cube, Waters Trizaic, Eksigent Nanoflex).

For experiments employing electrospray (>1 μL/min) or nanoelectrospray (<1 μL/min), a front-end quadrupole filter or ion trap is typically employed. A decade ago a triple quadrupole or an ion trap would have been used for the acquisition of both the full and product-ion mass spectra, and both are still used today. A hybrid instrument, however, is usually employed with the full mass spectrum recorded in the high mass-resolving power section of the instrument (e.g., the TOF in a Q-TOF or the ICR in an ion trap-FTICR mass spectrometer). High mass-resolving power allows for charge state determination, separation of peaks that are close in mass, and improved mass measurement accuracy. The mass accuracy of these instruments, typically better than 5 ppm, improves confidence in identification of peptides when using database searching algorithms (e.g., Sequest or Mascot). Several FT instruments (e.g., Thermo LTQ-FT, Thermo Orbitrap, Bruker Solarix) are able to record product-ion mass spectra in an ion trap simultaneously with full mass spectral acquisition by the FTMS.

For experiments in Q-TOF instruments, the TOF section measures all of the spectra. All ions pass through a collision cell, and their energy is reduced for the full mass spectrometric analysis and increased when fragmentation is desired. An advantage of the Q-TOF is its speed, as it can operate at 10 Hz or faster. Until recently, however, Q-TOF instruments did not produce MS^2 data comparable to that obtained from ion traps.

For experiments in hybrid ion trap-FTMS instruments, the duty cycle is approximately 1 Hz. In a little more than one second, the instrument can record a full mass spectrum at a mass resolving power of 100,000 (at $m/z = 400$) and simultaneously collect 5 to 10 product-ion spectra afforded by CAD in an ion trap. While the FTMS is recording the transient signal for the full mass spectrum in an ICR trap or an orbitrap, the ion trap is sequentially isolating and activating ions selected from a mass spectrum derived from a truncated portion of the transient. The mass spectrum is truncated for speed, even though the mass resolving power at short transient times is low. The ultimate mass spectrum is obtained from a longer transient, insuring narrower peaks and good mass accuracy.

For MALDI experiments, the sample can be spotted onto a multiwell plate (e.g., 96-, 192-, or 384-well plates); prior to spotting, a suitable matrix for peptide analysis (e.g., α-cyanohydroxycinnamic acid) is added. Repetition rates of 1 kHz are common; once a plate is spotted, the MALDI analysis is very fast. Given that MALDI is a pulsed ionization experiment, it couples well with time-of-flight instruments (in principle, it should also be compatible with FT instruments). MALDI also has the advantage that an immobilized, dried sample that can be interrogated repeatedly. MALDI also can be coupled to HPLC, as demonstrated by several research groups [37,42,43].

Single-stage TOF instruments provide mass spectra, but post-source decay can be employed to get fragmentation data for MHC peptides. More sophisticated and expensive machines make use of Bradbury ion gates to create a TOF-TOF geometry that allows for selection of a precursor ion, collisional activation of the analyte precursor ions in a gas-filled cell [11], and product-ion analysis in a second TOF.

Many MHC peptide studies employ the DDA methods that are used in bottom-up proteomics. The peptides are readily separated by reverse-phase HPLC and eluted directly into a mass spectrometer using ESI or nanospray. Collisional activation of peptides produces a variety of ions from cleavages at the many peptide bonds in the molecule. The most important of these for peptide identification are the b- and y-ions that are *N*-terminal and *C*-terminal fragments, respectively, after amide bond cleavage (see Figure 13.2). These ions are used to determine the sequence of a peptide manually, so-called de novo sequencing, or automatically by using a database search (see also Chapter 2 by Lin and O'Connor in this volume).

Another fragmentation method that is becoming increasingly useful is electron activation. There are two methods for this, electron-capture dissociation (ECD) and electron-transfer dissociation (ETD). Both methods employ energetic electron capture by highly positive (multiply charged) peptides or proteins. The activation produces radical cations that fragment in a completely different way than when activated by the many low-energy collisions with an inert gas in an ion trap. In ECD, the source of electrons is usually a hot cathode, whereas for ETD the electron is transferred to the analyte from a carefully prepared anion (e.g., the fluoranthene radical anion). These methods give rise to cleavage of the N–C bonds along the peptide backbone producing c- and z-ions instead of b- and y-ions. The electron-capture methods add greater amounts of energy to the peptide during fragmentation but in a faster way; thus they are favored for analyzing peptides with labile groups (e.g., posttranslational modifications like phosphorylation, glycosylation) because the peptides still undergo peptide bond cleavage to give c- and z-ions rather than losing the modification. For more on activation methods, see Chapter 2 of this volume.

13.2.4 MHC Peptide Analysis in Practice—Data Analysis

Another characteristic that the mass spectrometry of MHC peptides shares with standard bottom-up proteomics experiments is large data sets. Mass spectrometers coupled to HPLCs can generate more than 5000 spectra an hour. Analysis of these data requires powerful computing and algorithms. The first step is usually database searching by using one of many available programs. Most researchers use Mascot and Sequest, but other search engines are now available, including Spectrum Mill, X! Tandem, and Phenyx. Many laboratories produce programs to search for specific features in datasets (e.g., disulfide bonds).

Both Mascot and Sequest compare experimental product-ion spectra with those calculated from a database. This database can be created by the user, but it is most often derived from genomic data that were used to calculate protein sequences. Examples of databases are the NCBI non redundant database or a subset of this database (e.g., one derived from the proteome of the mouse) and databases available from IPI. Modifications, whether biological, such as phosphorylation, or not, such as reduction and alkylation of disulfide bonds, can be included as variables, but modifications increase the computing time geometrically. These search engines look for similarity between the recorded product-ion spectrum and the calculated spectra, so there is always a best match, even for random noise. Matches are ranked according

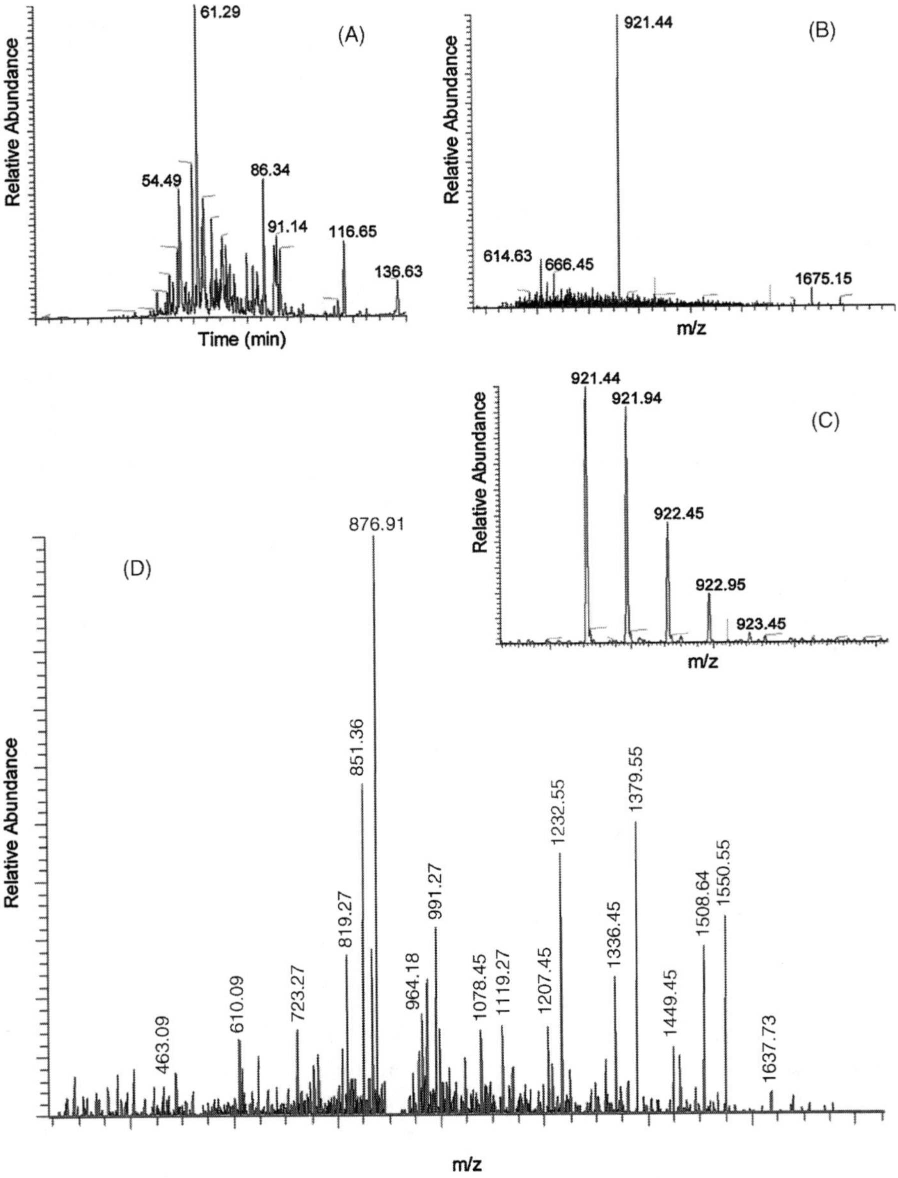

FIGURE 13.2 Proteomics applied to antigenic peptides. (*A*) The base peak chromatogram from a 150-min gradient separation of peptides eluted from the class II MHC IAg^7. (*B*) A full mass spectrum from 75.11 min in the chromatogram. (*C*) The isotopically resolved peaks in an expanded view of the mass spectrum shown in *B*. (*D*) The MS2 spectrum from the peak at m/z 921.44. A Mascot database search determined that best match to the spectrum was YQTIEE-NIKIFEEDA from the murine protein ITM2B. The b-ions are shown in red and the y-ions in blue. The other two ions are doubly charged *b*-ions. (*See the color version of this figure in Color Plates section.*)

to a scoring system (see Chapter 9 by Dasari and Tabb of this volume), and if several peptides are found for a single protein, this improves confidence that it is in the sample. This is problematic for MHC peptides because there may only be one peptide for a given protein. Other family members for a given motif improve the score somewhat, but there is rarely much coverage for the protein precursors that are the source for the presented peptides.

13.3 EXAMPLES OF THE APPLICATION OF MASS SPECTROMETRY TO ANTIGENIC PEPTIDE STUDY

In this section the work of some of the major contributors on antigen-presenting cells is highlighted. Some of the approaches they used, even as far back as 20 years ago, are still applicable, and the continual improvements to mass spectrometers make these methods more effective. The interplay of biomedical research and technology development drives this field forward. Most of the work mentioned below (i.e., that of Hunt, Unanue, Rammensee, and Allen) employs the immunocapture of the MHC molecule and mild acid elution of peptides pioneered by Rammensee. This sample preparation step is followed by Hunt's method of HPLC separation of peptides and direct elution into the mass spectrometer. The peptides are identified by database searching and de novo sequencing.

13.3.1 Work of D. Hunt

As was noted previously, the group of Donald Hunt was the first to apply mass spectrometry to the study of MHC peptides. This group has been active in the field ever since, and a review of their contributions was published in 2006 [17]. In addition to determining repertoires and motifs for several class I and class II MHC peptides, they continue to develop new analytical approaches.

In the "splitter method," class I MHC peptides from a melanoma cell line, DM6, were fractionated by multiple steps of chromatography [44]. To determine those ions whose peptide precursors are important in stimulating the T cells, a portion of the effluent was diverted to culture media for a T cell assay. Fractions that correspond to positive responses in the T cell assay were pooled. The peptides in these pooled fractions were again run through an HPLC separation step in which a sixth of the effluent was diverted and collected as fractions in cell media for a T cell assay. Most of the sample was directed into a triple quadrupole mass spectrometer so that the CTL assay can be linked to the mass spectrometry. In one example, Hunt and coworkers [45] found that a single peptide, YLEPGPVTA derived from a melanocyte protein Pmel 17, is the naturally occurring peptide that stimulates five different CTLs from melanoma patients. This experiment demonstrates the effectiveness of using the most sensitive readout, the T cell assay, to focus the mass spectrometric analysis. Hunt and his collaborators [46,47] also used this microfraction collection method in other studies.

In another example, the Hunt group used mass spectrometry to elucidate the role of tapasin in the processing of class I MHC peptides. The cell lines were tapasin deficient or contained either soluble or membrane-bound tapasin. Peptides from HLA-B8 were analyzed with a homemade FTMS and a Thermo LCQ Deca ion trap mass spectrometer. The repertoire of peptides expressed in the presence or absence of tapasin is distinctly different, four times as many peptides are bound when tapasin is present, and there is only a slight difference between the cells that expressed soluble or membrane-bound tapasin. When Hunt and coworkers tested a variety of peptides for binding, they found that tapasin-deficient cells bound peptides with higher affinities. On the basis of this and other work, they reasoned that tapasin plays a role in stabilizing the peptide-MHC complexes but does not selectivity edit peptides based on binding strength [48].

Mass spectrometric methods can be used to provide understanding of antigenic peptides that are posttranslationally modified. Hunt and coworkers [49] were the first to find naturally processed phosphopeptides, detecting them from eight class I MHCs. After immunoaffinity purification of the MHC-peptide complex, they added an enrichment step using immobilized metal affinity chromatography (IMAC) with iron cations to bind the phosphate groups of the modified peptides. These were eluted by using sodium phosphate and then desalted after being bound to C18 media. Because the phosphate group is labile under normal CID conditions, Hunt's group devised their own software to look for neutral losses of phosphoric acid in the spectra (i.e., loss of 49 and 98 from doubly and singly charged precursors, respectively). Product-ion spectra that showed such losses were then analyzed manually to determine sequence with the aid of other software tools (MS-Tag at www.prospector.ucsf.edu). In one example, between 9 and 122 candidates were identified in each of the eight MHCs by using neutral-loss software that Hunt's group developed. Hunt and his colleagues used a Thermo LCQ ion trap to sequence the phosphopeptides. In the example, no phosphopeptides in TAP-deficient cell lines were detected, showing that TAP is essential for their presentation. To add confidence, they used a T-cell assay to show that there is a specific T-cell response to one of these phosphopeptides. The example study also found evidence for peptides with more than one phosphorylation site and showed that the some MHC complexes displayed more phosphorylated peptides than others. Hunt and his colleagues [50,51] also found other PTM in antigenic peptides.

13.3.2 Work of E. Unanue

Emil Unanue and coworkers at Washington University in St. Louis continue to be active in antigenic peptide research, employing mass spectrometry in many of their projects. This group has developed cell lines that express foreign protein (e.g., hen egg lysozyme), and they use them to understand the peptide processing and the display by MHC of antigenic peptides that derive from this foreign protein. Their focus is on the autoimmune disease, type 1 diabetes mellitus (T1DM) [52,53].

Unanue and his colleagues have analyzed peptides derived from murine tumor cell lines expressing the class II MHC I-A^{g7} and a mutant, I-A^{g7PD}. I-A^{g7} is the MHC most responsible for T1DM in the nonobese diabetic (NOD) mouse. They found that I-A^{g7} showed a strong preference for peptides that have acidic residues in the P9 binding pocket, and that binding affinities can be increased when there are acidic residues beyond the P9 pocket. When they used a cell line that had mutated I-A^{g7} in which

I-A^{g7} has P→D at position 57 (a mutation in the binding pocket of the MHC), there was only 5% overlap in the peptide repertoire, and the preference for an acidic carboxyl terminal residues was lost. To be sure that the peptides were not in the mutant PD sample, extracted ion chromatograms showed absence of these peptides at the threshold of detection. Peptides were then synthesized on the basis of those found in the study and tested for binding to show that mutation to acidic residues outside the pocket has an impact on binding in I-A^{g7} [25].

These researchers [54] co-expressed I-A^{g7} and I-A^{d} in a cell line and compared the peptide repertoire with cell lines that expressed only I-A^{g7} or both I-A^{g7} and I-A^{g7PD}. They employed data-dependent analysis on a Thermo LCQ-Deca equipped with a nanospray source to identify dozens of families from each MHC and more than 50 peptides per MHC from each line. The large number of peptides allowed them to infer that even closely related MHCs display different peptides and have different motifs. In cells with more than one MHC, neither has much of an effect on the peptides displayed by the other. This finding is significant because other researchers were trying to determine why the presence of other MHCs both in the NOD mouse and in humans has a protective effect in T1DM. These results suggest that it is not due to an alteration in the presentation of antigen by the MHC.

They have also generated a cell line that expresses the human class II MHC DQ8. Both DQ8 and I-A^{g7} lack an asparagine at b57 in the binding pocket of the MHC, and this amino-acid residue is one determinant in developing T1DM. They sequenced a large number of peptides (301 for I-A^{g7} and 206 for DQ8) by using 2D chromatography (strong cation exchange and RP C18). They showed that DQ8 and I-A^{g7} share similar binding motifs. In particular, they both have a strong preference for acidic residues in the P9 pocket. They also found that the two MHCs shared several identical epitopes. EENIKIFEE (Table 13.1) is an example of a peptide motif with many family members that is shared by I-A^{g7} and DQ8 [55].

In another experiment, Unanue and coworkers [56] created a cell line that expresses I-A^{g7} from a β cell insulinoma. β islet cells in the pancreas are the target of autoreactive T cells. β cells, however, are not antigen-presenting cells (APCs). β cell peptides are presented by other cells that are in their vicinity. This system is difficult to reproduce in the laboratory. The cell line, NitCIITA, solved this problem; it was created by splicing the class II MHC machinery for I-A^{g7} into a β cell tumor line. 320 peptides from 120 distinct families were identified with the Thermo LCQ-Deca and a high-performance Thermo LTQ-FT. This study showed that the β cell has a diverse MHC peptidome for I-A^{g7}. An analysis of a subset 21 peptide families that are beta cell specific revealed that 19 of them display the expected binding motif, either aspartic or glutamic acid, in the P9 pocket.

The studies mentioned above also demonstrate the impact of steady improvement in mass spectrometers. The speed of the Fourier-transform hybrid instrument allows for more peptide identifications. The LTQ-FT can carry out up to 10 data-dependent MS2 events in the LTQ (i.e., the linear quadrupole ion trap) while measuring a full mass spectrum in the ICR trap. This is a significant improvement in both the speed of MS2 acquisition and the mass accuracy for parent ions, leading to more confident sequencing by database searching.

13.3.3 Work of H. Rammensee

Hans-Georg Rammensee and coworkers employ both MALDI-TOF and Q-TOF mass spectrometers for analysis of antigenic peptides. They investigated the presence of PTMs in both class I and class II MHCs. To detect phosphopeptides in HLA-DA from human tissue, from a human B lymphoblastoid cell line, and from a human melanoma cell line, they used a titanium dioxide phosphopeptide enrichment step, RP C18 chromatography to separate the peptides, and ESI analysis on a Waters Q-TOF Ultima. Using isotopic labeling to compare normal and cancerous renal tissue from a patient, they sequenced 16 class I phosphopeptides and identified that one of these was tumor restricted. They used the cell lines to look for class II phosphpeptides and sequenced 27 in the melanoma line and 20 in the lymphoblastoid line. Peptides derived from membrane proteins had between one and three phosphorylations [57].

The Rammensee group was the first to find class II peptides that had been glycosylated from a human B-cell lymphoblastoid line that expresses HLA-DR4. Signals corresponding to high *m/z* ions were seen with a MALDI TOF, and that determined candidates for further ESI MS studies. Given that the CAD of glycosylated peptides produces a facile cleavage of the glycan, in-source fragmentation was used to yield the peptide with only a GlcNac attached, and this was isolated for CAD. The two glycosylated peptides that were found are family members with the same *N*-linked glycan, and they originate from the protein CD53. Further experiments revealed that the glycan is a pentasaccharide core modified with one fucose [58].

As in Unanue's group, Rammensee's group is interested in autoimmune disease, and they recently identified both class I and class II MHC peptides from the central nervous system tissue of multiple sclerosis patients. Investigating the brain tissue from eight patients and using their standard method of immunocapture, they assayed HLA-A, -B, -C, and –DR and found approximately 34 peptides per patient, approximately two-thirds from class II MHC. They used the Mascot database searching program to find the peptides and the SYPEITHI algorithm to assess their binding affinity. In seven of eight brain samples they detected peptides from myelin basic protein, which is believed to be a target of the autoimmune response [59].

13.3.4 Work of P. Allen

The Allen group uses mass spectrometry to study the alloreactivity of peptide–MHC class II complexes. Alloreactivity occurs when T cells recognize different alleles than those that occur in the organism in which they developed. Alloresponses are of great clinical significance in graft versus host disease and in transplantation. The Allen group uses a 2.102 T cell known to be specific for the class II MHC I-E^k. To study the nature of alloreactivity, it is necessary to find peptides that participate in the alloresponse.

To that end, the Allen group [60] used MS to determine that the 2.102 T cell is alloreactive to I-E^p. They lysed B6P.C3 cells that expressed I-E^p and stimulated the 2.102 T cell. The peptides were separated using both 1D HPLC with reversed phase C18 media and a 2D approach with offline strong cation exchange followed by reverse-phase C18 HPLC. The researchers identified 295 peptides from 120 families and were able to establish the binding motif for I-E^p. They then used a bioinformatics approach to screen proteins for sequences that contained the appropriate binding sequence and residues for TCR contact derived from a mimotope that stimulated the 2.102 T cell. This led to the discovery of a self-peptide that triggered an alloresponse; the peptide is from residues 531–545 of the protein GPR128. They also learned that flanking residues beyond the P9 pocket are important to the interaction. The Allen group recently employed this strategy in the analysis of the alloreactivity of other MHCs [61].

13.3.5 Work of P. Thibault

Pierre Thibault and his colleagues at the University of Montreal use comparative proteomics to study the class I MHC repertoire. Rather than employ antibody capture of MHCs, they use two cell lines, a wild type (WT) EL4 and another in which b2-microglobulin is knocked out. The EL4 thymoma cell line expresses $H2D^b$, $H2K^b$, Qa1, and Qa2. The removal of b2-microglobulin eliminates the presentation of class I peptides in the mutant cell line.

Thibault and coworkers [62] collected peptides directly from the cells without lysis by using mild acid elution, purified them by using solid-phase extraction and ultrafiltration, and separated the mixtures with 2D chromatography, similar to MUDPIT [36], employing SCX and reversed-phase C12 media. They eluted the peptides directly through a nanospray column into a Thermo LTQ-Orbitrap mass spectrometer. They employed one full MS scan at 60,000 (at *m/z* 400) mass-resolving power and three CID scans in the LTQ by using data-dependent triggering and dynamic exclusion. They developed their own label-free quantitation software and applied accurate mass and time tagging to the features in the full MS. They focused on ions from peptides that showed a significant increase in the WT cell line, reasoning that features that appeared in both are contaminants (i.e., are not class I MHC peptides). Those ions that triggered MS2 scans and were identified by a Mascot search were manually verified. Studies of the EL4 line allowed the researchers to validate their methods and understand the global distribution of class I MHC peptides among the four MHCs. They compared the thymoma EL4 class I MHC repertoire to that captured from normal mouse thymocytes to determine the differential expression of these peptides in neoplastic cells.

13.4 FUTURE WORK

Although many human and animal model MHCs have been studied, much remains to be done, owing to both the diversity of the peptide repertoire and the genetic variation of the MHC. Allelic variations play a role in autoimmune disease and finding peptides that trigger autoimmune response in type 1 diabetes mellitus (T1DM), rheumatoid arthritis (RA), and multiple sclerosis will aid in understanding these conditions. In addition there are nonclassical MHCs that present other types of molecules (e.g.,

CD1a presents lipids). Although these have not yet drawn the level of attention of classical MHCs, they may be of comparable importance.

Issues of PTMs in antigenic peptides remain important. Recent work has demonstrated success employing derivatization of citrulline [63,64] or the use of a reporter fragment ion of citrulline [65]. Chemical modification may play a role in enrichment and detection of low-level modified MHC peptides. Electron-capture dissociation (ECD) and electron-transfer dissociation (ETD) are now widely available and have utility in the detection of PTM [66].

In addition to label-free quantitation, many proteomics researchers have employed *N*-terminal labeling to derive quantitative information in experiments that have controls and samples [67]. For targeted work, isotopically labeled peptides can be synthesized by several commercial suppliers, and these can be introduced into sample isolates and simultaneously measured for absolute quantitation.

Instruments continue to improve in speed and sensitivity. State-of-the-art Q-TOF platforms can fragment ions without isolation and link precursors and products using software to track their time profiles. This is called MS-to-the-E, and is available on Waters instruments. This approach may provide better coverage, although it is not yet clear whether this approach offers more sensitivity than ion trap instruments. Improvements (e.g., ion funnels) to front- end optics on ion traps, quadrupoles, and hybrids have increased sensitivity and speed (i.e., the requisite number of ions for a trapping experiment are collected in a shorter time). New techniques like ion mobility might have an impact (e.g., the use of drift time traces may allow discovery of new families of peptides).

New methods, such as intensity binning for label-free, directed, or targeted proteomics experiments [68,69] should improve sensitivity and coverage of MHC peptidomes. The application of standard MS methods like multiple reaction monitoring may come to play a role in the detection of specific important peptides or in the detection of diagnostic fragment ions (e.g., derived from modified peptides).

Software improvements directed specifically at MHC peptides have lagged. This is a small area of research compared with the vast applications of conventional bottom-up proteomics. Databases of peptides will continue to grow, and out of necessity, interested research groups will likely develop software for their own specific needs.

Although the sensitivity of MS continues to improve, the detection limit of immunological assays is far lower. Almost all researchers verify any critical MS data with an immunological test. Given that these biological assays afford both selectivity and sensitivity even in the midst of complicated backgrounds, they will not soon be displaced by mass spectrometry but rather will remain complementary. MS appears to be the only approach to give structural information for these low-level materials.

ACKNOWLEDGMENTS

Partial funding of this chapter comes from the WU Mass Spectrometry Research Resource, which is supported by the NCRR of the NIH (2P41RR000954). HWR thanks Dr. Manolo Plasencia for help in preparing the figures and Professor E. Unanue for his support for many years.

ABBREVIATIONS

AMT	Accurate Mass and Time
APC	Antigen Presenting Cell
CAD	Collisionally Activated Dissociation
CTL	Cytotoxic T Lymphocyte
DDA	Data-Dependent Analysis
ECD	Electron Capture Dissociation
ESI	ElectroSpray Ionization
ETD	Electron Transfer Dissociation
FT	Fourier Transform
HEL	Hen Egg Lysozyme
HLA	Human Leukocyte Antigen
HPLC	High-Performance Liquid Chromatography
ICR	Ion Cyclotron Resonance
IMAC	Immobilized Metal Affinity Chromatography
IPI	International Protein Index
MALDI	Matrix-Assisted Laser Desorption Ionization
MHC	Major Histocompatibility Complex
MS	Mass Spectrometry
MUDPIT	MUltiDimensional Protein Identification Technology
NCBI	National Center for Biological Information
PDB	Protein DataBank
Q-TOF	Quadrupole Time Of Flight
RA	Rheumatoid Arthritis
T1DM	Type 1 Diabetes Mellitus
TFA	TriFluoroacetic Acid
TOF	Time Of Flight
UPLC	Ultra–high-Performance Liquid Chromatography
WT	Wild Type

REFERENCES

1. Kinter, M., Sherman, N. E. (2000). *Protein Sequencing and Identification Using Tandem Mass Spectrometry*. Wiley, New York.
2. Ziegler, K., Unanue, E. R. (1981). Identification of a macrophage antigen-processing event required for I-region-restricted antigen presentation to T lymphocytes. *J Immunol 127*, 1869–1875.
3. Babbitt, B. P., Allen, P. M., Matsueda, G., Haber, E., Unanue, E. R. (1985). Binding of immunogenic peptides to Ia histocompatibility molecules. *Nature 317*, 359–361.

4. Bjorkman, P. J., Saper, M. A., Samraoui, B., Bennett, W. S., Strominger, J. L., Wiley D. C. (1987). Structure of the human class I histocompatibility antigen, HLA-A2. *Nature 329*, 506–512.
5. Falk, K., Rotzschke, O., Stevanovic, S., Jung, G., Rammensee, H. G. (1991). Allelespecific motifs revealed by sequencing of self-peptides eluted from MHC molecules. *Nature 351*, 290–296.
6. Jardetzky, T. S., Lane, W. S., Robinson, R. A., Madden, D. R., Wiley D. C. (1991). Identification of self peptides bound to purified HLA-B27. *Nature 353*, 326–329.
7. Rudensky, A., Preston-Hurlburt, P., Hong, S. C., Barlow, A., Janeway, C. A., Jr., (1991). Sequence analysis of peptides bound to MHC class II molecules. *Nature 353*, 622–627.
8. Nelson, C. A., Roof, R. W., McCourt, D. W., Unanue, E. R. (1992). Identification of the naturally processed form of hen egg white lysozyme bound to the murine major histocompatibility complex class II molecule I-Ak. *Proc Natl Acad Sci USA 89*, 7380–7383.
9. Hunt, D. F., Henderson, R. A., Shabanowitz, J., Sakaguchi, K., Michel, H., Sevilir, N., Cox, A. L., Appella, E., Engelhard, V. H. (1992). Characterization of peptides bound to the class I MHC molecule HLA-A2.1 by mass spectrometry. *Science 255*, 1261–1263.
10. Hunt, D. F., Michel, H., Dickinson, T. A., Shabanowitz, J., Cox, A. L., Sakaguchi, K., Appella, E., Grey, H. M., Sette, A. (1992). Peptides presented to the immune system by the murine class II major histocompatibility complex molecule I-Ad. *Science 256*, 1817–1820.
11. Purcell, A. W., Gorman, J. J. (2004). Immunoproteomics: Mass spectrometry-based methods to study the targets of the immune response. *Mol Cell Proteomics 3*, 193–208.
12. Fenn, J. B., Mann, M., Meng, C. K., Wong, S. F., Whitehouse, C. M. (1989). Electrospray ionization for mass spectrometry of large biomolecules. *Science 246*, 64–71.
13. Karas, M., Bachmann, D., Bahr, U., Hillenkamp, F. (1987). Matrix-assisted laser desorption of non-volatile compounds. *Int J Mass Spectrom Ion Processes 78*, 53–68.
14. Costello, C. E. (1997). Time, life ... and mass spectrometry: New techniques to address biological questions. *Biophys Chem 68*, 173–188.
15. Glish, G., Busch, K. L., McLuckey, S. (1989). *Mass Spectrometry-Mass Spectrometry: Techniques and Applications of Tandem Mass Spectrometry*. Wiley-VCH, New York.
16. Hillen, N., Stevanovic, S. (2006). Contribution of mass spectrometry-based proteomics to immunology, *Expert Rev Proteomics 3*, 653–664.
17. Engelhard, V. H. (2007). The contributions of mass spectrometry to understanding of immune recognition by T lymphocytes. *Int J Mass Spectrom 259*, 32–39.
18. Hager-Braun, C., Tomer, K. B. (2005). Determination of protein-derived epitopes by mass spectrometry. *Expert Rev Proteomics 2*, 745–756.
19. Burlet-Schiltz, O., Claverol, S., Gairin, J. E., Monsarrat, B. (2005). The use of mass spectrometry to identify antigens from proteasome processing. *Methods Enzymol 405*, 264–300.
20. Downard, K. M. (2000). Contributions of mass spectrometry to structural immunology. *J Mass Spectrom 35*, 493–503.
21. de Jong, A. (1998). Contribution of mass spectrometry to contemporary immunology. *Mass Spectrom Rev 17*, 311–335.
22. Amigorena, S., Savina, A. (2010). Intracellular mechanisms of antigen cross presentation in dendritic cells. *Curr Opin Immunol 22*, 109–117.

23. Houde, M., Bertholet, S., Gagnon, E., Brunet, S., Goyette, G., Laplante, A., Princiotta, M. F., Thibault, P., Sacks, D., Desjardins, M. (2003). Phagosomes are competent organelles for antigen cross-presentation. *Nature 425*, 402–406.

24. Velazquez, C., Vidavsky, I., van der Drift, K., Gross, M. L., Unanue, E. R. (2002). Chemical identification of a low abundance lysozyme peptide family bound to I-Ak histocompatibility molecules. *J Biol Chem 277*, 42514–42522.

25. Suri, A., Vidavsky, I., van der Drift, K., Kanagawa, O., Gross, M. L., Unanue, E. R. (2002). In APCs, the autologous peptides selected by the diabetogenic I-Ag7 molecule are unique and determined by the amino acid changes in the P9 pocket. *J Immunol 168*, 1235–1243.

26. Peters, B., Sidney, J., Bourne, P., Bui, H. H., Buus, S., Doh, G., Fleri, W., Kronenberg, M., Kubo, R., Lund, O., Nemazee, D., Ponomarenko, J. V., Sathiamurthy, M., Schoenberger, S., Stewart, S., Surko, P., Way, S., Wilson, S., Sette, A. (2005). The immune epitope database and analysis resource: from vision to blueprint. *PLoS Biol 3*, e91.

27. DeFranco, A. L., Locksley, R. M., Robertson, M. (2007). *Immunity*. New Science Press, London.

28. Murphy, K. M., Travers, P., Walport, M. (2007). *Janeway's Immunobiology*, 7 ed. Garland Science, New York.

29. Sompayrac, L. M. (2008). *How the Immune System Works*, 3rd ed., Wiley-Blackwell, Hoboken, NJ.

30. Delves, P., Martin, S., Burton, D., Roitt, I. (2006). *Roitt's Essential Immunology*, 11th ed. Wiley-Blackwell, Hoboken, NJ.

31. Paulovich, A. G., Billheimer, D., Ham, A. J., Vega-Montoto, L., Rudnick, P. A., Tabb, D. L., Wang, P., Blackman, R. K., Bunk, D. M., Cardasis, H. L., Clauser, K. R., Kinsinger, C. R., Schilling, B., Tegeler, T. J., Variyath, A. M., Wang, M., Whiteaker, J. R., Zimmerman, L. J., Fenyo, D., Carr, S. A., Fisher, S. J., Gibson, B. W., Mesri, M., Neubert, T. A., Regnier, F. E., Rodriguez, H., Spiegelman, C., Stein, S. E., Tempst, P., Liebler, D. C. (2010). Interlaboratory study characterizing a yeast performance standard for benchmarking LC-MS platform performance. *Mol Cell Proteomics, 9*, 242–254.

32. Rudnick, P. A., Clauser, K. R., Kilpatrick, L. E., Tchekhovskoi, D. V., Neta, P., Blonder, N., Billheimer, D. D., Blackman, R. K., Bunk, D. M., Cardasis, H. L., Ham, A. J., Jaffe, J. D., Kinsinger, C. R., Mesri, M., Neubert, T. A., Schilling, B., Tabb, D. L., Tegeler, T. J., Vega-Montoto, L., Variyath, A. M., Wang, M., Wang, P., Whiteaker, J. R., Zimmerman, L. J., Carr, S. A., Fisher, S. J., Gibson, B. W., Paulovich, A. G., Regnier, F. E., Rodriguez, H., Spiegelman, C., Tempst, P., Liebler, D. C., Stein, S. E. (2010). Performance metrics for liquid chromatography-tandem mass spectrometry systems in proteomics analyses. *Mol Cell Proteomics, 9*, 225–241.

33. Tabb, D. L., Vega-Montoto, L., Rudnick, P. A., Variyath, A. M., Ham, A. J., Bunk, D. M., Kilpatrick, L. E., Billheimer, D. D., Blackman, R. K., Cardasis, H. L., Carr, S. A., Clauser, K. R., Jaffe, J. D., Kowalski, K. A., Neubert, T. A., Regnier, F. E., Schilling, B., Tegeler, T. J., Wang, M., Wang, P., Whiteaker, J. R., Zimmerman, L. J., Fisher, S. J., Gibson, B. W., Kinsinger, C. R., Mesri, M., Rodriguez, H., Stein, S. E., Tempst, P., Paulovich, A. G., Liebler, D. C., Spiegelman, C. (2010). Repeatability and reproducibility in proteomic identifications by liquid chromatography-tandem mass spectometry. *J Proteome Res, 9*, 761–776.

34. Rotzschke, O., Falk, K., Wallny, H. J., Faath, S., Rammensee, H. G. (1990). Characterization of naturally occurring minor histocompatibility peptides including H-4 and H-Y. *Science 249*, 283–287.

35. Rotzschke, O., Falk, K., Deres, K., Schild, H., Norda, M., Metzger, J., Jung, G., Rammensee, H. G. (1990). Isolation and analysis of naturally processed viral peptides as recognized by cytotoxic T cells. *Nature 348*, 252–254.
36. Florens, L., Washburn, M. P. (2006). Proteomic analysis by multidimensional protein identification technology. *Meth Mol Biol 328*, 159–175.
37. Chen, X., Sans, M. D., Strahler, J. R., Karnovsky, A., Ernst, S. A., Michailidis, G., Andrews, P. C., Williams, J. A. (2010). Quantitative organellar proteomics analysis of rough endoplasmic reticulum from normal and acute pancreatitis rat pancreas. *J Proteome Res 9*, 885–896.
38. Thingholm, T. E., Larsen, M. R. (2009). The use of titanium dioxide micro-columns to selectively isolate phosphopeptides from proteolytic digests, *Meth Mol Biol 527*, 57–66, xi.
39. Bruins, A. P., Koch, K. D. (2007). Electrospray ionization: Principles and instrumentation. In Gross, M. L., and Caprioli, R. M., eds., *Ionization Methods*. Elsevier, Oxford, pp 415–421.
40. Van Berkel, G. J. (2007). Electrochemistry of the electrospray ionization source. In Gross, M. L., and Caprioli, R. M., eds., *Ionization Methods*. Elsevier, Oxford, pp 422–426.
41. Thomson, B. A. (2007). Micro and nano-electrospray ionization. In Gross, M. L., and Caprioli, R. M., eds., *Ionization Methods*. Elsevier, Oxford, pp 434–444.
42. Fugmann, T., Neri, D., Roesli, C. (2010). DeepQuanTR: MALDI-MS-based label-free quantification of proteins in complex biological samples. *Proteomics 10*, 2631–2643.
43. Maccarrone, G., Turck, C. W., Martins-de-Souza, D. (2010). Shotgun mass spectrometry workflow combining IEF and LC-MALDI-TOF/TOF. *Protein J, 29*, 99–102.
44. Cox, A. L., Skipper, J., Chen, Y., Henderson, R. A., Darrow, T. L., Shabanowitz, J., Engelhard, V. H., Hunt, D. F., Slingluff, C. L., Jr., (1994). Identification of a peptide recognized by five melanoma-specific human cytotoxic T cell lines. *Science 264*, 716–719.
45. Kittlesen, D. J., Thompson, L. W., Gulden, P. H., Skipper, J. C., Colella, T. A., Shabanowitz, J., Hunt, D. F., Engelhard, V. H., Slingluff, C. L., Jr., (1998). Human melanoma patients recognize an HLA-Al-restricted CTL epitope from tyrosinase containing two cysteine residues: Implications for tumor vaccine development. *J Immunol 160*, 2099–2106.
46. Henderson, R. A., Cox, A. L., Sakaguchi, K., Appella, E., Shabanowitz, J., Hunt, D. F., Engelhard, V. H. (1993). Direct identification of an endogenous peptide recognized by multiple HLA-A2.1-specific cytotoxic T cells. *Proc Nat Acad Sci USA 90*, 10275–10279.
47. den Haan, J. M., Sherman, N. E., Blokland, E., Huczko, E., Koning, F., Drijfhout, J. W., Skipper, J., Shabanowitz, J., Hunt, D. F., Engelhard, V. H., et al. (1995). Identification of a graft versus host disease-associated human minor histocompatibility antigen. *Science 268*, 1476–1480.
48. Zarling, A. L., Luckey, C. J., Marto, J. A., White, F. M., Brame, C. J., Evans, A. M., Lehner, P. J., Cresswell, P., Shabanowitz, J., Hunt, D. F., Engelhard, V. H. (2003). Tapasin is a facilitator, not an editor, of class I MHC peptide binding. *J Immunol 171*, 5287–5295.
49. Zarling, A. L., Ficarro, S. B., White, F. M., Shabanowitz, J., Hunt, D. F., Engelhard, V. H. (2000). Phosphorylated peptides are naturally processed and presented by major histocompatibility complex class I molecules in vivo. *J Exp Med 192*, 1755–1762.
50. Skipper, J. C., Hendrickson, R. C., Gulden, P. H., Brichard, V., Van Pel, A., Chen, Y., Shabanowitz, J., Wolfel, T., Slingluff, C. L., Jr., Boon, T., Hunt, D. F., Engelhard, V. H. (1996). An HLA-A2-restricted tyrosinase antigen on melanoma cells results from

posttranslational modification and suggests a novel pathway for processing of membrane proteins. *J Exp Med 183*, 527–534.

51. Mosse, C. A., Meadows, L., Luckey, C. J., Kittlesen, D. J., Huczko, E. L., Slingluff, C. L., Shabanowitz, J., Hunt, D. F., Engelhard, V. H. (1998). The class I antigen-processing pathway for the membrane protein tyrosinase involves translation in the endoplasmic reticulum and processing in the cytosol. *J Exp Med 187*, 37–48.
52. Lovitch, S. B., Walters, J. J., Gross, M. L., Unanue, E. R. (2003). APCs present A beta(k)-derived peptides that are autoantigenic to type B T cells. *J Immunol 170*, 4155–4160.
53. Calderon, B., Suri, A., Miller, M. J., Unanue, E. R. (2008). Dendritic cells in islets of Langerhans constitutively present beta cell-derived peptides bound to their class II MHC molecules. *Proc Nat Acad Sci USA 105*, 6121–6126.
54. Suri, A., Walters, J. J., Kanagawa, O., Gross, M. L., Unanue, E. R. (2003). Specificity of peptide selection by antigen-presenting cells homozygous or heterozygous for expression of class II MHC molecules: The lack of competition. *Proc Nat Acad Sci USA 100*, 5330–5335.
55. Suri, A., Walters, J. J., Gross, M. L., Unanue, E. R. (2005). Natural peptides selected by diabetogenic DQ8 and murine I-A(g7) molecules show common sequence specificity. *J Clin Invest 115*, 2268–2276.
56. Suri, A., Walters, J. J., Rohrs, H. W., Gross, M. L., Unanue, E. R. (2008). First signature of islet beta-cell-derived naturally processed peptides selected by diabetogenic class II MHC molecules. *J Immunol 180*, 3849–3856.
57. Meyer, V. S., Drews, O., Gunder, M., Hennenlotter, J., Rammensee, H. G., Stevanovic, S. (2009). Identification of natural MHC class II presented phosphopeptides and tumor-derived MHC class I phospholigands. *J Proteome Res 8*, 3666–3674.
58. Dengjel, J., Rammensee, H. G., Stevanovic, S. (2005). Glycan side chains on naturally presented MHC class II ligands. *J Mass Spectrom 40*, 100–104.
59. Fissolo, N., Haag, S., de Graaf, K. L., Drews, O., Stevanovic, S., Rammensee, H. G., Weissert, R. (2009). Naturally presented peptides on major histocompatibility complex I and II molecules eluted from central nervous system of multiple sclerosis patients. *Mol Cell Proteomics 8*, 2090–2101.
60. Felix, N. J., Suri, A., Walters, J. J., Horvath, S., Gross, M. L., Allen, P. M. (2006). I-Ep-bound self-peptides: identification, characterization, and role in alloreactivity. *J Immunol 176*, 1062–1071.
61. Felix, N. J., Donermeyer, D. L., Horvath, S., Walters, J. J., Gross, M. L., Suri, A., Allen, P. M. (2007). Alloreactive T cells respond specifically to multiple distinct peptide-MHC complexes. *Nat Immunol 8*, 388–397.
62. Fortier, M. H., Caron, E., Hardy, M. P., Voisin, G., Lemieux, S., Perreault, C., Thibault, P. (2008). The MHC class I peptide repertoire is molded by the transcriptome. *J Exp Med 205*, 595–610.
63. Holm, A., Rise, F., Sessler, N., Sollid, L. M., Undheim, K., Fleckenstein, B. (2006). Specific modification of peptide-bound citrulline residues. *Anal Biochem 352*, 68–76.
64. Stensland, M., Holm, A., Kiehne, A., Fleckenstein, B. (2009). Targeted analysis of protein citrullination using chemical modification and tendem mass spectrometry. *Rapid Commun Mass Spectrom 23*, 2754–2762.
65. Hao, G., Wang, D., Gu, J., Shen, Q., Gross, S. S., Wang, Y. (2009). Neutral loss of isocyanic acid in peptide CID spectra: A novel diagnostic marker for mass spectrometric identification of protein citrullination. *J Am Soc Mass Spectrom 20*, 723–727.

66. Mikesh, L. M., Ueberheide, B., Chi, A., Coon, J. J., Syka, J. E., Shabanowitz, J., Hunt, D. F. (2006). The utility of ETD mass spectrometry in proteomic analysis. *Biochim Biophys Acta 1764*, 1811–1822.

67. Weinzierl, A. O., Lemmel, C., Schoor, O., Muller, M., Kruger, T., Wernet, D., Hennenlotter, J., Stenzl, A., Klingel, K., Rammensee, H. G., Stevanovic, S. (2007). Distorted relation between mRNA copy number and corresponding major histocompatibility complex ligand density on the cell surface. *Mol Cell Proteomics 6*, 102–113.

68. Schmidt, A., Gehlenborg, N., Bodenmiller, B., Mueller, L. N., Campbell, D., Mueller, M., Aebersold, R., Domon, B. (2008). An integrated, directed mass spectrometric approach for in-depth characterization of complex peptide mixtures. *Mol Cell Proteomics 7*, 2138–2150.

69. Domon, B., Aebersold, R. (2010). Options and considerations when selecting a quantitative proteomics strategy. *Nat Biotechnol 28*, 710–721.

CHAPTER 14

Neuropeptidomics

JONATHAN V. SWEEDLER, FANG XIE, and ADRIANA BORA

14.1 INTRODUCTION

Endogenous cell–cell signaling peptides (SPs), a diverse group of compounds that includes neuropeptides, hormones, and cytokines, play key roles in brain physiology. Research on SPs has been accelerating as aspects of their function become established, even for those recently characterized [1–4]. Nevertheless, the function of many native SPs remains unknown. SPs and other neuropeptides are important biomarker candidates related to a variety of health concerns, including Alzheimer's and cardiovascular disease, schizophrenia, depression, and cancer [5–11]. So that we may understand their role in physiological processes and realize their potential as psychopharmaceutical targets, comprehensive analyses of the peptide content of specific cells, groups of cells, brain structures, or even entire brains are required. Thus the analysis of neuropeptides has important implications in drug discovery.

Two recent measurement strategies, proteomics and peptidomics, have comparable objectives and employ similar techniques, but there are key differences. Proteomics, the large-scale study of proteins in a sample, often uses enzymatic digestion to generate small peptides that can be analyzed by mass spectrometry (MS). When proteomics is used as a "bottom-up" approach, the goal is to identify a protein by identifying some of its constituent peptide fragments [12]. In contrast, peptidomics is the study of the "comprehensive suite" of endogenous peptides expressed within a specific cell, organ, or organism. These peptides are analyzed without enzymatic pre-treatment. The aim of an MS-based peptidomics experiment is to detect and characterize the peptides present, including their post-translational modifications (PTMs), as a function of time and often, in a spatially or temporally defined manner.

Protein and Peptide Mass Spectrometry in Drug Discovery, Edited by Michael L. Gross, Guodong Chen, and Birendra N. Pramanik.

Peptidomics has long been a field of research, even before 2001 [13–16] when the term was introduced [17–19]. Since these initial studies, many research groups have made spectacular progress using peptidomic methods for studying SPs in a range of brain samples, from organelle [20], neuronal cell [21,22], and nucleus [23] to brain structures [24–27], entire nervous systems [28,29], and cerebrospinal fluid [30,31]. It is challenging to characterize and measure SPs in a biological sample given the variety of bioactive peptides, the large number of PTMs, their high spatial heterogeneity, and their wide concentration dynamic range. At times they are localized in small, spatially segregated structures or under specific physiological conditions, increasing the analytical challenge. This chapter focuses on the sample preparation techniques and analytical methods for detecting and characterizing endogenous SPs in the central nervous system (CNS).

14.2 NEUROPEPTIDOMICS: CHARACTERIZING PEPTIDES IN THE BRAIN

Neuropeptidomics, the measurement of peptides in CNS-relevant samples, has been used by several research groups to identify neuropeptides and to measure changes in their expression levels [23–27,32]. There are a number of inherent challenges in neuropeptidomics; these include a dynamic range of endogenous peptides (even for those from the same prohormone) that is wider than the dynamic range of mass spectrometers, the post-mortem degradation of proteins and peptides obscuring the endogenous peptides, and the difficulty of detecting or sequencing neuropeptides of low abundance. Given these factors, proper sample preparation is often crucial to a successful neuropeptidomics experiment. As a result the extraction or direct tissue preparation procedures to maximize neuropeptidome coverage and facilitate the mass spectrometric detection of novel peptides must be optimized. Given the importance of sample preparation, this procedure will be discussed first.

SPs include neuropeptides, trophic factors, and other endogenous peptides produced and secreted by populations of cells in the brain; these peptides have roles in neuromodulation, cell–cell signaling, and neuronal network formation. After SPs are synthesized and released, they can exert their actions either locally or at long distance and hence directly impact a number of physiological parameters and even organism behavior. For example, oxytocin and vasopressin, secreted by the hypothalamic magnocellular neurons in the supraoptic nuclei, directly affect many physiological systems. Discovered and chemically identified in 1953 by Du Vingneaud [33], these two peptide hormones, conserved in vertebrates, have similar chemical structures but distinct physiological activities. In the peripheral nervous system, oxytocin is associated with reproductive functions, and vasopressin with water balance in the kidney [34]. In the brain they have robust effects on social and maternal behavior, as well as on pair bonding [35].

The biosynthesis of neuropeptides starts with the production of the preprohormone in the cellular cytosol. The preprohormone contains an *N*-terminal signal peptide, typically between 15 and 30 amino acids, which guides the forming preprohormone to the rough endoplasmic reticulum where it is cleaved by a signal endopeptidase [11,22]. The complete prohormone (now without the signal peptide) is transferred to the trans-Golgi network where it is sorted and packed into large, dense core vesicles or secretory granules. Here a number of processing enzymes become active and cleave the prohormone at specific sequence motifs (e.g., basic amino acid residues) (Figure 14.1) [36]. Additional modifications, such as *C*-terminal amidation, acetylation, phosphorylation, sulfation, and glycosylation, may also occur during this process [37]. The PTMs of endogenous neuropeptides provide resistance to enzymatic degradation and regulate binding affinity to receptors, and thus directly impact bioactivity [38]. After intracellular processing, the peptides are usually secreted at the axon termini (presynaptic release), the soma or dendrites [35,39–41]. Following secretion, at times with further extracellular processing, the peptides interact with their cognate receptors. They are then degraded in the extracellular space by peptidases or even taken up by cells. As an example, neprilysin and neurolysin are essential metalloenzymes in the formation and degradation of angiotensin and endothelin, respectively [42].

Present in the brain, spinal cord, cerebrospinal fluid, gastrointestinal tract, pancreas, and blood [30,31,43–45], SPs act as neurotransmitters, neuromodulators, or hormones [36,46]. They also may have other functions (e.g., cytokine or trophic activity [47]). A single gene encodes a single prohormone; however, the prohormone often encodes multiple bioactive SPs (Figure 14.1). For example, multiple peptides processed from the proopiomelanocortin hormone have distinct functions: adrenocorticotropic hormone stimulates the adrenal glands to release cortisol; melanocyte-stimulating hormone increases the pigmentation of skin by increasing melanin production in melanocytes; and beta-endorphin and Met-enkephalin are endogenous opiates with a broad spectrum of functions [36].

It is not necessarily a given, however, that all peptides derived from a prohormone will be bioactive. Peptides may be expressed in a tissue- and/or time-specific manner. A number of peptides are present in neurons at high levels under normal conditions and some remain at low or undetectable levels; others are present only during the early phase of development and then are significantly down-regulated after birth. For example, on the one hand, substance P in primary sensory neurons, galanin in hypothalamic neurons, and neuropeptide Y and vasoactive intestinal peptide in cortical neurons are present at high levels; on the other hand, these peptides are present at low levels in sensory neurons under normal conditions. Certain peptides are only transiently expressed, for example, during early development; such peptides include somatostatin, which is expressed widely in the CNS, substance P in spinal guiding neurons, and galanin in primary sensory neurons [37].

14.3 SAMPLE PREPARATION FOR MASS SPECTROMETRY

Several analytical techniques offer a number of performance specifications that make them well suited for peptide discovery and characterization. Specifically, the fast analysis speed, low detection limits, and excellent mass accuracy of current

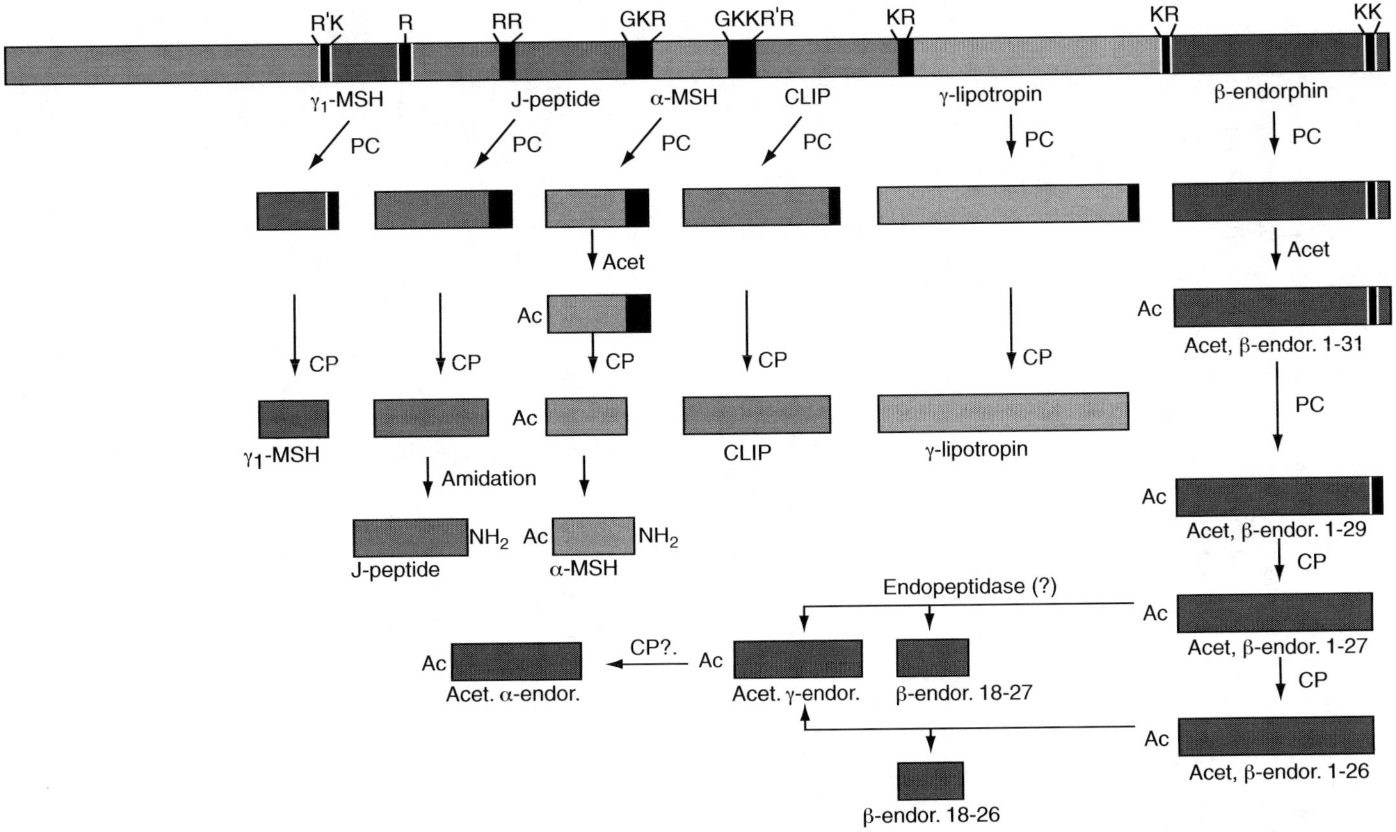

FIGURE 14.1 Processing of proopiomelanocortin (POMC). The top line shows the general organization of various peptides within POMC. Black rectangles represent the cleavage sites that contain basic amino acids. In most cases groups of basic amino acids are cleaved to the *C*-terminus of the group, although there are two exceptions, as indicated by the arrowheads. Abbreviations: Acet., acetyl or acetylation; CP, carboxypeptidase E and/or D; CP?, carboxypeptidase with specificity for *C*-terminal Leu (not CPE or CPD); endor., endorphin. Reproduced with permission from John Wiley & Sons, Ltd. [32].

mass spectrometers are particularly useful for peptide characterization. As was stated earlier, sample preparation also plays an important role in the success of an SP characterization. The objectives of the sample preparation process are to recover as large a fraction of the SP complement as possible, to minimize interference from protein degradation, and to match the appropriate separation/mass analyzer specifications to the analyte of interest. Presented here are various sample handling methodologies that lead to improved detection and identification of SPs from neural tissues using direct MS, as well as separations hyphenated to MS.

14.3.1 Direct Tissue Profiling

A variety of analytes in a wide range of biological samples can be measured directly by matrix-assisted laser desorption/ionization (MALDI) time-of-flight (TOF) MS. This method provides high sensitivity and enables the analysis of small specimens such as individual cellular organelles [20,48], single neurons [23,49,50], single nuclei [51], single-cell-sized tissue samples [52], and brain slices [51]. Sample preparation consists of three steps: isolating the tissue, mounting on the MALDI target, and introducing the MALDI matrix [51]. Often sample preparation for direct-tissue MALDI involves obtaining fresh frozen sections with a cryostat [51]. After mounting, the sample surface is usually sprayed with an acidic matrix solution, allowing analytes to be extracted locally and incorporated into matrix crystals [53]. A direct profiling approach for investigating neuropeptides may be useful [52]; a thin tissue section is placed on an array of 38-μm glass beads attached to an elastic, hydrophobic membrane (Figure 14.2). The membrane is stretched so that the tissue can be divided into single-cell-sized pieces. This method of direct tissue analysis will be described further in Section 14.3.4.

MALDI MS is an efficient tool for investigating the neuropeptidome of single cells in invertebrates [22,49] and vertebrates [50]. Sample preparation starts with selecting the cell(s) of interest. To study peptides from the invertebrate nervous system, one can use manual isolation of single cells or cell clusters; manual isolation is the easiest and fastest method (e.g., requires less than 1 min) [48]. Single-cell MALDI can be applied to mammalian cells [50] and hypothalamic neurons [23]. Once isolated, the cells are plated on a MALDI target and treated with the appropriate sized-matched volumes of matrix solution. One should also consider laser-capture microdissection, which is used to isolate cells from thin mammalian tissue sections [54].

Protocols [50,55] can be developed to analyze the peptide content of individual mammalian cells with sizes under 10 μm by employing micromanipulation to transfer the isolated cells onto a MALDI target and removing the extracellular solution to reduce the chemical background. Prior to this cell-transfer step, glycerol is used for stabilization to maintain cell morphology and prevent neuropeptide redistribution. Guided microdeposition of the MALDI matrix is the last step to create the optimal analyte-to-matrix ratio, thereby minimizing peptide dilution. These sampling protocols enable investigation of neurobiological systems with cells in the range of 8 to 100 μm and with analyte amounts at the amol level.

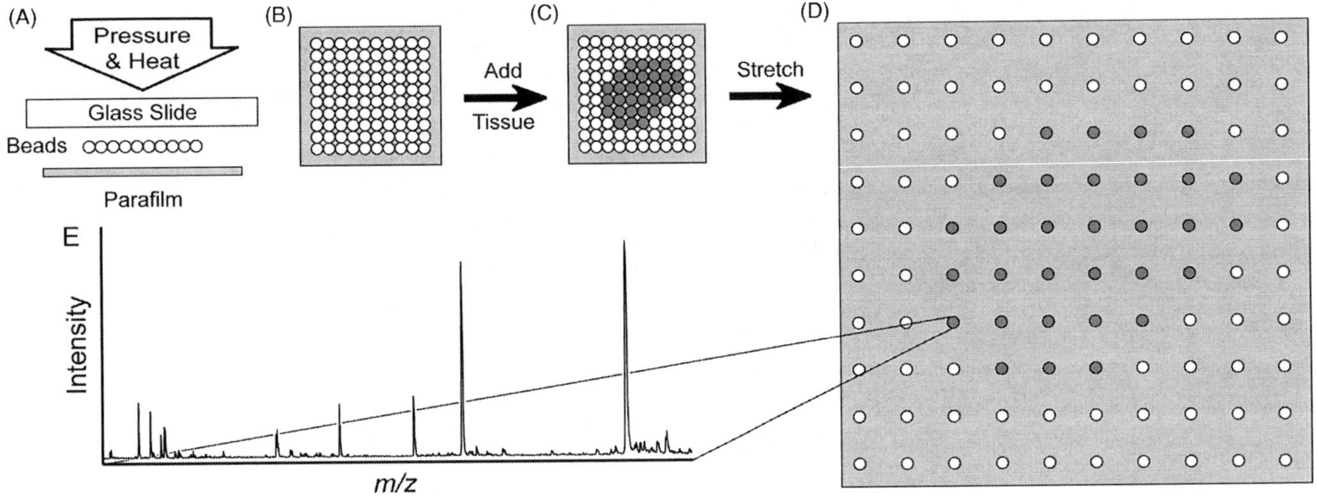

FIGURE 14.2 Schematic of the stretched sample preparation method for MS imaging. (*A*, *B*) Pressure and heat are used to form a layer of glass beads on a Parafilm M membrane surface. (*C*) A thin tissue slice is placed onto the glass bead layer. (*D*) The Parafilm M membrane is manually stretched. As a result the tissue slice is fragmented into thousands of spatially isolated pieces. (*E*) After MALDI matrix application, individual pieces of tissue may then be investigated with MALDI MS. Reproduced with permission from the American Chemical Society [52].

Direct-tissue profiling can also be achieved by applying mass spectrometry imaging (MSI) to the identification and localization of peptides in a sample. A typical preparation procedure includes mounting a section of frozen tissue (e.g., 10 μm thick) on a stainless steel target plate, coating it with a matrix solution, and drying it prior to MALDI-TOF MS analysis [56]. MSI can be used to study neuropeptides and proteins from well-defined regions of mouse brain and human brain tumor sections [56], and to locate specific compounds that are highly expressed in a treated or diseased sample relative to normal tissue [54,56]. This technique will also be discussed in more detail below.

14.3.2 Extraction-Based Strategies

Often the first step in an experimental protocol is to extract the peptides from the sample of interest into the appropriate solution. Various extraction-based protocols for post-mortem biological specimens are available, and their overall goal is to preserve the peptides as close as possible to their original state. Post-mortem degradation becomes a critical issue in neuropeptidomics owing to proteolytic enzyme activity that can rapidly degrade SPs in sample tissues. Deactivating these enzymes by thermal treatment appears effective. A variety of procedures, including focused microwave irradiation (before or after freezing) [57,58], rapid microwave irradiation of the entire brain [24], boiling the tissue immediately after decapitation [23], or a combination of snap-freezing, cryostat dissection, and boiling extraction buffer [25], improve the recovery of SPs for mass spectrometric characterization.

After protease deactivation, the next step is to extract the peptides. Fortunately, we now have a good understanding of various techniques that enhance the reproducible recovery of peptides and minimize their detection from protein degradation fragments [23,25,59,60]. Acidified organic extraction buffers, including acidified acetone, ethanol, and methanol, can be successfully employed to extract peptides from biological samples and deactivate proteases via denaturation and precipitation, and acidified extraction has proved effective for invertebrate samples [28]. For samples from mammals, dilute acetic acid affords the highest yield among the various extraction buffers evaluated (e.g., dilute acetic acid, acidified methanol, and sodium dodecyl sulfate with KCl) [25]. Samples treated with hot or boiling water followed by cold acidic extraction provide reproducible recoveries of many hypothalamic peptides, including a few novel compounds [23,25,59]. By combining techniques, such as sonication with detergent treatment [61], one may achieve a threefold increase in neuropeptide identifications. Subjecting samples to sonication and heated water at 70°C for 20 min followed by cold acid enables the efficient extraction of many neuropeptides without forming the protein degradation fragments seen with hot acid extractions.

A less complex extraction protocol, developed by Romanova et al. [62] for both invertebrate and vertebrate samples, also appears to be effective. Tissues are stored in saturated 2,5-dihydroxybenzoic acid (DHB) solution, leading to effective extraction of endogenous peptides and long-term preservation of tissue samples without freezing. Additional advantages of this protocol are its technical simplicity, reproducibility, and ease of remote sample preparation. Another novel extraction protocol [63] makes use of an 8-M urea solution, a well-known cell lysate factor and protein denaturant (Figure 14.3). A higher number of putative novel bioactive peptides can be characterized in a concentrated urea extract than in an acidic extract. Urea may solubilize the tissue more efficiently than does an acidic solution, liberating and dissolving the peptide content. An alternative method for sample preparation without microwave fixation is to select and isolate cellular subcompartments that have high levels of SPs (e.g., synaptosomes or even dense core vesicles).

When analyzing complex systems (e.g., the mammalian brain), samples of relatively large morphological structures are preferred, such as an entire hypothalamus, striatum, or hippocampus [25,64,65]. For MS analyses, obtaining adequate analyte amounts typically requires these large sample sizes. Fortunately, optimized sample-handling protocols are available to encompass a range of sample sizes, including single nuclei isolated from ex vivo brain slices via tissue punch [23]. Unlike the majority of previous mammalian peptidomic studies, these samples do not require immediate heating or microwaving. The preparation of brain slices is a well-established technique; the slices remain functional for more than 24 h ex vivo [66,67], allowing the recovery of altered neurosecretory vesicles and other cell organelles [68]. Furthermore the isolated tissue punches from freshly prepared hypothalamic slices are processed through a multistage peptide extraction protocol, including boiling as a primary extraction step and deactivation of proteolytic enzymes, followed by a two-stage acidic extraction solution of acidified acetone and acetic acid [23]. This multistage extraction, where the boiling water is followed by an organic and a slightly acidic solution, increases the recovery of the peptidome from a small brain region. Advantages of the method include the ability to combine functional tests of peptide release with peptidome measurement, and to delineate clearly small and anatomically well-defined brain structures, such as individual brain nuclei from fresh (non–heat-treated) samples [23].

As noted above, direct-tissue profiling with MS enables a number of unique studies, such as the classification of individual mammalian cells by peptide profiling, the localization of different compounds within a tissue, the elucidation of cell-specific prohormone processing (including PTMs of peptides), the study of neuronal transport by measuring peptides below the level of a single cell, and the discovery of novel signaling peptides on a cell-to-cell basis. However, extraction-based strategies, when combined with the appropriate separation and MS techniques, are essential methods for qualitative and quantitative measurement of the brain peptide complement.

14.3.3 Collecting Peptide Release

The release of endogenous neuropeptides represents a fundamental and dynamic component of cell–cell signaling within the brain. Neuropeptides are secreted locally at moderate concentrations but are quickly diluted after release. Both direct-tissue profiling and analysis of tissue extracts are methods that generate valuable information about the content of peptides and their locations in a biological sample. Although we are able to detect and identify the fully processed peptides and prohormone

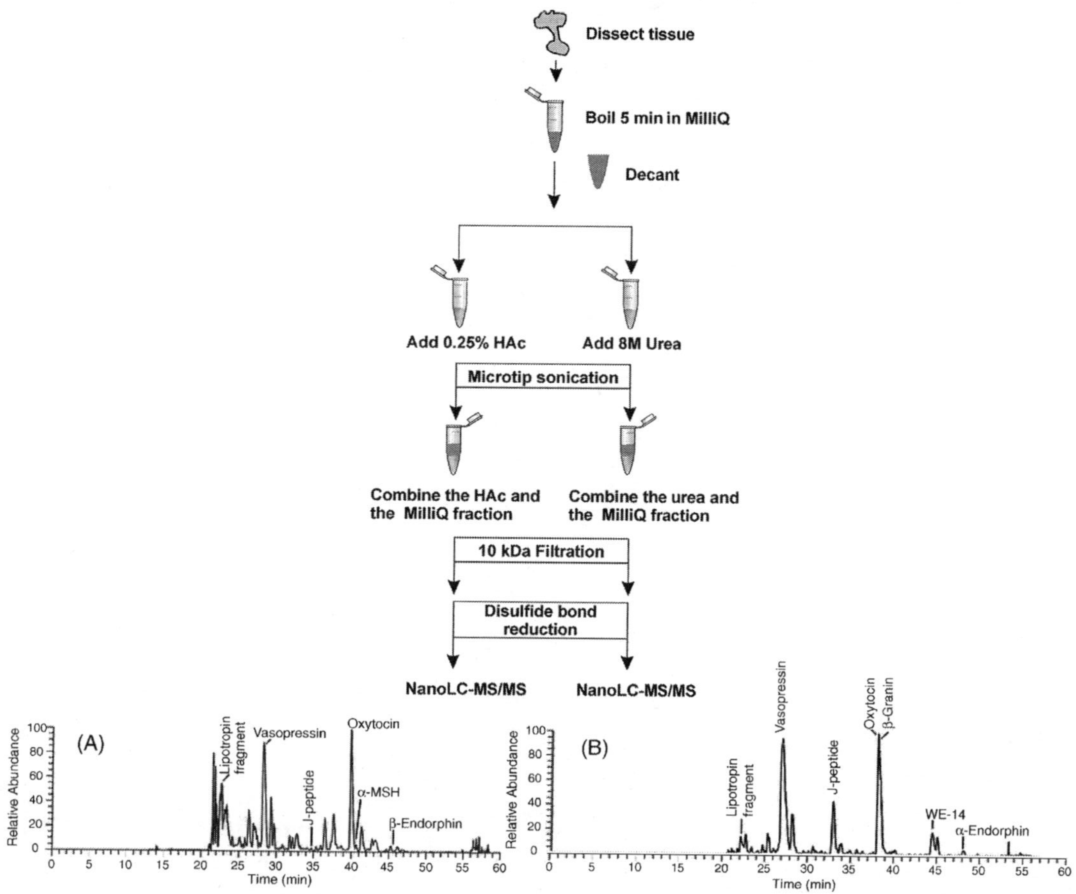

FIGURE 14.3 Overview of an extraction-based sample preparation approach used to analyze and identify peptides from murine pituitary tissue; total ion chromatograms of the extracted endogenous peptides from murine pituitary tissue using (*A*) HAc extraction and (*B*) urea extraction. The identity of some of the more abundant peptides, as identified by using LC MS/MS, is indicated. Adapted with permission from the American Chemical Society [63].

processing intermediates, these procedures do not provide direct information on their biological function. Peptides derived from a single prohormone may have diverse bioactivities, but only a small number of them may be involved in cell–cell signaling. Those that are involved in cellular communication are released in the extracellular space at different neuronal sites [39–41]. A number of approaches can be employed to monitor activity-dependent peptide release from the CNS.

Solid-phase extraction (SPE) collection, combined with electrophysiology and the appropriate separation and mass spectrometric techniques, may be successful in generating functional insights regarding peptides [1,69,70]. In addition to using micrometer-sized SPE beads as region-specific sampling probes [69,71,72], one should consider other methods to collect and concentrate releasate directly from hypothalamic brain slices by using SPE material embedded within pipettes [1] or in porous monoliths fabricated in capillaries [70] (Figure 14.4). In both cases the releasates can be eluted from the SPE material and characterized offline with MALDI-TOF MS. This approach to peptide collection using individual bead probes

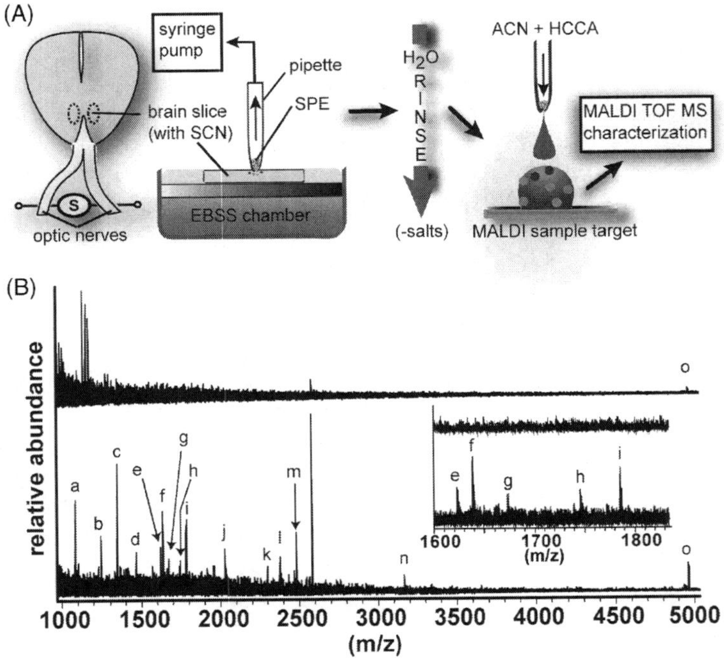

FIGURE 14.4 Schematic workflow of releasate collection and characterization. (*A*) Sample collection and preparation. (*B*) MS characterization of SCN releasates. Inset shows a zoomed mass range highlighting compounds not clearly observable in the expanded spectrum. Labeled analytes are as follows: (*a*) AVP, (*b*) proSomatostatin 89–100, (*c*) substance P, (*d*) PENK 219–229, (*e*) melanotropin α, (*f*) somatostatin 14, (*g*) pyro-glu neurotensin, (*h*) big LEN, (*i*) little SAAS, (*j*) unknown 2028.02 *m/z*, (*k*) PEN, (*l*) unknown 2380.10 *m/z*, (*m*) unknowns 2481.26/2481.77 *m/z*, (*n*) galanin, and (*o*) thymosin β-4. Reproduced with permission from the National Academy of Sciences [1]. (*See the color version of this figure in Color Plates section.*)

has a number of advantages. Not only can the beads be placed separately at discrete locations, they can also concentrate compounds present in the immediate extracellular region and provide a method for desalting, which is essential for quality MALDI MS. SPE material placed directly on intact nervous tissue before and after stimulation using chemical or electrical depolarization minimizes sample loss and facilitates removal of salts that unevenly suppress peaks in the mass spectra [1]. Spatially resolved neuropeptide sampling combined with electrophysiology and MS can measure site-specific peptide releasates within the neurohormonal system and single neurons [69,71]; detected peptides are then sequenced by tandem mass spectrometry (MS/MS). Following this strategy for mammalian models, ex vivo hypothalamic rat brain slices or in vitro rat pituitary, releasates can be screened spatially and temporally for multiple compounds [1,69]. The same assay can be used for functional testing of specific molecules. This method demonstrates the feasibility of collecting peptides from discrete locations of biological tissues and extracellular media in an activity-dependent manner. It also allows a close interface between biological release and SPE collection, thereby decreasing sample dilution resulting from analyte diffusion and degradation in the extracellular space [69].

Perhaps the most common approach for monitoring the release and metabolism of the brain is in vivo microdialysis. A semipermeable membrane is implanted in discrete brain regions of non-anesthetized, freely moving animals. Because there is a concentration gradient, the analytes present in the extracellular space diffuse into the probe and are collected into fractions for chemical analysis [73,74]. Neuropeptide measurement with microdialysis is usually performed with radioimmunoassay, a highly sensitive technique; however, it does require analyte preselection [75,76]. Owing to sample complexity and low concentration of peptides, the collected material is often separated by using nano-liquid chromatography (LC)/capillary electrophoresis (CE) followed by sensitive MS detection [73,74,77,78]. An important advantage of microdialysis over those of other methods is the ability to use living and behaving animals. The use of small-volume separations hyphenated to MS allows higher temporal and spatial resolution, affords higher sensitivity and selectivity, provides multi-analyte monitoring, and identifies peptide sequences along with their PTMs.

Compared to direct tissue profiling and extraction-based methods, measuring brain releasates with MS can also shed light on the putative functions of peptides because context-specific release patterns are obtained. These methods can have significant success in peptide detection and identification, but do keep in mind that each assay must be optimized for the specific analytes.

14.3.4 Sample Preparation for MSI

An important requirement for MSI sample preparation is to preserve the native spatial distribution of analytes within the specimen. A general sample preparation protocol for MALDI MSI involves freezing or drying the samples immediately after collection, slicing them into thin sections (typically 5–20 μm) on a cryomicrotome just before analysis, and then transferring the sections onto a MALDI target plate. Limited fixation, via either thaw-mounting or chemical approaches, is sometimes used.

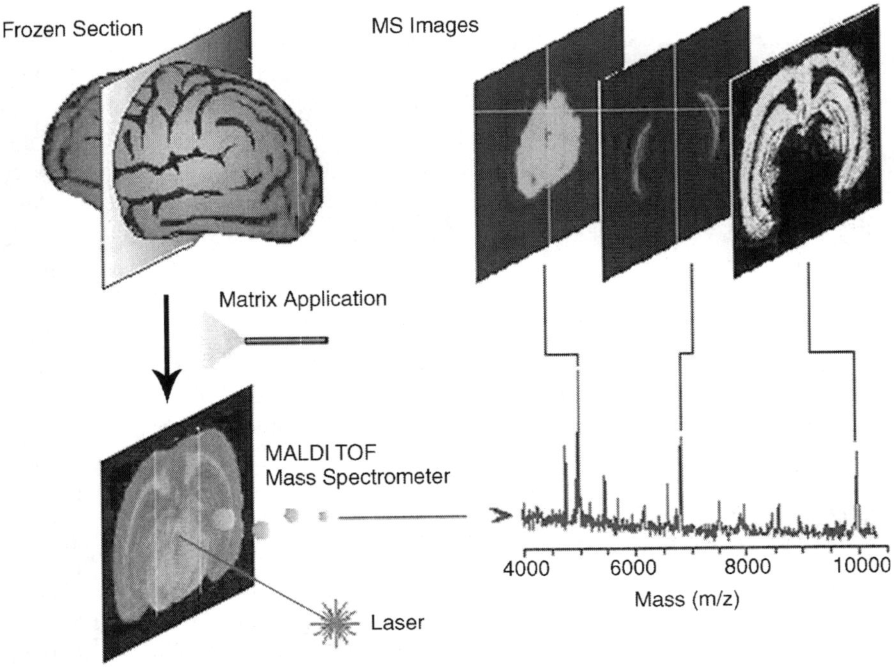

FIGURE 14.5 Schematic workflow of the spatial analysis by MALDI TOF mass spectrometry. A cell, tissue, or tissue slice is mounted on a metal plate, coated with an UV-absorbing matrix and placed in the mass spectrometer. A pulsed UV laser desorbs and ionizes analytes from the tissue and their *m/z* values are determined using a TOF analyzer. From a raster over the sample and measurement of the peak intensities over thousands of spots, mass spectrometric images are generated at specific molecular weight values. Reproduced with permission from Nature Publishing Group [56]. (*See the color version of this figure in Color Plates section.*)

Next sample sections are coated with a MALDI matrix solution; analytes are extracted and incorporated into the matrix as it crystallizes (Figure 14.5).

Choosing the appropriate matrix depends on the mass range and the analytes of interest. The commonly used matrix, α-cyano-4-hydroxycinnamic acid, works well for compounds under 3000 Da (e.g., neuropeptides). DHB is less used in MSI because it tends to form large heterogeneous crystals, decreasing resolution. It is commonly used in neuropeptide profiling experiments, however, because such profiling is a form of low spatial resolution mapping. Tissues are surrounded by physiological saline, in which case a washing step may be helpful before matrix deposition to ensure formation of uniform and small-sized crystals. Typically a brief 70% ethanol wash eliminates both salts and debris, followed by a subsequent 90% to 100% ethanol wash to complete tissue dehydration and temporarily fix the tissue [79]. The addition of glycerol can also be used for this purpose [48]. The matrix application step is critical for MSI analysis, and a balance must be achieved between extraction efficiency and spatial resolution. Specifically, the longer a sample is exposed to the matrix solution, the greater the extraction of analytes and the better the sensitivity. Longer exposure to the matrix solution, however, causes analytes to migrate further from their original positions, leading to decreased spatial resolution and blurred images. This is especially true for neuropeptides because they are small, hydrophilic analytes [48].

To address issues related to peptide migration versus analyte extraction, one can use a unique stretched sample method [52] (Figure 14.2). A thin tissue section is adhered to an array of glass beads that are mechanically attached to a Parafilm M membrane. The membrane is stretched, causing the tissue to be divided rapidly into thousands of cell-sized pieces, and the MALDI matrix is applied. The physical separation between the beads, along with the hydrophobic nature of the membrane, allows peptides to be extracted from the tissue into the matrix, without any horizontal redistribution.

Another way to minimize analyte migration is to optimize the MALDI matrix deposition technique, with the goal of delivering small droplets of matrix solution on the tissue to form a homogeneous layer of crystals. A variety of approaches are available including electrospray deposition [80], automated acoustic printing [81], and nebulizer spray coating [82]. The matrix can also be seeded on tissues before deposition, either by adding it to the washing solutions (e.g., ethanol), applying during a rinsing procedure [83] or, if the solid matrix can be mechanically ground into a fine power, spreading it across the tissue section [84]. Seeding results in greater crystal homogeneity, and thus the MS signals are more stable and uniform across the tissue sample. As for the choice of seeding matrix, sinapinic acid minimizes spatial redistribution [80]. An alternative deposition approach combines matrix seeding and high-density micro-spotting; a sprayed low-concentration matrix seed layer is followed by inkjet printing of a high-concentration matrix solution [85].

14.4 SEPARATIONS

In many cases sample fractionation is required prior to MS analysis owing to the complexity and wide dynamic range of the peptides present in tissue extracts. Although 2D gel electrophoresis is a standard method in proteomics [86,87], it is not as common in peptidomic analyses because most gels are optimized for analytes larger than 10 kDa [88]. At present, reversed-phase (RP) capillary LC is more common for separating endogenous peptides, especially those of low abundance [89,90]. Typically a nanoflow LC column (e.g., 75-μm inner diameter [i.d.]) is connected to a pre-column with low backpressure to achieve time-efficient loading via a column-switching system [90,91]. Moreover this functions as a desalting mechanism because the undesired, less-retained components (e.g., the small inorganic ions) of the sample matrix are driven to waste during sample loading. This approach can be utilized in peptidomic analyses, allowing numerous peptides to be identified efficiently in a single chromatographic run [92]. Even smaller columns (e.g., 25-μm i.d.) can be used [93]. For neuropeptidomic studies, capillary LC is typically either online coupled to electrospray ionization (ESI) MS, or offline coupled to MALDI MS (more details in Section 14.5.1) [94,95]. RP chromatography is well suited for peptide separations and easily interfaced to ESI MS. In the nanoliter flow rate range, the electrospray process becomes a more efficient ionization method [53]; the mobile phase additives are often chosen to achieve the best compromise between separation resolution and detection sensitivity. (The effect of different ion-pairing reagents on the MS signal intensities in peptide analysis was reviewed elsewhere [96].) At present, RPLC that employs a water-acetonitrile gradient containing trifluoroacetic acid of low concentration (e.g., 0.05–0.1%) is the most common choice for neuropeptidomic analyses [96,97].

Although one-dimensional capillary LC offers high separation efficiency, its peak capacity is insufficient for the complexity of many CNS samples. Thus multidimensional chromatography, which couples two or more LC columns with different separation mechanisms, is often used to further reduce sample complexity [98,99]. In neuropeptidomics, a common multidimensional LC approach is strong cation exchange (SCX) chromatography followed by RP chromatography [100]. This combination can be realized either online or offline. In the typical online scheme, peptides are trapped on an SCX column and eluted stepwise to an RP column by using salt fractions of increasing concentration. For each set of peptides, an RP linear gradient is used to elute them from the RP column. A subsequent salt fraction is then applied to elute the next set of peptides from the SCX column. The online approach allows automated analysis with minimal sample loss [101,102]. For protein identification, one can use a dual-phase column [103]. This column is packed with SCX beads followed by RP beads and functions essentially as an online 2D LC separation. In the offline approach, peptides are eluted from the SCX column with a linear salt gradient. A series of fractions are collected and then subjected to RP HPLC analysis. Larger sample quantities can be analyzed by using this scheme because a wider bore SCX column can be used as the first dimension [102]. The applications of SCX-RP in peptidomic studies of *Drosophila melanogaster* [101] and *Caenorhabditis elegans* [104] are successful examples of the enhanced detection of low-abundance peptides. Because many endogenous peptides are doubly or triply charged in solution, the majority of them elute on the SCX column within few and relatively narrow time windows [105].

Other 2D LC schemes, including size exclusion chromatography-RP [106–108], affinity chromatography-RP [64,109], RP-RP [25,110,111], and hydrophilic interaction chromatography (HILIC)-RP [100,112,113], can be utilized for endogenous peptide analysis; more details are available in a review by Gilar et al. [114] on the orthogonality of these different combinations. RP usually serves as the second dimension because it has high resolving power, and its desalting capability is compatible with an ESI MS interface. For example, Dowell et al. [25] reported that 2D RP-RP LC (different pH conditions in the first and second dimensions) increased successfully the neuropeptidome coverage of rat brain. 2D HILIC-RP LC can also be employed to study rat hypothalamus and rhinencephalon samples, and affords approximately three times more peaks as compared to the traditional 2D SCX-RP LC system [113]. This improvement results from the enhanced fractionation capability and the increased orthogonality of the HILIC combination. Affinity chromatography is also useful for the selection of subsets of peptides; for example, an anhydrotrypsin column can be used to purify neuropeptides with a *C*-terminal basic amino acid extension [64]. Because neuropeptidomics aims to identify as complete a set as possible of the naturally occurring peptides present in a cell, organ, or organism, this technique is less applicable for peptidomics [53]. An online 3D microcapillary LC system, RP1-SCX-RP2, may be an alternative because it can improve the peak capacity and resolution for peptide separation [115] and be useful for proteomic and neuropeptidomic studies.

As an alternative to capillary LC, CE can also be coupled to MS detection for peptide studies, using both online or offline arrangements. CE requires low analyte amounts, making it optimal for analyzing peptides, especially those from small cells. It provides some additional advantages, including high efficiency, fast analysis time, low reagent consumption, and great versatility [116]. Three main CE separation modes—capillary zone electrophoresis (CZE) [117], capillary electrochromatography (CEC) [118], and capillary isoelectric focusing (CIEF) [119])—can be coupled with a variety of mass spectrometric approaches to study peptides [116]. Of these, CZE is frequently used; in this mode, peptides are separated based on their electrophoretic migration rate, which depends on both their charge and size [120]. CEC, a hybrid of HPLC and CE, separates peptides via HPLC mechanisms, using electroosmotic flow as the driving force [116]. In CIEF, peptides are separated based on their pI values in a pH gradient formed by carrier ampholytes under an electric voltage [116]. The various properties of the peptides being analyzed, such as charge, size, conformation, hydrophobicity, solubility, and chemical stability, are all factors to consider when selecting the appropriate CE separation mode [121].

14.5 PEPTIDE CHARACTERIZATION VIA MASS SPECTROMETRY

Just as there are numerous sample fractionation techniques, there are a number of mass analyzers available, each with considerably different figures of merit. There is no one ideal instrumentation platform for peptide investigation. In what follows, several of the more common approaches for peptide characterization and measurement are described and categorized based on their applicability for qualitative or quantitative measurements. We also include a discussion of those approaches that either characterize tissues directly or provide spatial and temporal information.

14.5.1 Qualitative Analyses

As was noted above, it is often necessary to reduce the complexity of biological samples by using one or multiple stages of separation prior to MS analysis. In neuropeptidomic studies the separation method is usually coupled either online or offline with MS. A successful combination of soft ionization with improved mass analysis allows the detection and identification of peptides in the attomole range. The MS technologies typically used in neuropeptide discovery will now be considered.

Direct Analysis MALDI MS is commonly used for direct analysis of the peptidome of a cell organelle, a single cell, a specific tissue, or an organ. A sample is placed on a MALDI target, mixed with the appropriate organic matrix, and ionized in a mass spectrometer, generating mass profiles. MALDI is highly sensitive and relatively tolerant to salts and detergents. It is "soft" ionization that generates singly charged ions, facilitating the analysis of complex samples. In many commercial systems, the laser is focused to $> 10\,\mu m$, providing an effective tool for profiling cellular samples, and it can work with samples such as 1 to 2-μm-diameter dense core vesicles [20], 8-μm mammalian cells [23,50], single-cell-sized tissue samples [52], and cellular nuclei [51]. Although sample contaminants affect the sensitivity, mass accuracy, resolving power, and data reproducibility, MALDI allows the detection of known and unknown peptides without preselection or tagging.

The quadrupole ion trap (QIT), TOF, and Fourier-transform ion cyclotron resonance (FTICR) mass analyzers are widely used for peptide identification; they have different merits of mass accuracy, mass-resolving power, and fragmentation capabilities. One straightforward setup is a MALDI-TOF mass spectrometer, which can be used successfully for direct tissue profiling [48,50,52,54]. Fragmentation techniques such as collision-induced dissociation (CID) and post-source decay (PSD) are needed for peptide fingerprinting. As examples, both PSD and CID approaches are useful when profiling single molluskan neurons [22,122] whereas CID fragmentation with an in-cell accumulation technique is helpful for the identification of crustacean neuropeptides [123,124].

Secondary ion mass spectrometry (SIMS) has long been used in the bioanalytical investigation of cells and tissues. It is well suited for the direct analyses of elements, metabolites, lipids, pharmacological agents, and other low-molecular-weight analytes. Because of its limited high mass range, SIMS has thus far seen modest use in peptidomics, although it is capable of studying several physiologically relevant peptides when taking advantage of the advances in sample preparation and instrumentation (discussed in detail in Section 14.3). In SIMS a focused primary ion beam bombards the sample surface, and the ejected secondary ions are collected and analyzed. During primary impact, however, surface charging and analyte fragmentation can take place. One approach for resolving this issue during sample preparation is via surface metallization, which prevents charging and "softens" the primary ion impact by forming a protective shell over the surface. Another method is to coat tissues with a MALDI matrix in a similar manner to that used in MALDI MSI, so-called matrix-enhanced SIMS [125]. This technique can greatly increase the mass range (up to the 1-kDa range) and sensitivity of SIMS imaging, but the experimental spatial resolution becomes compromised and limited to 2 to 3 μm because the matrix crystal dimensions are finite [126]. As an example, the subcellular distribution of the neuropeptide APGWamide in sections of *Lymnaea stagnalis* cerebral ganglia, covered with a layer of DHB and investigated using a mass spectrometer with an $^{115}In^{+}$ liquid metal ion gun, can be visualized with about 3-μm lateral resolution.

Hyphenated Analysis Hyphenated techniques that combine separations and MS are used for analyzing complex biological tissue extracts. A key step is sample ionization; ESI has a soft ionization mechanism that usually requires the analyte to be in a solvent. The sample is dissolved in the selected solvent (volatile acids, bases, or buffers) and injected into the mass spectrometer through a small, high-voltage capillary. A noble gas or nitrogen is used to assist the production of solvent droplets. As the solvent evaporates, multiply charged ions are created by protonation or deprotonation. ESI, which can be combined with separation methods such as LC or CE, is compatible with multiple mass analyzers, including QIT, FTICR, and TOF. ESI MS provides high reproducibility because there is no need for crystallization, but it is not often used for molecular imaging.

Liquid Chromatography–Mass Spectrometry Various sampling methods, such as microdialysis, solid phase or solvent extraction, and separations, are widely used in combination with ESI or MALDI. Setups that combine a separation method (e.g., LC) with MS detection enable high-throughput analyses of peptides as well as other molecules in biological samples. Having described multiple separation methods in Section 14.4, we highlight next the coupling of LC with different mass spectrometers.

As previously mentioned, smaller flow rates improve sensitivity; micro-/nano-HPLC and MS systems are frequently interfaced by micro-/nano-ESI [53,78,90]. An online-coupled LC-ESI MS setup can analyze the eluting peptides directly. MALDI is a complementary ionization technique to ESI, and offline-coupled LC-MALDI-TOF MS is an alternative approach [23,27]. Here the LC eluent is spotted on a MALDI target plate and mixed with the appropriate matrix for detection via MS. Nano-HPLC (e.g., 50-μm i.d.) can be coupled offline with MALDI-TOF/TOF MS and employed for the identification and quantification of neuropeptides. An example is microwave-fixed rat brain tissue [27], where multiple compounds ($\sim$1 pM) can be detected and 10 peptides, including 3 novel peptides, were identified. Furthermore quantification of substance P revealed 6.8-fold higher levels than previously reported in the rat striatum [27]. One can use in vivo microdialysis followed by high sensitivity nano-flow capillary LC/microelectrospray MS to quantify and compare extracellular peptides from normal and KCl-stimulated discrete regions of the brain [73]. Endogenous Met-enkephalin [127] and neurotensin [73] can be measured in 10 μL microdialysate fractions with an ESI-quadropole (Q) mass spectrometer. Using an in vivo microdialysis system, consisting of fused-silica nano-capillary LC columns (25-μm i.d.) and integrated electrospray emitters (3-μm i.d.) interfaced to a QIT mass spectrometer, one can achieve high-sensitivity LC-MS/MS to monitor endogenous peptides at amol levels [93]. Specifically, the analysis of endogenous Met-enkephalin and Leu-enkephalin can be achieved by using nano-HPLC (25-μm i.d. column)-ESI-MS/MS in dialysates from the globus pallidus of rat brain; from injection of 1.8 μL of sample, detection at the 2 pM ($\sim$4 amol) level is achievable. These approaches permit the identification of multiple known and new putative bioactive neuropeptides [93].

A number of other innovative LC/MS interface designs are available. A micro-HPLC (25-μm i.d. column)-ESI-Q-TOF MS system, in combination with MALDI-TOF MS analysis, can be used, for example, to identify a novel neuropeptide in honey bee brain [128]. Neuropeptide FF-related peptides (NPFF) in rat and mouse spinal cord [129] can be identified with a micro-HPLC (300-μm i.d. column)-ESI MS system. In human, NPFF peptides can be identified in cerebrospinal fluid and characterized by nano-HPLC (75-μm i.d. column)-ESI-ion trap (IT) MS [130].

Another example utilizes a novel dissection protocol and an enhanced extraction approach, coupled with 2D RP-RP LC-ESI-QTOF MS/MS or MALDI-FTICR MS. This approach identified as many as 56 peptides from known prohormones and characterized 17 new peptides [25]. The high mass-resolving power and mass accuracy of the MALDI-FTICR MS and 2D LC-ESI-MS/MS approaches enable improved coverage of the rat brain peptidome. In another example, Altelaar et al. [63], using a similar approach, analyzed extracted peptide mixtures with high-resolution nanoLC-MS/MS and CID on an LTQ-Orbitrap™, and nanoLC electron transfer dissociation on a linear ion trap. This study characterized 147 peptides from mouse pituitary tissue; 48 were uniquely identified in urea extract and 11 novel peptides in acetic acid. These results demonstrate that coupling HPLC with MS provides a versatile and powerful toolset for the detection, identification, and quantitation of neuropeptides in brain tissues [58,73,74].

Capillary Electrophoresis–Mass Spectrometry To take advantage of CE-MS for peptidomic analyses, the limited sample loading volume of CE, caused by the small inner diameter of CE separation capillaries, must be addressed or the inherent sensitivity of MS is compromised. This is more of an issue when using a nanospray sheath flow interface because analytes are diluted by the sheath liquid (details about interfaces will be discussed later in this chapter). Sample preconcentration has proved useful in addressing these challenges, and several tested approaches can be utilized [131–135]. One interesting solution uses a microcartridge containing an SPE sorbent or a membrane impregnated with different chromatographic stationary phases at the inlet end of the separation capillary [136,137]. In another example, a monolithic column modified by Cu(II) iminodiacetic acid functional groups was used to preconcentrate low-abundance peptides from complex biological samples [138].

In addition to the chromatographic methods, electrophoretic approaches can be employed to preconcentrate samples. For instance, Yang et al. [134] optimized many aspects of the field-amplified sample injection procedure, including the type and concentration of the acid and organic solvent in the sample solution, the length of the water plug, and the sample injection time; as a result the detection sensitivity can be enhanced more than 3000-fold. This technique was successfully used in the analysis of low-concentration (i.e., picomolar range) bioactive peptide mixtures. Another strategy is to introduce a dynamic pH junction within the capillary, which narrows the sample zone [139]. This method has led to a nearly 1000-fold sensitivity improvement when analyzing peptides [140]. To minimize the sample loss in CE-MS that results from the absorption of peptides onto the inner wall, a number of covalent and physically adsorbed capillary coatings can be employed [133,141–143].

Unfortunately, the online hyphenation of ESI MS with CE is not as straightforward as with LC, for two main reasons. First, the CE effluent is typically in the nL/min flow rate range, which is too low to establish a stable spray. Second, the compounds used in the background electrolyte are typically nonvolatile, making it incompatible with the spray formation. These issues can be addressed with several interface schemes, categorized as sheath flow [144], sheathless [145], and liquid junction [146] designs. The sheath flow design normally consists of two coaxial tubes (one for the nebulization gas, and one for the sheath liquid) interfaced with the separation capillary, and is common in peptide analyses [147]. The sheath flow rate is higher than the CE flow rate, which allows the use of high salt concentrations in CE separation. As mentioned previously, this interface reduces sensitivity owing to the dilution of analytes. This issue is avoided with the sheathless interface where no additional flow is used to close the electrical contact. One sheathless design is to coat the fine tip of the capillary with a conductive metal, usually gold or silver, or with a polymer or graphite [148], to preserve the CE and ESI circuits; however, the coated tips have a reduced life span, and the manufacturing techniques are not yet well established [116,121]. Another design is to insert a conductive wire, such as a platinum electrode [149], directly into the capillary. A novel sheathless interface, developed and applied to peptide studies by Chen et al. [150], has a Nafion tubing junction to integrate an open tubular CE capillary and the electrospray emitter [150]. In a liquid junction interface, an additional electrolyte reservoir is used to provide the electrical connection. The spray is formed through an emitter capillary placed 10 to 20 μm away, next to the outlet of the separation capillary. For example, Fanali et al. [151] developed a pressurized liquid junction nanoflow interface, in which the capillary and the emitter were placed in the electrode vessel, and a stable spray was generated in the emitter tip by the hydrostatic pressure of the spray liquid. A variety of mass analyzers can be online-coupled to a CE setup to analyze peptides, such as single quadrupole, triple quadrupole, IT, TOF, Q-TOF, and FTICR.

Matrix-based MS has also been interfaced to CE for peptide analysis. Along with the many advantages of MALDI, such as generating mostly singly charged ions, matrixes with high acid concentrations afford higher tolerance to a greater range of CE buffers [121]. The online coupling of MALDI MS to CE usually requires reconstruction of the inlet of the mass spectrometer, although the sample handling changes can be minimized [152]. The offline interface of MALDI and CE is more popular, and various strategies are available. For example, the MALDI matrix can be added in the prepositioned droplets [153], through the CE running buffer [154,155], or crystallized onto the sample spots prior to fractionation [156]. To enhance sample collection and concentration, one may consider a MALDI matrix-precoated membrane target for the continuous deposition of effluent using solid-phase preconcentration [74]. This approach is applicable to analyses of rapidly degraded and complex mixtures of peptides found in high salt concentration solutions. The online solid-phase CE transfer allows maximum sample recovery, whereas the offline MS analysis permits each analytical technique to be optimized independently [74]. An example is the in vivo metabolism of neuropeptides in anesthetized rat [74,157]. One may also use a silicone-coated prestructured MALDI target plate specifically for CE fraction collection to achieve enhanced sensitivity in the CE-MALDI-TOF MS analysis of hydrophobic peptides [152]. Despite these opportunities, CE-MS has yet to be commonly used in neuropeptidomic studies; many reported applications deal with peptide standards or enzymatic digests of protein [116,119,120,132–135,141–146].

Mass Spectrometry Imaging Mass spectrometric methods offer unprecedented information on the constituents in a sample. In MSI a tissue is sequentially probed so that an image can be constructed. Owing to the high sensitivity and selectivity of MSI, it can offer unique capabilities when determining the localization of known and unknown analytes in a tissue slice. MALDI-based MSI is now commonly used to study proteins and peptides in the nervous system. Generally, the sample is coated with a homogeneous thin matrix layer, a series of mass spectra from an array of locations across the sample are collected, and finally, ion images of different peptides are generated. For any mass of interest, the intensities of signals corresponding to the selected mass-to-charge ratio (*m/z*) are plotted from each spatially resolved spectrum to create a 2D distribution map. MALDI MSI was applied in the direct spatial analysis of amyloid beta peptide in mouse brain sections, achieving 50-μm spatial resolution with attomole sensitivity [158]. It can also be used to visualize the distribution of multiple neuropeptides within the processes of single isolated and cultured neurons [48].

One must be careful, however, when interpreting peptide distributions because MS signal intensities do not necessarily reflect the amounts of analytes in the tissue. Some intensity distortions occur owing to the inherent heterogeneous nature of the tissue and/or the ion suppression effect [159]. Unlike extracted peptide samples used for separations, the chemical environment of the peptides in the tissue section will impact peptide extraction and MALDI crystallization occurring on the tissue, and hence MALDI performance. An illustration of this can be seen in spinal tissues; gray matter and white matter have different hydrophobicities, and so different peptide extraction efficiencies result. Thus the differences in the amount of peptides observed in the MSI of these two areas will be due to differences in peptide abundance as well as the factors discussed above.

In addition to the advances discussed in the sample preparation section, instrumentation has been improved to give better spatial resolution and acquisition speed. Notably cellular length scales can be attained using stigmatic imaging [160,161] and spatial resolution better than the laser beam diameter can be achieved via oversampling the raster pattern [162]. MS/MS with MSI enables de novo sequencing of novel peptides. For example, DeKeyser et al. [163] performed an organ-level MSI study of crustacean neuronal tissues with MALDI-TOF/TOF MS, followed by MS/MS to identify and confirm novel neuropeptides. This study illustrated not only the spatial relationships between multiple neuropeptide isoforms of the same family but also the distributions of different neuropeptide families. Similarly neuropeptide imaging of insect neurosecretory tissues can be carried out on an LTQ™ ion trap mass spectrometer equipped with an intermediate pressure MALDI source (vMALDI™) [164]; imaging the distribution of diagnostic fragment ions across a sample helped distinguish different compounds with the same nominal molecular mass.

The SIMS imaging approach, mentioned previously, was developed earlier than MALDI. Its application, however, has traditionally been limited to atoms and small molecules because its ionization mechanism is energetic and can cause decomposition of large molecules. SIMS uses primary-ion impacts to desorb and ionize analytes from a sample, and thus large molecules are often not ionized intact. The development of "softer" primary-ion sources, however, along with advances in sample preparation, can significantly increase the mass range and sensitivity of SIMS and expand its application to neuropeptide imaging. Cluster ion sources, such as Au_x^+, Bi_x^+, and C_{60}^+ [165–167], make possible the imaging of large biomolecules in tissues and even isolated cells. Other approaches to large molecules include optimization of primary-ion flux and ion-beam size [84]. To date, however, SIMS has not been used as much as MALDI MS in the field of neuropeptide imaging.

Nonetheless, both SIMS and MALDI MS accomplish the same ultimate goal of attaining spatially resolved, high-sensitivity imaging of peptides. They do, however, start with different performance specifications. SIMS achieves a higher spatial resolution over a lower mass range (<1000 Da), whereas MALDI MS covers a higher and much broader mass range ($\sim$500 Da to >25 kDa), but with a lower spatial resolution [126,159]. For example, the spatial distribution of multiple neuropeptides in rat spinal cord sections can be analyzed using both SIMS and MALDI MS [168].

14.5.2 Relative Quantitative Analyses

MS is not often used for absolute quantitation without using standards for each analyte of interest. For peptides this is due to the nonlinear relationship between each peptide concentration and its signal intensity and also the dependence of the signal intensity on other sample constituents [88]. Internal standards that contain heavy stable isotopes (e.g., ^2H, ^{13}C, ^{15}N, or ^{18}O) can be used for accurate measurement of the absolute quantity of a particular peptide [14,169,170]. It is impractical, however, to use this approach for every peptide present in a biological sample.

Although absolute quantitative analysis may be desirable, the ease of obtaining relative changes of mass spectral signal intensities makes this an important aspect of many neuropeptidomic studies. Often relative changes in peptide amounts between two different samples (e.g., knockout vs. wild type) are measured; the incorporation of isotopic labels into peptides allows a global-scale measurement of the variations of each peptide level. In general, an isotopic label can be introduced into peptides either metabolically or chemically. A metabolic method is to feed a cell culture or an organism a diet that contains isotope-labeled amino acids, so that the isotopes will be incorporated into the peptides during biosynthesis [171–173]. An example is the work of Che et al. [171], who reported labeling neuroendocrine cell lines with *L*-leucine containing ten deuterium residues (D_{10}-Leu) and successfully determined the peptide biosynthesis rate. The drawback of this approach is that it can only be used to study living cells in vitro. The discussion below focuses on analyses of peptides labeled with isotopic reagents after they are extracted from tissue samples [174].

Subsequent to labeling with light or heavy tags, samples are mixed and subjected to LC-MS analysis. A variety of mass spectrometric approaches can be used for peptide quantitation; examples are ESI-IT, ESI-Q-TOF, and MALDI-TOF. The relative amount of a peptide is revealed from the ratio of the peak intensity or peak area of the two isotopic forms in the mass spectrum (Figure 14.6). We see an

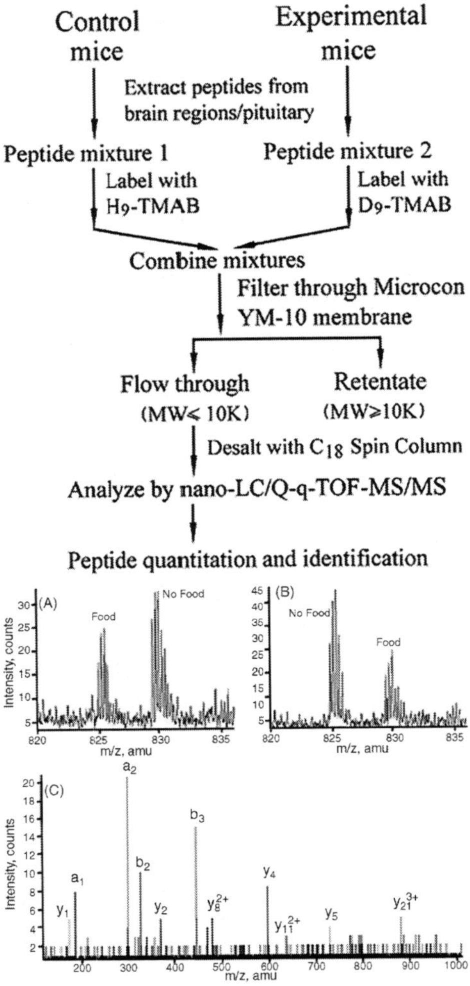

FIGURE 14.6 General strategy for peptide quantification with stable isotopes and mass spectrometry. Peptides are extracted from two groups of animals, and labeled with either the heavy (deuterated) or light (hydrogenated) form of a label such as TMAB. After labeling, the two samples are pooled, filtered, desalted, and analyzed by nano-LC/MS/MS. For accurate quantification, it is important to repeat the labeling scheme above with pools of tissues from distinct animals, and to reverse the heavy and light labels so that the control tissue receives the heavy label and the experimental tissue the light label. (*A*) and (*B*) mass spectra of peak pairs of a peptide with two TMAB tags. (*A*) Normal labeling scheme in which the experimental group was labeled with the heavy tag and the control group with the light tag; (*B*) reverse labeling scheme. Quantitation of the relative levels of the peptide can be done by calculating the ratio of the peak height or peak area of the two groups. (*C*) The product-ion (MS/MS) spectrum of the peptide tagged with two TMAB groups. Adapted from with permission from Wiley Periodicals, Inc., A Wiley Company [26].

example in the work of DeKeyser et al. [175], who developed a novel quantitation method using a Fourier transform mass spectrometer equipped with a MALDI ion source. Analyte ions in two differentially labeled and separately spotted samples were combined in the ion cyclotron resonance cell prior to detection. A single mass spectrum containing isotopically paired peaks was generated for quantitative comparison of relative ion abundances. Quantitation via this in-cell combination method has potential for quantitative comparisons of neuropeptides directly from neurons or tissues.

There are several factors to consider when choosing a stable isotopic reagent [26]. First, the reagent should react with a functional group common to peptides. For example, an isotope-coded affinity tag [176], which reacts with the sulfhydryl group of cysteines, is widely used in quantitative proteomics [177,178]; however, it has limited use in peptidomics because few peptides contain cysteine [26]. (More details about quantitative proteomics are discussed in Chapter 4 by Galan, Iliuk, and Tao in this volume) Amine and carboxyl groups are found on most peptides. H_3/D_3-methanol can be used to label free carboxyl groups (i.e., aspartic and glutamic acid side chains, and the *C*-terminal carboxyl group) [174], whereas H_2/D_2-formaldehyde can target primary amines (i.e., lysine side chains and the *N*-terminal amine groups) [179]. Because some common neuropeptide PTMs (e.g., *C*-terminal amidation, *N*-terminal pyroglutamic acid formation, and *N*-terminal acetylation) cause issues when such schemes are employed, many common neuropeptides are not detectable. A second criterion is that the reagent should be sufficiently reactive to achieve complete peptide labeling, and thus provide reproducible results. It is possible that the light and heavy forms of labeling reagents may react differently with the analyte peptides; to avoid errors resulting from such differences, a reversed labeling is usually employed in parallel [24,180,181]. Third, once incorporated into the peptides, the isotopic tags should be stable during the analysis, including the ionization processes of MS. Fourth, both the light and heavy forms of the labeled peptides should co-elute from the LC system so that they can be analyzed simultaneously by MS; otherwise, the ratio of the light and heavy forms may not be accurate for relative quantitation [182–184]. When the interaction with the stationary phase is via H/D bonds, peptides labeled with such will elute at slightly different times [90,185]. Finally, the isotopic tag should not interfere with MS detection, and the *m/z* values of the two isotopic forms should be well separated in the mass spectrum; the required separation then depends on the capabilities of the mass analyzer and sample complexity. For instance, H_6/D_6-acetic anhydride (H_6/D_6–Ac_2O) and H_4/D_4-succinic anhydride (H_4/D_4–Suc_2O) labels can be compared in a relative quantitation of the neuropeptide levels. Using different mouse pituitary extracts [180,185], more peptides were resolved in the mass spectrum after succinylation than after acetylation, resulting from the larger *m/z* separation between the two isotopic forms of Suc_2O-labeled than for Ac_2O-labeled peptides (4 vs. 3 Da). As a practical matter the reagent should be easy to synthesize or commercially available.

A wide range of isotopic labeling approaches are available for quantitative neuropeptidomic applications. Both H_4/D_4–Suc_2O and H_6/D_6–Ac_2O are commercially available amine-based labeling reagents; compared to Ac_2O labeled peptides, Suc_2O-labeled peptides are more readily separated, and they show improved co-elution from RP HPLC. Both approaches, however, suffer because the labeling reagents create analytes with less positive charge, reducing the MS signal intensity of many peptides when analyzed in the positive-ion mode [180]. If a peptide does not contain histidine or arginine residues, it will be poorly detected in the positive-ion mode owing to reduced positive charge. The drawback of this and other amine-based labeling reagents can be addressed by using trimethylammonium butyrate (TMAB) [183]. Although TMAB is not commercially available, it is relatively easy to synthesize from inexpensive reagents (i.e., gamma-amino butyric acid and methyl iodide) [183]. For example, TMAB was successfully used in the quantitative studies of endogenous mouse pituitary peptides [180]. Besides the excellent separation of 9 Da between the light and heavy forms, TMAB contains a permanent positive charge, allowing the positive charge of the peptides to be preserved during MS analysis. Moreover the positive quaternary amine group significantly reduces the hydrogen/deuterium bonding between the hydrogen or deuterium atoms and the reverse phase resin; as a result TMAB-labeled peptides have better co-elution from RP columns. A shortcoming of TMAB labeling, however, can be the loss of the trimethylamine group in MALDI-TOF MS and ESI-IT MS, but not in ESI-QTOF MS [26,180]. This drawback also pertains to other isotopic labels, such as trimethylammonium acetyl, trimethylammonium propyl and methylnicotinic acid [26]. It is possible that the use of multiple labeling reagents will lead to better peptidome coverage in the biological sample [26,180].

Differential peptidomics reveals the putative function of endogenous peptides in a specific physiological state. The emergence of new stable isotopic labeling reagents that have multiple labels in one set makes differential peptidomics more promising because multiple samples can be analyzed and compared in a single peptidomics experiment. An illustration of this is the work of Morano and coworkers [186] who synthesized four different forms of TMAB labels (H_9/H_6D_3/H_3D_6/D_9-TMAB) and utilized multivariate analysis in a single LC-MS experiment [187]. The iTRAQ MS strategy is another promising method for quantitative peptidomics studies. It can have up to eight differently labeled samples in one analysis [188]; however, this technique is not yet common in neuropeptidomics.

14.5.3 Data Analysis with Bioinformatics

The plethora of MS-based techniques described here underscores the capability of MS to provide valuable information about peptide function. A typical peptidomics experiment generates large amounts of data for analysis; consequently bioinformatics plays a critical role in extracting and decoding the relevant data and translating it into neuropeptide identities. Depending on the experimental methodology, instrument capability, and the species of interest, a number of different bioinformatics tools can be used for neuropeptide identification.

To many proteomics practitioners, a peptidomics experiment may simply appear to be a proteomics experiment after the peptides have been digested by the cells; however, the prohormone processing is actually quite different from exogenous enzymes. As was mentioned previously, the processing of neuropeptide prohormones into biologically active neuropeptides requires multiple enzymatic processing steps, so that one cannot normally predict the final products without measuring them via MS or some other approach. Several bioinformatics toolsets can be used to predict prohormone cleavages and hence, the resulting neuropeptides. In utilizing a variety of available resources, such as expert systems and logistic regression models, these tools facilitate prediction of cleaved and noncleaved basic sites in a new prohormone [189–195]. Consider that in mammals only about 30% of the basic amino-acid sites are cleaved; how do you determine which ones are or are not? Using models trained on the neuropeptide cleavages characterized in a number of common neuronal animal models, we can make well-informed predictions about the expected neuropeptides from a novel prohormone. These prediction tools are accessible via the NeuroPred application suite at http://neuroproteomics.scs.illinois.edu/neuropred.html [190]. The resulting mass list of expected neuropeptides is useful when directly profiling a cell or a tissue. Many peptides from a specific prohormone can be detected simultaneously provided that one has accurate mass data. Accurate mass enhances the ability to assign peptide identities confidently without MS/MS data in a manner similar to the improvement in the identification of a protein that arises from detecting multiple peptides from its tryptic processing [196]. Neuropeptidomic examination of a single cell with direct MALDI MS measurement is an excellent example in this category; at times almost complete characterization of the peptides from a prohormone can be accomplished.

When dealing with larger samples, such as extracts or bigger brain regions that have many more prohormones, it is critical to obtain the sequence information of the peptides, typically via MS/MS. Database searching is the simplest and fastest way to identify neuropeptides. There are a variety of database search engines currently available; examples are SEQUEST [197], Mascot [198], and X!Tandem [199], although these programs were originally developed for protein identifications. When using such programs, the proteins in the database are virtually digested into peptides by specified enzymes (e.g., trypsin, Asp-N protease, Lys-C protease). The product-ion (MS/MS) spectrum is then predicted and compared with those of the virtual peptides. The disadvantage associated with applying such programs to neuropeptide identification is that there is no virtual enzyme that has the same cleavage sequences or creates the same suite of PTMs as the endogenous suite of enzymes (e.g., convertases that cleave the neuropeptide prohormones). Moreover neuropeptides are relatively short and often contain PTMs, which further impair the possibility of obtaining good MS/MS data for a significant identification [200]. One potential way to identify neuropeptides with PTMs is via sequence tag searching, which does not rely on the parent masses but uses short, contiguous amino-acid strings to match the sequences in the database. Both Mascot and MS-Seq (available at http://prospector.ucsf.edu) offer this function.

To enhance the confidence of neuropeptide assignment, database search requires as complete a list of neuropeptide prohormones as possible. Searching against a standard protein database, such as the UniProt Knowledgebase or the nonredundant protein sequence collection from the National Center for Biotechnology Information, will reduce the sensitivity of peptide identification owing to the high possibility of obtaining false-positive hits. Currently there are several peptide databases available. For example, the EROP-Moscow oligopeptide database contains bioactive peptides extracted directly from literature; no peptide in EROP is larger than 50 amino-acid residues in length [201]. The SwePep database is designed for the identification of endogenous peptides and small proteins (below 10 kDa) by using MS-derived data [200]. Most peptides in SwePep are collected from Uniprot and stored in three categories: (1) biologically active peptides, such as neuropeptides and peptide hormones, (2) potential biologically active peptides, and (3) uncharacterized peptides. There are three databases that consist of currently known peptides, peptide precursors, and peptide motifs in Metazoa, respectively [202]. These databases (available at http://peptides.statik.be) help identify new members of a particular peptide family and provide information on the evolutionary relationship of a group of organisms that share peptide homologues. Moreover incorporating SwePep peptide identification into this database suite should have utility for characterizing bioactive peptides [202].

Besides using the currently available databases, one can build a neuropeptide precursor database if the genome of the species of interest is sequenced. One way is to use sequences of known neuropeptides in other species as homologs and use BLAST to search the sequenced genome of that species [203,204]. Another way is to search for the conserved patterns or motifs of neuropeptides or their precursors in the sequenced genome [202]. Neuropeptide genes (if present) not discovered previously can be found via such bioinformatics methods [53]. Having a neuropeptide precursor database helps in deducing the correct amino-acid residues of peptides, which otherwise can be problematic when it comes to differentiating Leu and Ile by using MS detection alone.

Identifying a neuropeptide via database searching is only feasible for those species with complete genome (protein) databases. If no such database is available for the species of interest, de novo sequencing is often required to interpret the MS/MS data. This method uses linear amino-acid combinations to derive the best possible peptide sequence from the MS/MS data. Several programs perform automated de novo sequencing: LuteFisk [205], PEAKS [206], PepNovo [207], and Mascot Distiller [198] are among the widely used approaches. With experience, de novo sequencing can be done manually as well, but this is time-consuming and only effecive when high-quality product-ion spectra are available.

Although de novo sequencing does not always generate full sequences from MS/MS data, partial sequences can often be deduced. These partial sequences (sequence tags) can then be used to search against a protein database of a closely related species; homology searches are usually tolerant to possible sequencing errors and homolog mutations [53]. In fact homologs of the peptide query, rather than the exact and complete sequences, are often identified in this manner, and the results shed light on the evolutionary origins and potential functions of the partially sequenced peptides.

Pattern-finding programs, such as SPIDER (Software for Protein Identification from Sequence Tags) and MEME (Multiple EM For Motif Elicitation) [208,209], provide improved partial sequence homology searches over the traditionally used BLAST.

14.6 CONCLUSIONS

The rapid development of improved mass spectrometric approaches that offer high mass-resolving power and sequencing capabilities, coupled with advances in separation systems, are significantly accelerating the pace of SP discovery. Direct MS analysis is the method of choice for peptide profiling of brain regions, clusters of neurons, single cells, and even cell organelles, although interfacing MS to LC or CE provides powerful modalities for detecting peptides from complex samples. Besides their chemical identity, the location and time dependence of the peptides in the biological system can also be obtained. Owing to the advances in microdialysis that help monitor the dynamic mosaic of released peptides, we now have a greater understanding of the fundamental mechanisms underlying the complex aspects of physiology and behavior. The development of MS imaging, including improved sample-preparation procedures and instrument and data-acquisition methods, provides us with unique opportunities to investigate the chemical information of brain peptides directly from complex environments. Moreover we can determine their distribution and relative abundance over a broad range of analytes in a spatially resolved manner.

14.7 FUTURE PERSPECTIVES

Despite the accelerating neuropeptide discovery rate of the past few decades, a substantial number of signaling peptides may still be unidentified. In fact, given that the number of novel neuropeptides discovered has increased with the introduction of every technological innovation that affords higher sensitivity, the need for further identification will continue. Thus further developments in analytical instrumentation and methodologies are required, including sample preparation protocols, separation techniques, MS detection, and bioinformatics tools. Exploiting a combination of improved sample-handling protocols and new analysis approaches, we will be able to identify unknown neuropeptides, especially those of low abundance. New in vivo sampling approaches will allow the detection of SPs released in an activity-dependent manner; these peptides are more likely to function as neurotransmitters, neuromodulators, or hormones. To overcome further the challenge of sample complexity, improvements in small-volume separations are needed and may be realized via miniaturizing LC and CE using lab-on-a-chip approaches, creating novel LC packing materials such as monolithic columns, and developing multidimensional separations. Another challenge is the limited sensitivity of current detection techniques, and this may be addressed by new or improved ionization techniques.

Presently MS imaging techniques can localize neuropeptides with greater spatial resolution and without the preselection required by immunocytochemistry and in situ hybridization approaches. Improved spatial resolution and sensitivity, however, are important. Current MSI measurements only provide snapshots of the peptides present in a tissue at a specific physiological phase. If the dynamic functional imaging of tissue slices can be coupled with MSI snapshots at specific time points, information regarding the function of specific peptides will result.

These methodological improvements, coupled with more genome-sequencing information and the associated increase in transcriptomic data, will drive the discovery of new neuropeptides. The availability of the genome sequence expedites the analysis of the neuropeptidome (obviously only for Metazoan for which the complete genome is available). As discussed in this review, however, the integration of bioinformatics into neuropeptidomics still stands as the bottleneck for many applications, leaving considerable room for improvements.

Finally, although the traditional biochemical and physiological assays of neuropeptides will not be replaced in the near future, neuropeptidomic approaches will lead to more profound insights into cell–cell communication in the animal kingdom. Some of these insights may have implications in drug discovery.

ACKNOWLEDGMENTS

This work was supported by Award. P30 DA018310 from the National Institute on Drug Abuse (NIDA) and by Award. 5R01NS031609 from the National Institute of Neurological Disorders and Stroke (NINDS). The content is solely the responsibility of the authors and does not necessarily represent the official views of the National Institutes of Health, NIDA, or NINDS.

REFERENCES

1. Hatcher, N. G., Atkins, N., Jr., Annangudi, S. P., Forbes, A. J., Kelleher, N. L., Gillette, M. U., Sweedler, J. V. (2008). Mass spectrometry-based discovery of circadian peptides. *Proc Natl Acad Sci USA 105*, 12527–12532.
2. Jing, J., Vilim, F. S., Horn, C. C., Alexeeva, V., Hatcher, N. G., Sasaki, K., Yashina, I., Zhurov, Y., Kupfermann, I., Sweedler, J. V., Weiss, K. R. (2007). From hunger to satiety: Reconfiguration of a feeding network by Aplysia neuropeptide Y. *J Neurosci 27*, 3490–3502.
3. Wei, S., Feng, Y., Che, F. Y., Pan, H., Mzhavia, N., Devi, L. A., McKinzie, A. A., Levin, N., Richards, W. G., Fricker, L. D. (2004). Obesity and diabetes in transgenic mice expressing proSAAS. *J Endocrinol 180*, 357–368.
4. Thiele, T. E., Marsh, D. J., Ste Marie, L., Bernstein, I. L., Palmiter, R. D. (1998). Ethanol consumption and resistance are inversely related to neuropeptide Y levels. *Nature 396*, 366–369.
5. Ekman, R., Gobom, J., Persson, R., Mecocci, P., Nilsson, C. L. (2001). Arginine vasopressin in the cytoplasm and nuclear fraction of lymphocytes from healthy donors and patients with depression or schizophrenia. *Peptides 22*, 67–72.

6. Holsboer, F. (2003). Corticotropin-releasing hormone modulators and depression. *Curr Opin Investig Drugs 4*, 46–50.
7. Holsboer, F. (2003). The role of peptides in treatment of psychiatric disorders. *J Neural Transm* (Suppl), 17–34.
8. Karl, J., Borgya, A., Gallusser, A., Huber, E., Krueger, K., Rollinger, W., Schenk, J. (1999). Development of a novel, *N*-terminal-proBNP (NT-proBNP) assay with a low detection limit. *Scand J Clin Lab Invest 230* (Suppl), 177–181.
9. McCullough, P. A., Omland, T., Maisel, A. S. (2003). B-type natriuretic peptides: A diagnostic breakthrough for clinicians. *Rev Cardiovasc Med 4*, 72–80.
10. Schrader, M., Selle, H. (2006). The process chain for peptidomic biomarker discovery. *Dis Markers 22*, 27–37.
11. Strand, F. L. (2003). Neuropeptides: General characteristics and neuropharmaceutical potential in treating CNS disorders. *Prog Drug Res 61*, 1–37.
12. Gygi, S. P., Aebersold, R. (2000). Mass spectrometry and proteomics. *Curr Opin Chem Biol 4*, 489–494.
13. Desiderio, D. M. (1992). Quantitative analytical mass spectrometry of endogenous neuropeptides in human pituitaries. *Life Sci 51*, 169–176.
14. Desiderio, D. M., Zhu, X. (1998). Quantitative analysis of methionine enkephalin and beta-endorphin in the pituitary by liquid secondary ion mass spectrometry and tandem mass spectrometry. *J Chromatogr A794*, 85–96.
15. Hokfelt, T. (1991). Neuropeptides in perspective: The last ten years. *Neuron 7*, 867–879.
16. Mutt, V. (1990). Recent developments in the chemistry of gastrointestinal peptides. *Eur J Clin Invest 20* (Suppl 1), S2–9.
17. Clynen, E., Baggerman, G., Veelaert, D., Cerstiaens, A., Van der Horst, D., Harthoorn, L., Derua, R., Waelkens, E., De Loof, A., Schoofs, L. (2001). Peptidomics of the pars intercerebralis-corpus cardiacum complex of the migratory locust, *Locusta migratoria*. *Eur J Biochem 268*, 1929–1939.
18. Schrader, M., Schulz-Knappe, P. (2001). Peptidomics technologies for human body fluids. *Trends Biotechnol 19*, S55–60.
19. Verhaert, P., Uttenweiler-Joseph, S., de Vries, M., Loboda, A., Ens, W., Standing, K. G. (2001). Matrix-assisted laser desorption/ionization quadrupole time-of-flight mass spectrometry: An elegant tool for peptidomics. *Proteomics 1*, 118–131.
20. Rubakhin, S. S., Garden, R. W., Fuller, R. R., Sweedler, J. V. (2000). Measuring the peptides in individual organelles with mass spectrometry. *Nat Biotechnol 18*, 172–175.
21. Jimenez, C. R., van Veelen, P. A., Li, K. W., Wildering, W. C., Geraerts, W. P., Tjaden, U. R., van der Greef, J. (1994). Neuropeptide expression and processing as revealed by direct matrix-assisted laser desorption ionization mass spectrometry of single neurons. *J Neurochem 62*, 404–407.
22. Li, L., Garden, R. W., Sweedler, J. V. (2000). Single-cell MALDI: A new tool for direct peptide profiling. *Trends Biotechnol 18*, 151–160.
23. Bora, A., Annangudi, S. P., Millet, L. J., Rubakhin, S. S., Forbes, A. J., Kelleher, N. L., Gillette, M. U., Sweedler, J. V. (2008). Neuropeptidomics of the supraoptic rat nucleus. *J Proteome Res 7*, 4992–5003.
24. Che, F. Y., Lim, J., Pan, H., Biswas, R., Fricker, L. D. (2005). Quantitative neuropeptidomics of microwave-irradiated mouse brain and pituitary. *Mol Cell Proteomics 4*, 1391–1405.

25. Dowell, J. A., Heyden, W. V., Li, L. (2006). Rat neuropeptidomics by LC-MS/MS and MALDI-FTMS: Enhanced dissection and extraction techniques coupled with 2D RP-RP HPLC. *J Proteome Res 5*, 3368–3375.
26. Fricker, L. D., Lim, J., Pan, H., Che, F. Y. (2006). Peptidomics: Identification and quantification of endogenous peptides in neuroendocrine tissues. *Mass Spectrom Rev 25*, 327–344.
27. Wei, H., Nolkrantz, K., Parkin, M. C., Chisolm, C. N., O'Callaghan, J. P., Kennedy, R. T. (2006). Identification and quantification of neuropeptides in brain tissue by capillary liquid chromatography coupled off-line to MALDI-TOF and MALDI-TOF/TOF-MS. *Anal Chem 78*, 4342–4351.
28. Baggerman, G., Cerstiaens, A., De Loof, A., Schoofs, L. (2002). Peptidomics of the larval *Drosophila melanogaster* central nervous system. *J Biol Chem 277*, 40368–40374.
29. Skold, K., Svensson, M., Kaplan, A., Bjorkesten, L., Astrom, J., Andren, P. E. (2002). A neuroproteomic approach to targeting neuropeptides in the brain. *Proteomics 2*, 447–454.
30. Schulte, I., Tammen, H., Selle, H., Schulz-Knappe, P. (2005). Peptides in body fluids and tissues as markers of disease. *Expert Rev Mol Diagn 5*, 145–157.
31. Yuan, X., Desiderio, D. M. (2005). Human cerebrospinal fluid peptidomics. *J Mass Spectrom 40*, 176–181.
32. Che, F. Y., Biswas, R., Fricker, L. D. (2005). Relative quantitation of peptides in wild-type and Cpe(fat/fat) mouse pituitary using stable isotopic tags and mass spectrometry. *J Mass Spectrom 40*, 227–237.
33. Du Vigneaud, V., Ressler, C., Trippett, S. (1953). The sequence of amino acids in oxytocin, with a proposal for the structure of oxytocin. *J Biol Chem 205*, 949–957.
34. Acher, R., Chauvet, J., Chauvet, M. T. (1995). Man and the chimaera: Selective versus neutral oxytocin evolution. *Adv Exp Med Biol 395*, 615–627.
35. Ludwig, M., Leng, G. (2006). Dendritic peptide release and peptide-dependent behaviours. *Nat Rev Neurosci 7*, 126–136.
36. Strand, F. L. (1999). *Neuropeptides: Regulators of Physiological Processes*. MIT Press, Cambridge.
37. Hokfelt, T., Broberger, C., Xu, Z. Q., Sergeyev, V., Ubink, R., Diez, M. (2000). Neuropeptides—An overview. *Neuropharmacology 39*, 1337–1356.
38. Cuttitta, F. (1993). Peptide amidation: Signature of bioactivity. *Anat Rec 236*, 87–93. 172–173; discussion 193–175.
39. Ludwig, M. (1998). Dendritic release of vasopressin and oxytocin. *J Neuroendocrinol 10*, 881–895.
40. Pow, D. V., Morris, J. F. (1989). Dendrites of hypothalamic magnocellular neurons release neurohypophysial peptides by exocytosis. *Neuroscience 32*, 435–439.
41. Sabatier, N., Caquineau, C., Douglas, A. J., Leng, G. (2003). Oxytocin released from magnocellular dendrites: A potential modulator of alpha-melanocyte-stimulating hormone behavioral actions? *Ann NY Acad Sci 994*, 218–224.
42. Kim, S. I., Grum-Tokars, V., Swanson, T. A., Cotter, E. J., Cahill, P. A., Roberts, J. L., Cummins, P. M., Glucksman, M. J. (2003). Novel roles of neuropeptide processing enzymes: EC3.4.24.15 in the neurome. *J Neurosci Res 74*, 456–467.

43. Boonen, K., Baggerman, G., D'Hertog, W., Husson, S. J., Overbergh, L., Mathieu, C., Schoofs, L. (2007). Neuropeptides of the islets of Langerhans: A peptidomics study. *Gen Comp Endocrinol 152*, 231–241.

44. Schulz-Knappe, P., Zucht, H. D., Heine, G., Jurgens, M., Hess, R., Schrader, M. (2001). Peptidomics: The comprehensive analysis of peptides in complex biological mixtures. *Comb Chem High Throughput Screen 4*, 207–217.

45. Tammen, H., Peck, A., Budde, P., Zucht, H. D. (2007). Peptidomics analysis of human blood specimens for biomarker discovery. *Expert Rev Mol Diagn 7*, 605–613.

46. Salio, C., Lossi, L., Ferrini, F., Merighi, A. (2006). Neuropeptides as synaptic transmitters. *Cell Tissue Res 326*, 583–598.

47. Kandel, E., Schwarts, J., Jessell, T. (2000). *Principles of Neural Science*. McGraw-Hill, New York.

48. Rubakhin, S. S., Greenough, W. T., Sweedler, J. V. (2003). Spatial profiling with MALDI MS: Distribution of neuropeptides within single neurons. *Anal Chem 75*, 5374–5380.

49. Jimenez, C. R., Li, K. W., Dreisewerd, K., Spijker, S., Kingston, R., Bateman, R. H., Burlingame, A. L., Smit, A. B., van Minnen, J., Geraerts, W. P. (1998). Direct mass spectrometric peptide profiling and sequencing of single neurons reveals differential peptide patterns in a small neuronal network. *Biochemistry 37*, 2070–2076.

50. Rubakhin, S. S., Churchill, J. D., Greenough, W. T., Sweedler, J. V. (2006). Profiling signaling peptides in single mammalian cells using mass spectrometry. *Anal Chem 78*, 7267–7272.

51. Fournier, I., Day, R., Salzet, M. (2003). Direct analysis of neuropeptides by in situ MALDI-TOF mass spectrometry in the rat brain. *Neuro Endocrinol Lett 24*, 9–14.

52. Monroe, E. B., Jurchen, J. C., Koszczuk, B. A., Losh, J. L., Rubakhin, S. S., Sweedler, J. V. (2006). Massively parallel sample preparation for the MALDI MS analyses of tissues. *Anal Chem 78*, 6826–6832.

53. Boonen, K., Landuyt, B., Baggerman, G., Husson, S. J., Huybrechts, J., Schoofs, L. (2008). Peptidomics: The integrated approach of MS, hyphenated techniques and bioinformatics for neuropeptide analysis. *J Sep Sci 31*, 427–445.

54. Xu, B. J., Caprioli, R. M., Sanders, M. E., Jensen, R. A. (2002). Direct analysis of laser capture microdissected cells by MALDI mass spectrometry. *J Am Soc Mass Spectrom 13*, 1292–1297.

55. Rubakhin, S. S., Sweedler, J. V. (2007). Characterizing peptides in individual mammalian cells using mass spectrometry. *Nat Protoc 2*, 1987–1997.

56. Stoeckli, M., Chaurand, P., Hallahan, D. E., Caprioli, R. M. (2001). Imaging mass spectrometry: A new technology for the analysis of protein expression in mammalian tissues. *Nat Med 7*, 493–496.

57. Theodorsson, E., Stenfors, C., Mathe, A. A. (1990). Microwave irradiation increases recovery of neuropeptides from brain tissues. *Peptides 11*, 1191–1197.

58. Kennedy, R. T., Watson, C. J., Haskins, W. E., Powell, D. H., Strecker, R. E. (2002). In vivo neurochemical monitoring by microdialysis and capillary separations. *Curr Opin Chem Biol 6*, 659–665.

59. Che, F. Y., Zhang, X., Berezniuk, I., Callaway, M., Lim, J., Fricker, L. D. (2007). Optimization of neuropeptide extraction from the mouse hypothalamus. *J Proteome Res 6*, 4667–4676.

60. Svensson, M., Skold, K., Nilsson, A., Falth, M., Svenningsson, P., Andren, P. E. (2007). Neuropeptidomics: Expanding proteomics downwards. *Biochem Soc Trans 35*, 588–593.

61. Parkin, M. C., Wei, H., O'Callaghan, J. P., Kennedy, R. T. (2005). Sample-dependent effects on the neuropeptidome detected in rat brain tissue preparations by capillary liquid chromatography with tandem mass spectrometry. *Anal Chem 77*, 6331–6338.

62. Romanova, E. V., Rubakhin, S. S., Sweedler, J. V. (2008). One-step sampling, extraction, and storage protocol for peptidomics using dihydroxybenzoic acid. *Anal Chem 80*, 3379–3386.

63. Altelaar, A. F., Mohammed, S., Brans, M. A., Adan, R. A., Heck, A. J. (2009). Improved identification of endogenous peptides from murine nervous tissue by multiplexed peptide extraction methods and multiplexed mass spectrometric analysis. *J Proteome Res 8*, 870–876.

64. Che, F. Y., Yan, L., Li, H., Mzhavia, N., Devi, L. A., Fricker, L. D. (2001). Identification of peptides from brain and pituitary of Cpe(fat)/Cpe(fat) mice. *Proc Natl Acad Sci USA 98*, 9971–9976.

65. Svensson, M., Skold, K., Svenningsson, P., Andren, P. E. (2003). Peptidomics-based discovery of novel neuropeptides. *J Proteome Res 2*, 213–219.

66. Hatton, G. I., Doran, A. D., Salm, A. K., Tweedle, C. D. (1980). Brain slice preparation: Hypothalamus. *Brain Res Bull 5*, 405–414.

67. Gillette, M. U. (1986). The suprachiasmatic nuclei: circadian phase-shifts induced at the time of hypothalamic slice preparation are preserved in vitro. *Brain Res 379*, 176–181.

68. Fiala, J. C., Kirov, S. A., Feinberg, M. D., Petrak, L. J., George, P., Goddard, C. A., Harris, K. M. (2003). Timing of neuronal and glial ultrastructure disruption during brain slice preparation and recovery in vitro. *J Comp Neurol 465*, 90–103.

69. Hatcher, N. G., Richmond, T. A., Rubakhin, S. S., Sweedler, J. V. (2005). Monitoring activity-dependent peptide release from the CNS using single-bead solid-phase extraction and MALDI TOF MS detection. *Anal Chem 77*, 1580–1587.

70. Iannacone, J. M., Ren, S., Hatcher, N. G., Sweedler, J. V. (2009). Collecting peptide release from the brain using porous polymer monolith-based solid phase extraction capillaries. *Anal Chem 81*, 5433–5438.

71. Hatcher, N. G., Sweedler, J. V. (2008). Aplysia bag cells function as a distributed neurosecretory network. *J Neurophysiol 99*, 333–343.

72. Jakubowski, J. A., Hatcher, N. G., Xie, F., Sweedler, J. V. (2006). The first gamma-carboxyglutamate-containing neuropeptide. *Neurochem Int 49*, 223–229.

73. Andren, P. E., Caprioli, R. M. (1999). Determination of extracellular release of neurotensin in discrete rat brain regions utilizing in vivo microdialysis/electrospray mass spectrometry. *Brain Res 845*, 123–129.

74. Zhang, H., Stoeckli, M., Andren, P. E., Caprioli, R. M. (1999). Combining solid-phase preconcentration, capillary electrophoresis and off-line matrix-assisted laser desorption/ionization mass spectrometry: Intracerebral metabolic processing of peptide E in vivo. *J Mass Spectrom 34*, 377–383.

75. Lindefors, N., Brodin, E., Ungerstedt, U. (1987). Microdialysis combined with a sensitive radioimmunoassay. A technique for studying in vivo release of neuropeptides. *J Pharmacol Meth 17*, 305–312.

76. Maidment, N. T., Brumbaugh, D. R., Rudolph, V. D., Erdelyi, E., Evans, C. J. (1989). Microdialysis of extracellular endogenous opioid peptides from rat brain in vivo. *Neuroscience 33*, 549–557.

77. Li, Q., Zubieta, J. K., Kennedy, R. T. (2009). Practical aspects of in vivo detection of neuropeptides by microdialysis coupled off-line to capillary LC with multistage MS. *Anal Chem 81*, 2242–2250.

78. Haskins, W. E., Watson, C. J., Cellar, N. A., Powell, D. H., Kennedy, R. T. (2004). Discovery and neurochemical screening of peptides in brain extracellular fluid by chemical analysis of in vivo microdialysis samples. *Anal Chem 76*, 5523–5533.

79. Goodwin, R. J., Pennington, S. R., Pitt, A. R. (2008). Protein and peptides in pictures: Imaging with MALDI mass spectrometry. *Proteomics 8*, 3785–3800.

80. Kruse, R., Sweedler, J. V. (2003). Spatial profiling invertebrate ganglia using MALDI MS. *J Am Soc Mass Spectrom 14*, 752–759.

81. Aerni, H. R., Cornett, D. S., Caprioli, R. M. (2006). Automated acoustic matrix deposition for MALDI sample preparation. *Anal Chem 78*, 827–834.

82. Chaurand, P., Norris, J. L., Cornett, D. S., Mobley, J. A., Caprioli, R. M. (2006). New developments in profiling and imaging of proteins from tissue sections by MALDI mass spectrometry. *J Proteome Res 5*, 2889–2900.

83. Chaurand, P., Schwartz, S. A., Reyzer, M. L., Caprioli, R. M. (2005). Imaging mass spectrometry: Principles and potentials. *Toxicol Pathol 33*, 92–101.

84. Zimmerman, T. A., Monroe, E. B., Tucker, K. R., Rubakhin, S. S., Sweedler, J. V. (2008). Imaging of cells and tissues with mass spectrometry: Adding chemical information to imaging. In Correia, D. J. J., Detrich, H. W. I. (eds.), *Methods Cell Biol.*, Academic Press, San Diego, pp. 361–390.

85. Sugiura, Y., Shimma, S., Setou, M. (2006). Two-step matrix application technique to improve ionization efficiency for matrix-assisted laser desorption/ionization in imaging mass spectrometry. *Anal Chem 78*, 8227–8235.

86. Pandey, A., Mann, M. (2000). Proteomics to study genes and genomes. *Nature 405*, 837–846.

87. Gorg, A., Weiss, W., Dunn, M. J. (2004). Current two-dimensional electrophoresis technology for proteomics. *Proteomics 4*, 3665–3685.

88. Paulson, L., Persson, R., Karlsson, G., Silberring, J., Bierczynska-Krzysik, A., Ekman, R., Westman-Brinkmalm, A. (2005). Proteomics and peptidomics in neuroscience. Experience of capabilities and limitations in a neurochemical laboratory. *J Mass Spectrom 40*, 202–213.

89. Hu, L., Ye, M., Jiang, X., Feng, S., Zou, H. (2007). Advances in hyphenated analytical techniques for shotgun proteome and peptidome analysis—A review. *Anal Chim Acta 598*, 193–204.

90. Saz, J. M., Marina, M. L. (2008). Application of micro- and nano-HPLC to the determination and characterization of bioactive and biomarker peptides. *J Sep Sci 31*, 446–458.

91. Sinnaeve, B. A., Van Bocxlaer, J. F. (2004). Evaluation of nano-liquid chromatography-tandem mass spectrometry in a column switching setup for the absolute quantification of peptides in the picomolar range. *J Chromatogr A1058*, 113–119.

92. Hummon, A. B., Richmond, T. A., Verleyen, P., Baggerman, G., Huybrechts, J., Ewing, M. A., Vierstraete, E., Rodriguez-Zas, S. L., Schoofs, L., Robinson, G. E., Sweedler, J. V.

(2006). From the genome to the proteome: Uncovering peptides in the Apis brain. *Science 314*, 647–649.

93. Haskins, W. E., Wang, Z., Watson, C. J., Rostand, R. R., Witowski, S. R., Powell, D. H., Kennedy, R. T. (2001). Capillary LC-MS2 at the attomole level for monitoring and discovering endogenous peptides in microdialysis samples collected in vivo. *Anal Chem 73*, 5005–5014.
94. Ericson, C., Phung, Q. T., Horn, D. M., Peters, E. C., Fitchett, J. R., Ficarro, S. B., Salomon, A. R., Brill, L. M., Brock, A. (2003). An automated noncontact deposition interface for liquid chromatography matrix-assisted laser desorption/ionization mass spectrometry. *Anal Chem 75*, 2309–2315.
95. Husson, S. J., Landuyt, B., Nys, T., Baggerman, G., Boonen, K., Clynen, E., Lindemans, M., Janssen, T., Schoofs, L. (2009). Comparative peptidomics of *Caenorhabditis elegans* versus *C. briggsae* by LC-MALDI-TOF MS. *Peptides 30*, 449–457.
96. Garcia, M. C. (2005). The effect of the mobile phase additives on sensitivity in the analysis of peptides and proteins by high-performance liquid chromatography-electrospray mass spectrometry. *J Chromatogr B Anal Technol Biomed Life Sci 825*, 111–123.
97. Bulau, P., Meisen, I., Schmitz, T., Keller, R. Peter-Katalinic, J. (2004). Identification of neuropeptides from the sinus gland of the crayfish Orconectes limosus using nanoscale on-line liquid chromatography tandem mass spectrometry. *Mol Cell Proteomics 3*, 558–564.
98. Issaq, H. J., Chan, K. C., Janini, G. M., Conrads, T. P., Veenstra, T. D. (2005). Multidimensional separation of peptides for effective proteomic analysis. *J Chromatogr B Anal Technol Biomed Life Sci 817*, 35–47.
99. Cottingham, K. (2007). Overcoming the challenges of mammalian neuropeptidomics. *Anal Chem 79*, 12.
100. Holm, A., Storbraten, E., Mihailova, A., Karaszewski, B., Lundanes, E., Greibrokk, T. (2005). Combined solid-phase extraction and 2D LC-MS for characterization of the neuropeptides in rat-brain tissue. *Anal Bioanal Chem 382*, 751–759.
101. Baggerman, G., Boonen, K., Verleyen, P., De Loof, A., Schoofs, L. (2005). Peptidomic analysis of the larval Drosophila melanogaster central nervous system by two-dimensional capillary liquid chromatography quadrupole time-of-flight mass spectrometry. *J Mass Spectrom 40*, 250–260.
102. Canas, B., Pineiro, C., Calvo, E., Lopez-Ferrer, D., Gallardo, J. M. (2007). Trends in sample preparation for classical and second generation proteomics. *J Chromatogr A1153*, 235–258.
103. Wolters, D. A., Washburn, M. P., Yates, J. R., III., (2001). An automated multidimensional protein identification technology for shotgun proteomics. *Anal Chem 73*, 5683–5690.
104. Husson, S. J., Clynen, E., Baggerman, G., De Loof, A., Schoofs, L. (2005). Discovering neuropeptides in Caenorhabditis elegans by two dimensional liquid chromatography and mass spectrometry. *Biochem Biophys Res Commun 335*, 76–86.
105. Boersema, P. J., Divecha, N., Heck, A. J., Mohammed, S. (2007). Evaluation and optimization of ZIC-HILIC-RP as an alternative MudPIT strategy. *J Proteome Res 6*, 937–946.
106. Stroink, T., Wiese, G., Teeuwsen, J., Lingeman, H., Waterval, J. C., Bult, A., de Jong, G. J., Underberg, W. J. (2002). On-line coupling of SEC and RP-LC for the determination of

structurally related enkephalins in cerebrospinal fluid. *J Pharm Biomed Anal 30*, 1393–1401.

107. Hu, L., Li, X., Jiang, X., Zhou, H., Jiang, X., Kong, L., Ye, M., Zou, H. (2007). Comprehensive peptidome analysis of mouse livers by size exclusion chromatography prefractionation and nanoLC-MS/MS identification. *J Proteome Res 6*, 801–808.
108. Tian, R., Ren, L., Ma, H., Li, X., Hu, L., Ye, M., Wu, R., Tian, Z., Liu, Z., Zou, H. (2008). Selective enrichment of endogenous peptides by chemically modified porous nanoparticles for peptidome analysis. *J Chromatogr A 1216*, 1270–1278.
109. Fricker, L. D., McKinzie, A. A., Sun, J., Curran, E., Qian, Y., Yan, L., Patterson, S. D., Courchesne, P. L., Richards, B., Levin, N., Mzhavia, N., Devi, L. A., Douglass, J. (2000). Identification and characterization of proSAAS, a granin-like neuroendocrine peptide precursor that inhibits prohormone processing. *J Neurosci 20*, 639–648.
110. Gilar, M., Olivova, P., Daly, A. E., Gebler, J. C. (2005). Two-dimensional separation of peptides using RP-RP-HPLC system with different pH in first and second separation dimensions. *J Sep Sci 28*, 1694–1703.
111. Rogatsky, E., Stein, D. T. (2006). Two-dimensional reverse phase-reverse phase chromatography: A simple and robust platform for sensitive quantitative analysis of peptides by LC/MS. Hardware design. *J Sep Sci 29*, 538–546.
112. Mihailova, A., Karaszewski, B., Faergestad, E. M., Hauser, R., Nyka, W. M., Lundanes, E., Greibrokk, T. (2008). Two-dimensional LC-MS/MS in detection of peptides in hypothalamus of the rat subjected to hypoxic stress. *J Sep Sci 31*, 468–479.
113. Mihailova, A., Malerod, H., Wilson, S. R., Karaszewski, B., Hauser, R., Lundanes, E., Greibrokk, T. (2008). Improving the resolution of neuropeptides in rat brain with on-line HILIC-RP compared to on-line SCX-RP. *J Sep Sci 31*, 459–467.
114. Gilar, M., Olivova, P., Daly, A. E., Gebler, J. C. (2005). Orthogonality of separation in two-dimensional liquid chromatography. *Anal Chem 77*, 6426–6434.
115. Wei, J., Sun, J., Yu, W., Jones, A., Oeller, P., Keller, M., Woodnutt, G., Short, J. M. (2005). Global proteome discovery using an online three-dimensional LC-MS/MS. *J Proteome Res 4*, 801–808.
116. Herrero, M., Ibanez, E., Cifuentes, A. (2008). Capillary electrophoresis-electrospray-mass spectrometry in peptide analysis and peptidomics. *Electrophoresis 29*, 2148–2160.
117. Simo, C., Cifuentes, A. (2003). Capillary electrophoresis-mass spectrometry of peptides from enzymatic protein hydrolysis: Simulation and optimization. *Electrophoresis 24*, 834–842.
118. Gucek, M., Gaspari, M., Walhagen, K., Vreeken, R. J., Verheij, E. R., van der Greef, J. (2000). Capillary electrochromatography/nanoelectrospray mass spectrometry for attomole characterization of peptides. *Rapid Commun Mass Spectrom 14*, 1448–1454.
119. Kuroda, Y., Yukinaga, H., Kitano, M., Noguchi, T., Nemati, M., Shibukawa, A., Nakagawa, T., Matsuzaki, K. (2005). On-line capillary isoelectric focusing-mass spectrometry for quantitative analysis of peptides and proteins. *J Pharm Biomed Anal 37*, 423–428.
120. Cifuentes, A., Poppe, H. (1997). Behavior of peptides in capillary electrophoresis: Effect of peptide charge, mass and structure. *Electrophoresis 18*, 2362–2376.
121. Kasicka, V. (2008). Recent developments in CE and CEC of peptides. *Electrophoresis 29*, 179–206.

122. Li, L., Garden, R. W., Romanova, E. V., Sweedler, J. V. (1999). In situ sequencing of peptides from biological tissues and single cells using MALDI-PSD/CID analysis. *Anal Chem 71*, 5451–5458.

123. Stemmler, E. A., Provencher, H. L., Guiney, M. E., Gardner, N. P., Dickinson, P. S. (2005). Matrix-assisted laser desorption/ionization fourier transform mass spectrometry for the identification of orcokinin neuropeptides in crustaceans using metastable decay and sustained off-resonance irradiation. *Anal Chem 77*, 3594–3606.

124. Kutz, K. K., Schmidt, J. J., Li, L. (2004). In situ tissue analysis of neuropeptides by MALDI FTMS in-cell accumulation. *Anal Chem 76*, 5630–5640.

125. Wittmaack, K., Szymczak, W., Hoheisel, G., Tuszynski, W. (2000). Time-of-flight secondary ion mass spectrometry of matrix-diluted oligo- and polypeptides bombarded with slow and fast projectiles: Positive and negative matrix and analyte ion yields, background signals, and sample aging. *J Am Soc Mass Spectrom 11*, 553–563.

126. McDonnell, L. A., Piersma, S. R., MaartenAltelaar, A. F., Mize, T. H., Luxembourg, S. L., Verhaert, P. D., van Minnen, J., Heeren, R. M. (2005). Subcellular imaging mass spectrometry of brain tissue. *J Mass Spectrom 40*, 160–168.

127. Emmett, M. R., Andren, P. E., Caprioli, R. M. (1995). Specific molecular mass detection of endogenously released neuropeptides using in vivo microdialysis/mass spectrometry. *J Neurosci Methods 62*, 141–147.

128. Takeuchi, H., Yasuda, A., Yasuda-Kamatani, Y., Kubo, T., Nakajima, T. (2003). Identification of a tachykinin-related neuropeptide from the honeybee brain using direct MALDI-TOF MS and its gene expression in worker, queen and drone heads. *Insect Mol Biol 12*, 291–298.

129. Bonnard, E., Burlet-Schiltz, O., Frances, B., Mazarguil, H., Monsarrat, B., Zajac, J. M., Roussin, A. (2001). Identification of neuropeptide FF-related peptides in rodent spinal cord. *Peptides 22*, 1085–1092.

130. Burlet-Schiltz, O., Mazarguil, H., Sol, J. C., Chaynes, P., Monsarrat, B., Zajac, J. M., Roussin, A. (2002). Identification of neuropeptide FF-related peptides in human cerebrospinal fluid by mass spectrometry. *FEBS Lett 532*, 313–318.

131. Neususs, C., Pelzing, M., Macht, M. (2002). A robust approach for the analysis of peptides in the low femtomole range by capillary electrophoresis-tandem mass spectrometry. *Electrophoresis 23*, 3149–3159.

132. Kato, M., Sakai-Kato, K., Jin, H., Kubota, K., Miyano, H., Toyo'oka, T., Dulay, M. T., Zare, R. N. (2004). Integration of on-line protein digestion, peptide separation, and protein identification using pepsin-coated photopolymerized sol-gel columns and capillary electrophoresis/mass spectrometry. *Anal Chem 76*, 1896–1902.

133. Larsson, M., Lutz, E. S. (2000). Transient isotachophoresis for sensitivity enhancement in capillary electrophoresis-mass spectrometry for peptide analysis. *Electrophoresis 21*, 2859–2865.

134. Yang, Y., Boysen, R. I., Hearn, M. T. (2006). Optimization of field-amplified sample injection for analysis of peptides by capillary electrophoresis-mass spectrometry. *Anal Chem 78*, 4752–4758.

135. Janini, G. M., Zhou, M., Yu, L. R., Blonder, J., Gignac, M., Conrads, T. P., Issaq, H. J., Veenstra, T. D. (2003). On-column sample enrichment for capillary electrophoresis sheathless electrospray ionization mass spectrometry: Evaluation for peptide analysis and protein identification. *Anal Chem 75*, 5984–5993.

136. Liu, Z., Pawliszyn, J. (2006). Online coupling of solid-phase microextraction and capillary electrophoresis. *J Chromatogr Sci 44*, 366–374.

137. Sandra, K., Lynen, F., Devreese, B., Van Beeumen, J., Sandra, P. (2006). On-column sample enrichment for the high-sensitivity sheath-flow CE-MS analysis of peptides. *Anal Bioanal Chem 385*, 671–677.

138. Zhang, L., Zhang, L., Zhang, W., Zhang, Y. (2005). On-line concentration of peptides and proteins with the hyphenation of polymer monolithic immobilized metal affinity chromatography and capillary electrophoresis. *Electrophoresis 26*, 2172–2178.

139. Monton, M. R., Imami, K., Nakanishi, M., Kim, J. B., Terabe, S. (2005). Dynamic pH junction technique for on-line preconcentration of peptides in capillary electrophoresis. *J Chromatogr A 1079*, 266–273.

140. Imami, K., Monton, M. R., Ishihama, Y., Terabe, S. (2007). Simple on-line sample preconcentration technique for peptides based on dynamic pH junction in capillary electrophoresis-mass spectrometry. *J Chromatogr A 1148*, 250–255.

141. Amon, S., Plematl, A., Rizzi, A. (2006). Capillary zone electrophoresis of glycopeptides under controlled electroosmotic flow conditions coupled to electrospray and matrix-assisted laser desorption/ionization mass spectrometry. *Electrophoresis 27*, 1209–1219.

142. Hu, A., Chen, C. T., Tsai, P. J., Ho, Y. P. (2006). Using capillary electrophoresis-selective tandem mass spectrometry to identify pathogens in clinical samples. *Anal Chem 78*, 5124–5133.

143. Sassi, A. P., Andel, F., III, Bitter, H. M., Brown, M. P., Chapman, R. G., Espiritu, J., Greenquist, A. C., Guyon, I., Horchi-Alegre, M., Stults, K. L., Wainright, A., Heller, J. C., Stults, J. T. (2005). An automated, sheathless capillary electrophoresis-mass spectrometry platform for discovery of biomarkers in human serum. *Electrophoresis 26*, 1500–1512.

144. Nilsson, S. L., Bylund, D., Jornten-Karlsson, M., Petersson, P., Markides, K. E. (2004). A chemometric study of active parameters and their interaction effects in a nebulized sheath-liquid electrospray interface for capillary electrophoresis-mass spectrometry. *Electrophoresis 25*, 2100–2107.

145. Kele, Z., Ferenc, G., Klement, E., Toth, G. K., Janaky, T. (2005). Design and performance of a sheathless capillary electrophoresis/mass spectrometry interface by combining fused-silica capillaries with gold-coated nanoelectrospray tips. *Rapid Commun Mass Spectrom 19*, 881–885.

146. Foret, F., Zhou, H., Gangl, E., Karger, B. L. (2000). Subatmospheric electrospray interface for coupling of microcolumn separations with mass spectrometry. *Electrophoresis 21*, 1363–1371.

147. Schmitt-Kopplin, P., Englmann, M. (2005). Capillary electrophoresis–mass spectrometry: Survey on developments and applications 2003–2004. *Electrophoresis 26*, 1209–1220.

148. Sanz-Nebot, V., Balaguer, E., Benavente, F., Barbosa, J. (2005). Comparison of sheathless and sheath-flow electrospray interfaces for the capillary electrophoresis-electrospray ionization-mass spectrometry analysis of peptides. *Electrophoresis 26*, 1457–1465.

149. Chang, Y. Z., Her, G. R. (2000). Sheathless capillary electrophoresis/electrospray mass spectrometry using a carbon-coated fused-silica capillary. *Anal Chem 72*, 626–630.

150. Chen, Z., Boggess, B., Chang, H. C. (2007). Open-tubular capillary electrochromatography-mass spectrometry with sheathless nanoflow electrospray ionization for analysis of amino acids and peptides. *J Mass Spectrom 42*, 244–253.

151. Fanali, S., D'Orazio, G., Foret, F., Kleparnik, K., Aturki, Z. (2006). On-line CE-MS using pressurized liquid junction nanoflow electrospray interface and surface-coated capillaries. *Electrophoresis 27*, 4666–4673.

152. Jacksen, J., Redeby, T., Emmer, A. (2006). Capillary electrophoretic separation and fractionation of hydrophobic peptides onto a pre-structured matrix assisted laser desorption/ionization target for mass spectrometric analysis. *J Sep Sci 29*, 288–295.

153. Ojima, N., Shingaki, T., Yamamoto, T., Masujima, T. (2001). Droplet electrocoupling between capillary electrophoresis and matrix assisted laser desorption/ionization-time of flight-mass spectroscopy and its application. *Electrophoresis 22*, 3478–3482.

154. Page, J. S., Rubakhin, S. S., Sweedler, J. V. (2002). Single-neuron analysis using CE combined with MALDI MS and radionuclide detection. *Anal Chem 74*, 497–503.

155. Rejtar, T., Hu, P., Juhasz, P., Campbell, J. M., Vestal, M. L., Preisler, J., Karger, B. L. (2002). Off-line coupling of high-resolution capillary electrophoresis to MALDI-TOF and TOF/TOF MS. *J Proteome Res 1*, 171–179.

156. Johnson, T., Bergquist, J., Ekman, R., Nordhoff, E., Schurenberg, M., Kloppel, K. D., Muller, M., Lehrach, H., Gobom, J. (2001). A CE-MALDI interface based on the use of prestructured sample supports. *Anal Chem 73*, 1670–1675.

157. Zhang, H., Caprioli, R. M. (1996). Capillary electrophoresis combined with matrix-assisted laser desorption/ionization mass spectrometry;continuous sample deposition on a matrix-precoated membrane target. *J Mass Spectrom 31*, 1039–1046.

158. Stoeckli, M., Staab, D., Staufenbiel, M., Wiederhold, K. H., Signor, L. (2002). Molecular imaging of amyloid beta peptides in mouse brain sections using mass spectrometry. *Anal Biochem 311*, 33–39.

159. Rohner, T. C., Staab, D., Stoeckli, M. (2005). MALDI mass spectrometric imaging of biological tissue sections. *Mech Ageing Dev 126*, 177–185.

160. Altelaar, A. F., Luxembourg, S. L., McDonnell, L. A., Piersma, S. R., Heeren, R. M. (2007). Imaging mass spectrometry at cellular length scales. *Nat Protoc 2*, 1185–1196.

161. Altelaar, A. F. M., Taban, I. M., McDonnell, L. A., Verhaert, P. D. E. M., de Lange, R. P. J., Adan, R. A. H., Mooi, W. J., Heeren, R. M. A., Piersma, S. R. (2007). High-resolution MALDI imaging mass spectrometry allows localization of peptide distributions at cellular length scales in pituitary tissue sections. *Int J Mass Spectrom 260*, 203–211.

162. Jurchen, J. C., Rubakhin, S. S., Sweedler, J. V. (2005). MALDI-MS imaging of features smaller than the size of the laser beam. *J Am Soc Mass Spectrom 16*, 1654–1659.

163. DeKeyser, S. S., Kutz-Naber, K. K., Schmidt, J. J., Barrett-Wilt, G. A., Li, L. (2007). Imaging mass spectrometry of neuropeptides in decapod crustacean neuronal tissues. *J Proteome Res 6*, 1782–1791.

164. Verhaert, P. D., Conaway, M. C. P., Pekar, T. M., Miller, K. (2007). Neuropeptide imaging on an LTQ with vMALDI source: The complete 'all-in-one' peptidome analysis. *Int J Mass Spectrom 260*, 177–184.

165. Fletcher, J. S., Lockyer, N. P., Vaidyanathan, S., Vickerman, J. C. (2007). TOF-SIMS 3D biomolecular imaging of Xenopus laevis oocytes using buckminsterfullerene (C60) primary ions. *Anal Chem 79*, 2199–2206.

166. Tempez, A., Schultz, J. A., Della-Negra, S., Depauw, J., Jacquet, D., Novikov, A., Lebeyec, Y., Pautrat, M., Caroff, M., Ugarov, M., Bensaoula, H., Gonin, M., Fuhrer, K., Woods, A. (2004). Orthogonal time-of-flight secondary ion mass spectrometric analysis of peptides using large gold clusters as primary ions. *Rapid Commun Mass Spectrom 18*, 371–376.

167. Touboul, D., Kollmer, F., Niehuis, E., Brunelle, A., Laprevote, O. (2005). Improvement of biological time-of-flight-secondary ion mass spectrometry imaging with a bismuth cluster ion source. *J Am Soc Mass Spectrom 16*, 1608–1618.

168. Monroe, E. B., Annangudi, S. P., Hatcher, N. G., Gutstein, H. B., Rubakhin, S. S., Sweedler, J. V. (2008). SIMS and MALDI MS imaging of the spinal cord. *Proteomics 8*, 3746–3754.

169. Desiderio, D. M. (1996). Mass spectrometric quantification of neuropeptides. *Methods Mol Biol 61*, 57–65.

170. Kusmierz, J. J., Sumrada, R., Desiderio, D. M. (1990). Fast atom bombardment mass spectrometric quantitative analysis of methionine-enkephalin in human pituitary tissues. *Anal Chem 62*, 2395–2400.

171. Che, F. Y., Yuan, Q., Kalinina, E., Fricker, L. D. (2004). Examination of the rate of peptide biosynthesis in neuroendocrine cell lines using a stable isotopic label and mass spectrometry. *J Neurochem 90*, 585–594.

172. Veenstra, T. D., Martinovic, S., Anderson, G. A., Pasa-Tolic, L., Smith, R. D. (2000). Proteome analysis using selective incorporation of isotopically labeled amino acids. *J Am Soc Mass Spectrom 11*, 78–82.

173. Zhu, H., Pan, S., Gu, S., Bradbury, E. M., Chen, X. (2002). Amino acid residue specific stable isotope labeling for quantitative proteomics. *Rapid Commun Mass Spectrom 16*, 2115–2123.

174. Goodlett, D. R., Keller, A., Watts, J. D., Newitt, R., Yi, E. C., Purvine, S., Eng, J. K., von Haller, P., Aebersold, R., Kolker, E. (2001). Differential stable isotope labeling of peptides for quantitation and de novo sequence derivation. *Rapid Commun Mass Spectrom 15*, 1214–1221.

175. DeKeyser, S. S., Li, L. (2006). Matrix-assisted laser desorption/ionization Fourier transform mass spectrometry quantitation via in cell combination. *Analyst 131*, 281–290.

176. Gygi, S. P., Rist, B., Gerber, S. A., Turecek, F., Gelb, M. H., Aebersold, R. (1999). Quantitative analysis of complex protein mixtures using isotope-coded affinity tags. *Nat Biotechnol 17*, 994–999.

177. Julka, S., Regnier, F. (2004). Quantification in proteomics through stable isotope coding: A review. *J Proteome Res 3*, 350–363.

178. Tao, W. A., Aebersold, R. (2003). Advances in quantitative proteomics via stable isotope tagging and mass spectrometry. *Curr Opin Biotechnol 14*, 110–118.

179. Hsu, J. L., Huang, S. Y., Chow, N. H., Chen, S. H. (2003). Stable-isotope dimethyl labeling for quantitative proteomics. *Anal Chem 75*, 6843–6852.

180. Che, F. Y., Fricker, L. D. (2005). Quantitative peptidomics of mouse pituitary: Comparison of different stable isotopic tags. *J Mass Spectrom 40*, 238–249.

181. Lim, J., Berezniuk, I., Che, F. Y., Parikh, R., Biswas, R., Pan, H., Fricker, L. D. (2006). Altered neuropeptide processing in prefrontal cortex of Cpe (fat/fat) mice: Implications for neuropeptide discovery. *J Neurochem 96*, 1169–1181.

182. Zhang, R., Regnier, F. E. (2002). Minimizing resolution of isotopically coded peptides in comparative proteomics. *J Proteome Res 1*, 139–147.

183. Zhang, R., Sioma, C. S., Thompson, R. A., Xiong, L., Regnier, F. E. (2002). Controlling deuterium isotope effects in comparative proteomics. *Anal Chem 74*, 3662–3669.

184. Zhang, R., Sioma, C. S., Wang, S., Regnier, F. E. (2001). Fractionation of isotopically labeled peptides in quantitative proteomics. *Anal Chem 73*, 5142–5149.

185. Che, F. Y., Fricker, L. D. (2002). Quantitation of neuropeptides in Cpe(fat)/Cpe(fat) mice using differential isotopic tags and mass spectrometry. *Anal Chem 74*, 3190–3198.

186. Ross, P. L., Huang, Y. N., Marchese, J. N., Williamson, B., Parker, K., Hattan, S., Khainovski, N., Pillai, S., Dey, S., Daniels, S., Purkayastha, S., Juhasz, P., Martin, S., Bartlet-Jones, M., He, F., Jacobson, A., Pappin, D. J. (2004). Multiplexed protein quantitation in Saccharomyces cerevisiae using amine-reactive isobaric tagging reagents. *Mol Cell Proteomics 3*, 1154–1169.

187. Morano, C., Zhang, X., Fricker, L. D. (2008). Multiple isotopic labels for quantitative mass spectrometry. *Anal Chem 80*, 9298–9309.

188. Pierce, A., Unwin, R. D., Evans, C. A., Griffiths, S., Carney, L., Zhang, L., Jaworska, E., Lee, C. F., Blinco, D., Okoniewski, M. J., Miller, C. J., Bitton, D. A., Spooncer, E., Whetton, A. D. (2008). Eight-channel iTRAQ enables comparison of the activity of six leukemogenic tyrosine kinases. *Mol Cell Proteomics 7*, 853–863.

189. Amare, A., Hummon, A. B., Southey, B. R., Zimmerman, T. A., Rodriguez-Zas, S. L., Sweedler, J. V. (2006). Bridging neuropeptidomics and genomics with bioinformatics: Prediction of mammalian neuropeptide prohormone processing. *J Proteome Res 5*, 1162–1167.

190. Southey, B. R., Amare, A., Zimmerman, T. A., Rodriguez-Zas, S. L., Sweedler, J. V. (2006). NeuroPred: A tool to predict cleavage sites in neuropeptide precursors and provide the masses of the resulting peptides. *Nucl Acids Res 34*, W267–272.

191. Southey, B. R., Rodriguez-Zas, S. L., Sweedler, J. V. (2006). Prediction of neuropeptide prohormone cleavages with application to RFamides. *Peptides 27*, 1087–1098.

192. Southey, B. R., Sweedler, JV and Rodriguez-Zas, SL. (2008). A python analytical pipeline to identify prohormone precursors and predict prohormone cleavage sites. *Front Neuroinformatics 2*, 7.

193. Southey, B. R., Sweedler, J. V., Rodriguez-Zas, S. L. (2008). Prediction of neuropeptide cleavage sites in insects. *Bioinformatics 24*, 815–825.

194. Tegge, A. N., Southey, B. R., Sweedler, J. V., Rodriguez-Zas, S. L. (2008). Comparative analysis of neuropeptide cleavage sites in human, mouse, rat, and cattle. *Mamm Genome 19*, 106–120.

195. Hummon, A. B., Hummon, N. P., Corbin, R. W., Li, L., Vilim, F. S., Weiss, K. R., Sweedler, J. V. (2003). From precursor to final peptides: A statistical sequence-based approach to predicting prohormone processing. *J Proteome Res 2*, 650–656.

196. Zhang, W., Chait, B. T. (2000). ProFound: An expert system for protein identification using mass spectrometric peptide mapping information. *Anal Chem 72*, 2482–2489.

197. Sadygov, R. G., Cociorva, D., Yates, J. R., III. (2004). Large-scale database searching using tandem mass spectra: Looking up the answer in the back of the book. *Nat Methods 1*, 195–202.

198. Perkins, D. N., Pappin, D. J., Creasy, D. M., Cottrell, J. S. (1999). Probability-based protein identification by searching sequence databases using mass spectrometry data. *Electrophoresis 20*, 3551–3567.

199. Craig, R., Beavis, R. C. (2004). TANDEM: Matching proteins with tandem mass spectra. *Bioinformatics 20*, 1466–1467.

200. Falth, M., Skold, K., Norrman, M., Svensson, M., Fenyo, D., Andren, P. E. (2006). SwePep, a database designed for endogenous peptides and mass spectrometry. *Mol Cell Proteomics 5*, 998–1005.

201. Zamyatnin, A. A., Borchikov, A. S., Vladimirov, M. G., Voronina, O. L. (2006). The EROP-Moscow oligopeptide database. *Nucl Acids Res 34*, D261–266.

202. Liu, F., Baggerman, G., Schoofs, L., Wets, G. (2008). The construction of a bioactive peptide database in Metazoa. *J Proteome Res 7*, 4119–4131.

203. Hewes, R. S., Taghert, P. H. (2001). Neuropeptides and neuropeptide receptors in the *Drosophila melanogaster* genome. *Genome Res 11*, 1126–1142.

204. Vanden Broeck, J. (2001). Neuropeptides and their precursors in the fruitfly, *Drosophila melanogaster*. *Peptides 22*, 241–254.

205. Taylor, J. A., Johnson, R. S. (1997). Sequence database searches via de novo peptide sequencing by tandem mass spectrometry. *Rapid Commun Mass Spectrom 11*, 1067–1075.

206. Ma, B., Zhang, K., Hendrie, C., Liang, C., Li, M., Doherty-Kirby, A., Lajoie, G. (2003). PEAKS: Powerful software for peptide de novo sequencing by tandem mass spectrometry. *Rapid Commun Mass Spectrom 17*, 2337–2342.

207. Frank, A., Pevzner, P. (2005). PepNovo: De novo peptide sequencing via probabilistic network modeling. *Anal Chem 77*, 964–973.

208. Han, Y., Ma, B., Zhang, K. (2005). SPIDER: Software for protein identification from sequence tags with de novo sequencing error. *J Bioinform Comput Biol 3*, 697–716.

209. Bailey, T. L., Williams, N., Misleh, C., Li, W. W. (2006). MEME: Discovering and analyzing DNA and protein sequence motifs. *Nucl Acids Res 34*, W369–373.

CHAPTER 15

Mass Spectrometry for the Study of Peptide Drug Metabolism

PATRICK J. RUDEWICZ

15.1 INTRODUCTION

The drug discovery and development process is a time and labor-intensive endeavor. The time from identification of a biological target to the registration of a new drug at the conclusion of clinical trials may take on average 12 years and cost 800 to 900 million dollars [1]. Drug discovery is often described as a funnel where hundreds of thousands of compounds are initially screened for potency against a biological target by using high-throughput screening. At this early stage, chemical hits are ligands having a large affinity for a particular biological target. The hits, however, do not necessarily have the pharmaceutical properties desired in a drug. It is in the hit-to-lead and lead optimization stages that pharmaceutical scientists optimize absorption, distribution, metabolism, and excretion (ADME) qualities. At this juncture of drug discovery, drug metabolism scientists have the largest impact through the judicious use of screening and mechanistic studies so that the resulting compounds not only have the required in vitro and cellular potency but also enter systemic circulation and reach optimal concentrations at the therapeutic target.

When a drug is administered orally, it becomes necessary to overcome several barriers to reach levels in blood required to have the desired pharmacological effect. First, the drug undergoes dissolution within the stomach where the compound must possess adequate solubility and stability at low pH. As it passes through the intestines, the drug needs to possess the necessary permeability to be absorbed. Permeability may be described as either passive diffusion through the lipid bilayer or active uptake by transporter proteins anchored in the lipid bilayer. Once absorbed, the drug passes through the portal vein into the liver where it may be

Protein and Peptide Mass Spectrometry in Drug Discovery, Edited by Michael L. Gross, Guodong Chen, and Birendra N. Pramanik.

metabolized by enzymes located in hepatocytes including the cytochrome (CYP) P450 enzymes which are mainly responsible for what is referred to as Phase I oxidative metabolism.

Phase II metabolic reactions are conjugation reactions including glucuronidation, sulfation, and amino acid conjugation. Conjugation of small endogenous molecules may occur directly to the parent drug molecule or to a Phase I metabolic product such as a hydroxylated drug molecule. These reactions increase the polarity of the drug molecule so that it may more easily be excreted via the kidneys or in bile. Apart from the liver, metabolizing enzymes are located in blood and other organs and peripheral tissues, including the intestines. Similarly uptake and efflux transporter proteins are located in many tissues, including hepatocytes, and there exists an interplay between transporters and P450 enzymes. For example, a drug molecule may be effluxed out of an enterocyte in the small intestine by a P-gp transporter and then metabolized by CYP3A4 [2].

The goal in drug metabolism and pharmacokinetics (DMPK) groups is to eliminate compounds that are poor candidates for pharmaceutical development. This is achieved by implementing metabolism screening assays that have a relatively high throughput, meaningful endpoints, and low compound requirements. Typical in vitro screens include P450 inhibition, permeability, metabolic stability, transporter, including P-lycoprotein interaction, and plasma protein binding. In vivo screens often include IV/PO pharmacokinetic screening in mouse and rats.

15.2 PEPTIDE DRUG METABOLISM

The metabolism of peptide and protein drugs is governed by hydrolysis of peptide bonds by peptidases found in blood and in various tissues [3–5]. These metabolizing enzymes can restrict the half-life of peptides administered orally to range from a few minutes to an hour; hence the drug may not reach the desired concentration at the target tissue. Peptide bond hydrolysis is catalyzed by peptidases, which are ubiquitous in body tissue. Metabolism of peptide drugs can be assessed in vitro using microsomes, hepatocytes and Caco-2 cells [6–8]. Some endopeptidase activity might be lost in cryopreserved compared to fresh hepatocytes. Rat brain synaptosomes can be used for the evaluation of synaptic metabolism of investigational neuropeptides [9]. In vivo identification of peptide drug metabolites is often performed using samples from plasma, urine, bile, and feces after an investigational drug is dosed in animal species. In addition to metabolism, due to the hydrophilic nature of most peptides, renal excretion also contributes to the rapid clearance of peptide drugs.

As an example, the peptide drug metkephamid undergoes extensive hydrolysis by peptidases at the surface of endothelial cells in the intestine and also at the surface of hepatocytes, limiting its overall bioavailability to 30% [10]. Certain peptides can also undergo phase I oxidative metabolism typical of other small molecule drugs. The metabolism of cyclosporine A, a cyclic undecapeptide, occurs principally in the liver and yields mono- and di-hydroxlylated metabolites as well as *N*-demethylated metabolites [11,12].

Absorption of peptide drugs across intestinal cell membranes is limited by their hydrophilicity and molecular mass. Foger et al. [6] reported the correlation of in vitro models for studying peptide drug absorption and showed that the molecular mass of peptides correlates linearly with permeability in rat in vitro models. In addition some peptide drugs may be substrates for P-glycoprotein efflux transporters that further limit their absorption across intestinal and blood–brain membranes [6].

Several approaches can be employed to improve the stability of peptide drugs toward proteolytic enzyme degradation as well as enhance membrane permeability. These include the inclusion of *D* or unnatural amino acids, chemical modification of *N* or *C* termini, creating cyclic analogues, the synthesis of peptidomimetics, and the introduction of structural constraints. These approaches often lead to peptides that possess improved metabolic properties and better bioavailability. The plasma half-life of somatostatin was improved from a few minutes to 1.5 h by shortening the number of amino acids and replacing selective *L*-amino acids with *D*-amino acids [5]. The addition of polyethylene glycol (PEG) or polymers of *N*-acetylneuramic acid (polysialic acids) to peptide or protein drugs can also dramatically improve half-life [13]. This is accomplished by providing a steric barrier to proteolytic enzymatic cleavage and by increasing molecular mass, thereby reducing renal excretion. Several PEGylated protein drugs are on the market including PEGylated alpha-interferon for the treatment of hepatitis C. PEGylated alpha-interferon has a half-life of 50 h, whereas native alpha-interferon has a half-life of only 5 h [13]. Cyclic peptides (both head-to-tail and side-chain to side-chain) are known to be more resistant to peptidase hydrolysis and consequently may possess improved half-lives for metabolism when compared to their linear analogues [14]. For example, cyclic growth regulating factor (GRF) (1–29)-NH_2 has a half-life of 2 h whereas the half-life of the linear analogue is only 13 min [5]. Cyclic peptides also display improved membrane permeability compared to that of linear analogues [14–15]. This may be explained not only by their increased resistance to proteolytic cleavage but also by facilitation of internal hydrogen-bonded conformations and by the reduction of charged termini. Branched peptides have also been shown to resist proteolytic cleavage by proteases and peptidases in vivo [16].

15.3 LC-MS/MS FOR METABOLITE IDENTIFICATION

Metabolite identification of drug candidates is done at various stages of the drug discovery and development process. In the lead optimization stage, structural elucidation of metabolites is often sought to identify metabolism "hot spots" that may lead to poor pharmacokinetic properties, including high clearance and short half-life clearance. This is usually accomplished by using samples from both in vitro experiments (e.g., hepatocyte incubations) and in vivo studies in rodents [17]. Another important reason for performing metabolite identification in drug discovery is to identify and minimize the formation of any reactive metabolites that may covalently bind to macromolecules and potentially form organ toxicity or idiosyncratic drug reactions [18,19].

In preclinical development, metabolite identification aids in the selection of the proper rodent and non-rodent species for toxicology studies [20]. The purpose is to ensure that the biotransformation pathways in the preclinical toxicology species are similar to humans and that there are no human-specific metabolites. As part of the preclinical development package, mass balance studies are performed [21]. These are done in rat and also in a non-rodent species (e.g., dog or monkey). The goal is to gain an in-depth understanding of the metabolic fate and clearance of a drug. The studies are performed by administering radiolabeled drug, usually ^{14}C, to animals and collecting plasma, urine, feces, and sometimes bile at various time intervals after dosing. Radiometric scintillation detection is used to determine the percent of the dose eliminated by liver or kidneys as unchanged drug or metabolites. LC/MS/MS with flow scintillation detection is used to gather a profile of the metabolic fate of the drug and to identify metabolite structures. Given that authentic standards of metabolites are not always available, quantitation is accomplished by using radiolabel detection. At this stage the contributions of metabolites to clearance, efficacy, and toxicity may also be evaluated.

In clinical development, metabolite identification is carried out as part of the human absorption, metabolism, and excretion study in which a radiolabeled drug is administered to humans. Evaluation and identification of all metabolites formed in humans is essential to ensure that they are also formed by the preclinical toxicology species at an adequate level.

Metabolite identification is done using LC/MS/MS with electrospray ionization in conjunction with various types of mass spectrometers. In the 1980s, tandem MS was introduced for metabolite identification of Phase I metabolites [22]. This approach was extended to on-line LC-MS/MS with neutral loss and precursor ion scans for rapid identification and structural elucidation of secondary drug conjugates, including glucuronides and aryl sulfate esters [23].

Triple quadrupoles mass spectrometers with electrospray ionization are still used today for metabolite identification in biological matrices. Other mass analyzers, however, now supplement the triple quadrupole. Ion trap mass spectrometers have the advantages of increased sensitivity in the full-scan product ion mode relative to triple quadrupole mass spectrometers [24]. Another advantage is the capability of MS^n production scanning to enhance structural information for unknown metabolites. Disadvantages include in-trap space charging, which limits dynamic range in quantitative analysis, and the lack of conventional neutral loss and precursor ion scanning. Hybrid triple quadrupole-linear ion trap mass spectrometers, introduced in 2002, are now used extensively for both quantitative analysis and metabolite identification [25]. They maintain all the functions of a triple quadrupole mass spectrometer and hence can be used not only for routine quantitation but also for acquisition of enhanced sensitivity product-ion spectra of metabolites. Hybrid quadrupole-linear ion traps can also be employed for simultaneous metabolite identification and quantitation of drugs and their metabolites in plasma, thereby improving throughput for the metabolite identification process [26].

High mass-resolving power mass spectrometry is an important tool for metabolite identification; it aids in the assignment of structurally informative fragment ions and also helps to differentiate between isobaric molecular ions. The advent of high resolving power hybrid orthogonal quadrupole time-of-flight (QqTOF) instruments makes accurate mass measurements more accessible to pharmaceutical metabolism laboratories [27]. These instruments are routinely used at mass resolving powers of 30,000 allowing accurate masses to be assigned within 3 ppm mass accuracy. A more recent development that has also had a large impact in metabolite identification is the introduction of the LTQ-Orbitrap [28]. This instrument combines a linear ion trap with an Orbitrap mass analyzer with Fourier-transform detection. With the Orbitrap, online LC/MS/MS experiments may be done with a mass-resolving power of 100,000 and concomitant improved accurate mass measurements.

15.4 QUANTITATIVE ANALYSIS

Quantitation in drug metabolism and pharmacokinetic studies is most often performed by using triple quadrupole mass spectrometers in the selected reaction monitoring (SRM) mode with electrospray ionization. In the SRM mode the protonated molecule of the compound of interest is selected with the first quadrupole and undergoes collisional-induced dissociation in the second quadrupole to yield product ions. Usually one or two of the most abundant fragment ions are monitored by using the third quadrupole. This approach has been routinely applied in drug metabolism laboratories since the early 1990s. Analytical standards are used to generate a calibration standard curve, and the amount of drug is determined by back calculation from a linear regression.

Quantitative determination of larger molecular weight peptides is also done using enzyme-linked immunosorbent assays (ELISA). The technology is very sensitive with lower limits of quantitation of 1 to 10 pg/mL, and sample analysis is fast and amenable to automation. Method development for an ELISA assay, however, may take months owing to the time required to generate and screen antibodies. Furthermore ELISA assays have limited dynamic range, and the lack of selectivity is also a concern. Endogenous antibodies directed against the therapeutic agent can interfere with ELISA assays. Because of these limitations, ELISA assays for peptide drugs are often replaced by more specific LC/MS/MS assays using either conventional sample preparation such as solid-phase extraction or immuno-capturing of the analyte [29].

Quantitation in drug metabolism studies is also done by using ^{14}C or ^{3}H radioisotope labeled drug [30]. The use of radiolabeled compound obviates the need to synthesize analytical standards for each drug metabolite to be used for normal "cold" LC/MS/MS assays. The radiolabeled atoms, however, need to be incorporated into the molecule in such a fashion that the majority of the metabolites also contain the radiolabel. This approach presents a limitation in the metabolism of peptide drugs that easily hydrolyze to release the labeled part of the molecule; thus, major metabolic pathways may be missed during metabolic profiling when using radio-chemical detectors. Hence dual radiolabels are sometimes employed for drugs that contain peptide bonds so that the metabolism may be tracked at different sides of the molecule [31].

15.5 CASE STUDY: IL-1β PROTEASE INHIBITORS

Interleukin-1β (IL-1β) converting enzyme (ICE) is a cytoplasmic cysteine protease that is responsible for the conversion of inactive 31 kDa precursor IL-1β to the active 17.5 kDa IL-1β by cleavage of the peptide bond between Asp 116 and Ala 117. The 17.5 kDa active form is secreted into extracellular fluid and has systemic effects. Elevated levels of IL-1β are present in patients with infections, inflammation, rheumatoid arthritis, and other diseases [32]. The goal of this particular drug discovery program was to produce an orally available compound with the possibility of reducing IL-1β processing by inhibiting ICE. The rationale was that inhibition of IL-1β processing might allow modulation of its pathological effects while leaving its immunological functions intact.

The compounds that were initially screened for activity as ICE inhibitors were small di- and tri-peptides that also contained various leaving groups necessary for in-vitro enzyme and cellular activity against ICE (see Figure 15.1 for the structures of three such ICE inhibitors). One is a tripeptide, referred to as Z-Val-Ala-Asp-TPP, and two others are dipeptides, Z-Val- Asp-TPP and Z-Val-Asp-DPP.

As part of the later stage selection of a lead compound, oral bioavailability studies are run, as they were here in the dog. The assessment of bioavailability is a key step in the design of an orally active therapeutic molecule. For non-rodent species, like dogs and monkeys, often the drug candidate is dosed to an animal intravenously (IV) and

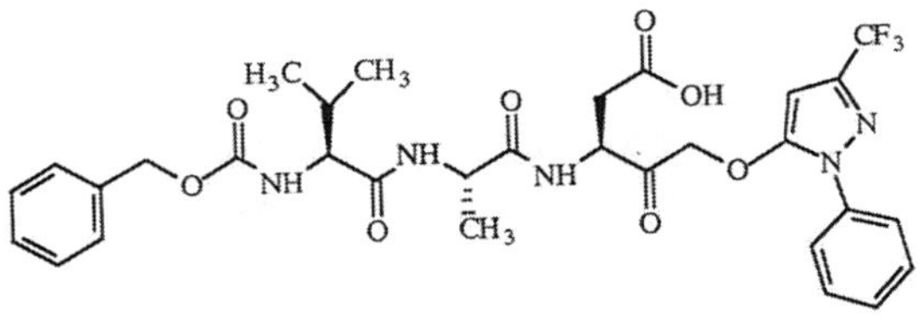

Z-Val-Ala-Asp-TPP

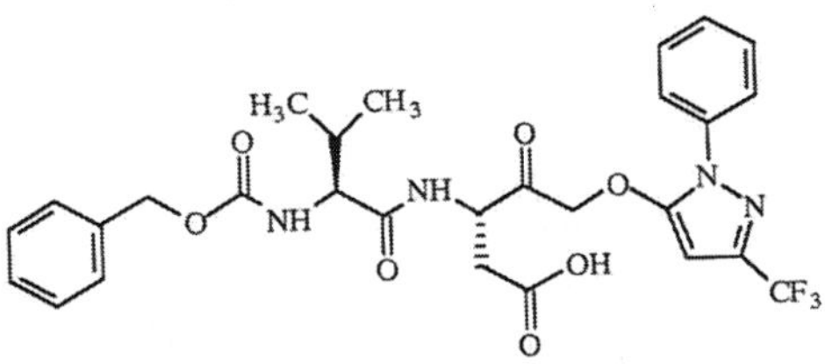

Z-Val-Asp-TPP

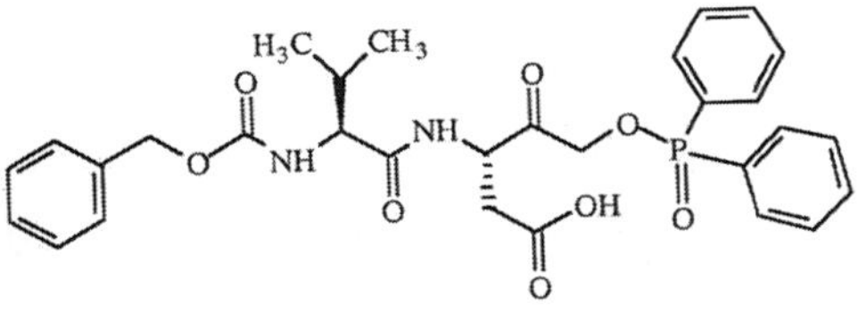

Z-Val-Asp-DPP

FIGURE 15.1 Structures of IL-1β protease inhibitors.

then after a washout period, the compound is administered to the same animal orally. The ratio of the area under the curve of the concentration time profile after an oral dose (AUC_{oral}) to the AUC_{iv} normalized to any difference in dose is defined as bioavailability (%F). In this study the IV and oral doses were given 3 h apart in the same animal as a means to reduce intra-animal variability. Any contributions of AUC_{iv} to AUC_{oral} were deconvoluted by using Monte Carlo calculations [33].

In addition compounds of interest can be administered together as a "cassette" rather than individually administered on separate occasions. The cassette approach has been utilized to enhance throughput of in vivo PK assessments in drug discovery. For these studies the three compounds were administered as a mixture both IV (2 mg/kg each) and then 2.5 h later, orally (5 mg/kg each). Plasma samples were collected using EDTA as the anticoagulant at various time points out to 24 h. Protein precipitation with acetonitrile was used for sample processing. Sample introduction was accomplished with a fast 5-min gradient on a C18 HPLC column. The analytes were quantitated by using ionspray on a triple quadrupole mass spectrometer. The use of a triple quadrupole mass spectrometer in the SRM mode added specificity to this analysis and allowed successful calibration from 1 to 1000 ng/mL.

The resulting IV/PO PK profiles for each compound (Figure 15.2) show that the tri-peptide Z-Val-Ala-Asp-TPP and the di-peptide Z-Val-Asp-DPP had poor bioavailability of less than 5%. The dipeptide Z-Val-Asp-TPP, however, had a much better bioavailability of 47%. As shown in Figure 15.3, the plasma versus concentration profiles are similar when administered either as a cassette or individually.

To investigate the mechanistic reasons for the large difference in %F for Z-Val-Ala-Asp-TPP and Z-Val-Asp-TPP a discovery excretion/ mass balance study was performed in dog using ^{14}C radiolabeled compounds. The radiolabel was located on the methylene carbon of the Z group. The labeled compound was administered to dogs at 2 mg/kg oral, and urine and feces were collected at various time intervals out to

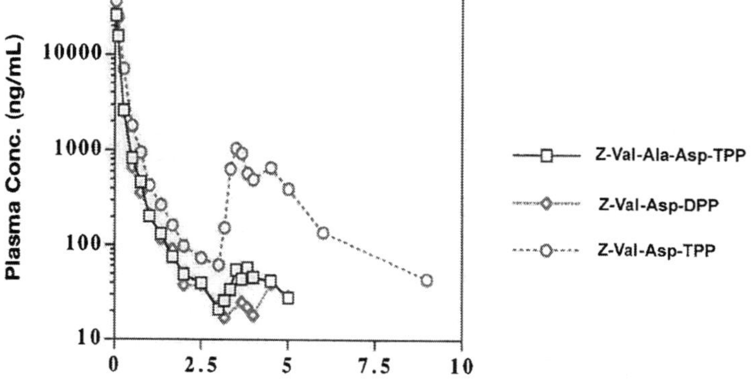

FIGURE 15.2 Plasma concentration versus time profiles.

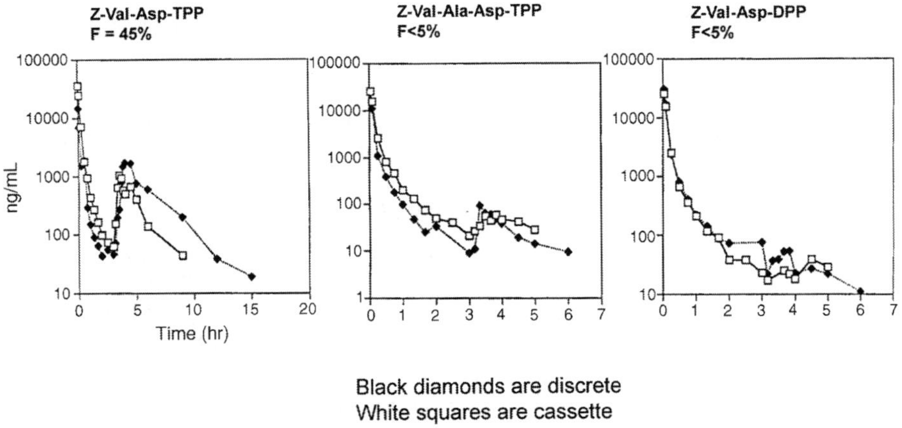

FIGURE 15.3 Comparison of plasma concentration versus time profiles for discrete versus cassette dosing.

120 h; plasma was collected at 1 and 4 h. Total radioactivity was measured in feces, plasma, and urine by using a scintillation counter. Metabolite profiles and LC/MS/MS spectra for metabolite identification were recorded by splitting the LC effluent and directing 33% of the flow to the mass spectrometer and 67% to a flow scintillation detector (Figure 15.4). The plasma radio chromatogram for Z-Val-Ala-Asp-TPP from a 4-h post oral dose time point is shown in Figure 15.5.

To ascertain the structure of the metabolites, product-ion spectra were obtained for all radiochromatogram metabolites as well as for the parent drug. The product ion spectrum for the parent compound, Z-Val-Ala-Asp-TPP (Figure 15.6) shows the protonated molecule at *m/z* 662. Major product ions are produced by the loss of the

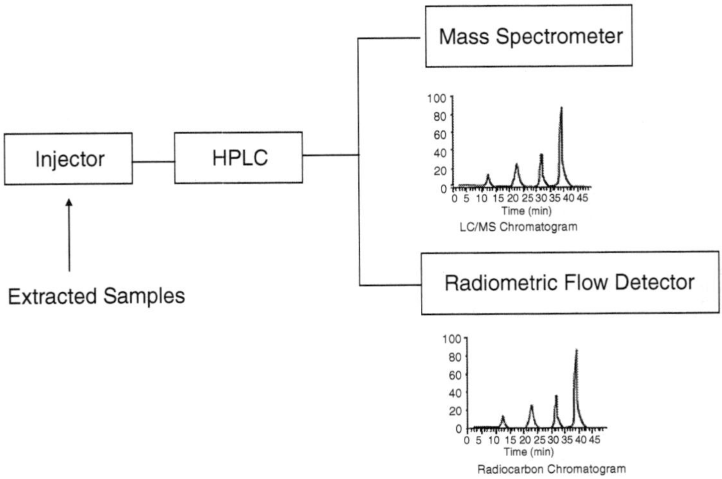

FIGURE 15.4 Schematic for radio-profiling/metabolite identification experiment.

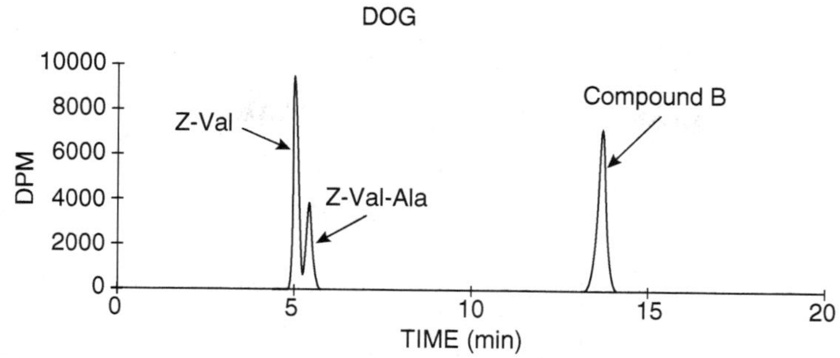

FIGURE 15.5 Radiochromatogram of dog plasma for Z-Val-Ala-Asp-TPP.

TPP (ion of *m/z* 434), the loss of the Z group and TPP to give an ion of *m/z* 344, and loss of CO_2 from both ions of *m/z* 434 and 344. From the product ion spectra, it was determined that the largest circulating metabolite in plasma, with a retention time of 5 min, is Z-Val, which is formed by peptide-bond hydrolysis. Similarly the smaller radiolabeled metabolite is Z-Val-Ala. The dog urine radiochromatogram (Figure 15.7) shows only one peak corresponding to Z-Val indicating complete peptide-bond hydrolysis and excretion of Z-Val as the major metabolite. As expected, the major

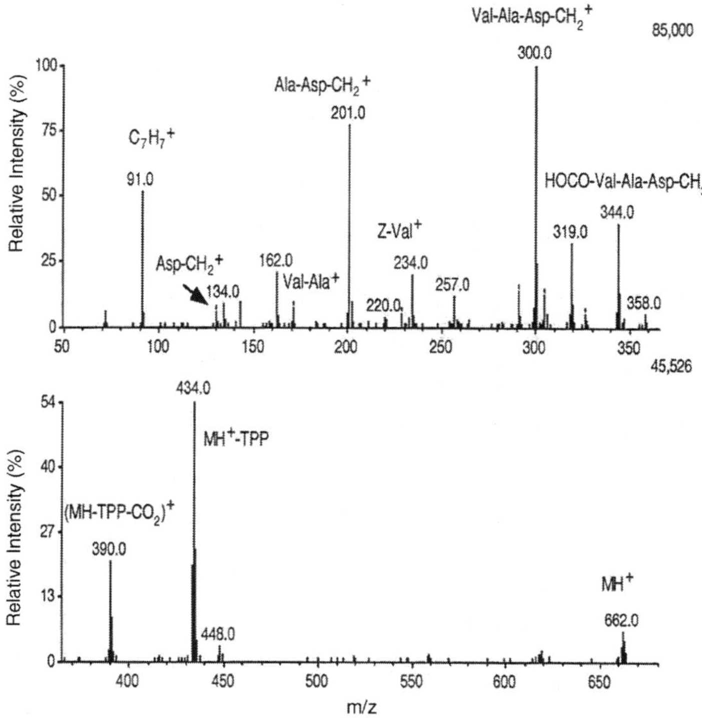

FIGURE 15.6 Product ion spectrum for Z-Val-Ala-Asp-TPP.

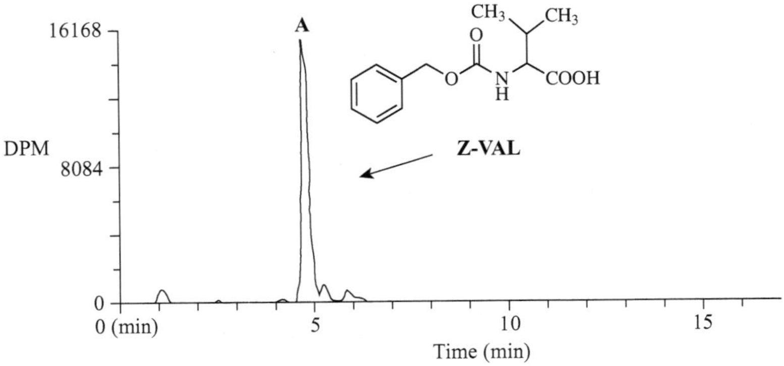

FIGURE 15.7 Radiochromatogram of dog urine for Z-Val-Ala-Asp-TPP.

biotransformation pathway for Z-Val-Ala-Asp-TPP is peptide bond hydrolysis, resulting in low oral bioavailability (5%) for this compound (Figure 15.8).

To summarize the conclusions for the dog excretion/mass balance study for Z-Val-Ala-Asp-TPP, the major route of metabolism was hydrolysis of peptide bonds to yield, Z-Val (major) and Z-Val-Ala (minor). The major plasma metabolite at both 1- and 4-h post oral dose is Z-Val. The major metabolite in urine is also Z-Val.

The results of the mass/balance study for the di-peptide Z-Val-Asp-TPP showed that Z-Val-Asp-TPP was excreted intact in feces, indicating that biliary excretion was

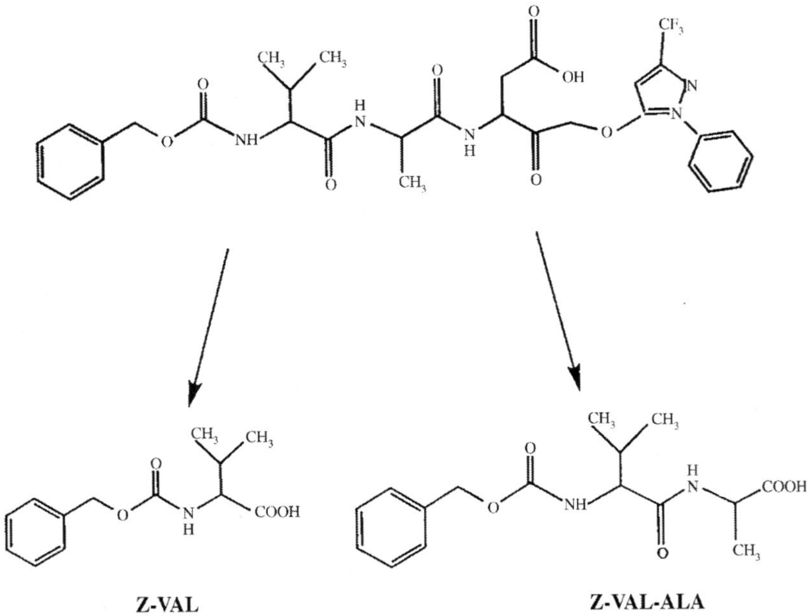

FIGURE 15.8 Metabolic pathway for Z-Val-Ala-Asp-TPP in dog.

a major mechanism of clearance (data not shown). Interestingly peptide bond hydrolysis is not a major route of metabolism for this compound. Presumably transporters assist in the uptake of this compound across the gut endothelial cells and result in improved absorption into the portal vein and higher levels in systemic circulation. The result is better oral bioavailability of approximately 45%.

15.6 FUTURE DIRECTIONS

One area where MS will play a larger role in the future is in the quantitative analysis of protein and antibody drugs. As was previously mentioned, this is presently most often done using ELISA assays that are easily automated and have excellent sensitivity with limits of detection in the low pg/mL range. MS is more specific, and method development is faster because antibodies do not need to be prepared and characterized. Recent reports in the literature describe the analogous use of MS for protein quantitation by using LC/SRM-MS for the detection of specific signature peptides released by enzymatic digestion [34,35].

MS will become increasingly integrated into labs that now use ELISA, at the very least to verify specificity during optimization of ELISA assays and also to evaluate novel biomarkers. The sensitivity of MS detection with conventional (nonimmuno-affinity sample preparation) has reached the low ng/mL range; hence at this time, MS is most useful for preclinical toxicology study support. Micro-chromatography and nano-ESI will be exploited to achieve better sensitivity for target protein quantitation. Continuous advances in MS instrumentation will enhance sensitivity and selectivity so that MS will begin to play a larger role not only in peptide quantitation and metabolism but also for proteins.

REFERENCES

1. Rawlins, M. D. (2006). Cutting the cost of drug development. *Nat Rev Drug Discov 3*, 360–364.
2. Benet, L. Z. (2009). The drug transporter-metabolism alliance: Uncovering and defining the interplay. *Mol Pharmaceut 6*, 6 1631–1643.
3. Testa, B., Mayer, J. M. (2003). *Hydrolysis in Drug and Prodrug Metabolism, Chemistry, Biochemistry and Enzymology*, Verlag Helvetica Chimica Acta, Zurich.
4. Bernkop-Schnurch, A., Schmitz, T. (2007). Presystemic metabolism of orally administered peptide drugs and strategies to overcome it. *Curr Drug Metab 8*, 509–517.
5. Werle, M., Bernkop-Schnurch, A. (2006). Strategies to improve plasma half life of peptide and protein drugs. *Amino Acids 30*, 351–367.
6. Foger, F., Kopf, A., Loretz, B., Albrecht, K., Bernkop-Schnurch, A. (2008). Correlation of in vitro and in vivo models for the absorption of peptide drugs. *Amino Acids 35*, 223–241.
7. Augustijns, P. F., Borchardt, R. T. (1995). Transport and metabolism of delta sleep-inducing peptide in cultured human intestinal epithelial cell monolayers. *Drug Metab Disp. 23*, 1372–1378.

8. Cao, X., Gibbs, S. T., Fang, L., Miller, H. A., Landowski, C. P., Shin, H. C., Lennernas, H., Zhong, Y., Amidon, G. L., Yu, L. X., Sun, D. (2006). Why is it challenging to predict intestinal drug absorption and oral bioavailability in human using rat model. *Pharmaceut Res 23*, 1675–1686.
9. Prokai, L., Zharikova, A. D. (1998). Identification of synaptic metabolites of dynorphin A (1–8) by electrospray ionization and tandem mass spectrometry. *Rapid Commun Mass Spectrom 12*, 1796–1800.
10. Taki, Y., Sakane, T., Nadai, T., Sezaki, H., Amidon, G. L., Langguth, P., Yamashita, S. (1998). First-pass metabolism of peptide drugs in rat perfused liver. *J Pharm Pharmacol 50*, 1013–1018.
11. Wallemaq, P. E., Lhoest, G., Dumont, P. (1989). Isolation, purification and structure elucidation of cyclosporin A metabolites in rabbit and man. *Biomed Environ Mass Spectrom 18*, 48–56.
12. Fabre, G., Bertault-Peres, P., Fabre, I., Maurel, P., Just, S., Cano, J.-P. (1987). Metabolism of cyclosporin A. I. Study in freshly isolated rabbit hepatocytes. *Drug Metab Disp 15*, 384–390.
13. Webster, R., Didier, E., Harris, P., Siegel, N., Stadler, J., Tilbury, L., Smith, D. (2007). PEGylated proteins: Evaluation of their safety in the absence of definitive metabolism studies. *Drug Metab Disp 35*, 9–16.
14. Rezai, T., Yu, B., Millhauser, G. L., Jacobson, M. P., Lokey, R. S. (2006). Testing the conformational hypothesis of passive membrane permeability using synthetic cyclic peptide diasereomers. *J Am Chem Soc 128*, 2510–2511.
15. Rezai, T., Bock, J. E., Zhou, M. V., Kalyanaraman, C., Lokey, R. S., Jacobson, M. P., (2006). Conformational flexibility, internal hydrogen bonding and passive membrane permeability: successful in silico prediction of the relative permeabilities of cyclic peptides. *J Am Chem Soc 128*, 14073–14080.
16. Pini, A., Falciani, C., Bracci, L. (2008). Branched peptides as therapeutics. *Curr Protein Peptide Sci 9*, 468–477.
17. Zhang, D., Zhu, M., Humphreys, W. G., (eds.), (2008). Drug Metabolism in Drug Design and Development. John Wiley and Sons, Inc. New Jersey.
18. Baillie, T. A., Cayen, M. N., Fouda, H., Gerson, R. J., Green, J. D., Grossman, S. J., Klunk, L. J., LeBlanc, B., Perkins, D. G., Shipley, L. A. (2002). Drug metabolites in safety testing. *Toxicol Appl Pharmacol 182*, 188–196.
19. Ma, S., Subramanian, R. (2006). Detecting and characterizing reactive metabolites by liquid chromatography/tandem mass spectrometry. *J Mass Spectrom 41*, 1121–1139.
20. Chowdhury, S. K. (ed.), (2005) *Progress in Pharmaceutical and Biomedical Analysis Volume 6: Identification and Quantification of Drugs, Metabolites and Metabolizing Enzymes by LC-MS*. Elsevier, New York.
21. Roffey, S. J., Obach, R. S., Gedge, J. I., Smith, D. (2007). What is the objective of the mass balance study? A retrospective analysis of data in animal and human excretion studies employing radiolabeled drugs. *Drug Metab Rev 39*, 17–43.
22. Perchalski, R. J., Yost, R. A., Wilder, B. J. (1982). Structural elucidation of drug metabolites by triple-quadrupole mass spectrometry. *Anal Chem 54*, 1466–1471.
23. Rudewicz, P., Straub, K. M. (1986). Rapid structure elucidation of catecholamine conjugates with tandem mass spectrometry. *Anal Chem 58*, 2928–2934.
24. March, R. E. (1997). An introduction to quadrupole ion trap mass spectrometry. *J Mass Spectrom 32*, 351–369.

25. Hager, J. W. (2002). A new linear ion trap mass spectrometer. *Rapid Commun Mass Spectrom 16*, 512–526.

26. Hopfgartner, G., Husser, C., Zell, M. (2003). Rapid screening and characterization of drug metabolites using a new quadrupole-linear ion trap mass spectrometer. *J Mass Spectrom 38*, 138–150.

27. Chernushevich, I. G., Loboda, A. V., Thomson, B. A. (2001). An introduction to quadrupole-time-of-flight mass spectrometry. *J Mass Spectrom 36*, 849–865.

28. Hu, Q., Noll, R. J., Li, H., Makarov, A., Hardman, M., Cooks, G. (2005). The orbitrap: A new mass spectrometer. *J Mass Spectrom 40*, 430–443.

29. Ackermann, B. L., Berna, M. J. (2007). Coupling immunoaffinity techniques with MS for quantitative analysis of low-abundance protein biomarkers. *Expert Rev Proteomics 4*, 175–186.

30. Knadler, M. P., Ackerman, B. L., Coutant, J. E., Hurst, G. H. (1992) Metabolism of the anticoagulant peptide, MDL 28,050, in rats. *Drug Metab Disp 20*, 89–95.

31. Zheng, K., Lubman, D. M., Rossi, D. T., Nordblom, G. D., Barksdale, C. M. (2000). Elucidation of peptide metabolism by on-line immunoaffinity liquid chromatography mass spectrometry. *Rapid Commun Mass Spectrom 14*, 261–269.

32. Singer, I. I., Scott, S., Chin, J., Bayne, E. K., Limjuco, G., Weidner, J., Miller, D. K., Chapman, K., Kostura, M. J. (1995). The Interleukin-1β-converting enzyme (ICE) is localized on the external cell surface membranes and in the cytoplasmic ground substance of human monocytes by immuno-electron microscopy. *J Exp Med 182*, 1447–1459.

33. Karlsson, M. O., Bredburg, U. (1990). Bioavailability estimation by semisimultaneous drug administrtaion: A Monte Carlo simulation study. *J Pharmacokinet Biopharmaceut 18*, 103–120.

34. Heudi, O., Barteau, S., Zimmer, D., Schmidt, J., Bill, K., Lehmann, N., Bauer, C., Kretz, O. (2008). Towards absolute quantification of therapeutic monoclonal antibody in serum by LC-MS/MS using isotope-labeled antibody standard and protein cleavage isotope dilution mass spetcrometry. *Anal Chem 80*, 4200–4207.

35. Yang, Z., Hayes, M., Fang, X., Daley, M. P., Ettenberg, S., Tse, F. L. S., (2007) LC-MS/MS approach for quantification of therapeutic proteins in plasma using a protein internal standard and 2D-solid-phase extraction cleanup. *Anal Chem 79*, 9294–9301.

INDEX

Protein and Peptide Mass Spectrometry in Drug Discovery, Edited by Michael L. Gross, Guodong Chen, and Birendra N. Pramanik.